Encyclopedia of
Environmental Issues

ENCYCLOPEDIA OF ENVIRONMENTAL ISSUES

REVISED EDITION

Volume 3
Land clearance—Sewage treatment and disposal

Editor
Craig W. Allin
Cornell College

SALEM PRESS
Pasadena, California   Hackensack, New Jersey

Editor in Chief: Dawn P. Dawson

Editorial Director: Christina J. Moose *Photo Editor:* Cynthia Breslin Beres
Development Editor: R. Kent Rasmussen *Production Editor:* Joyce I. Buchea
Project Editor: Judy Selhorst *Graphics and Design:* James Hutson
Acquisitions Manager: Mark Rehn *Layout:* William Zimmerman
Research Supervisor: Jeffry Jensen

Cover photo: ©Naluphoto/Dreamstime.com

Library of Congress Cataloging-in-Publication Data

Encyclopedia of environmental issues / editor, Craig W. Allin. — Rev. ed.
     Planned in 4 v.
     Includes bibliographical references and index.
     ISBN 978-1-58765-735-1 (set : alk. paper) — ISBN 978-1-58765-736-8 (vol. 1 : alk. paper) —
ISBN 978-1-58765-737-5 (vol. 2 : alk. paper) — ISBN 978-1-58765-738-2 (vol. 3 : alk. paper) —
ISBN 978-1-58765-739-9 (vol. 4 : alk. paper)
     1. Environmental sciences—Encyclopedias. 2. Pollution—Encyclopedias.
I. Allin, Craig W. (Craig Willard)
     GE10.E523 2011
     363.7003—dc22
                                                                                            2011004176

Contents

Complete List of Contents

Volume 1

Volume 2

Volume 3

Volume 4

ENCYCLOPEDIA OF
ENVIRONMENTAL ISSUES

Land clearance

CATEGORIES: Ecology and ecosystems; land and land use

DEFINITION: Removal of plant life, stones, and other obstacles from a land surface to increase the area available for farming or for construction of buildings

SIGNIFICANCE: Although humans have modified land surfaces for food production and habitation since prehistoric times, the process of land clearance escalated tremendously from the mid-twentieth century onward, causing dramatic changes in ecosystems around the world. These changes have raised environmental concerns regarding soil, water, and air pollution and the loss of animal habitats.

Land clearance increased significantly when human beings transitioned from hunter-gatherer to agricultural societies. As agriculture developed, extensive land areas were cleared for crop growth. With the advancement of technology, the process of land clearance escalated as humans using heavy equipment became capable of not only rapidly clearing but also reshaping the landscape. Often, areas of cleared land that have become less productive than they were previously are abandoned in favor of newly cleared land that is more supportive of agricultural endeavors. These deserted areas often become barren wastelands devoid of human and animal habitation.

THE AMAZON RAIN FOREST

An example of land clearance of epic proportions is the Amazon rain forest. The Amazon River basin is the largest rain forest on the planet; it covers about 40 percent of the South American continent and is roughly the size of the contiguous forty-eight U.S. states. The biodiversity of the rain forest is unequaled: It is home to some 40,000 plant species, 2.5 million insect species, and more than 2,000 bird and animal species.

Deforestation began in the Amazon basin during the 1960's, when access to the forest's interior was first allowed. Large areas were cleared for the planting of crops, and when the soil became depleted, the farmers moved to adjacent areas. These practices resulted in extensive damage to the forest. In 1991, the area of deforestation was 415,000 square kilometers (160,230 square miles); by 2000, it had grown to 587,000 square kilometers (226,640 square miles). From 2000 to 2005

the deforestation rate was 22,392 square kilometers (8,646 square miles per year). If this rate continues, about 40 percent of the rain forest will have vanished by 2030.

NEGATIVE CONSEQUENCES

Land clearance and the agricultural activities that follow have major impacts on climate through the release of greenhouse gases into the atmosphere. (Greenhouse gases include water vapor, carbon dioxide, methane, nitrous oxide, ozone, and chlorofluorocarbons.) Soil disturbance results in the release of carbon dioxide, and methane is also produced by soil disturbance, wetland drainage, rice paddies, and cattle flatus. Nitrous oxide is produced by nitrogen fertilizers, irrigation, the cultivation of nitrogen-fixing plants, and the burning of plant materials, or biomass combustion. Sulfur dioxide and particulate matter are also produced by biomass combustion.

Another effect of land clearance is related to the alteration of the reflection of sunlight from land surfaces, known as the albedo effect (albedo is a measure of the amount of incident radiation, such as sunlight, that a surface reflects). Altered vegetation (such as crops replacing rain forest) and the construction of structures change the land's heat balance by changing albedo, affecting the amount of moisture evaporated from plants (transpiration) and altering air circulation by changing the surface roughness (for example, planted areas interspersed with structures versus a smooth rain-forest canopy). An example of this alteration is the fact that urban areas have generally warmer temperatures than do rural areas in a given region.

When a forest is cleared for agricultural use, animal and plant life within that forest is immediately lost. When plants eaten by the herbivores (plant eaters) native to an area no longer exist, the herbivore population dies out or migrates; the carnivores, which feed on the herbivores, thus lose their food source as well. Forested areas adjacent to cleared areas are also affected. When contiguous areas of forest are replaced by "islands" of forest interspersed with cultivated areas, the movements of forest animals are restricted, and some may not be able to reach food sources or mating grounds. Land clearance can thus markedly reduce the populations of species or even result in species extinction.

Cleared land becomes vulnerable to soil erosion by wind and water. As topsoil erodes, the land's fertility decreases and particulate matter is released into the air and into surface water, such as streams and rivers. When land is cleared by burning, pollutants are immediately discharged into the atmosphere. Plants and animals are negatively affected by the contamination of air and water with particulate matter, and runoff from agricultural irrigation, containing nitrogen and phosphorus from fertilizers as well as toxins from pesticides, further pollutes the water supply.

Some areas that have been cleared for agricultural become nonproductive over time owing to overgrazing or soil depletion. These areas may be simply abandoned or they may be taken over by commercial and residential structures, particularly in regions where populations are increasing. The urbanization of former agricultural lands often adds to pollution, as the numbers of motor vehicles in these areas increase.

Countering the Negative Consequences

The negative environmental impacts of land clearance can be reduced through the implementation of sustainable agriculture practices. Such practices include crop rotation to avoid nutrient depletion of soils and contour farming (in which crops are planted in furrows that follow natural land contours rather than in straight rows) to reduce erosion. The management of agricultural lands must allow for a constant or growing supply of food and other resources to human populations. It also must incorporate the use of less toxic pesticides and the control of runoff into the water supply. Sustainable agriculture practices do not solve the problem of the contribution of land clearance to climate change, however.

To counter the impacts of land clearance on biodiversity, some nations have developed conservation reserves where animals and plant species are protected within their natural habitats. In some areas previously subjected to land clearance, native plants have been reintroduced and their growth supported by both government and volunteer efforts. Another strategy used to promote biodiversity is the establishment of corridors connecting habitat areas where these have been separated by cleared land; this greatly expands the area available to animal species and preserves the migratory habits of many animals.

Robin L. Wulffson

Further Reading

Baden, John. *The Vanishing Farmland Crisis: Critical Views of the Movement to Preserve Agricultural Land.* Lawrence: University of Kansas Press, 1984.

Chew, Sing C. *World Ecological Degradation: Accumulation, Urbanization, and Deforestation, 3000 B.C.-A.D. 2000.* Walnut Creek, Calif.: AltaMira Press, 2001.

Houghton, John. *Global Warming: The Complete Briefing.* 4th ed. New York: Cambridge University Press, 2009.

Knox, Paul. *Urbanization: An Introduction to Geography.* 2d ed. Upper Saddle River, N.J.: Prentice Hall, 2005.

Levine, Marvin J. *Pesticides: A Toxic Time Bomb in Our Midst.* Westport, Conn.: Praeger, 2007.

London, Mark. *The Last Forest: The Amazon in the Age of Globalization.* New York: Random House, 2005.

Martine, George, et al., eds. *The New Global Frontier: Urbanization, Poverty, and Environment in the Twenty-first Century.* Sterling, Va.: Earthscan, 2008.

Palmer, Charles, and Stefanie Engel, eds. *Avoided Deforestation: Prospects for Mitigating Climate Change.* New York: Routledge, 2009.

SEE ALSO: Agricultural chemicals; Air pollution; Amazon River basin; Biodiversity; Carbon dioxide; Cattle; Deforestation; Erosion and erosion control; Forest management; Rain forests; Slash-and-burn agriculture; Soil conservation; Sustainable agriculture.

Land pollution

CATEGORIES: Land and land use; waste and waste management

DEFINITION: Contamination of soil and other land components with substances that could damage human health or the environment

SIGNIFICANCE: Land may be polluted by a variety of substances, including industrial chemicals, oils, wastes from water treatment processes, solid wastes, and polluted water. These contaminants in turn contribute to the pollution of freshwater sources and the oceans. The problem of land pollution demonstrates the importance of the proper disposal of wastes of all kinds.

Land pollution can result from human activities large and small, from the improper disposal of urban and industrial wastes to individual acts of littering. As urbanization and industrialization have increased, human beings have generated an ever larger array of wastes that promote high levels of land pollution.

Sources

Land can become polluted in many different ways. One common source of land pollution is the improper disposal of wastes of all kinds, including hazardous industrial wastes, municipal solid wastes, construction and demolition debris, and the wastes generated by mining operations. Additional sources of land pollution are sediment pollutants from improperly managed construction sites, salts from agricultural irrigation practices, oil from urban runoff and energy production, petroleum products deposited by leaking vehicles, bacteria from improperly operated septic systems, pet wastes, and ground deposits from air pollution, such as particulate matter produced by power plants, motor vehicle exhaust, forest fires, and some industries, as well as lead, carbon monoxide, sulfur dioxide, and nitrogen oxides. Land contamination can also be caused by accidental spills that occur during industrial or commercial activities.

Other sources of land pollution are wastes from forestlands and agriculture (including animal manure and the residues of chemical pesticides and fertilizers), solids from sewage treatment, garbage (moist, biodegradable food wastes) and other wastes left behind on land after human activities such as recreation, wastes washed ashore from boats and sewage outlets, and nuclear wastes released on open land. Soil can also become polluted as the result of agricultural practices, such as the use of chemicals fertilizers, pesticides, and herbicides. Agricultural irrigation can result in soil salinization.

Negative Impacts

Among the many negative impacts of land pollution are destruction of the habitats of terrestrial and aquatic animals; reduction of light penetration in rivers, creeks, and lakes from sediments and excessive algal growth; reduction of soil oxygen levels caused by the decomposition of improperly disposed-of organic materials (such as cleared vegetation); and acid rain, which damages trees and other plants. Fossil-fuel use and fertilizer runoff on land can lead to the pollution of massive areas in the oceans.

Pollutants from acid mine drainage and other toxic chemicals released by mining activities can damage both land and aquatic wildlife species and their habitats. It has been estimated that more than 405 hectares (1,000 acres) of land are disturbed in some fashion by the average mining operation.

Some of the chemicals in polluted land are potentially toxic to humans and other living organisms and can reduce soil fertility. Exposure to contaminated soils can cause or exacerbate skin and respiratory diseases and has been linked to various kinds of cancers and birth defects. Toxic materials in polluted soils can enter human bodies through direct skin contact, through inhalation of soil particles, through the consumption of fruits and vegetables grown in the soils, and through the consumption of water contaminated by the soils.

Some chemical pesticides and herbicides are not easily biodegradable and can therefore have long-lasting effects on soil. Land-polluting organophosphate insecticides, which include the highly toxic parathion and about forty other insecticides, are used in agriculture in the United States and around the world. Fungicides may contain copper and mercury, which are toxic to plants and fish. Organomercury compounds, used to kill sedges, can accumulate in the central nervous system; these compounds are difficult to remove from the soil.

Control Measures

Among the measures taken to protect critical land areas and wildlife habitats from pollutants are the recycling and reclamation of used materials (such as paper, metals, glass, and plastic) and the proper disposal of domestic as well as industrial waste. Solid wastes must be properly buried in managed landfills, incinerated in a controlled manner, or recycled.

When soils contain pesticides or metals, off-site disposal is recommended. Soils that are polluted by large amounts of buried construction and demolition debris may need to undergo special screening. Industrial effluents and other forms of industrial and commercial solid and liquid waste (polluted water and chemicals) must be properly treated and disposed of.

Land pollution can be reduced through the practice of organic farming and gardening methods, which strictly limit the use of chemical pesticides and fertilizers. Individuals can help to reduce land pollution by buying biodegradable products and products with minimal packaging materials whenever possible; storing all liquid chemicals and chemical wastes in spill-proof containers and disposing of these containers only at designated toxic waste disposal sites; repairing automobile oil leaks; and disposing of used motor oil at designated sites rather than by dumping it on the ground.

In the United States, government regulations limit the uses of many chemicals and other materials that have been found to be detrimental to the health of humans and other living organisms, and strict laws have been passed to ensure the proper disposal of all kinds of wastes in the environment. Additional measures that have been taken to reduce land pollution include educational campaigns aimed at informing the public of the environmental harms caused by the improper dumping of wastes.

Samuel V. A. Kisseadoo

FURTHER READING

Boulding, J. Russell, and Jon S. Ginn. *Practical Handbook of Soil, Vadose Zone, and Ground-Water Contamination: Assessment, Prevention, and Remediation.* 2d ed. Boca Raton, Fla.: CRC Press, 2003.

Hilgenkamp, Kathryn. "Land." In *Environmental Health: Ecological Perspectives.* Sudbury, Mass.: Jones and Bartlett, 2006.

Meyer, Peter B., Richard H. William, and Kristen R. Yount. *Contaminated Land: Reclamation, Redevelopment, and Reuse in the United States and the European Union.* Northampton, Mass.: Edward Elgar, 1995.

Wise, Donald L., and Debra Trantolo. *Bioremediation of Contaminated Soils.* Boca Raton, Fla.: CRC Press, 2000.

Zehnder, Alexander J. B. *Soil and Groundwater Pollution: Fundamentals, Risk Assessment, and Legislation.* New York: Springer, 1995.

SEE ALSO: Acid deposition and acid rain; Agricultural chemicals; Brownfields; Electronic waste; Groundwater pollution; Hazardous waste; Incineration of waste products; Irrigation; Landfills; Nuclear and radioactive waste; Oil spills; Pesticides and herbicides; Plastics; Recycling; Septic systems; Sewage treatment and disposal; Soil salinization; Strip and surface mining; Waste management.

Land-use planning

CATEGORIES: Land and land use; urban environments
DEFINITION: Systematic development of a vision and public policy that guides the future use of land, facilities, and resources
SIGNIFICANCE: Land-use planning affects the livability and sustainability of a community by establishing social, economic, and environmental policies. Land-use plans provide a basis for the efficient use of land, for community improvements and economically viable development, and for the preservation and management of the environment and existing resources.

Governments engage in land-use planning for many reasons. Land-use plans support the adoption of legitimate land-use regulations that will withstand scrutiny in a court of law. Land-use planning results in proposals and policies that guide future land development and the installation of infrastructure to support such uses. Infrastructure planning requires the projection of the locations of roads and transportation networks, sewers, water lines, and other utilities and the consideration of future budgeting to allow for the construction of such improvements. Moreover, as development that is based on a land-use plan becomes a reality, a community is able to budget in advance for capital improvement facilities such as schools, fire stations, and municipal buildings. By adopting land-use plans, cities and towns are able to achieve sustainable development through the integration of existing and future land uses and natural systems with existing and future infrastructure and major projects. Ideally, land-use planning promotes order and economic change while preserving the environment.

A major specialty in land-use planning is urban planning, which considers the built environment and attempts to address some kinds of social problems, such as by finding ways to discourage crime. Land-use planning for urban environments has existed since ancient times, when the first human settlements were designed and built to achieve order, efficiency, aesthetics, and safety and to meet the functional needs of society, including community services, water, sanitation, transportation, and military defense. Modern urban land-use planning may involve urban renewal proposals that encourage new investment and development in decaying and declining areas of a city.

SUSTAINABLE DEVELOPMENT

The United Nations has been in the forefront of the sustainable development movement, which involves the efficient marriage of economic development, environmental protection, and sociopolitical policy. It is quite challenging to respect the carrying capacity of the natural environment while at the same time encouraging economic development, and land-use planning can be a viable tool for achieving sustainable development that will meet both the present and the future needs of a society.

To contribute to the achievement of sustainable development, those involved in land-use planning must consider whether the natural environment will be able to withstand certain social policies, such as bringing jobs to a community through industrial development, and whether certain environmental policies, such as the adoption of stringent environmental regulations, will thwart economic development. In addition, economic development must not deplete natural resources faster than they can be replenished, and sociopolitical policies should achieve equitable development that does not benefit one segment of society at the expense of another.

THE PLANNING PROCESS

Land-use planning is multidisciplinary and brings together many divergent views of practitioners such as architects, landscape architects, regional planners, urban designers, engineers, transportation planners, and environmentalists. Before a land-use plan becomes final, planning experts conduct numerous background studies and analyses. Examples include analyses of the suitability and carrying capacity of land for various land uses. A suitability study might consider not only the soils and topography of an area but also the availability of supportive infrastructure for future development. Another goal of land-use planning may be to conserve land and protect valuable ecosystems while allowing the public to make some use of

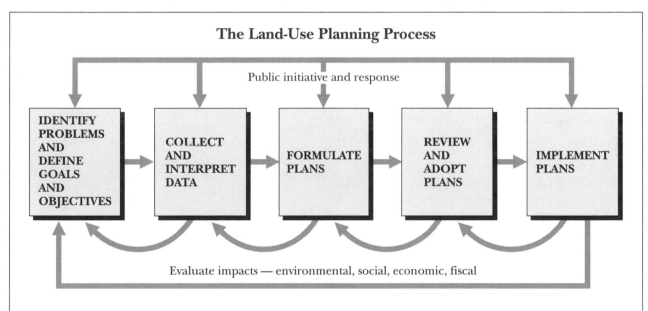

The Land-Use Planning Process

Public initiative and response

IDENTIFY PROBLEMS AND DEFINE GOALS AND OBJECTIVES → COLLECT AND INTERPRET DATA → FORMULATE PLANS → REVIEW AND ADOPT PLANS → IMPLEMENT PLANS

Evaluate impacts — environmental, social, economic, fiscal

COLLECT DATA

Earth science and other
   information
Background studies
   Existing land use
   Transportation
   Economic
   Political
   Social
Land capability studies

FORMULATE PLANS

Land use
Watershed
Natural resources
Hazard mitigation
Open space
Waste management
Public facilities

IMPLEMENT PLANS

Zoning and subdivision regulations
Erosion and sedimentation control
   ordinances
Building and housing codes
Environmental impact statements
Capital improvement programs
Health and information codes

these community assets. Planners may thus be concerned with environmental planning and might conduct ecological assessments of flora and fauna habitats. Planners may also consider watershed and groundwater management while concomitantly proposing viable locations for water-related infrastructure and facilities to support future development. Other important components of a land-use plan concern the management of sewers, stormwater, and solid waste.

Planners develop land-use plans and policies and present them to legislative bodies for adoption. As a plan evolves, land-use planners may take advantage of new technologies, such as digital media and geospatial data and information systems, that enable them to gain a better understanding of the interplay between the area's natural and developed land systems and the environmental science necessary to mitigate the impacts of past and existing land-use practices. Gaining approval for land-use plans requires planners to engage in negotiation with both government officials and interested members of the public to arrive at acceptable proposals supported by all parties. Land-use plans are the basis for many land-use regulations, including zoning, subdivision control, building codes, and environmental regulations. Often governments must rely on land-use plans to overcome legal challenges concerning the constitutionality of development regulations.

Carol A. Rolf

FURTHER READING

Berke, Philip R., and David R. Godschalk. *Urban Land Use Planning*. 5th ed. Champaign: University of Illinois Press, 2006.

Haar, Charles M., and Michael A. Wolf. *Land Use Planning and the Environment*. Washington, D.C.: Environmental Law Institute, 2009.

Juergensmeyer, Julian Conrad, and Thomas E. Roberts. *Land Use Planning and Development Regulation Law*. 2d ed. Eagan, Minn.: Thomson West, 2006.

Randolph, John. *Environmental Land Use Planning and Management*. Washington, D.C.: Island Press, 2003.

Silberstein, Jane, and Chris Maser. *Land-Use Planning for Sustainable Development*. Boca Raton, Fla.: CRC Press, 2000.

SEE ALSO: Conservation easements; Greenbelts; Open spaces; Planned communities; Road systems and freeways; Urban parks; Urban planning; Urban sprawl; Zoning.

Land-use policy

CATEGORY: Land and land use

DEFINITION: The way in which a society organizes, plans, and manages social and physical activities on the landscape

SIGNIFICANCE: Policy makers around the world have become increasingly aware of the potential environmental impacts of government and community decisions regarding land use. Since the late twentieth century, approaches to land-use planning have trended toward ecological conservation and preservation.

Human societies are organized to survive in particular environments. The adaptive nature of culture allows a society to respond to changes in the environment and to cause even more changes. Once human beings began to establish cities, where their lifestyles became relatively sedentary, formal land-use policies arose, and decisions about how to manage land became critical to survival. Even during the earliest periods of the Mayan, Egyptian, and other city-state civilizations, land-use control was implemented for religious, agricultural, hunting, and residential purposes. Religious proscriptions dictated the appropriate appearances of structures. Plagues, warfare, and resource distribution demonstrated the need to isolate structures as well as groups of people. Crowding and cultural conflicts made it necessary for groups to create processes to make government decisions about who would get to use particular resources. Land-use policies are even more necessary in the contemporary world of competing interests, growing populations, and diminishing natural resources. Increased scientific knowledge about environmental impacts has fueled the need to ensure that appropriate land-use decisions are made in the interests of survival through long-term resource management.

ROOTS OF LAND-USE POLICY

The earliest land-use policies took the form of religious prohibitions and mandates as the priest class interpreted the needs of the gods for sacred spaces. Kings exercised a divine mandate in interpreting just what could be allowed in sacred areas, while market forces dictated the use of less important secular space. Later, after kings began losing their divine authority, they could still regulate the use of secular space in the name of promoting social order. Social class structure continued to support the allegiance to class-based authority.

After the Renaissance, European societies believed that while God no longer directly intervened in day-to-day activities, his presence was felt in the need to maintain social order through hierarchies. It was believed that this social order must also be reflected in physical space, or in the way in which a landscape was arranged. Accordingly, it was seen as only proper that buildings, towns, and landscapes reflect particular patterns in form and ownership. In the American colonies that became the United States, examples of kingly intervention in land-use decisions can be found in the marking of potential mast pine trees in New England with the king's broad arrow, the designation of village squares, and the reservation of lots for the king's agents. However, the rise of a democratic society provided a shift in decision-making authority to elected representatives.

By the mid-nineteenth century, early planning laws in the United States began to regulate urban tenement housing and prohibit "obnoxious uses." In 1893 the World's Columbian Exposition in Chicago promoted the exchange of planning and design concepts in exhibits by landscape architect Frederick Law Olmsted, artist Saint Gaudens, and others. By 1895 Los Angeles had an ordinance prohibiting the locating of steam shoddying plants (plants that reprocessed scrap wool) within 30.5 meters (100 feet) of a church. In addition to ordinances and the designation of parks, planned communities were initiated. The publication of Ebenezer Howard's *Tomorrow: A Peaceful Path to Real Reform* (1898) launched the garden city movement. In an 1899 decision, the Massachusetts Supreme Court upheld a building height limitation, and by 1909 zones of limits for buildings were upheld in the U.S. Supreme Court.

By the early twentieth century, organized planning efforts were under way in some large cities. Hartford, Connecticut, established a planning board in 1907. By 1909 Wisconsin had passed the first state enabling act for planning, and Los Angeles provided the first American use of zoning to direct future development in a series of multiple zoning ordinances. In 1913 Massachusetts became the first state to make planning mandatory for local governments. Newark hired the first full-time city planner in 1914, and the first comprehensive zoning code in the United States was enacted in New York City in 1916. By 1925 Cincinnati, Ohio, had become the first major U.S. city to adopt a comprehensive plan, and Burlington, Vermont, had authorized a municipal planning commission.

POLICIES AFTER WORLD WAR II

In the 1940's planning for war and postwar public housing brought the U.S. government to review the land-use planning process. The mustering of resources for World War II demonstrated the value of planning, and many towns implemented town plans as postwar prosperity began. In 1962 the Chicago Area Transit Study showed the applicability of cost-benefit studies to planning for suitable development.

In the 1970's Americans became increasingly concerned with the need to control development. Performance standards were upheld in the courts as one mechanism to manage growth through land-use policy. In 1971 the concept of transferable development rights (TDRs) was introduced to help preserve urban landmarks. Creative land-use tools such as conservation easements, controlled access rights, planned unit development (PUD) density credits, and overlay districts were increasingly used in the 1970's as communities expanded their regulatory schemes. In their zeal to manage growth, some communities enacted land-use policies that were discriminatory. In an early decision on discriminatory land-use regulation, in 1975 the New Jersey Supreme Court struck down a restrictive Mount Laurel zoning ordinance on the basis that it did not allow a regional "fair share" of low- and middle-income housing.

By the late 1980's a sufficient number of court cases existed to reduce the likelihood that communities would enact discriminatory ordinances, but a slump in the U.S. economy fueled the challenge of some land-use ordinances and policies as unauthorized "takings"—the erosion or loss of landowner rights through excessive regulation without due compensation. Yet the vast majority of land-use regulations are not generally found to be true takings because at least some economic use of land is allowed. Still, the issue of a taking is frequently raised when regulations deprive an owner of one or more desired uses. The economic climate has a strong influence on the reaction of a population to particular issues such as takings and to land-use regulation in general.

Land-use regulation is generally viewed as the control of two categories: subdivision (physical size or boundary change) and development (physical use or alteration within the boundary). Changing landownership or subdivision is a land use because it affects the management of resources on the land and can fragment habitat. Most land that has been subdivided stays that way; it is unusual for such land to revert to an original larger tract. Land that is used or developed through construction, clearing, or other alteration, including various forms of land management such as agriculture, also undergoes a change to the natural path of succession. In fact, sufficient past alteration through introduction of new species or direct physical action has so altered some landscapes that it is almost impossible to determine their true "natural" condition. It is this perspective that, along with concern regarding environmental impacts from yet-to-occur changes, is most commonly used to justify land-use policies.

CONVENTIONAL AND CONSERVATION PLANNING

Conventional land-use planning assumes the desirability of economic growth through new development and therefore tends to favor revenue generators. Tools such as zoning are the main techniques for conventional planning. A government land-use plan is implemented through a series of development regulations that differ according to zones. A series of base maps are prepared after the completion of natural, social, and infrastructural resource inventories. The maps can be viewed as both opportunities for and constraints on growth. Individual base maps, when combined with the community's or region's goals and objectives for growth, result in a land-use map containing zones. Each zone reflects a particular category: commercial, industrial, residential, recreational, historical, governmental, agricultural, special, and others that are suggested by the inventory and goal processes. Each category can contain a variety of subcategories based on lot size and range of allowable uses.

The size, configuration, and pattern of the arrangement of lots can all affect growth. For example, large lots reduce the number of houses, while small lots increase the number of houses, reduce the amount of open space, and increase the fragmentation of landownership. The community uses the planning process to agree upon the development capacity of given areas. By using various tools and approaches, a community can seek a balance between achieving appropriate densities and maintaining natural and aesthetic resources; the objective is to achieve sustainability.

Conservation planning is one approach to improv-ing land use within conventional planning. In conservation planning, structures and uses such as septic systems are located away from valued or critical natural resources on a tract. Conventional planning, even if conservation-oriented, can still lead to checkerboard or highway strip patterns of development that are harmful to open space. Clustering is a technique that goes one step further by attempting to preserve or conserve open space by treating it as a natural resource. Other cited benefits of clustering are the fostering of a sense of community, reduction of urban sprawl, and the presentation of traditional village appearances.

In clustering, dwelling units (or commercial structures) are grouped together to allow a larger uninterrupted area of open space (generally 25 to 30 percent), which is often maintained in a residential or commercial subdivision as common land under shared ownership. Some communities encourage clustering through policies that allow a greater density of units or square feet of construction. Thus a 20-hectare (50-acre) tract that might be approved for construction of five houses in a traditional checkerboard subdivision of ten lots might be allowed to contain up to fifteen units of housing if the houses (or condominiums) are clustered and if a specified percentage of the parcel is preserved as open space. The open space might be a separate 8-hectare (20-acre) lot that is deeded to the prospective unit owners as shared or common land to be managed in a certain way.

Conservation planning can be employed on a city- or statewide level through the administration of tiered levels of permits and other development review processes coupled with a system of land-use planning in which designated areas are conserved. An example is Oregon's establishment of greenline boundaries, which limit growth at the edges of cities. Trails and greenbelt corridors also reflect conservation planning, but these require considerable coordination and cooperation when more than one community or state is involved.

ECOLOGICAL PLANNING

Ecological land-use planning expands on conventional land-use planning by taking a more integrative perspective of ecosystem dynamics and applying it over a greater period of time than the normal five-year period. To take this comprehensive approach, planners require significant amounts of data about a wide variety of resources and a fairly stable set of goals and objectives in a relatively constant sociopolitical setting. Ecological planning provides the benefit of long-range dynamic planning while attempting to prevent, rather than remedy, problems. However, it is more costly than conventional planning in initial expenditures. The Netherlands and some other northern European countries have a long history of ecological land-use planning, particularly in response to increased population pressures in areas of finite land resources.

The decrease in numbers of rural inhabitants and the growing numbers of suburbanites in the United States in the latter half of the twentieth century caused an increase in the consideration of various planning techniques and tools. Vermont adopted a statewide land-use policy law in which individual development decisions are made by a regional volunteer citizen panel at quasi-judicial hearings in which dispute resolution techniques and consensus building are encouraged. Vermont has found this case-by-case process to work quite well despite the lack of a comprehensive statewide plan and despite an expanding population. Oregon uses a similar process. Florida's Environmental Land and Water Management Act of 1972 allows the state to designate areas of critical interest that local governments must consider when enacting local policies. It also requires state review of development projects that are large or have regional impacts.

The cumulative adverse environmental effects of many small subdivision and construction projects can significantly outweigh the effects of larger projects that are more intensely regulated. By the end of the twentieth century most U.S. states had begun to recognize this and had increased local control of land use. Most had also implemented statewide natural resource programs that reflected ecological understanding of the interactions between changing land use and the need to manage wildlife and other natural resources.

Forms of Land-Use Regulation

Land use can be controlled through incentives or restrictive processes. Incentive-based land-use control approaches include direct funding, grants, tax abatements, trade-offs such as credits for clustering, and other forms of positive feedback. Restrictive land-use control is the form more commonly recognized and employed, notably through the use of permits, li-

censes, environmental assessments, taxes, and direct prohibitions. Restrictions may be imposed by government or by landowners via covenants or easements. Government restrictive regulations have two forms: prescriptive, in which the objectives and specifications are precisely articulated; and subscriptive, in which the outcomes are specified but the individual means of achieving them are left to the discretion of the owner or community. Critical land uses and issues, such as matters of public health, are more likely to be prescriptively regulated. Less critical matters might be handled by subscriptive means, also called performance-based planning. Even tax structuring can be performance-based when land is taxed based on actual use rather than potential use. This can reduce the pressure for commercial development of high-value or expensive properties.

Most communities and governments use a combination of the two forms of land-use control and the two forms of regulations. Much of land-use policy concerns the manner in which the combination is achieved. Issues of land-use control can become highly politicized, as in the wise-use movement or the controversy involving the northern spotted owl and logging policy in the northwestern United States. In such cases, reaching sufficient community consensus to support consistent regulation can be difficult to achieve because of the differing cultural and social values held by the parties involved.

Although the United States and Canada have contributed greatly to the literature on land-use planning and the development of innovative techniques, North America has relatively weak land-use controls, as does Australia. Northern Europe and countries such as Japan are known for their comprehensive land-use controls. In the United States, as in many countries, the redistribution of population and wealth, together with the restructuring of public lands, necessitates constant reexamination of land-use policies and regulations. For developing nations, land-use policies become particularly critical in the evaluation of trade-offs between natural and land-based resources on one hand and economic well-being on the other. Pressures exist for these countries to exploit their natural resources while striving to achieve the prosperity they see in more developed countries such as those in North America and Northern Europe. As global markets have expanded, the need for international dialogue on land-use issues has also grown.

Robert M. Sanford and Hubert B. Stroud

FURTHER READING

Arendt, Randall. *Rural by Design: Maintaining Small Town Character.* Chicago: American Planning Association, 1994.

Arnold, Craig Anthony. *Fair and Healthy Land Use: Environmental Justice and Planning.* Chicago: American Planning Association, 2007.

Goetz, Stephan J., James S. Shortle, and John C. Bergstrom, eds. *Land Use Problems and Conflicts: Causes, Consequences, and Solutions.* New York: Routledge, 2005.

Johnston, Robert J., and Stephen K. Swallow, eds. *Economics and Contemporary Land Use Policy: Development and Conservation at the Rural-Urban Fringe.* Washington, D.C.: Resources for the Future, 2006.

Marsh, William M. *Landscape Planning: Environmental Applications.* 5th ed. Hoboken, N.J.: John Wiley & Sons, 2010.

Porter, Douglas R. *Managing Growth in America's Communities.* 2d ed. Washington, D.C.: Island Press, 2008.

Sargent, Frederic O., et al. *Rural Environmental Planning for Sustainable Communities.* Washington, D.C.: Island Press, 1991.

SEE ALSO: Conservation policy; Forest and range policy; Greenbelts; Open spaces; Planned communities; Urban planning.

Landfills

CATEGORY: Waste and waste management

DEFINITION: Sites reserved for the disposal of solid wastes

SIGNIFICANCE: Modern sanitary landfills are designed to minimize the hazards to public health and safety that can arise from improper disposal of wastes, in particular the pollution of groundwater or surface water that can occur when chemicals produced by wastes leach into water systems.

Prior to the 1930's most solid waste in the United States was discarded in unsightly, open areas known as dumps. The dumps attracted animal pests, and the chemicals and bacteria in the wastes deposited could easily move into nearby water sources. In the 1930's an increasing number of American communities began disposing of solid wastes in sanitary landfills; by 1960 more than fourteen hundred cities had adopted this method of dealing with waste materials. Subsequently, government regulations mandated the use of sanitary landfills to avoid the pollution and other problems posed by open dumps.

The solid waste placed in sanitary landfills in the United States is about 40 percent paper and 20 percent yard trimmings, with the remaining 40 percent consisting of glass, metal, plastics, wood, food, and other materials. The ratios of the various materials in landfills differ from country to country; for example, one Asian city has a landfill composed of mostly vegetable matter.

At a sanitary landfill, bulldozers compact the trash accumulated at the end of each day, then cover it with a layer of soil 15 to 30 centimeters (6 to 12 inches) thick to keep out animals and air. In areas where the water table (a zone in the ground below which all the pore space contains water) is close to the surface, the waste may be placed directly on the ground in flat areas and covered with soil each day. In places with a lower water table, the waste is placed in a trench before it is covered with soil. In sloping areas, the waste is placed on the slope and covered with material taken from the base of the slope. The most favorable sites for landfills are dry areas far from population centers, but such placement of landfills may not be possible, as transporting waste over long distances is quite expensive. Dry areas are best for landfills because they have little or no water to percolate into the solid waste.

LEACHATE

As water moves through solid waste in a landfill, it dissolves many of the constituent elements in the waste to high concentrations, producing a corrosive liquid called leachate. Ideally, solid waste in landfills should be placed in areas that contain rocks, such as mudstone or igneous rocks, that are impermeable to the movement of the landfill leachate to groundwater or surface water. Burial in mudstone is especially desirable, as the fine-grained clay minerals in the mudstone act as a natural filter and adsorb many of the dissolved or suspended constituents in the leachate. Ideally, a minimum of about 10 meters (33 feet) of impermeable rock is desirable between the base of the landfill and the water table. Landfills placed in permeable sediment such as sand or gravel with the water table close to the surface favor rapid movement of leachate into the groundwater system.

In most modern landfills, a double thickness of an

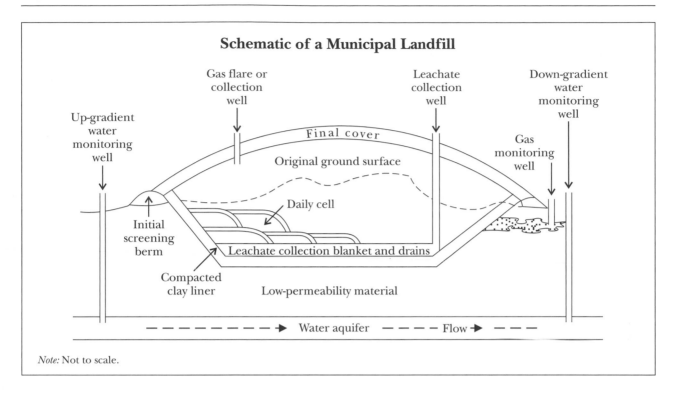

Schematic of a Municipal Landfill

Note: Not to scale.

impermeable liner—made of a material such as plastic—is placed between the landfill material and the impermeable rock, and a monitoring system positioned between the two impermeable layers is used to check for leakage of the leachate. In addition, landfills must have a way to remove the leachate to minimize the amount of waste that can potentially move into the groundwater. Finally, monitor wells adjacent to landfills are periodically sampled for evidence of any contamination that may be moving into the groundwater.

As water slowly moves laterally through waste material, bacteria and many organic and inorganic substances are incorporated into the leachate and can reach high concentrations. The concentration of constituents in the leachate depends on the composition and grain size of the solid waste, the pH (negative log of the hydrogen ion concentration), the amount of water, contact time of the solid waste and leachate, and composition of the gas produced in the landfill. For example, most of the biodegradable solid wastes are in the organic material called carbohydrates. The carbohydrates, with the help of bacteria in an oxygen-free environment, break down into methane and carbon dioxide gas as heat is produced. A variety of organic compounds, such as sugar, acetic acid, butyric acid, and proprionic acid, have been detected in leachate. Also, the acidic leachate helps to dissolve many inorganic compounds out of the solid waste. Thus fresh leachates from landfills in the United States contain relatively high concentrations of constituents such as lead (0.4 milligrams per liter), zinc (22 milligrams per liter), copper (0.1 milligrams per liter), and iron (540 milligrams per liter). Over time, the acidity of the leachate decreases, and the concentration of most constituents tends to decrease correspondingly (for example, lead drops to 0.1 milligrams per liter).

CLOSURE OF LANDFILLS

When a landfill is closed in the United States, it is usually covered with impermeable material and then topped by a layer of soil up to 1 meter (3.3 feet) thick. The layer of soil is graded and planted, and the land may then be used for a park or other recreational facility. In Virginia Beach, Virginia, for example, a landfill had been placed on the surface because the water table there was close to the surface. As waste was added to the landfill over time, a hill more than 280 meters (920 feet) by 120 meters (395 feet) by 20 meters (65 feet) high was built above the land surface. When the landfill was eventually closed, the land became a recreational area.

Buildings should not be constructed on repurposed landfill sites, because the waste below the surface may continue to be compacted gradually over time, and any buildings atop it may subside, resulting in structural damage. In addition, gases such as methane could potentially move from the trapped waste below into any buildings above.

Robert L. Cullers

FURTHER READING

Christensen, Thomas H., Raffaello Cossu, and Rainer Stegmann, eds. *Sanitary Landfilling: Process, Technology, and Environmental Impact.* San Diego, Calif.: Academic Press, 1989.

Melosi, Martin V. *Garbage in the Cities: Refuse, Reform, and the Environment.* Rev. ed. Pittsburgh: University of Pittsburgh Press, 2005.

Reinhart, Debra R., and Timothy G. Townsend. *Landfill Bioreactor Design and Operation.* Boca Raton, Fla.: Lewis, 1998.

Rogers, Heather. *Gone Tomorrow: The Hidden Life of Garbage.* New York: New Press, 2005.

Tammemagi, Hans. *The Waste Crisis: Landfills, Incinerators, and the Search for a Sustainable Future.* New York: Oxford University Press, 1999.

Westlake, Kenneth. *Landfill Waste Pollution and Control.* Chichester, West Sussex, England: Albion, 1995.

SEE ALSO: Electronic waste; Garbology; Groundwater pollution; Hazardous waste; Leachates; Solid waste management policy; Waste management; Water pollution.

Landscape architecture

CATEGORIES: Land and land use; urban environments

DEFINITION: Profession devoted to bringing the natural environment and human beings into greater accord through aesthetic and functional planning and design

SIGNIFICANCE: Landscape architects employ creativity and scientific principles to integrate the natural and constructed environments for sustainable human use and enjoyment. Urban landscape architecture serves to increase the livability of cities by including natural elements in these densely populated and heavily human-built environments.

Landscape architects work to manage and enhance the natural environment, and in so doing they provide many valuable services. Landscape plans and designs usually commence with analysis and research, including feasibility studies and environmental impact assessments. Landscape architects work not only with natural features but also with the built environment; they may be involved in designing for circulation and site security, for the preservation of historic and natural features, or for the restoration of blighted and decaying urban areas. Landscape architects have knowledge of plant materials, construction materials (such as woods, metals, masonry, asphalt, and concrete), soils, civil engineering, and structural landscape design.

HISTORY

From the times of the earliest civilizations—the Sumerians in the Fertile Crescent, the Egyptians in the Nile Valley, the Chinese and Japanese in the Far East, and, later, pre-Columbian societies in the Americas—human beings have developed techniques for manipulating the landscape, including soil, water, and plants, for the purposes of sustenance and shelter. Societies quickly progressed from growing life-sustaining crops to developing plant varieties not only for food but also for ornamentation. Renaissance Europe saw the creation of complex landscape designs as palaces were surrounded by grounds featuring extensive geometric and formal gardens, fountains and reflecting pools, and distinctive topiary, from the pope's summer retreat at Tivoli, Italy, to the palace of the king and queen of France at Versailles.

In England during the late seventeenth and eighteenth centuries gardens were designed that utilized natural landforms and included the mixing of plant materials and water features that served functional purposes and resulted in a more naturalistic appearance. Inigo Jones, who brought Renaissance architectural design to England, became one of the most famous designers of English gardens and is generally recognized as the first landscape architect. In North America, and particularly in the United States, Jones's influence played a larger role than did the more formal approach of the Italian and French designers of palace gardens.

The first American to be recognized as a landscape architect was Frederick Law Olmsted, who became known as the father of American landscape architecture. Along with his partner, architect Calvert Vaux,

Olmsted designed many urban parks, including Central Park in New York City and the Emerald Necklace chain of parks in Boston. Olmsted's park designs became famous for their use of diverse plant materials, undulating landforms, and constructed landscape features that might appear along curved walkway vistas. Olmsted's firm also utilized landscape architecture in the design of the 1893 Chicago World's Fair and of college campuses, including Stanford University.

Landscape architecture advanced significantly in the United States during the twentieth century. During the Great Depression of the 1930's, agencies of the U.S. government such as the National Park Service and the Forest Service engaged landscape architects to design and construct public facilities at sites from the Blue Ridge Parkway in the East to the Grand Canyon in the West. Landscape architects began to be employed as town planners, and, especially during the 1960's and 1970's, some became social reformers and advocates for the establishment of parks and green spaces to revitalize cities and enhance suburban development. In addition, landscape architects have played a major role in helping to write and gain passage of legislation aimed at preserving, protecting, and managing natural resources, including the National Parks Act of 1916 and much of the environmental legislation of the 1970's.

THE WORK OF LANDSCAPE ARCHITECTS

Landscape architects design, develop, and preserve many features in the built and natural landscape, including highways, pedestrian trails, parks and forests, scenic rivers, and historic sites. They are involved in all phases of design and construction, including analysis of the suitability of sites for specific land uses and the formulation of master plans that may become part of a community's land-use plan for growth management. The site plans that landscape architects prepare include detailed graphic design specifications with written explanations and cost estimates for all elements, such as grading, drainage, irrigation, and the construction of retaining walls, decks, signage, lighting, pools, fountains, paving, and other human-made landscape features. In their role as aesthetic and functional designers, landscape architects collaborate with many other professionals, including architects and engineers, in the design and construction of roads and bridges and the arrangement and placement of buildings and structures within the natural environment.

Carol A. Rolf

FURTHER READING

Booth, Norman K., and James E. Hiss. *Residential Landscape Architecture.* 5th ed. Upper Saddle River, N.J.: Prentice Hall, 2007.

Hopper, Leonard J., ed. *Landscape Architectural Graphic Standards.* Hoboken, N.J.: John Wiley & Sons, 2007.

McLeod, Virginia. *Detail in Contemporary Landscape Architecture.* London: Laurence King, 2008.

Simonds, John O., and Barry Starke. *Landscape Architecture: A Manual of Land Planning and Design.* 4th ed. New York: McGraw-Hill, 2006.

Swaffield, Simon, ed. *Theory in Landscape Architecture: A Reader.* Philadelphia: University of Pennsylvania Press, 2002.

Vranckx, Bridget. *Urban Landscape Architecture.* Beverly, Mass.: Rockport, 2007.

SEE ALSO: Balance of nature; Community gardens; Forest Service, U.S.; Greenbelts; Land-use planning; Mumford, Lewis; Olmsted, Frederick Law; Open spaces; Planned communities; Road systems and freeways; Urban parks; Urban planning.

Law of the Sea Treaty. *See* United Nations Convention on the Law of the Sea

Leachates

CATEGORIES: Pollutants and toxins; waste and waste management; water and water pollution

DEFINITION: Liquids produced when fluids dissolve the constituent elements in landfill wastes

SIGNIFICANCE: Some of the potentially toxic materials found in landfills include mercury, lead, cadmium, chromium, arsenic, many organic compounds, and even pathogenic organisms. Because leachates may contain high concentrations of such materials, care is taken in modern landfills to prevent leachates from escaping and contaminating groundwater supplies.

Leachate fluids can vary a great deal depending on the composition of material in the landfill, the age of the leachate, and the speed of the addition and removal of water in the landfill. The flow of water depends on the amount of rainfall, the permeability of the garbage and surrounding rocks (that is, how readily water flows through the material), the kind of liner placed at the base of the waste material, and whether or not a cover has been placed over the landfill to prevent water inflow.

COMPOSITION

Underground water moves slowly and may take years to flow through a landfill. The water thus has a lot of time to dissolve materials such as motor oil, paint, batteries, chlorinated hydrocarbons, pesticides, and other industrial wastes. Many of these materials have long been banned from landfills in the United States, but tens of thousands of old landfills still contain such materials. In fact, many old waste disposal sites were simply depressions in the ground in which almost anything was placed. The U.S. military was one of the worst offenders in creating poor landfills in the past, as military facilities generate more than 500,000 tons of waste per year.

The composition of leachates varies from landfill to landfill and even within a single landfill. The upper concentrations of certain constituents in some leachates can greatly exceed the maximum concentration of those constituents allowed in drinking water. For example, lead concentrations have been found to exceed 5 milligrams per liter in some leachates; drinking water in the United States is required to have less than 0.01 milligram of lead per liter. Zinc concentrations have been found to exceed 1,300 milligrams per liter in some leachates; drinking water is required to have less than 5 milligrams of zinc per liter.

The concentrations of constituents in the leachate in a given landfill may also change over time. For example, large amounts of carbon dioxide are initially produced in the landfill by organisms reacting with biodegradable waste. The carbon dioxide reacts with water to produce acidic waters so that inorganic materials can more readably dissolve. The initial acidic leachate gradually becomes less acid over time as it dissolves the inorganic materials. The total amount of dissolved materials in leachates has been found to be as high as 40,000 milligrams per liter (most drinking water has less than 500 milligrams per liter).

A number of toxic and carcinogenic organic constituents have also been found in landfills, such as dioxin compounds. Dioxins are a group of chlorinated hydrocarbons that are not very soluble in water, but they concentrate in the fatty tissues of animals. They are chemically stable, so they concentrate upward in the food chain. Dioxins are toxic and can cause problems with the immune and reproductive systems; they may also cause some cancers. Other toxic organic constituents that may be present in landfill leachate include certain alcohols, chlorobenzene, acetone, methylene chloride, and toluene.

MANAGEMENT

The goal of leachate management is to prevent leachates from escaping landfills and polluting groundwater supplies. Methods have been developed to achieve this goal and to collect and treat leachates to remove dangerous impurities.

Modern landfills are lined with impermeable and durable materials that prevent leachates from moving into the groundwater. Some of these liners consist of clay minerals that are compacted; others are geomembranes, such as polyethylene sheets. The clay minerals used to line landfills are fine-grained minerals formed naturally during the weathering of coarser-grained minerals. Some of the clay minerals used, such as smectites, may also adsorb many of the metal ions from leachate so the metals do not move with the fluid.

Polyethylene liners are useful in landfills because they are durable and easy to install in addition to being relatively impermeable to most fluids. Such liners can, however, be torn or otherwise damaged by some materials placed into landfills. The most effective system for lining landfills to prevent the movement of leachates out of the landfills and into groundwater combines the two kinds of liners, alternating layers of compacted clay minerals with polyethylene sheets.

Landfills are ideally placed in impermeable sedimentary rocks, such as mudrocks (which also contain abundant clay minerals), that are above the water table rather than in permeable materials such as sand or sandstone. A dry climate also is desirable, but this may not be possible since transportation costs to move material to a landfill may be too high if the landfill is too far from the communities that use it.

Modern landfills also install systems to collect leachate and move it to sites where the worst impurities and contaminants can be removed through chemical and biological treatment. Sometimes leachate may be recycled through a landfill again so that bacteria can further reduce certain impurities. Leachate may also be placed into an oxygen-rich lagoon so that other bacteria may oxidize some of the dissolved materials contained in the leachate. Chemical treatment of leachate can further reduce some kinds of undesirable materials. For example, the precipitation of calcium or sodium hydroxide solids can remove many dissolved metals.

Robert L. Cullers

FURTHER READING

Bagchi, Amalendu. *Design of Landfills and Integrated Solid Waste Management.* 3d ed. Hoboken, N.J.: John Wiley & Sons, 2004.

Bedient, Philip B., Hanadi S. Rifai, and Charles J. Newell. *Ground Water Contamination: Transport and Remediation.* 2d ed. Upper Saddle River, N.J.: Prentice Hall, 1999.

Rogers, Heather. *Gone Tomorrow: The Hidden Life of Garbage.* New York: New Press, 2005.

Tammemagi, Hans. *The Waste Crisis: Landfills, Incinerators, and the Search for a Sustainable Future.* New York: Oxford University Press, 1999.

Westlake, Kenneth. *Landfill Waste Pollution and Control.* Chichester, West Sussex, England: Albion, 1995.

SEE ALSO: Arsenic; Carcinogens; Dioxin; Environmental Protection Agency; Groundwater pollution; Hazardous waste; Heavy metals; Landfills; Lead; Mercury; Methyl parathion; Pesticides and herbicides; Water pollution.

Lead

CATEGORY: Pollutants and toxins

DEFINITION: Heavy and malleable bluish-white metallic element that occurs naturally, at low levels, in the environment

SIGNIFICANCE: Excessive, long-term exposure to lead—which has been used in the manufacture of gasoline, paint, and water pipes, among other things—may lead to severe health problems and even death.

Lead is a relatively rare element that has been mined and used for centuries. The concentration of lead in the earth's crust is only 12.5 parts per million (ppm), and most of this lead is found in the ore mineral galena. Evidence of the use of lead dates back at least 5,000 years, and the environmental impacts of smelting and using lead can be detected in ancient bog deposits in England and Spain that date back at least 2,800 years. Lead was used by ancient European and Chinese civilizations for plumbing, drinking vessels, food and wine storage containers, weights, and ornaments. Considerable evidence from human skeletons indicates that lead poisoning was a significant problem in ancient Roman society.

Lead is now the fifth most commonly used metal in the world. The majority of lead is used for the manufacture of lead-acid batteries, but much of the lead mined in the twentieth century was used for the production of antiknock compounds (such as tetraethyl lead) that were added to gasoline. Lead has also been used in the manufacture of some paints, ceramic glazes, ammunition, and solder, and it is the preferred material for preventing unwanted exposure to X rays. Because of the widespread burning of gasolines with tetraethyl lead for many years, which caused the transport of lead compounds throughout the atmosphere, lead may be the most widely distributed of the heavy metals. Measured levels of lead in remote areas have been found to range from three to five times higher than natural background levels.

About 60 percent of lead exposure for adults comes from food, both fresh and canned. Lead is naturally taken into plants, but it also accumulates in food because of dry deposition from the atmosphere and storage in cans. Approximately 30 percent of the lead in humans comes from the inhalation of air, although this exposure has decreased dramatically since leaded gasoline was phased out of use for most vehicles in the United States in the late 1970's. Lead is also released into the atmosphere through the burning of solid waste, coal, and oil. Tobacco smoke is an additional source of airborne lead exposure. The remaining 10 percent of lead in humans comes from water. Much of the exposure through water is the result of lead pipes, fixtures, and solders used in older plumbing. An additional significant exposure route for children is through ingestion of chips or flakes of the lead-based paint that was used in many homes in the past; children may also ingest dirt containing lead paint residues.

LEAD POISONING

Lead is a toxic element that can cause both acute effects from short-term high-dosage exposure and chronic effects from long-term exposures at lower levels. Children and pregnant women are at particularly high risk with regard to lead exposure. Children may ingest higher levels of lead from soil, and the effects of lead poisoning in children begin at lower blood levels. Pregnant women are at high risk because lead can substitute for calcium in bones and may be mobilized during periods of calcium deficiency, such as pregnancy. Lead released into a pregnant woman's blood can cross the placenta and cause damage to the fetus or even miscarriage.

At high levels, lead poisoning can cause severe brain damage, gastrointestinal disorders, kidney damage, and even death. At lower levels, the symptoms of lead poisoning are not as severe; they include constipation, vomiting, abdominal pains, and loss of muscular coordination. Because these symptoms may also result from other causes, it is not always clear when lead poisoning has occurred.

The long-term chronic effects of low-level lead exposure may lead to many health problems, primarily with the circulatory and central nervous systems. Most of the lead that is absorbed ends up in the blood, and its general residence time is between two and three weeks. Long-term exposure can lead to disruption of the formation of heme in the blood and cause other enzymatic disorders. Anemia can result, with

the symptoms arising at much lower blood lead levels in children.

Lead is also a neurotoxin that affects primarily workers in lead-related mining and manufacturing, as well as children. In children, exposure to lead may lead to lowered mental capabilities, but it is not yet clear to what extent this can be established. It is known that lead poisoning can cause hearing disorders and even slow the growth of children. Chronic exposure to lead can cause blood-pressure problems and interference with vitamin D metabolism. Reproductive problems such as low sperm counts and spontaneous abortions have also been linked to lead exposure.

It is not yet clear whether lead is a carcinogen. Studies on workers in lead industries have been inconclusive with regard to links between cancer and lead exposure, but animal studies have shown an increase in cancers of the kidney. The U.S. Environmental Protection Agency (EPA) considers lead to be a probable human carcinogen and has classified it as a Group B2 carcinogen.

Because of its highly toxic nature, lead has been ex-

Villagers wait to have their children's blood tested near the Dongling Lead and Zinc Smelting Co. plant in China's Shaanxi Province in 2009. Hundreds of children living nearby were apparently poisoned by lead from the plant. (AP/Wide World Photos)

tensively studied and closely regulated in the United States and many other countries. Lead is one of the six primary pollutants regulated by the Clean Air Act, and the EPA is required to establish ambient outdoor air-quality standards for this pollutant. The removal of leaded gasoline from the U.S. market reduced the concentrations of lead in the atmosphere by 93 percent between 1979 and 1988. In addition, the removal of most lead-based paints has reduced the risk of accidental ingestion of lead by children. As a result of these and other regulations, the lead exposure of most adults and children has been significantly reduced. However, lead levels in the blood of many children indicate that exposure is still a great concern.

Jay R. Yett

FURTHER READING

Casdorph, H. Richard, and Morton Walker. *Toxic Metal Syndrome: How Metal Poisonings Can Affect Your Brain.* Garden City Park, N.Y.: Avery, 1995.

Timbrell, John. "Environmental Pollutants." In *Introduction to Toxicology.* 3d ed. New York: Taylor & Francis, 2002.

Warren, Christian. *Brush with Death: A Social History of Lead Poisoning.* Baltimore: The Johns Hopkins University Press, 2000.

Yu, Ming-Ho. *Environmental Toxicology: Biological and Health Effects of Pollutants.* 2d ed. Boca Raton, Fla.: CRC Press, 2005.

SEE ALSO: Biomagnification; Clean Air Act and amendments; Environmental illnesses; Hazardous and toxic substance regulation; Heavy metals.

League of Conservation Voters

CATEGORIES: Organizations and agencies; activism and advocacy; preservation and wilderness issues

IDENTIFICATION: American nonprofit organization that promotes proenvironmental policies and works to elect political candidates who support such policies

DATE: Founded in 1970

SIGNIFICANCE: The League of Conservation Voters has played an important role in influencing the voting public in the United States with its campaigns for or against candidates for national office based on those persons' support or lack of support for environment-related legislation.

In 1969 environmental leader David Brower created the organization Friends of the Earth (FOE), which later grew to become one of the world's largest networks of grassroots environmental organizations. In 1970, after the first Earth Day, Brower split off some of FOE's staff in Washington, D.C., to establish the League of Conservation Voters (LCV), an organization that would focus specifically on politics, tracking the actions of the U.S. Senate and the House of Representatives and campaigning for proenvironmental candidates and issues. Early issues addressed by LCV included nuclear power, the Trans-Alaska Pipeline, and the protection of endangered species. By the twenty-first century, the group had expanded its range of concerns to include global climate change and clean energy.

Every year since 1970, LCV has issued its National Environmental Scorecard, which rates each member of Congress according to his or her voting record on bills concerning environmental, public health, and energy issues. The Scorecard reports on the results of Senate and House votes on certain bills and gives each member of Congress a percentage score. Representatives from approximately twenty environmental and conservation groups work together to identify which votes should be considered and to determine the percentage score for each member of Congress. Several states have their own branches of LCV, and these track members of their state legislatures in the same way. LCV also issues an annual Presidential Report Card, and during election campaigns it releases information about candidates' environmental records.

Since 1996, LCV has released a list that it calls the Dirty Dozen during election years; this list identifies the twelve sitting members of Congress who have most consistently voted against environmental protection and who are up for reelection. In selecting the legislators who will appear on the list, LCV targets candidates in races that are projected to be close enough that an oppositional campaign could make a difference in the outcome; the organization then actively campaigns against these incumbents, emphasizing their environmental voting records. The candidates selected may be members of the Senate or of the House of Representatives and may belong to any political party, although most have been Republicans. LCV also endorses candidates who take strong proenvironmental stances and supports their campaigns with funding through its LCV Action Fund.

LCV set up the LCV Education Fund in 1985 to support environmental education efforts at the local, state, and national levels. The fund provides small grassroots environmental organizations with training, research assistance, and technology that they could not develop on their own; it also helps local organizations conduct opinion polls and run get-out-the-vote programs.

Cynthia A. Bily

FURTHER READING

Duffy, Robert J. *The Green Agenda in American Politics: New Strategies for the Twenty-first Century.* Lawrence: University Press of Kansas, 2003.

Gibson, James William. *A Reenchanted World: The Quest for a New Kinship with Nature.* New York: Macmillan, 2009.

League of Conservation Voters. *National Environmental Scorecard '05.* Washington, D.C.: Author, 2006.

SEE ALSO: Alternative energy sources; Antinuclear movement; Brower, David; Conservation; Conservation policy; Friends of the Earth International; Public opinion and the environment.

Environmental philosopher Aldo Leopold. (AP/Wide World Photos)

Leopold, Aldo

CATEGORIES: Activism and advocacy; preservation and wilderness issues

IDENTIFICATION: American wilderness conservationist and environmental philosopher

BORN: January 11, 1887; Burlington, Iowa

DIED: April 21, 1948; near Baraboo, Sauk County, Wisconsin

SIGNIFICANCE: Leopold, who has been called the father of modern wildlife management and ecology, applied his insightful concepts of ethics and philosophy to conservation strategies and thus helped raise awareness of environmental issues.

Aldo Leopold developed an interest in wildlife while observing waterfowl and animals living in the Mississippi River marshes near his childhood home. After completing graduate studies at the Yale School of Forestry in 1909, he worked for the U.S. Forest Service in the Arizona Territory and New Mexico. With his colleague Arthur Carhart, Leopold demanded that the Forest Service preserve wilderness areas in national forests for their aesthetic and recreational value. As a result, in June, 1924, the Forest Service set aside 305,500 hectares (755,000 acres) in New Mexico as the Gila Wilderness Area, initiating the protection of millions of forest acres for environmentalists to study and enjoy.

Leopold then initiated a game protection movement in the southwestern United States. An avid sportsman, he initially promoted the hunting of predators to protect game. When a deer herd on the Arizona Kaibab Plateau suffered starvation because predator-control measures, among other factors, had resulted in a great increase in the deer population, Leopold realized that the predator-prey balance is crucial to a healthy, stable environment. As game consultant for the Sporting Arms and Ammunition Manufacturers' Institute, he published one of the first studies of American game populations, titled *Report on Game Survey of the North Central States* (1931). He traveled to Europe to assess the game management techniques used in various countries, and he developed a national game management policy for the American Game Protective Association.

In 1933 Leopold published *Game Management,* a popular textbook in which he discussed wildlife population dynamics and habitat protection, emphasizing the preservation of ecosystems. He considered the wilderness as a community to be shared rather than a commodity to be controlled by humans, stressing that nature should be treated ethically and not appropriated for economic gain. Leopold accepted the chair in game management, created especially for him, at the University of Wisconsin in 1933. He built a cabin by the Wisconsin River, planted trees, experimented with land restoration, and observed the wildlife of Sauk County. Active in conservation groups, he helped organize the Wilderness Society and the Wildlife Society. President Franklin D. Roosevelt appointed Leopold to the Special Committee on Wild Life Restoration in 1934.

Leopold wrote a number of essays about environmental ethics, which were compiled in *A Sand County Almanac, and Sketches Here and There* (1949) after his death. He outlined his concept of a land ethic in which all wilderness residents have a vital role. He believed that humans are a part of nature and that all persons have a responsibility to understand, protect, and live in harmony with a healthy environment. He argued that humans should respect nature and "preserve the integrity, stability, and beauty of the biotic community." Because of its literary appeal, Leopold's book introduced a broader audience to the conservation movement. When the book was reissued in the 1960's, a new generation of environmental readers embraced Leopold's passion for nature, praising his conservation initiatives and wilderness wisdom.

Elizabeth D. Schafer

FURTHER READING

Knight, Richard L., and Suzanne Riedel, eds. *Aldo Leopold and the Ecological Conscience.* New York: Oxford University Press, 2002.

Leopold, Aldo. *A Sand County Almanac, and Sketches Here and There.* 1949. Reprint. New York: Oxford University Press, 1987.

Lorbiecki, Marybeth. *Aldo Leopold: A Fierce Green Fire.* 1996. Reprint. Guilford, Conn.: Globe Pequot Press, 2005.

SEE ALSO: Audubon, John James; Conservation; Environmental ethics; Gila Wilderness Area; Kaibab Plateau deer disaster; Nature preservation policy; Predator management; Wilderness areas; Wilderness Society; Wildlife management.

Life-cycle assessment

CATEGORY: Resources and resource management

DEFINITION: Technique for investigating and evaluating the environmental consequences arising from the provision of a product or service

SIGNIFICANCE: Life-cycle assessment is a method of considering the "cradle-to-grave" environmental impacts a product has during its entire life cycle—from raw material acquisition through production, use, consumption, and reuse or final disposal. The technique has the advantage of revealing hidden environmental impacts.

The term "life-cycle assessment" (LCA) is applied to a whole family of environmental assessment tools. The driving force behind the first LCA was an environmental debate about waste and packaging in the United States during the late 1960's. The first LCAs compared the resource consumption and emissions of beverage cans, plastic bottles, and refillable glass bottles. During the global oil crises of 1973 and 1979 the focus of LCAs shifted to the analysis of different petroleum substitutes. With growing environmental awareness LCA became an attractive methodology for product-oriented environmental policies in several countries.

During the 1990's the first attempt to establish an international standard for LCAs was undertaken by the Society of Environmental Toxicology and Chemistry (SETAC). This effort outlined the way for a systematic application of LCA as a decision support tool used in industry, government, and nongovernmental organizations. During the late 1990's the International Organization for Standardization (ISO) released its first international guidelines for LCA practitioners.

According to ISO standards, an LCA study consists of four distinct phases: goal and scope definition, inventory analysis, impact assessment, and interpretation. In the goal and scope phase the product and the purpose of the study is set. Reasons for carrying out LCAs vary; an LCA may be undertaken to compare the environmental impacts of two different products, for example, or to support product development. According to the goal and scope, the system boundaries (boundaries in relation to natural or technical systems, geographic boundaries, time horizon) for the assessment are determined. The second phase, the life-cycle inventory (LCI) analysis, is a process of quantifying the environmentally relevant flows for the entire life cycle of the product. Inputs include energy, raw materials, and ancillary requirements; outputs are atmospheric emissions, waterborne emissions, solid wastes, and other releases. Inputs and outputs are represented in a flowchart. The third phase, called life-cycle impact assessment (LCIA), consists of the evaluation of potential environmental consequences of the inputs and outputs quantified in the inventory analysis.

The general categories of impacts considered in an LCA are ecological consequences, human health effects, and resource use. For practical reasons these three categories are normally divided into more specific impact categories, such as global warming, ozone depletion, resource depletion, acidification, photochemical smog, human health, land and water use, or terrestrial toxicity. The LCI parameters are sorted and assigned to the various categories. For comparison between LCI results within a category, equivalency factors are used. In the global warming category, for example, different greenhouse gases can be expressed in terms of carbon dioxide equivalents. In a forth and final step, the results of the LCI and LCIA are analyzed, and conclusions are drawn regarding the environmental impacts of the investigated product or service.

The application of LCA is widespread and ranges from the analysis of waste management to production processes and different sorts of consumer goods. Despite the popularity of LCA and the international standardization of the process, LCA has received some criticism for a lack of consistency. Possible sources of inconsistency are variations in system boundaries and cutoff criteria, as well as the quality and availability of data. Critics have also noted that any assessment of the social implications of products is generally lacking in LCA.

Claudia Reitinger

FURTHER READING

Brown, Lester R., and Hal Kane. *Full House: Reassessing the Earth's Population Carrying Capacity*. New York: W. W. Norton, 1994.

Dietz, Thomas, and Paul C. Stern, eds. *Public Participation in Environmental Assessment and Decision Making*. Washington, D.C.: National Academies Press, 2008.

Hendrickson, Chris T., Lester B. Lave, and H. Scott Matthews. *Environmental Life Cycle Assessment of Goods and Services: An Input-Output Approach*. Washington, D.C.: Resources for the Future, 2006.

Levy, Geoffrey M., ed. *Packaging, Policy, and the Environment*. New York: Aspen, 2000.

SEE ALSO: Carbon footprint; Ecological footprint; Energy policy; Environmental economics; Green marketing; Recycling; Resource recovery; Sustainable development.

Light pollution

CATEGORIES: Energy and energy use; urban environments

DEFINITION: Human-caused illumination of areas beyond where light is intended and wanted

SIGNIFICANCE: In addition to its adverse effects on the work of astronomers and the wasted energy it represents, light pollution can have deleterious impacts on animals, including humans.

The most obvious effect of light pollution is aesthetic. Stray light from human-made lighting sources makes the sky bright at night, which makes stars and constellations less easily visible. The field of astronomy is adversely affected by light pollution for this reason; in fact, in many highly populated areas, astronomers are unable to see most galaxies and nebulae, even with the use of telescopes. Because of the impact of this situation on their field of study, astronomers have been among the most vocal critics of light pollution, but environmentalists have increasingly joined in efforts to address the problem, noting that it has negative impacts on the environment as well.

Light pollution is often an unintended consequence of intentional lighting. The light from human-made sources, if not properly directed, often goes beyond the areas that are meant to be illuminated. Many people who install outdoor lighting fixtures do not even think about how the light they generate affects others. In this way, light pollution is much like noise pollution. Light pollution is also like noise pollution in that for many years it was thought of solely as a nuisance issue, a problem without any real consequences for the environment.

ENVIRONMENTAL EFFECTS

One of the biggest environmental impacts associated with light pollution is the waste of energy represented by light that goes beyond where it is intended to go. When light shines beyond the area that needs to be illuminated, the light source is using more energy than is really necessary to do the intended job. It has been estimated that the energy used for outdoor lighting could be cut in half if all lighting fixtures were appropriately shielded to direct their light more precisely. Aside from the economic benefits, such a reduction in energy consumption would have a significant impact on the need for electricity generation, which often has negative environmental effects.

Light pollution has also been found to be detrimental to wildlife. Nocturnal animals are adapted to life in the dark, and the lack of full darkness in areas with high levels of light pollution can influence the activities of such animals; for example, it can affect the relationship between predator and prey, as the prey cannot hide in darkness as they normally would. Some studies have shown that migratory animals, particularly birds, are also affected by light pollution, and other research has found that increased light affects the breeding practices of many animals. Excess nighttime lighting can affect plant growth as well.

Humans also appear to be physically affected by light pollution. An increasing body of evidence indicates that the human circadian rhythm developed to include a certain number of hours of darkness per day. Disruptions of a person's circadian rhythm can result in serious health problems, ranging from sleeplessness to irritability and depression.

MITIGATION

Outdoor lighting is essential for safety and security, but not all outdoor lighting generates the same level of light pollution. Reducing light pollution can be as easy as replacing conventional outdoor lighting fixtures with fixtures that are shielded so that they direct light where it is wanted and limit the amount of light that goes in other directions. Such fixtures can achieve the same level of illumination in intended areas with lower-wattage bulbs than are needed in conventional fixtures, resulting in energy savings in addition to reductions in stray light. Shielded light fixtures tend to cost more than unshielded fixtures, but the savings in electricity costs may offset the greater initial expense over time. Retrofitting existing lights with shielding on a large scale, however, can be prohibitively expensive.

The biggest impediment to the reduction of light pollution is a general lack of awareness and understanding of the effects of light pollution among the public. The deleterious effects of air and water pollution are generally well understood, but the effects of light pollution—other than on astronomy—are often not immediately recognized. Because of this lack of awareness, outdoor lighting is often installed by nonexperts. Municipalities, businesses, and home owners could reduce light pollution by consulting with lighting engineers who can design lighting plans to ensure sufficient illumination of desired areas with minimal waste of light and energy.

Raymond D. Benge, Jr.

FURTHER READING

Bakich, Michael E. "Can We Win the War Against Light Pollution?" *Astronomy*, February, 2009, 56-59.

Gallaway, Terrel, Reed N. Olsen, and David M. Mitchell. "The Economics of Global Light Pollution." *Ecological Economics* 69, no. 3 (2010): 658-665.

Klinkenborg, Verlyn. "Our Vanishing Night." *National Geographic*, November, 2008, 102-123.

Luginbuhl, Christian B., Constance E. Walker, and Richard J. Wainscoat. "Lighting and Astronomy." *Physics Today*, December, 2009, 32-37.

Mizon, Bob. *Light Pollution: Responses and Remedies.* New York: Springer, 2002.

Rich, Catherine, and Travis Longcore, eds. *Ecological Consequences of Artificial Night Lighting.* Washington, D.C.: Island Press, 2006.

SEE ALSO: Ecological footprint; Energy conservation; Noise pollution; Urban ecology; Urban planning.

Limited Test Ban Treaty

CATEGORIES: Treaties, laws, and court cases; nuclear power and radiation

THE TREATY: International agreement to halt nuclear test explosions in the atmosphere, under the sea, and in outer space

DATE: Opened for signature on August 5, 1963

SIGNIFICANCE: The Limited Test Ban Treaty opened a dialogue between the United States and the Soviet Union that helped to ease Cold War tensions. It represented a fundamental step toward a global policy of arms control and disarmament, and it also had a positive ecological impact, particularly in the atmosphere, where it led to reduced radioactive contamination.

The atomic bomb was developed by the United States during World War II. The first test explosion of the bomb was carried out in July, 1945, in the desert of New Mexico. Less than one month later, the U.S. military used atomic bombs to destroy the Japanese cities of Hiroshima and Nagasaki, bringing the war to an end. Other countries soon developed their own bomb technology. The Soviet Union exploded an atomic bomb in 1949, followed by Great Britain in 1952 and France in 1960. During the 1950's both the United States and the Soviet Union embarked on military programs to build intimidating nuclear arsenals.

In 1954 the United States detonated a powerful hydrogen bomb at Bikini Atoll in the South Pacific. A Japanese fishing boat that was some 160 kilometers (100 miles) away from the blast was contaminated by radioactive fallout when the wind unexpectedly shifted, and the twenty-three sailors aboard suffered radiation sickness. This incident led to worldwide protests against further nuclear weapons testing. Antinuclear rallies in the United States and Europe mobilized public opinion against the escalating arms race.

In 1958 U.S. president Dwight D. Eisenhower and Soviet premier Nikita Khrushchev agreed to a moratorium on nuclear weapons testing in the atmosphere. It lasted for almost three years, until renewed tensions between the two superpowers—caused by the U-2 spy plane incident and the Berlin Crisis of 1961—led both countries to resume testing. The most powerful bomb in history was a 58-megaton nuclear device detonated by the Soviet Union in October, 1961.

Nuclear explosions in the atmosphere create radioactive particles that are spread around the world by prevailing winds and return to earth with precipitation. Radioactive cesium and iodine are two notably harmful materials that have contaminated grass pastures, causing grazing cows to produce radioactive milk. A study of baby teeth in the early 1960's showed the presence of radioactivity, primarily from milk consumption. The amount of radioactivity was not enough to cause radiation sickness, but even small doses of radiation have a statistical probability of increasing the likelihood of the development of cancers. Researchers for the Brookings Institution esti-

Soviet premier Nikita S. Khrushchev, right, toasts the signing of the Limited Test Ban Treaty on August 5, 1963, with, from left, U.S. senators J. William Fulbright and Hubert Humphrey and United Nations secretary-general U Thant. (AP/Wide World Photos)

mated that, worldwide, the testing of atomic bombs resulted in at least seventy thousand more cases of cancer than would otherwise have occurred.

The Cuban Missile Crisis of 1962 brought the world to the brink of nuclear war. This confrontation between the United States and the Soviet Union led to the mutual realization that the nuclear arms race could escalate into annihilation for both sides. Subsequent negotiations resulted in the signing of the Limited Test Ban Treaty on August 5, 1963, by the United States, the United Kingdom, and the Soviet Union. The treaty, which went into effect on October 10, 1963, prohibited the signatory nations from conducting nuclear weapons tests in the atmosphere, under the ocean, and in outer space; it still permitted underground explosions, however.

The agreement opened a dialogue between the United States and the Soviet Union and significantly eased Cold War tensions. It represented a fundamental step toward a global policy of arms control and disarmament. The treaty also had a positive ecological impact, particularly in the atmosphere, where it led to reduced radioactive contamination; as a result, human health problems linked with nuclear testing and radioactive fallout decreased substantially.

By 1992, 125 countries had signed and become parties to the Limited Test Ban Treaty; by 2008, the number of nations that had signed or acceded to the treaty had risen to 133, although some of them had yet to ratify it. A voluntary ban on underground explosions took effect in 1992, and negotiations for the Comprehensive Nuclear-Test-Ban Treaty were initiated at the United Nations in the late 1990's; that treaty was adopted by the General Assembly of the United Nations in September, 1996, but by the end of the first decade of the twenty-first century, it still had not entered into force. In the meantime, nuclear testing continued as France exploded bombs at the Mururoa atoll in the Pacific Ocean as late as 1995, and as Pakistan and India conducted underground nuclear testing in 1998.

Hans G. Graetzer

FURTHER READING

Barrett, Scott. "International Environmental Agreements." In *Environment and Statecraft: The Strategy of Environmental Treaty-Making.* New York: Oxford University Press, 2003.

Hansen, Keith A. "Early Efforts to Limit Nuclear Testing." In *The Comprehensive Nuclear Test Ban Treaty: An Insider's Perspective.* Stanford, Calif.: Stanford University Press, 2006.

SEE ALSO: Antinuclear movement; Bikini Atoll bombing; Nuclear testing; Nuclear weapons production; Radioactive pollution and fallout.

Limits to Growth, The

CATEGORIES: Ecology and ecosystems; resources and resource management; population issues

IDENTIFICATION: Report on a study, sponsored by the Club of Rome, that used computer modeling to examine global systems

DATE: Published in 1972

SIGNIFICANCE: *The Limits to Growth* marked the general public's first exposure to computer modeling of complex systems, such as the world ecosystem. The book ignited a great debate concerning the long-term sustainability of existing demographic, economic, and resource-use trends.

The Limits to Growth, written by Donella H. Meadows, Dennis L. Meadows, Jørgen Randers, and William W. Behrens III, is sometimes called the Club of Rome study because that organization sponsored the research effort. The authors used a computer modeling technique developed by Massachusetts Institute of Technology (MIT) professor Jay Forrester to trace the interaction of five major variables in the global ecosystem: population, pollution, nonrenewable resources, per-capita food production, and industrial output.

The researchers used the power of the computer to track the behavior of these variables as they responded to the interaction of multiple influences through feedback loops. For example, population change is influenced by birthrate relative to death rate. Both of these rates are influenced by the amount of pollution, which is, in turn, influenced by the population level. The complexity of the multiple interactions quickly exceeds the ability of the human mind to anticipate the behavior of the system. Using the calculating power of a computer, the model could track the performance of the variables. The model was based on five major assumptions: There is a finite stock of exploitable nonrenewable resources; there is a finite amount of arable land; there are limits to the ability of the environment to absorb pollutants; there is a limit to the amount of food that can be produced on each unit of arable land; and population, pollution, resource use, and industrial output grow exponentially, but technological change is incremental and noncontinuous.

The baseline scenario, based on the rates of change of each of the five variables during the twentieth century, projected a rapidly growing global population until the middle of the twenty-first century. According to the model, population growth would greatly overshoot the earth's long-term carrying capacity. At some point during the twenty-first century, a massive population collapse would occur as the depletion of resources, exponential increases in pollution, and dramatic decreases in food per capita would combine to increase death rates and decrease birthrates. When one or more of the variables were altered—for example, when resources were assumed to be unlimited—the timing of the collapse changed, but a catastrophic outcome was not avoided.

Although the authors took great pains to explain that the results of their computer modeling were projections rather than predictions, critics attacked *The Limits to Growth* as a "doomsday" report. Critics noted that the results of the simulation would change radically—there would be no population collapse—if a variable incorporating continual technological change were included in the model. Because ongoing technological change is a common feature of industrialized society, critics argued that this omission was fatal to the model. Despite the criticism, the discussion that surrounded publication of *The Limits to Growth* had a significant impact on the emergence of the notion of sustainable development.

Allan Jenkins

FURTHER READING

Bindé, Jérôme, ed. *Making Peace with the Earth: What Future for the Human Species and the Planet?* New York: UNESCO, 2007.

Meadows, Donella H., Jørgen Randers, and Dennis L. Meadows. *Limits to Growth: The Thirty-Year Update.* White River Junction, Vt.: Chelsea Green, 2004.

Pirages, Dennis, and Ken Cousins, eds. *From Resource Scarcity to Ecological Security: Exploring New Limits to Growth.* Cambridge, Mass.: MIT Press, 2005.

SEE ALSO: Climate models; Club of Rome; Population growth; Renewable resources; Resource depletion; Sustainable development.

Logging and clear-cutting

CATEGORY: Forests and plants

DEFINITIONS: Logging is the harvesting of timber from forestlands with the intention of using it for specific purposes, such as lumber, fuelwood, or the production of pulp or chemicals; clear-cutting is a logging technique in which all the timber is removed from a stand at the same time

SIGNIFICANCE: Improper logging and clear-cutting can pose a variety of threats to the environment, including the destruction of wildlife habitat and the disturbance of soil in ways that can lead to erosion and flooding. Environmentally sensitive logging practices can minimize such negative impacts.

Although some people may think of logging and clear-cutting as practically synonymous, the two are not the same. Similarly, many people may have the impression that commercial logging is responsible for all losses of forestland, but, particularly in tropical areas, many hectares of forestland are cleared annually for other purposes. Rain forests in Amazonia, for example, are often bulldozed to create pastureland for cattle. The timber is not harvested; rather, it is simply pushed into piles and burned at the site.

Logging and clear-cutting, if improperly done or motivated by short-term economic goals, can pose significant threats to the environment. Logging always involves some disturbance to soil and wildlife. If performed in environmentally sensitive areas, it can destroy irreplaceable habitat, contribute to problems with erosion and flooding, and worsen the threat of global warming. Heavy equipment can compact soil, leaving ruts that may persist for many years, while clear-cutting hillsides can lead to erosion, stream siltation, and devastating floods. In Asia, for example, clear-cutting in the mountains of Nepal and India resulted in disastrous floods in Bangladesh. Even when logging does not inflict long-term damage on the immediate environment, the simple removal of trees can contribute to global warming. The burning of slash (waste material) at logging sites pumps greenhouse gases into the atmosphere, and the loss of forest means that there are fewer trees to break those gases down into oxygen and organic compounds.

Regardless of whether a logger is cutting only one tree or one thousand, logging involves four basic steps: selecting the timber to be harvested, felling the trees, trimming away waste material, and removing the desired portion of the tree from the woods. Equipment used in logging ranges from simple hand tools, such as axes and crosscut saws, to multifunction harvesting machines costing hundreds of thousands of dollars each. A mechanized feller buncher, for example, can fell a tree, trim off the branches, cut the stem into logs of the desired length, and stack the logs to await removal from the forest. The choice of equipment utilized in harvesting any specific stand of timber depends on factors such as the terrain, the type of timber to be logged, and whether the logger intends to harvest only selected trees or to clear-cut the site.

Loggers are more likely to clear-cut, or remove all the standing timber from a section of land, if the timber is plantation grown and of a uniform age and size. Clear-cutting also occurs in forests where the desired species of trees need large amounts of sunlight to regenerate. Many conifers, such as Douglas fir, are shade-intolerant. Landowners will occasionally decide to change the dominant species on a tract and so will clear-cut existing timber to allow for replanting with new, more commercially desirable trees. Clear-cutting can be an acceptable practice in sustainable forestry when plantation stands are harvested in rotation.

Selective harvesting, in contrast with clear-cutting, leaves trees standing on the tract. Selective harvesting can be utilized with even-age plantation stands as a thinning technique. More commonly, it is used in mixed and uneven-age stands to harvest only trees of desired species or sizes. In cutting hardwood for use as lumber, for example, 30 centimeters (12 inches) may be considered the minimum diameter of a harvestable tree. Trees smaller than that will be left in the woods to continue growing.

An individual, noncommercial woodcutter may fell only a few trees per year on small parcels of land. Commercial loggers, in contrast, annually harvest hundreds of thousands of trees and operate on large parcels of land. Nonetheless, a significant number of hectares of forestland are cleared annually by people who rely on noncommercial logging for wood for their own individual needs, such as fuel for cooking or heating their homes. Although some woodcutters may cut more than they need for their own use and then sell the surplus, fuelwood for individual house-

A clear-cut mountainside in Canada. (©Charles Dyer/Dreamstime.com)

holds is usually gathered by members of the households that use the wood. Other examples of noncommercial logging include farmers cutting trees for use as fencing or building materials on their own property.

From an environmental viewpoint, the biggest difference between commercial and noncommercial logging would seem to be one of scale, but even this is not always true. Improper logging on a small parcel can have more of an impact on a watershed or ecosystem than professional harvesting of a large stand. Even if no single household's logging practices pose a problem, the gathering of fuelwood or other timber by multiple households in an area can be devastating. With no guidance from professional foresters, woodcutters tend to harvest trees based on convenience for themselves rather than on principles of sustainable forestry or watershed management. Many nations have developed programs in which professional foresters provide advice to small property owners on environmentally sound timber harvesting practices and improvement of timber stands, but the availability of such help varies widely from country to country.

Nancy Farm Männikkö

FURTHER READING

Bevis, William W. *Borneo Log: The Struggle for Sarawak's Forests.* Seattle: University of Washington Press, 1995.

Colfer, Carol J. Pierce. *The Equitable Forest: Diversity, Community, and Resource Management.* Washington, D.C.: Resources for the Future, 2005.

Davis, Lawrence S., et al. *Forest Management: To Sustain Ecological, Economic, and Social Values.* 4th ed. Boston: McGraw-Hill, 2001.

Stenzel, George, Thomas A. Walbridge, Jr., and J. Kenneth Pearce. *Logging and Pulpwood Production.* 2d ed. New York: John Wiley & Sons, 1985.

Tacconi, Luca, ed. *Illegal Logging: Law Enforcement, Livelihoods, and the Timber Trade.* Sterling, Va.: Earthscan, 2007.

Walker, Laurence C., and Brian P. Oswald. *The Southern Forest: Geography, Ecology, and Silviculture.* Boca Raton, Fla.: CRC Press, 2000.

U.S. Lumber Consumption
(billions of board feet)

	1995	2000	2003	2005	2007
Species group					
Softwoods	47.6	54.0	56.5	64.4	50.5
Hardwoods	11.7	12.2	10.5	11.2	10.3
End use					
New housing	15.9	20.6	24.0	28.6	26.2
Residential	14.3	16.4	18.3	20.6	20.9
New nonresidential construction	5.8	5.1	4.4	4.3	3.9
Manufacturing	5.5	—	8.1	7.7	7.3
Shipping	8.5	7.4	7.5	7.6	7.7
Other	9.3	16.1	4.7	7.0	—

Source: U.S. Forest Service, *U.S. Timber Production, Trade, Consumption, and Price Statistics, 1965-1999,* 2001; and U.S. Department of Commerce, *Statistical Abstract of the United States, 2009,* 2009.

SEE ALSO: Deforestation; Forest management; National forests; Old-growth forests; Rain forests; Slash-and-burn agriculture; Sustainable forestry.

London Convention on the Prevention of Marine Pollution

CATEGORIES: Treaties, laws, and court cases; water and water pollution

THE CONVENTION: International agreement intended to halt the reckless dumping of wastes in marine waters

DATE: Opened for signature on December 29, 1972

SIGNIFICANCE: While not the largest source of marine pollution, dumping of wastes at sea was the norm for the decades leading up to the Convention on the Prevention of Marine Pollution. With a burgeoning global population and new technology generating greater amounts of more potent waste, the need to limit this practice was and remains crucial.

The Convention on the Prevention of Marine Pollution by Dumping of Wastes and Other Matter, also known as the London Convention, was drafted and opened for signature in 1972 and entered into force in 1975. The aim of the convention is to prevent indiscriminate dumping of wastes at sea that could pose a threat to human health, damage marine life, or interfere with other uses of the sea. The convention applies to all marine waters other than a nation's internal waters; it excludes land-based dumping. By 2010 eighty-six nations had become parties to the convention.

The convention employs a so-called black list/gray list system. The black list specifies a number of dangerous pollutants that the parties to the convention are banned from dumping into marine waters, except in some cases where there are only trace levels of the pollutant or it is "rapidly rendered harmless." Examples of pollutants found on the black list include radioactive waste, persistent plastics, mercury, cadmium, organohalides, and materials for chemical or biological warfare. In contrast, pollutants found on the gray list can be dumped with a permit and under conditions specified by the convention.

Signatory nations are responsible for setting up their own laws to implement the convention's policies, and dumping figures are self-reported. The practice of self-reporting has at times hampered the convention's effectiveness, such as during the 1980's, when the Soviet Union regularly failed to report its dumping of radioactive waste. The convention, however, includes arbitration procedures for those times when parties disagree on reporting. When issuing permits for dumping, member nations are required to record the types and amounts of materials dumped, when and where the dumping takes place, and the condition of the seas of the area where the materials are dumped.

The contracting parties to the convention meet regularly at annual consultative meetings at which issues are raised and addressed and policies are shaped. The meetings are organized by the International Maritime Organization, which acts as the secretariat of the convention. Generally these meetings have widened the scope of the convention, adding new pollutants to the black list and setting tougher standards for permitting. The parties to the convention receive scientific and other expert input in making their decisions from a number of groups. The Scientific Group on Dumping, the convention's principal advisory board, is made up of experts chosen by the contracting parties; the group provides advice on issues related to the implementation of the convention's policies based on the latest scientific information.

In 1996 the London Protocol, a new agreement aimed at modernizing and eventually replacing the convention, was drafted. The protocol takes a much stricter stance on marine dumping, employing both the precautionary principle and the polluter pays principle. Regardless of the absence of conclusive evidence, the protocol regards all pollutants as detrimental to the marine environment. Thus, instead of black and gray lists, this agreement uses a "reverse list" that notes the pollutants permissible for dumping with a permit and bans all others. Also, parties are banned from exporting their waste to noncontracting parties for the purpose of dumping at sea. The London Protocol went into effect in 2006, and by 2010 thirty-seven nations had become parties to the agreement.

Daniel J. Connell

FURTHER READING

Guruswamy, Lakshman D., with Kevin L. Doran. "Dumping." In *International Environmental Law in a Nutshell.* 3d ed. St. Paul, Minn.: Thomson/West, 2007.

Louka, Elli. "Marine Environment." In *International Environmental Law: Fairness, Effectiveness, and World Order.* New York: Cambridge University Press, 2006.

SEE ALSO: Environmental law, international; Marine debris; Ocean dumping; Ocean pollution; Polluter pays principle; Precautionary principle; United Nations Convention on the Law of the Sea; Waste management.

London smog disaster

CATEGORIES: Disasters; atmosphere and air pollution

THE EVENT: Incident in which a lethal combination of fog, smoke, and pollutants settled over London, England, for several days

DATES: December 4-8, 1952

SIGNIFICANCE: After London's "killer smog" incident resulted in thousands of illnesses and deaths, the British government undertook to pass strong legislation that would address the problem of air pollution.

On the evening of December 4, 1952, weather conditions caused a "killer smog"—a lethal combination of fog, smoke, and pollutants—to settle over

Heavy smog in London's Piccadilly Circus on December 6, 1952. (Hulton Archive/Getty Images)

London, England; the heavy smog did not clear until December 9. Transportation was severely disrupted by low visibility during the disaster, and many outdoor sporting events had to be canceled. The chief culprit in the London smog disaster was deemed to be Great Britain's heavy dependence on coal, which was used by industry and burned in almost every household in London, since the country's wood supplies had long been depleted. Particular blame was placed on the popular "nutty slack," a low-grade soft coal used in most households that burned inefficiently and gave off noxious smoke and odor. Authorities estimated that approximately one-half of the smoke emitted during the killer smog came from household hearths.

While the smog was present and especially in the period immediately after the smog abated, the health effects of breathing such polluted air for days began to be seen. Thousands were hospitalized with circulatory and respiratory problems, including bronchitis, influenza, and pneumonia. During the last three weeks of December, the death rate in London climbed dramatically. The minister of health later announced that during a five-week period ending January 3, 1953, more than fifteen thousand deaths were registered in Greater London, compared with approximately nine thousand during the same period one year previously. It has been estimated that at least four thousand deaths were directly attributable to the smog, with most victims being babies under one year old and persons over the age of fifty-five. The death rate for this period exceeded that of previous disasters in London's history, including the worst periods of the cholera epidemic of 1866 and the great fog of December, 1873.

The British Clean Air Act of 1956, which represented an attempt to correct the smog problem, was a direct result of the 1952 disaster. The act banned the emission of black smoke from locomotives, vessels, and chimneys, and it required that all new furnaces be capable, as far as practical, of not producing smoke. In addition, the act banned the emission of grit or dust from all furnaces, new or old. It also created the Clean Air Council to advise government ministers in regard to clean air policy. One of the act's most important provisions gave cities the power to establish "smoke-control zones" to combat the problems caused by the burning of coal in home fireplaces and at local factories. To assist in this last objective, the national government offered financial grants to cover 40 percent of the cost of converting home appliances so that they could burn smokeless fuel; local government authorities were required to contribute another 30 percent, so that the appliance owners had to pay only 30 percent of the total cost of conversion.

Even with the measures initiated under the Clean Air Act, it was estimated that it would take up to fifteen years for the full impact of the improvements to be felt. London's smog problem was eventually overcome, however, thanks to this legislation and to the long-term trend of factories, railroads, and households converting to oil, gas, nuclear, and electrical energy.

David C. Lukowitz

FURTHER READING

Benton-Short, Lisa, and John R. Short. *Cities and Nature.* New York: Routledge, 2008.

Kessel, Anthony. *Air, the Environment, and Public Health.* New York: Cambridge University Press, 2006.

Vallero, Daniel. *Fundamentals of Air Pollution.* 4th ed. Boston: Elsevier, 2008.

SEE ALSO: Air pollution; Black Wednesday; Coal-fired power plants; Donora, Pennsylvania, temperature inversion; Environmental illnesses; Smog.

Los Angeles Aqueduct

CATEGORIES: Preservation and wilderness issues; water and water pollution

IDENTIFICATION: System of canals, tunnels, and pipes built to divert water from the Owens River to Los Angeles

DATE: Completed on November 5, 1913

SIGNIFICANCE: The Los Angeles Aqueduct was constructed to meet the growing demand for water in the city of Los Angeles, but the diversion of water from the Owens River led to the drying of the Owens River Valley and the subsequent collapse of that region's agricultural industry.

The Los Angeles Aqueduct extends 378 kilometers (235 miles) from the Owens River a few miles north of Independence, California, to San Fernando, California, on the north side of Los Angeles. Begun in 1908 and completed in 1913, it is a complex system of unlined, lined, and lined and covered canals, tunnels, and steel pipes. For seventy-three years the entire flow of the Owens River at the aqueduct intake point, an average of about 984 million liters (260 million gallons) per day, was diverted from the Owens Valley to Los Angeles. Beginning in 1986, a portion of the original flow was restored to the Owens River below the intake point. The aqueduct was extended north to the Mono Lake basin in 1940, and a parallel aqueduct was completed in 1970.

In 1900 the population of Los Angeles was about 200,000 and was rapidly increasing. The city's water supply came from the Los Angeles River and a few wells and local springs. A substantial supply of water was required for the city to continue to grow, and there was no local source. To solve the problem, Fred Eaton, an engineer and former mayor of Los Angeles, conceived the Los Angeles Aqueduct and discussed his idea with William Mulholland, superintendent of the Los Angeles Department of Water. Mulholland spent forty days surveying the proposed aqueduct route and discussed the proposal with the Los Angeles Board of Water Commissioners. All this was done in secret to avoid a burst of land speculation. Eaton had already obtained options to buy most of the private land along the aqueduct route and later agreed to sell the options to Los Angeles at cost.

One obstacle remained. The U.S. Reclamation Service (which was created in 1902 and renamed the Bureau of Reclamation in 1923) was in the process of

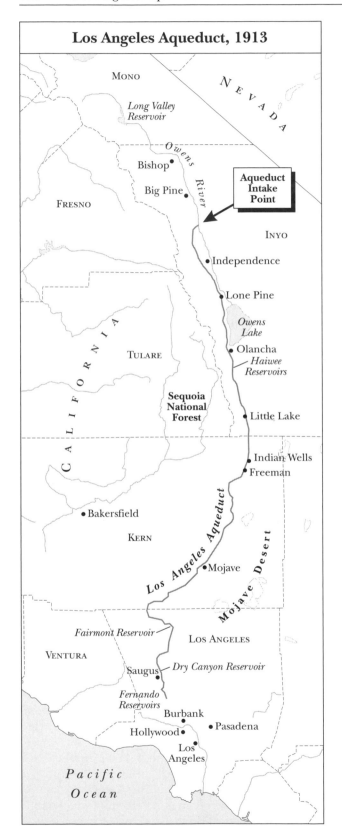

Los Angeles Aqueduct, 1913

planning a major irrigation project in the Owens Valley and had withdrawn the public lands in the area from claim. Part of the aqueduct route passed through these public lands. The residents of the valley were enthusiastically in favor of the irrigation project.

Plans for the aqueduct were made public, and a committee from the Los Angeles Chamber of Commerce met with President Theodore Roosevelt. To the disappointment of the Owens Valley residents, President Roosevelt reached the conclusion that the Owens River water would be much more beneficial to Los Angeles than to Owens Valley, and in 1906 the U.S. Congress granted the necessary right-of-way for the aqueduct. To ensure its right to the Owens River water and increase the available supply, the city of Los Angeles began buying the irrigated ranches and farms above the aqueduct intake in addition to properties below the intake that were now useless for lack of water. It was this maneuver, rather than the building of the aqueduct itself, that angered Owens Valley residents.

The Owens River Valley lies between the Sierra Nevada mountain range to the west and the White Mountains and Inyo range to the east. It is about 16 kilometers (10 miles) wide and 160 kilometers (100 miles) long. In its pristine state, the valley was a desert. The Owens River, flowing south from the Sierra Nevada, supported a fringe of willows and other riverside vegetation; the rest of the valley was thinly covered with cactus, chaparral, and sagebrush. The river ended at Owens Lake, an alkaline lake with no outlet. Because of evaporation over many thousands of years, the lake water was highly mineralized, primarily with sodium bicarbonate, but also with sulfates, chlorides, and other salts. The upper end of the lake was a freshwater marsh, which provided good habitat for waterfowl, as did the river north of the lake. The river, its fringe of vegetation undoubtedly home for many native birds, was on a major migratory pathway for several species of birds. Irrigation ditches serving farms, ranches, and orchards extended as far as 8 kilometers (5 miles) from the river. Carp, an exotic fish imported from Europe, swam in the irrigation ditches.

On completion of the Los Angeles Aqueduct in November, 1913, the lower 85 kilometers (53 miles) of the Owens River channel became dry; partial flow was not restored until 1986. The narrow fringe of riverside vegetation died, the marshes at the head of Owens Lake dried, and, eventually, the lake itself dried. Windstorms crossing the dry lake bed carried irritating alkali dust through the valley and often far beyond. The irrigated farms and ranches almost entirely disappeared, and Los Angeles, which owned most of the valley, allowed it to return to desert. The economy of the valley changed from one based on agriculture to one based on tourism and outdoor recreation. In purely economic terms, the income from tourism and outdoor recreation has vastly exceeded what might have been expected from an expansion of irrigation-based agriculture in the region.

By 1970 Los Angeles was using large quantities of well water to supplement aqueduct flow, lowering groundwater levels and drying local springs. In addition, the water level at Mono Lake was declining, threatening island breeding grounds of California gulls by creating land bridges that coyotes could cross.

If the aqueduct had not been built, the Reclamation Service probably would have constructed massive irrigation systems that might have been much more damaging to the environment of Owens Valley. Indeed, a Sierra Club spokesman once said, "We recognize that Los Angeles is probably the savior of the valley. Our goal is to save the valley as it is now."

Robert E. Carver

FURTHER READING

Deverell, William, and Greg Hise, eds. *Land of Sunshine: An Environmental History of Metropolitan Los Angeles.* Pittsburgh: University of Pittsburgh Press, 2005.

Fradkin, Philip L. *The Seven States of California: A Natural and Human History.* New York: Henry Holt, 1995.

Hundley, Norris, Jr. *The Great Thirst: Californians and Water—A History.* Rev. ed. Berkeley: University of California Press, 2001.

Mulholland, Catherine. *William Mulholland and the Rise of Los Angeles.* Berkeley: University of California Press, 2000.

Nadeau, Remi. *The Water Seekers.* 4th ed. Santa Barbara, Calif.: Crest, 1997.

Wood, Richard Coke. *The Owens Valley and the Los Angeles Controversy: Owens Valley as I Knew It.* Stockton, Calif.: University of the Pacific, 1973.

SEE ALSO: Aqueducts; Irrigation; Riparian rights; Roosevelt, Theodore; Water rights; Water use; Wetlands.

Love Canal disaster

CATEGORIES: Disasters; human health and the environment; waste and waste management

THE EVENT: Devastation of a community in Niagara Falls, New York, as the result of the burial of toxic chemical wastes in the area some years before

DATES: 1976-1980

SIGNIFICANCE: The events that took place at Love Canal demonstrate the environmental damage and dangers to human health posed by the improper disposal of toxic wastes. The Love Canal case also shows how informed and active citizens can influence legislators and policy makers to address environmental problems.

The discovery and identification of dangerous chemical wastes in the Love Canal neighborhood of Niagara Falls, New York, in 1976 transformed a community where the residents' livelihoods depended on the chemical companies long established in the area. Houses were boarded up and abandoned, a school was left empty and falling down, warning signs were posted, and the entire area was fenced off. The completeness of the human and ecological devastation at Love Canal made it the standard against which all subsequent chemical waste disasters have been compared. The toxic terror generated by Love Canal was caused by chemical waste buried at a time when the term "pollution" was not yet part of the American vocabulary.

In 1892 entrepreneur William Love arrived in Niagara Falls with plans to construct a navigable power canal between the upper and lower portions of the Niagara River and use the 90-meter (300-foot) drop in water level to generate electric energy. Digging began in May, 1894, but the depressed state of the economy at the time resulted in a withdrawal of investment capital, which ended the project with the canal less than one-half finished. Love Canal became the property of the Niagara Power and Development Company, which gave the Hooker Electrochemical Corporation permission in 1942 to dump wastes in the canal. The site was considered to be ideal for disposal of chemical wastes as the walls were lined with clay, which has a low level of permeability. Hooker purchased the property in 1947.

Between 1942 and 1952, Hooker disposed of 21,800 tons of chemicals at the site, burying them at depths of 6 to 8 meters (20 to 25 feet). The majority of

Love Canal, New York

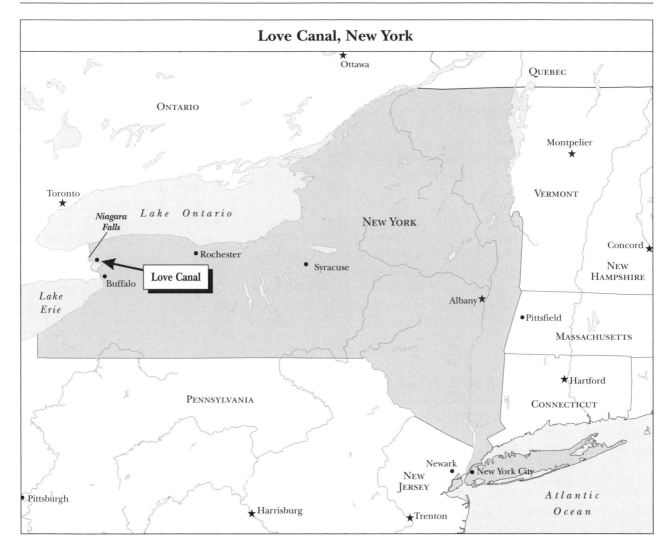

the chemicals were contained in metal and fiber barrels, although some waste was reputedly dumped directly into the canal. The customary method for disposal of chemical wastes throughout the United States in the 1950's was to dump them directly into unlined pits, lagoons, rivers, lakes, or surface impoundments; wastes were also sometimes burned. Apart from disposal into bodies of water, these methods were legal up until 1980.

In 1953 Hooker filled the canal and topped it with a 0.5-meter (2-foot) clay cap. Beneath lay 43.6 million pounds of eighty-two different chemical residues. Included were benzene, a chemical known to cause anemia and leukemia; lindane, exposure to which results in convulsions and excess production of white blood cells; chloroform, a carcinogen that attacks respiratory, nervous, and gastrointestinal systems; trichloro-

ethylene, a carcinogen that attacks genes, livers, and nervous systems; and methylene chloride, which can cause recurring respiratory distress and death. The most dangerous chemical in the waste, however, was dioxin, a component of the 200 tons of trichlorophenol dumped in the canal. Dioxin has been described as one of the most powerful known carcinogens.

Shortly after the canal was filled, the Niagara Falls Board of Education purchased the canal property for the token sum of one dollar on the condition that the board warn future owners of the buried chemicals, use the land only as a park with the school in close proximity, and build no houses on the property. Shortly after taking possession, however, the school board removed 13,000 cubic meters (17,000 cubic yards) of topsoil from the canal for grading at the site of the 99th Street School and the surrounding area. City workers then installed a sewer that punctured both the canal walls and the clay covering to facilitate adjacent housing tracts. Breaks in the clay cap and walls of the canal created openings through which the toxic chemicals eventually flowed.

During the winter of 1975-1976, heavy snowfall caused the groundwater level to rise in the Love Canal area, filling the uncovered canal. Portions of the land-fill subsided, and waste storage drums surfaced in a number of locations. Surface water, heavily contaminated with chemicals, was found in the backyards of houses bordering the canal. Residents complained of discomfort and illness caused by unpleasant chemical odors coming from the canal.

Lois Gibbs, president of the Love Canal Home-owners Association, united the community and became an effective and persuasive advocate for families seeking government aid. Gibbs involved herself in December, 1977, when her son, Michael, began to experience asthma and seizures just four months after he entered kindergarten at the 99th Street School. She went door-to-door and questioned other residents about their health in an attempt to discover the full extent of con-tamination. In 1979 Gibbs traveled to Washing-ton, D.C., where she testified before Congress on behalf of the Love Canal home owners.

In 1977 studies conducted by Niagara Falls city and New York state health officials showed exten-sive pollution affecting 57 percent of the homes at the southern end of the canal and a "moderate excess of spontaneous abortions and low birth weight infants occurring in households on 99th Street bordering the landfill." In August, 1978, the New York state health commissioner closed the 99th Street School and evacuated 240 families living within two blocks of the canal. In October, 1980, after the release of the alarming results of a health study conducted by the U.S. Environmen-tal Protection Agency (EPA), President Jimmy Carter ordered a total evacuation of the commu-nity.

The chemical disaster at Love Canal left be-hind a legacy of lawsuits and bitterness. In Octo-ber, 1983, a tentative settlement of the billions of dollars in lawsuits was reached by lawyers of the Hooker Electrochemical Corporation, the city of Niagara Falls, the Niagara Falls Board of Educa-tion, Niagara County, and former residents of the Love Canal area. Claims against Hooker and the pub-lic agencies totaled $16 billion. On February 19, 1984, former residents of Love Canal received payments ranging from $2,000 to $400,000.

In November, 1980, Congress passed legislation to deal with the cleanup of toxic wastes. The Compre-hensive Environmental Response, Compensation, and Liability Act (CERCLA), commonly referred to as Superfund, established a $1.6 billion fund for the cleanup of hazardous substances to be administered by the EPA. The money was to be used when "no re-sponsible party could be identified or when the re-sponsible party refuses to or is unable to pay for such a cleanup."

Peter Neushul

Mark Zanatian, one of the children endangered by the toxic chemicals buried under the site of the 99th Street School in the Love Canal neighbor-hood, waves a banner in protest during a neighborhood demonstration in August, 1978. (AP/Wide World Photos)

FURTHER READING

Blum, Elizabeth D. *Love Canal Revisited: Race, Class, and Gender in Environmental Activism.* Lawrence: University Press of Kansas, 2008.

Gibbs, Lois. *Love Canal: My Story.* Albany: State University of New York Press, 1982.

_____. *Love Canal: The Story Continues . . . .* Stony Creek, Conn.: New Society, 1998.

_____. "What Happened at Love Canal." 1982. In *So Glorious a Landscape: Nature and the Environment in American History and Culture,* edited by Chris J. Magoc. Wilmington, Del.: Scholarly Resources, 2002.

Layzer, Judith A. "Love Canal: Hazardous Waste and the Politics of Fear." In *The Environmental Case: Translating Values into Policy.* Washington, D.C.: CQ Press, 2002.

Levine, Adeline Gordon. *Love Canal: Science, Politics, and People.* Lexington, Mass.: Lexington Books, 1982.

SEE ALSO: Côte d'Ivoire toxic waste incident; Dioxin; Gibbs, Lois; Hazardous and toxic substance regulation; Hazardous waste; Superfund.

Lovejoy, Thomas E.

CATEGORIES: Activism and advocacy; preservation and wilderness issues

IDENTIFICATION: American tropical biologist

BORN: August 21, 1941; New York, New York

SIGNIFICANCE: Lovejoy is recognized for his contributions to conservation policy making. He is best known for developing creative solutions to issues of scientific concern, such as debt-for-nature swaps.

In 1971 Thomas E. Lovejoy earned a doctor of philosophy degree in biology from Yale University. His early research efforts included a long-range study of birds in the Brazilian Amazon. From 1973 to 1987 Lovejoy served as program director for the World Wildlife Fund (WWF) and was responsible for the organization's projects in the Western Hemisphere and those involving tropical forests. He is credited with be-

Tropical biologist Thomas E. Lovejoy. (©Lynn Goldsmith/CORBIS)

ing the first person to use the term "biological diversity," or "biodiversity," in 1980. Lovejoy also began speaking of global extinction rates, and in 1980 his first projection of such rates appeared in *The Global 2000 Report,* submitted to U.S. president Jimmy Carter. This marked the beginning of his many contributions to the policy arena.

It was during his tenure as executive vice president of WWF from 1985 to 1987 that Lovejoy originated the concept of the debt-for-nature swap, which involves forgiveness of foreign debts incurred by developing nations in exchange for those nations' agreements to protect the fragile ecosystems within their borders. Debt-for-nature swaps have been used to preserve wilderness in countries such as Costa Rica, Bolivia, Ecuador, the Philippines, and Madagascar. While working for WWF, Lovejoy increased public interest in the issue of preserving tropical forests.

Lovejoy is recognized for his practical approach to solving environmental problems. This is illustrated by his input for the Minimum Critical Size of Ecosystems Project, also known as the Biological Dynamics of Forest Fragments Project. This was a joint research project undertaken by the Smithsonian Institution and Brazil's National Institute of Amazonian Research. The project was one of the earliest attempts to define minimum sizes for national parks and biological reserves and to formulate management strategies for such protected areas. In recognition of his conservation work in Brazil, Lovejoy became the first environmentalist to receive the Order of Rio Branco, which the Brazilian government awarded him in 1988.

In 1987 Lovejoy created the Public Broadcasting Service television series *Nature*, for which he served as adviser for many years. In the same year he was appointed assistant secretary for environmental and external affairs at the Smithsonian Institution. By 1994 Lovejoy had risen to become counselor to the secretary for biodiversity and environmental affairs at the Smithsonian. While still associated with the Smithsonian Institution, he became chief biodiversity adviser to the World Bank in 1998.

While continuing his association with the Smithsonian, Lovejoy also served in a variety of advisory positions. In 1993 he was chosen to be science adviser by the U.S. secretary of the interior and helped set up the National Biological Survey. From 1994 to 1997 Lovejoy served as scientific adviser to the executive director of the United Nations Environment Programme (UNEP), and he later became senior adviser to the president of the United Nations Foundation. His professional contributions have included serving as chair of the Yale Institute for Biospheric Studies, as chair of the U.S. Man and the Biosphere Program, as president of the Society for Conservation Biology, and as president of the American Institute of Biological Sciences. Among the honors Lovejoy has received are the 2001 Tyler Prize for Environmental Achievement and the 2002 Lindbergh Award, which is presented by the Lindbergh Foundation in recognition of contributions made to maintaining a balance between technology and preservation of the environment.

Lovejoy has published numerous articles and several books on environmental conservation, including two coedited volumes: *Global Warming and Biological Diversity* (1992; with Robert L. Peters) and *Climate Change and Biodiversity* (2005; with Lee Hannah).

Michele Zebich-Knos

FURTHER READING

Lovejoy, Thomas E. "Biological Diversity." In *Life Stories: World-Renowned Scientists Reflect on Their Lives and on the Future of Life on Earth*, edited by Heather Newbold. Berkeley: University of California Press, 2000.

_____. "Climate Change and Prospects for Sustainability." In *Foundations of Environmental Sustainability: The Coevolution of Science and Policy*, edited by Larry L. Rockwood, Ronald E. Stewart, and Thomas Dietz. New York: Oxford University Press, 2008.

Lovejoy, Thomas E., and Lee Hannah, eds. *Climate Change and Biodiversity*. New Haven, Conn.: Yale University Press, 2005.

SEE ALSO: Biodiversity; Climate change and oceans; Debt-for-nature swaps; *Global 2000 Report, The*; International Union for Conservation of Nature.

Lovelock, James

CATEGORY: Ecology and ecosystems
IDENTIFICATION: English environmentalist and inventor
BORN: July 26, 1919; Letchworth, Hertfordshire, England
SIGNIFICANCE: Lovelock is best known for his Gaia hypothesis, which suggests that the earth itself is the source of life and that all living things on the planet have coevolved and therefore are inextricably intertwined.

James Lovelock was educated at the University of London and Manchester University and earned a doctoral degree in medicine. He taught in the United States at Yale University and at the Baylor College of Medicine, and he was a Rockefeller Fellow at Harvard University. He was elected a fellow of the Royal Society of London in 1974. He was president of the Marine Biological Association of the United Kingdom from 1986 to 1990, and in 1994 he accepted the position of Honorary Visiting Fellow of Green Templeton College, Oxford. Among the honors he has received are the Tswett Medal, the Dr. A. H. Heineken Prize for the Environment, the Royal Geographical Society Discovery Lifetime Award, and the Wollaston Medal, which is the highest honor bestowed by the Geological Society of London. He was named a Commander of the British Empire in 1990, and in 2003 he was named a Companion of Honour.

During the 1970's, Lovelock served as a consultant for the National Aeronautics and Space Administration (NASA) and was also known as an independent scientist and inventor. While at NASA he developed the electron capture detector, which has been utilized to study and analyze atmospheric gases as they exhibit themselves on other planets such as Mars and Venus. Data gained from his electron capture studies strongly suggest the uniqueness of Earth with respect to the fostering of life.

During his tenure as a NASA consultant, Lovelock developed the Gaia hypothesis, which suggests that as the earth's early atmosphere cooled, the mixing of gases and gradual increases of water on the planet provided the necessary nutrients for the evolution of microorganisms that began to utilize the earth's toxic gases. Those that could perform photosynthesis began to contribute heavily to the oxygen levels of the earth's troposphere. Eventually, other nonphotosynthetic microorganisms evolved that returned carbon dioxide to the atmosphere. In a continuous cycle, the microscopic biota cycled and recycled the oxygen and carbon dioxide gases to provide the atmosphere as it is recognized today.

Lovelock named his hypothesis after the Greek goddess of the earth, Gaia, upon the advice of his friend and colleague William Golding (author of the 1954 novel *Lord of the Flies*). The powerful notion that life itself contributes to and continually modifies the atmosphere was not readily accepted at the time. Skeptics argued that the sum of the biota could not purposefully maintain an entire atmosphere. Yet Gaia has remained a viable theory that continues to explain not only the earth's early evolution and the importance of early life-forms but also the importance of life in maintaining atmospheric balance. Current problems of global warming may also be related in that macroscopic organisms—namely, humans—may be upsetting this delicate balance.

Lovelock's first two major books on the Gaia hypothesis, *Gaia: A New Look at Life on Earth* (1979) and *The Ages of Gaia: A Biography of Our Living Earth* (1988), have continued to draw attention, and Lovelock has followed them up with several others, including *The Revenge of Gaia: Earth's Climate in Crisis and the Fate of Humanity* (2006) and *The Vanishing Face of Gaia: A Final Warning* (2008).

Kathleen Rath Marr

The Usefulness of the Gaia Hypothesis

At the end of the introductory chapter in Gaia: A New Look at Life on Earth *(1979), James Lovelock explains the value of the Gaia hypothesis and his purpose in writing the book:*

The Gaia of this book is a hypothesis but, like other useful hypotheses, she has already proved her theoretical value, if not her existence, by giving rise to experimental questions and answers which were profitable exercises in themselves. If, for example, the atmosphere is, among other things, a device for conveying raw materials to and from the biosphere, it would be reasonable to assume the presence of carrier compounds for elements essential in all biological systems, for example, iodine and sulphur. It was rewarding to find evidence that both were conveyed from the oceans, where they are abundant, through the air to the land surface, where they are in short supply. The carrier compounds, methyl iodide and dimethyl sulphide respectively, are directly produced by marine life. Scientific curiosity being unquenchable, the presence of these interesting compounds in the atmosphere would no doubt have been discovered in the end and their importance discussed without the stimulus of the Gaia hypothesis. But they were actively sought as a result of the hypothesis and their presence was consistent with it.

If Gaia exists, the relationship between her and man, a dominant animal species in the complex living system, and the possibly shifting balance of power between them, are questions of obvious importance. . . . [T]his book is written primarily to stimulate and entertain. The Gaia hypothesis is for those who like to walk or simply stand and stare, to wonder about the Earth and the life it bears, and to speculate about the consequences of our own presence here. It is an alternative to that pessimistic view which sees nature as a primitive force to be subdued and conquered. It is also an alternative to that equally depressing picture of our planet as a demented spaceship, forever travelling, driverless and purposeless, around an inner circle of the sun.

FURTHER READING

Lovelock, James. *The Ages of Gaia: A Biography of Our Living Earth*. Rev. ed. New York: W. W. Norton, 1995.

_____. *Gaia: A New Look at Life on Earth*. 1979. Reprint. New York: Oxford University Press, 2000.

_____. *The Revenge of Gaia: Earth's Climate in Crisis and the Fate of Humanity*. New York: Basic Books, 2006.

_____. *The Vanishing Face of Gaia: A Final Warning*. New York: Basic Books, 2008.

Schneider, Stephen H., James R. Miller, Eileen Crist, and Penelope J. Boston, eds. *Scientists Debate Gaia: The Next Century*. Cambridge, Mass.: MIT Press, 2004.

SEE ALSO: Biosphere; Gaia hypothesis.

Lovins, Amory

CATEGORIES: Activism and advocacy; energy and energy use

IDENTIFICATION: American physicist

BORN: November 13, 1947; Washington, D.C.

SIGNIFICANCE: Lovins, cofounder of the Rocky Mountain Institute, has worked to promote the use of sustainable and clean energy, particularly as a means to attain global stability and security.

A native of Washington, D.C., Amory Lovins was a student at Harvard University and Magdalen College, Oxford, England, during the mid- to late 1960's. He received his master's degree from Oxford University in 1971 and thereafter began working as a consultant physicist specializing in energy concerns and promoting the use of energy sources that are economical, efficient, diverse, sustainable, and environmentally sound. Lovins advocates an approach that maximizes energy efficiency and minimizes environmental impact as the best way to provide for the world's long-term energy needs without incurring the security risks inherent in the use of nuclear power plants or reliance on foreign oil.

Lovins has lectured extensively, has held several visiting academic chairs, and has published hundreds of papers. Among his many books are *World Energy Strategies: Facts, Issues, and Options* (1975), *Soft Energy Paths: Toward a Durable Peace* (1977), and *Non-Nuclear Futures: The Case for an Ethical Energy Strategy* (1980; with John H. Price). Lovins has been the recipient of numerous U.S. honorary doctorates and a 1993 MacArthur Fellowship. He has consulted for utilities, industries, and governments worldwide, has briefed heads of state, and has served on the U.S. Department of Energy's senior advisory board.

In 1979 Lovins married L. Hunter Sheldon, one of the cofounders (and for six years the assistant director) of the urban forestry and environmental educa-

Physicist Amory Lovins. (AP/Wide World Photos)

tion group TreePeople. In 1982 the Lovinses co-founded the Rocky Mountain Institute, a nonprofit resource policy center in Old Snowmass, Colorado, that promotes resource efficiency and global security, with Hunter as the institute's president and executive director and Amory as its vice president and director of research. The institute's stated mission is to "foster the efficient and sustainable use of resources as a path to global security." Its projects focus on such areas as transportation, green design and development, greenhouse gas reduction, water-efficient technologies, economic renewal, and corporate sustainability.

Amory and Hunter Lovins coauthored several books, even after they separated in 1989 (they divorced in 1999), including *Energy/War: Breaking the Nuclear Link* (1980), *Brittle Power: Energy Strategy for National Security* (1982), *Energy Unbound: A Fable for America's Future* (1986; with Seth Zuckerman), *Climate: Making Sense and Making Money* (1997), and *Natural Capitalism: Creating the Next Industrial Revolution* (1999; with Paul Hawken). The couple shared a number of honors, including the 1982 George and Cynthia Mitchell International Prize for Sustainable Development, the 1983 Right Livelihood Award (often called

the "alternative Nobel Prize"), the 1989 Delphi Prize (one of the world's top environmental awards, presented by the Onassis Foundation), and the 1993 Nissan Prize at the International Symposium on Automotive Technology and Automation, Europe's largest car technology conference. In a 2009 list published by *Time* magazine, Lovins was recognized as one of the world's one hundred most influential people.

Karen N. Kähler

FURTHER READING

Hawken, Paul, Amory B. Lovins, and L. Hunter Lovins. *Natural Capitalism: The Next Industrial Revolution.* New ed. Sterling, Va.: Earthscan, 2005.

Heintzman, Andrew, and Evan Solomon, eds. *Fueling the Future: How the Battle over Energy Is Changing Everything.* Toronto: House of Anansi Press, 2003.

Inslee, Jay, and Bracken Hendricks. *Apollo's Fire: Igniting America's Clean-Energy Economy.* Washington, D.C.: Island Press, 2008.

SEE ALSO: Alternative energy sources; Alternative fuels; Alternatively fueled vehicles; Renewable energy; Solar energy; Wind energy.

M

Maathai, Wangari

CATEGORIES: Activism and advocacy; preservation and wilderness issues

IDENTIFICATION: Kenyan environmentalist and social activist

BORN: April 1, 1940; Ihithe village, Nyeri District, Kenya

SIGNIFICANCE: A visionary and activist in the fight against deforestation in Africa and beyond, Maathai has spearheaded various initiatives that have resulted in the planting of billions of trees and have brought global attention to this critical environmental issue.

Even before she won the Nobel Peace Prize in 2004, Wangari Maathai was well known for the positive impacts of her efforts to raise environmental awareness, particularly in regard to the issue of deforestation. Born Wangari Muta, Maathai grew up in Kenya as the daughter of subsistence farmers. She received scholarships that enabled her to earn bachelor's and master's degrees in biological science in the United States, and she also spent some time studying in Germany.

Upon her return to Kenya, Maathai studied anatomy at the University of Nairobi, but her interests turned to the intertwined issues of poverty and deforestation, partly because she realized how much the practice of deforestation had changed Kenya's landscape in her absence. In 1969 she married Mwangi Mathai, with whom she would eventually have three children, and in 1971 she became the first woman in central and eastern Africa to earn a Ph.D. Her husband became politically active, and in 1974 he campaigned for a seat in the Kenyan parliament. In part to help him keep a campaign promise to create new jobs, Maathai founded a business called Envirocare, which paid people to raise tree seedlings

in nurseries for eventual transplantation across Kenya. Maathai's progressive political views strained her marriage, however, and Mwangi Mathai eventually sued for divorce. When the marriage ended, he asserted his legal right to demand that she change her surname, and, as she relates in her memoir, *Unbowed* (2006), instead of changing her surname completely Maathai instead chose to insert an extra "a."

Maathai's personal difficulties did not lessen her desire to have a positive influence on environmental outreach and activism. In 1977 Maathai renamed Envirocare the Green Belt Movement and gained the

Wangari Maathai speaks at the 2009 United Nations Summit on Climate Change. (AP/Wide World Photos)

support of Kenya's National Council of Women. The organization adopted the motto of "One person, one tree," which led to the goal of planting fifteen million trees, one for each person in Kenya. The group far exceeded the goal, planting more than twice that number of trees by the early twenty-first century.

At the same time, Maathai expanded her activities into other areas of the environmental movement. She successfully campaigned against the building of a planned skyscraper in Nairobi's Uhuru Park and opposed the government's attempts to sell valuable forestland to developers. Her work met with much opposition and she was imprisoned several times, but her international influence increased to the point that it became difficult for the authorities to detain her without cause.

In 2002 Maathai was elected to Kenya's parliament, and she was appointed the following year to the post of assistant minister of the environment, natural resources, and wildlife. Her tireless work was recognized worldwide when she became the first African woman to win the Nobel Peace Prize in 2004. Maathai then helped the United Nations Environment Programme launch the Billion Tree Campaign. The campaign's initial goal of planting one billion trees was reached more quickly than expected, and a new goal of planting more than seven billion trees by the end of 2009 was also exceeded. The significance of Maathai's contribution and inspiration to this campaign, which was carried out by workers and volunteers in more than 170 counties, is clear.

Amy Sisson

FURTHER READING

Maathai, Wangari. *The Green Belt Movement: Sharing the Approach and the Experience.* Rev. ed. New York: Lantern Books, 2003.

_____. *Unbowed: A Memoir.* New York: Alfred A. Knopf, 2006.

SEE ALSO: Africa; Deforestation; Forest management; Global ReLeaf; Land clearance; Soil conservation; Sustainable forestry; United Nations Environment Programme.

McToxics Campaign

CATEGORIES: Activism and advocacy; waste and waste management

THE EVENT: Consumer boycott against McDonald's restaurants

DATES: 1989-1990

SIGNIFICANCE: The McToxics Campaign prompted the huge McDonald's fast-food restaurant chain to replace the polystyrene food containers it had been using with more environmentally friendly packaging.

In 1989, the Environmental Defense Fund (EDF) inaugurated a grassroots campaign against McDonald's restaurants. By convincing consumers to boycott McDonald's, EDF hoped to persuade the fast-food giant to stop packaging its food in polystyrene "clamshells," which were cheap and efficient food insulators but did not readily biodegrade and were suspected of emitting chlorofluorocarbons (CFCs) harmful to the earth's ozone layer. EDF members worked to educate children about the environmental consequences of polystyrene packaging, then employed the young activists in their protests. Ronald McDonald, the company's well-known advertising character designed to appeal to young consumers, was redubbed "Ronald McToxic" by the campaign's directors.

Predictably, the children's participation drew media attention, and the protest took effect as consumers began to boycott McDonald's. Some wrote angry letters; others mailed empty clamshells to the company's offices. As the corporation's profits dropped, McDonald's lawyers began urging the company to undertake the packaging changes demanded by EDF and its supporters. On August 1, 1990, McDonald's came to a tentative agreement with EDF, and on November 1, 1990, McDonald's announced its decision to abandon the use of polystyrene packaging.

Nevertheless, the agreement was not without detractors. Some scientists expressed doubts that the changeover from polystyrene to paper packaging would produce environmental benefits, questioning whether the use of clamshells represented a legitimate environmental hazard. In addition, some environmental extremists objected to EDF's accommodation of corporate culture. Nevertheless, most observers agreed that the McToxics campaign and its conclusion represented a welcome example of how environmental and economic interests could be made to coexist.

Alexander Scott

See also: Chlorofluorocarbons; Environmental ethics; Green marketing; Waste management.

Mad cow disease

CATEGORIES: Human health and the environment; animals and endangered species

DEFINITION: Cattle disease affecting the central nervous system

SIGNIFICANCE: The sheep disease scrapie is generally believed to have jumped the species barrier to become mad cow disease. It breached the species barrier again to infect humans who ate beef. Introduction of the pathogen into the human food chain has been attributed to the use of scrapie-infected meat-and-bone meal in cattle feed.

Mad cow disease, formally known as bovine spongiform encephalopathy (BSE), is believed to be related to new-variant Creutzfeldt-Jakob disease (nvCJD), a fatal human illness. It is marked by the deterioration of brain tissue and progressive degeneration of the central nervous system, which cause symptoms such as impaired physical coordination, staggering, unusual aggression, and other abnormal behavior. Since it was first identified in Great Britain in 1986, BSE has infected cattle throughout the European Union. Scattered cases have also been found in the United States, Canada, the Falkland Islands, Israel, Oman, and Japan.

BSE is a transmissible spongiform encephalopathy (TSE), a class of diseases so named because of the brain damage that characterizes them—the tissue is left riddled with small holes, like a sponge. Other species affected by TSEs include deer, elk, and antelope (chronic wasting disease), sheep and goats (scrapie), mink (transmissible mink encephalopathy), cats (feline spongiform encephalopathy, or FSE), and humans (kuru, Gerstmann-Sträussler-Scheinker syndrome, fatal familial insomnia, and Creutzfeldt-Jakob disease, or CJD). All TSEs have long incubation periods that range from months to years. Diagnosis is confirmed only through autopsy and examination of the brain tissue. All known TSEs are incurable, untreatable, and fatal.

BSE is believed to have arisen because of an infectious agent present in cattle feed during the 1970's and 1980's. Meat-and-bone meal, a protein concentrate produced by rendering plants (facilities that process animal carcasses and slaughterhouse wastes into commercial products), was added to cattle feed as a nutritional supplement with increasing frequency during the 1980's. Among the rendered wastes from which the supplement was made were the carcasses of sheep that had died from scrapie. Through infected protein supplement, scrapie is thought to have crossed the species barrier from sheep into cattle and become a new TSE. An increase in Great Britain's sheep population during the late 1970's and early 1980's elevated the number of scrapie cases, and thus more infected material was rendered. Coincidentally, at about this time Britain's rendering industry made changes to the rendering process—eliminating the solvent extraction of fats and a subsequent steam-stripping treatment—that may have allowed more infectious agent to pass into the finished product.

THE BRITISH BSE EPIDEMIC

Not long after it was first identified in 1986, BSE reached epidemic proportions in the United Kingdom. The epidemic peaked in 1992, with 37,280 confirmed cases. Mandatory reporting of BSE cases and destruction of symptomatic animals began in 1988, as did a ban on the feeding of ruminants (cattle, sheep, goats, and deer) with ruminant-derived protein. In 1989 the British government banned the human consumption of cattle brains and other "specified offals" believed to be possible carriers of infection. In 1990 a British domestic cat was diagnosed as the first victim of FSE, causing public concern over the possible spread of BSE beyond livestock and a temporary drop in the nation's beef consumption. That year, the British government established a system for reporting and tracking CJD to monitor possible BSE effects on human beings.

In 1993 the British National CJD Surveillance Unit began to receive reports of unusual CJD cases. While CJD typically affects one person in one million between the ages of fifty-five and eighty, CJD was appearing in younger patients, with uncharacteristic patterns of brain damage, and with atypical frequency. By early 1996 eight cases of nvCJD in people under forty-two years of age had been identified. Global attention turned to Britain's mad cow crisis in March, 1996, when British health secretary Stephen Dorrell announced that ten cases of nvCJD had been confirmed and that they were most likely the result of exposure to BSE-contaminated beef before the 1989 specified offal ban. The announcement triggered a nationwide panic in Britain, caused a sharp drop in British beef sales, and exacerbated political and economic friction between Britain and the rest of the European Union.

Researchers have since found more conclusive evidence linking BSE and nvCJD, but they have not determined the exact cause of these and other TSEs. One of the more widely accepted theories is that the illnesses are induced by tiny proteinaceous infectious particles, or prions. Prions resemble an abnormally folded version of prion protein, a harmless, naturally occurring protein found in the brains of all mammals. Prions do not break down in mammalian digestive systems and provoke no immune response. Lacking genetic material (unlike more familiar infectious agents, such as bacteria, viruses, and fungi), prions are believed to spread by invading cells and converting normal prion protein into the aberrant form. As the prions accumulate, they damage brain cells and ultimately kill the organism.

Another theory is that BSE originated as a cattle TSE called bovine amyloidotic spongiform encephalopathy (BASE). Researchers found that the prions associated with BASE, first identified in 2003 during BSE testing in Italy, could convert to BSE prions in lab tests. Yet another hypothesis is that BSE's ability to manifest as multiple strains suggests instead a virus-like pathogen with nucleic acids that can transmit genetic information. Whatever the agent that causes BSE, it is highly resistant to ultraviolet light, pH and temperature extremes, ionizing radiation, and many chemical disinfectants.

Responses

British efforts to contain a livestock epidemic that was only partially understood while protecting human health and allaying public fears about beef consumption created new problems. In a government-instituted culling program, some 4.4 million animals considered at risk were slaughtered. Neither the designated rendering plants that processed the remains nor the designated incinerators that burned the rendered product were able to keep up with the slaughter rate. The resulting backlog of hundreds of thousands of tons of waste beef and rendered products was warehoused pending disposal. Incineration plans proved unpopular with Britons, many of whom were unconvinced that burning cull material was safe. Public concern also arose over the possible groundwater contamination risk posed by cattle carcasses landfilled and liquid rendering wastes discharged to the environment during the early years of the epidemic.

Another matter of concern was the possible person-to-person transmission of the disease through medical procedures such as blood transfusions, human growth hormone treatments, and cornea transplants. The first patient believed to have contracted nvCJD from infected blood died in 2003, nine years after receiving the suspect transfusion. The donor had given blood in 1996, before precautions to protect blood supplies were implemented; he died in 1999. No way has yet been found to screen blood for the nvCJD pathogen; this, combined with a disease incubation period that may be as long as thirty years, makes the likelihood of bloodborne transmission an unknowable and disturbing risk. Fears of nvCJD transmission have led many countries to ban British blood donations.

The number of BSE cases in the United Kingdom declined sharply after 1993 and by 2008 was down to thirty-seven. Cases reported outside the United Kingdom have generally decreased since 2005. Countries that have learned from the British experience have used feed bans, monitoring programs, precautionary recalls, and other measures to contain the spread of the disease and keep BSE pathogens out of the human food chain. Perceptions that the United States does not employ sufficient precaution against BSE has led many countries to implement partial or full restrictions on the import of American beef.

As of March, 2010, a total of 213 people had died from nvCJD since the disease was first identified: 172 in the United Kingdom and Ireland, 36 in Europe, 4 in North America, and 1 in Japan. Three of those who had died in the United Kingdom were known to have contracted the disease through blood transfusions. An additional six patients were known to be living with nvCJD, four of whom were in the United Kingdom, one in Italy, and one in Saudi Arabia. Of the world's nvCJD cases, 178 involved people who had lived in the United Kingdom for more than six months between 1980 and 1996; the one Japanese patient had spent twenty-four days in the United Kingdom during that period. The number of nvCJD cases has generally declined since the year 2000, but the disease's long incubation period makes it impossible to forecast its future trends.

Karen N. Kähler

FURTHER READING

Becker, Geoffrey S., Curtis W. Copeland, and Sarah A. Lister, eds. *Mad Cow Disease (Bovine Spongiform Encephalopathy)*. New York: Nova Science, 2008.

Lyman, Howard F., and Glen Merzer. *Mad Cowboy: Plain Truth from the Cattle Rancher Who Won't Eat Meat.* 1998. Reprint. New York: Touchstone, 2001.

Nunnally, Brian K., and Ira S. Krull, eds. *Prions and Mad Cow Disease.* New York: Marcel Dekker, 2004.

Prusiner, Stanley B. *Prion Biology and Diseases.* 2d ed. Cold Spring Harbor, N.Y.: Cold Spring Harbor Laboratory Press, 2004.

Rampton, Sheldon, and John Stauber. *Mad Cow U.S.A.* Updated ed. Monroe, Maine: Common Courage Press, 2004.

Rhodes, Richard. *Deadly Feasts: The "Prion" Controversy and the Public's Health.* New York: Simon & Schuster, 1997.

Schwartz, Maxime. *How the Cows Turned Mad: Unlocking the Mysteries of Mad Cow Disease.* Berkeley: University of California Press, 2004.

Yam, Philip. *The Pathological Protein: Mad Cow, Chronic Wasting, and Other Deadly Prion Diseases.* New York: Copernicus Books, 2003.

SEE ALSO: Brucellosis; Cattle; Environmental illnesses; Food chains; Food quality regulation; Vegetarianism.

Malathion

CATEGORY: Pollutants and toxins

DEFINITION: An organophosphorus pesticide used against mosquitoes, fleas, and other insects

SIGNIFICANCE: Although no adverse health effects have been found from low-level exposure to malathion, the use of this pesticide in aerial spraying programs has been controversial.

Pesticides have been used to control crop loss and prevent the spread of disease for hundreds of years. Until the middle of the twentieth century, pesticides were either toxic metals (such as arsenic or lead) or natural products (such as nicotine sulfate). With the discovery of dichloro-diphenyl-trichloroethane (DDT) in 1939, new human-made pesticides became available for use. However, it was soon found that DDT and related chlorinated hydrocarbon pesticides persist in the environment for years, causing damage to birds, fish, and other wildlife. Other pesticides, including organophosphate compounds, were therefore developed as alternatives to DDT.

Malathion is an organophosphate compound that is chemically similar to some types of nerve gas but less toxic than most other organophosphate pesticides. Unlike chlorinated hydrocarbons, malathion breaks down in the environment, transforming into carbonic and phosphoric acid over a period of days to a few weeks. Because of its relatively short lifetime, malathion does not bioaccumulate in aquatic organisms or contaminate groundwater. Following its introduction by the American Cyanamid Company in 1952, malathion quickly became a popular substitute for DDT in controlling mosquitoes, fleas, lice, and other insects. In 1956 malathion was first used to counteract fruit fly infestations.

In the following decades, the use of malathion continued to increase. In the 1980's, however, the spraying of malathion in California to control the Mediterranean fruit fly (also known as the Medfly) became a source of controversy. Those critical of the use of malathion in populated areas noted that commercial preparations of malathion contain trace impurities that are potentially toxic and that the initial breakdown products from malathion include malaoxin, a compound with acute toxicity forty times that of malathion itself. Critics also suggested that malathion might be carcinogenic or could weaken the immune system, making people more susceptible to disease. However, extensive studies of human populations exposed to malathion through the spraying program found no adverse health effects from low levels of exposure. Similarly, no harmful effects have been observed from low-level exposure to malathion in laboratory studies on animals. Despite the controversy, malathion continued to be used to prevent damage to citrus crops in California, Florida, and other states, and to be available in many commercial products.

By the end of the twentieth century, controversy over the use of synthetic pesticides such as malathion had convinced many people that the release of such compounds into the environment should be greatly reduced, but malathion continued to be used in some areas into the twenty-first century: Both New York City and the city of Winnipeg in Manitoba, Canada, used malathion spraying (in 2000 and 2005, respectively) as part of their attempts to eradicate insects, particularly mosquitoes, that might be carriers of West Nile virus. Critics of human-made pesticides have continued to call for more intelligent use of pesticides and the development of alternative methods of controlling insect populations.

Jeffrey A. Joens

FURTHER READING

Hamilton, Denis, and Stephen Crossley, eds. *Pesticide Residues in Food and Drinking Water: Human Exposure and Risks.* Hoboken, N.J.: John Wiley & Sons, 2004.
Levine, Marvin J. "Pesticides in Food." In *Pesticides: A Toxic Time Bomb in Our Midst.* Westport, Conn.: Praeger, 2007.
Manahan, Stanley E. *Fundamentals of Environmental Chemistry.* 3d ed. Boca Raton, Fla.: CRC Press, 2009.

SEE ALSO: Biomagnification; Biopesticides; Dichlorodiphenyl-trichloroethane; Medfly spraying; Pesticides and herbicides.

Maldive Islands

CATEGORIES: Places; weather and climate
IDENTIFICATION: Small island nation in the central Indian Ocean southwest of Sri Lanka
SIGNIFICANCE: Because of their very low elevation, the Maldive Islands are especially vulnerable to the rises in sea level that are projected to accompany progressive global warming.

The Maldives or Maldive Islands, formally known as the Republic of Maldives, comprise 1,192 very small islands, many of which are little more than coral reefs resting on the tips of an undersea volcanic ridge running north-south in the Indian Ocean some 700 kilometers (435 miles) southwest of Sri Lanka. The landmass of the islands amounts to 300 square kilometers (116 square miles) spread over an ocean area of some 90,000 square kilometers (35,000 square miles). The islands are protected by reefs that form lagoons around them, shielding them from the ocean waves, especially during the monsoon storms that rage annually from April to October. Some 200 of the islands are inhabited, supporting a population of more than 300,000. The republic's capital is Male, with a population of just over 100,000, on the North Male atoll. More than half a million tourists arrive in the Maldives every year, passing through the large international airport situated near the capital.

Until the mid-twentieth century, the population of the Maldives was approximately 100,000 and was largely supported by fishing, with a small amount of agriculture. Around 1970 the islands were discovered by the tourist industry, and by 2010 some 30 percent of the islands' gross domestic product and 90 percent of government revenues were accounted for by tourism. Approximately one hundred resorts have been developed on some of the previously uninhabited islands, offering scuba diving, surfing, fishing, cruising, and sunbathing on the white sandy beaches as their main attractions. The native population has grown and has been supplemented by large numbers of workers from mainland India. Food and freshwater resources of the islands have been put under increasing strain, and competition for suitable living space has grown with the population.

However, the Maldives' dependence on the tourist industry and the population pressures the islands have experienced as a result are overshadowed as threats to the nation by the projected rise in sea levels that could be caused by the melting of the earth's ice shelves and glaciers as global warming progresses. The average ground level in the islands is only 1.5 meters (5 feet) above sea level, and the highest point is only 2.3 meters (7.5 feet). Scientists have estimated that a rise in sea levels from 0.5 to 1.4 meters (1.7 to 4.6 feet) could occur during the twenty-first century. The worst-case scenario for the island nation of Maldives is complete inundation.

In 2008 the president of the Maldives, Mohamed Nasheed, proposed taxing tourists to create a sovereign wealth fund that the nation could use to buy land in India, Sri Lanka, and Australia in the event that the islands are flooded and their inhabitants become ecological refugees. This radical plan was accepted, but the Maldives also joined with other members of the Alliance of Small Island States to pressure larger nations to reduce their greenhouse gas emissions (which have been linked to global warming) and to provide help and resources to island states affected by rising sea levels.

Another environmental problem for the Maldives is the damage that was done to the islands' coral reefs by the Asian tsunami of 2004. A massive earthquake in the Indian Ocean produced waves more than 4 meters (13 feet) high, engulfing the nation and causing widespread destruction, including to the fragile reef ecosystems.

David Barratt

FURTHER READING

Hockly, T. W. *The Two Thousand Isles: A Short Account of the People, History, and Customs of the Maldive Archipelago*. New Delhi: Asian Educational Service, 2003.

Masters, Tom. *The Maldives*. 7th ed. London: Lonely Planet, 2009.

Phandmis, D. *Maldives: Wind of Change in an Atoll State*. New Delhi: South Asian Publishers, 1995.

Russell, Cristine. "First Wave." *Science News*, February 28, 2009, 24-30.

SEE ALSO: Alliance of Small Island States; Climate change and oceans; Coral reefs; Global warming; Kyoto Protocol; Sea-level changes.

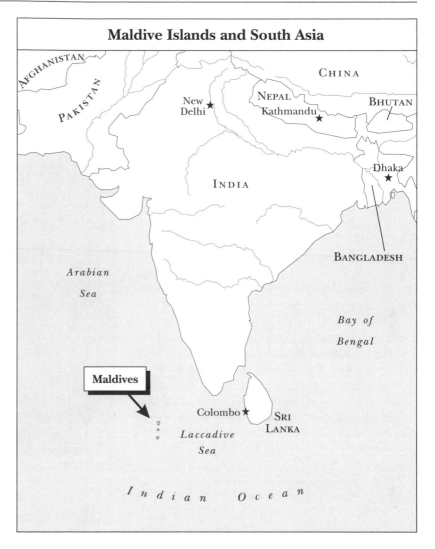

Maldive Islands and South Asia

Malthus, Thomas Robert

CATEGORIES: Population issues; philosophy and ethics

IDENTIFICATION: English political economist

BORN: February 13, 1766; The Rookery, near Dorking, Surrey, England

DIED: December 23, 1834; St. Catherine, near Bath, Somerset, England

SIGNIFICANCE: The author of *An Essay on the Principle of Population, as It Affects the Future Improvement of Society* (1798), Malthus provided the direst explanations of the causes and consequences of population growth.

Thomas Robert Malthus was born to a well-off and intellectual family. His father, Daniel Malthus, was a friend of the philosophers Jean-Jacques Rousseau and David Hume and was a minor figure in popularizing the utopian ideas of Rousseau and William Godwin. Thomas Malthus entered Jesus College Cambridge in 1784, where he was named a fellow in 1793. He was ordained in the Church of England in 1788. In 1804 Malthus married, left his fellowship at Cambridge, and was appointed chair of modern history and political economy at the East India Company College (now Haileybury). The marriage and his professorship lasted the rest of his days.

Malthus's earliest intellectual influence was his father. The two argued nonstop, with the older Malthus defending Godwin's optimistic ideas that human beings and human society can be perfected. In response, Malthus developed an account of human society and progress so pessimistic that economics has since been called the dismal science. The Malthusian theory of population growth and resource consumption is simple. Population grows geometrically: A population of two individuals will grow to four, then eight, then sixteen in just three generations. Available resources grow arithmetically, and the amount of available resources can increase only by a fixed amount. Starting with five units of resources and allowing an increase of two units in each generation, the amount of available resources will grow to seven, then nine, and then eleven in three generations. The amount of resources thus doubles in three generations, but the population grows eightfold. A given area's population will always grow to exceed the carrying capacity of the area, absent war or disease, and this overpopulation will result in famine and death. Malthus believed, however, that the planet's carrying capacity can grow infinitely, just not as rapidly as the population. The result, over time, is that the planet will support an ever larger population in misery and squalor.

The utopians thought that human character could be improved, and that people could learn to limit population growth and improve the cultivation of natural resources. A sustainable population could enjoy an ever-increasing standard of living. In the face of utopian optimism, Malthus was a relentless pessimist. Malthus denied that the human race could successfully curb the reproductive impulse. He believed that the utopians were wrong on another count too. Malthus thought that people who are not miserable have no incentive to improve their standard of living. A small population would not have the incentive to improve the cultivation of natural resources. According to Malthus, with overpopulation there is starvation and misery, and with limited population there is economic stagnation.

Malthus's ideas gained influence during the mid-twentieth century. In the twenty-first century, neo-Malthusians such as Paul R. Ehrlich and others who deny the infinite carrying capacity of the earth see the overconsumption of limited resources that results from large populations as a very serious problem arising from population growth. They assert that the consumption of resources that accompanies continued population growth will result in total environmental collapse, not just Malthus's persistent misery. In response, cornucopians such as Julian L. Simon respond that human beings never consume all of a re-

Political economist Thomas Robert Malthus. (Library of Congress)

source: As the resource becomes scarce the price increases, and people seek other ways to satisfy their desires. Total environmental degradation is prohibitively expensive. Other activists reject Malthusian thinking altogether, arguing that from an environmental standpoint, the problem is not overpopulation in the developing world but rather the overconsumption of resources by the relatively few wealthy.

Kevin Guilfoy

FURTHER READING

Ehrlich, Paul R., and Anne H. Ehrlich. *The Population Explosion.* New York: Simon & Schuster, 1990.

Elwell, Frank W. *A Commentary on Malthus' 1798 Essay on Population as Social Theory.* Lewiston, N.Y.: Edwin Mellen Press, 2001.

Huggins, Laura E., and Hanna Skandera, eds. *Population Puzzle: Boom or Bust?* Stanford, Calif.: Hoover Institution Press, 2004.

Malthus, Thomas. *An Essay on the Principle of Population, as It Affects the Future Improvement of Society.*

1798. Reprint. New York: Oxford University Press, 2008.

Simon, Julian L. *The Ultimate Resource 2*. Rev. ed. Princeton, N.J.: Princeton University Press, 1996.

See also: Ehrlich, Paul R.; *Limits to Growth, The*; Population Connection; Population-control and one-child policies; Population-control movement; Population growth.

Manabe, Syukuro

Category: Weather and climate

Identification: Japanese meteorological scientist

Born: September 21, 1931; Shingu-Mura, Uma-Gun, Ehim e-Ken, Japan

Significance: Manabe's research using computer modeling has improved humankind's understanding of the role that the oceans play in the global climate.

Syukuro Manabe is a meteorological scientist whose research area involves modeling the earth's atmosphere in order to predict global climate change. Modelers construct a system of mathematical equations that can be used, with a computer, to mimic the responses of a particular segment of nature to possible events. In Manabe's case, the equations center on the heat balance of the earth's atmosphere; the goal is to use data from previous years to arrive at accurate predictions of global weather patterns for future years. Even an approximate model allows its users to find the long-term effects of short-term environmental atmospheric inputs. This allows a better understanding of the causes and effects of environmental disturbances such as global warming.

Manabe completed his formal education in his home country of Japan, finishing a bachelor's degree in 1953, a master's degree in 1955, and a doctoral degree in 1958, all from Tokyo University. With a doctorate in meteorology, he moved to the United States and spent the years between 1958 and 1997 working his

Meteorological scientist Syukuro Manabe. (©Jim Sugar/CORBIS)

way up from the position of research meteorologist at the U.S. Weather Bureau to senior scientist at the National Oceanic and Atmospheric Administration (NOAA).

Manabe spent the majority of his time in the United States up to 1997 as a researcher with NOAA, located in Princeton, New Jersey. During this tenure, he contributed his expertise to many national and international committees and panels exploring the global climate. He also held an appointment as professor in the Program in Atmospheric and Oceanic Sciences at Princeton University. In 1997 Manabe returned to Tokyo as the director of the Global Warming Research Division of the Frontier Research System for Global Change. He returned to Princeton in 2002 as a visiting research collaborator in the Program in Atmospheric and Oceanic Sciences.

Manabe's research, the details of which have appeared in many publications, has brought him the international recognition of his peers, as is demonstrated by the numerous honors he has received. A survey of the citations that have accompanied these awards suggests the Manabe is without peer among scientists working to understand the global climate by using physical-mathematical models. He is considered a pioneer in the development of atmosphere-ocean circulation models and their application to the analysis of climate disturbances involving greenhouse gases. This research improves humankind's understanding of the role that the oceans play in the global climate. Manabe's work aids in the analysis of the interplay between dynamics and chemistry that controls the distribution of stratospheric ozone. His research also provides the intellectual underpinning for the increasing concern that many scientists and others have expressed regarding the future of the natural environment.

Kenneth H. Brown

FURTHER READING

Kiehl, J. T., and V. Ramanathan, eds. *Frontiers of Climate Modeling*. New York: Cambridge University Press, 2006.

McGuffie, Kendal, and Ann Henderson-Sellers. *A Climate Modelling Primer*. 3d ed. Hoboken, N.J.: John Wiley & Sons, 2005.

Weart, Spencer R. "Public Warnings." In *The Discovery of Global Warming*. Rev. ed. Cambridge, Mass.: Harvard University Press, 2008.

SEE ALSO: Climate change and oceans; Climate change skeptics; Climate models; Global warming; Greenhouse effect; National Oceanic and Atmospheric Administration; Ozone layer.

Marine debris

CATEGORIES: Waste and waste management; pollutants and toxins

DEFINITION: Solid litter discarded by humans that finds its way into the oceans

SIGNIFICANCE: The presence of large amounts of litter in the world's oceans is a matter of major environmental concern. Such debris can threaten marine wildlife and ecosystems in numerous ways, such as by entangling animals, by blocking the digestive systems of animals that ingest nonfood objects, and by contaminating ocean waters with dangerous chemical compounds.

Much of human-made material that is discarded after use ends up in the oceans to become marine debris. Of this debris, plastic products are among the most troublesome because they are produced in vast amounts, are lightweight and hydrophobic (water repelling), and resist degradation. In large parts of the ocean, plastic debris can be more abundant than the natural prey of marine animals. Animals that ingest these items are exposed to high levels of chemicals and may suffer blockages in their digestive systems. Large pieces of debris can entangle and choke large marine animals, including endangered turtles and marine mammals.

FATES OF DIFFERENT KINDS OF DEBRIS

The stability and the density of marine debris are the most important properties that affect the fates of these materials. Degradable materials such as paper and wood do not remain marine debris for long because microbes readily break down these materials. Nonsoluble and nondegradable materials such as rubber, plastic, most metals, and concrete remain in water for years and therefore represent the majority of marine debris. The density of the debris determines whether it sinks to the bottom or remains in the surface waters. Dense materials such as metals, concrete, ceramic, glass, and some plastics sink to the bottom of the sea, unless their shapes enable them to float. Materials that are less dense than seawater or equal to seawater in density, such as polystyrene and light plastics, drift in the surface currents of the oceans and often wash up on the shore with waves.

Large ocean gyres are vast areas where surface waters come together and downwell. This water movement causes light, nondegradable debris to concentrate in these areas. The amount of debris in these waters often exceeds the amount of living organisms, a condition known elsewhere only in landfills and some urban environments. At the North Pacific Gyre, the large accumulation of debris has been dubbed the Great Pacific Garbage Patch. Because the plastic materials have been broken down to small particles that are not easily seen from satellites, the size of this debris field is difficult to measure, but it is estimated to be on the order of 1 to 10 million square kilometers (0.39 to 3.9 million square miles). The debris is so abundant that some boat captains refuse to travel in the area because of the risk that their vessels could become entangled. In April, 2010, the Associated Press reported that scientists had discovered similar plastic garbage patches in the Atlantic Ocean.

EFFECTS ON MARINE WILDLIFE

Large pieces of debris provide structures and serve as habitats for marine organisms. In fact, old bridges, barges, and boats are intentionally discarded in some places to create artificial reefs for the purpose of increasing the productivity and diversity of marine life. Scientific research regarding the colonization and utilization of artificial reefs has revealed how the arrangement and shapes of debris can optimize its utility as habitat. Even large floating objects such as plastic buckets, buoys, and crates are used as habitat in much the same way as sargassum seaweed is used as natural habitat.

Most of the effects of debris on marine wildlife are negative. Marine debris can entangle, strangle, and choke some kinds of creatures. "Ghost nets" are fishing nets abandoned or lost in the sea that continue to function as they were designed to function, entangling animals. Fish, reptiles, mammals, and even birds that become entangled in ghost nets usually die from starvation, stress, injuries, or suffocation. Debris that gets wrapped around the throats of animals can cause strangulation. Ingested debris can block the digestive tracts of animals, interfering with their ability to feed, digest food, or pass waste. These animals usually die of starvation.

Predatory marine animals often consume plastic debris because it resembles their natural prey. Floating plastic bags and balloons are mistaken for jellyfish by sea turtles. Small plastic nurdles (resin pellets from which larger plastic items are made) are mistaken for fish eggs or zooplankton. Because most of the nurdles and small particles that have been found in debris samples are blue, black, white, or green, researchers have hypothesized that yellow and red plastic particles are selectively eaten by marine organisms.

Even if ingested plastic does not kill animals by choking their digestive systems, it can still have significant sublethal effects on organisms of all size. Plastic particles concentrate hydrophobic compounds on their surfaces. These compounds interfere with the endocrine systems of animals, creating reproductive, developmental, and immunological problems for these animals. By concentrating these compounds by many orders of magnitude, even small plastic particles can negatively affect large animals by increasing their exposure to endocrine-disrupting compounds, a problem that is already serious owing to the biomagnification of compounds in long marine food webs.

PREVENTION AND MITIGATION

The 1973 International Convention for the Prevention of Pollution from Ships (along with the 1978 protocol that amended it, collectively known as MARPOL 73/78) and the Convention on the Prevention of Marine Pollution by Dumping of Wastes and Other Matter (also known as the London Convention) are international laws that aim to limit marine debris by prohibiting the disposal of certain kinds of wastes in the open oceans. Enforcing such laws has proven to be difficult, however.

Marine debris is removed in a variety of ways, from individual scuba divers gathering litter by hand to boat-mounted skimmers picking debris from surface waters and teams of people and machines picking up the trash that washes up on beaches. Given that approximately 80 percent of marine debris comes from continental sources, strainers placed on storm drains and across streams to retain litter are effective preventive measures to combat marine debris. One unique program meant to reduce the number of ghost nets in the world's oceans is Hawaii's Nets-to-Energy Program, which collects more than 80 metric tons of derelict fishing nets every year and uses them to produce enough electricity to power thirty-four homes.

Greg Cronin

FURTHER READING

Andrady, Anthony L., ed. *Plastics and the Environment.* Hoboken, N.J.: John Wiley & Sons, 2003.

Hill, Marquita K. "Solid Waste." In *Understanding Environmental Pollution.* 3d ed. New York: Cambridge University Press, 2010.

Laws, Edward A. *Aquatic Pollution: An Introductory Text.* 3d ed. New York: John Wiley & Sons, 2000.

Moore, Charles. "Trashed: Across the Pacific Ocean, Plastics, Plastics, Everywhere." *Natural History,* November, 2003.

Ocean Conservancy. *Trash Travels: From Our Hands to the Sea, Around the Globe, and Through Time.* Washington, D.C.: Author, 2010.

United Nations Environment Programme. *Marine Litter: A Global Challenge.* Nairobi: Author, 2009.

Wolf, Nancy, and Ellen Feldman. *Plastics: America's Packaging Dilemma.* Washington, D.C.: Island Press, 1990.

SEE ALSO: Biomagnification; Gill nets and drift nets; London Convention on the Prevention of Marine Pollution; Marine Mammal Protection Act; Ocean dumping; Ocean pollution; Plastics; Seabed disposal; Solid waste management policy; United Nations Convention on the Law of the Sea; Water pollution.

Marine Mammal Protection Act

CATEGORIES: Treaties, laws, and court cases; animals and endangered species

THE LAW: U.S. federal legislation intended to provide for safe environments for marine mammals

DATE: Enacted on October 21, 1972

SIGNIFICANCE: The provisions of the Marine Mammal Protection Act contributed to improvements in the populations of various species, including seals, sea otters, and whales, but debates continue regarding the law's enforcement and its effects.

The Marine Mammal Protection Act (MMPA) prohibits ownership or importation of any marine mammal or any products of marine mammals. However, it does allow a limited catch by Alaska Natives and Native Americans for purposes of material survival or for reasons related to cultural heritage. The act was amended in 1994 to restructure jurisdiction and enforcement of the provisions of the law and to establish guidelines for transportation of marine mammals. Controversies exist among fishing interests, environmentalists, and members of indigenous cultures as to the interpretation and effects of the MMPA.

When the MMPA became law in 1972, the U.S. Fish and Wildlife Service (FWS), which is part of the Department of the Interior, became responsible for manatees, dugongs, polar bears, walruses, and sea otters. The National Marine Fisheries Service (NMFS), a division of the Department of Commerce, was assigned management of whales, dolphins, sea lions, fur seals, elephant seals, monk seals, true northern seals, and southern fur seals. The 1994 amendments stipulated stronger fishing regulations, especially the use of improved equipment to reduce the number of accidental killings of marine mammals and to exclude the bycatch of turtles, nontarget fish, and undersized fish of the targeted species.

Before the amendments, jurisdiction over care and transport of captive marine mammals was shared by the NMFS and the Department of Agriculture's Animal and Plant Health Inspection Service (APHIS). The amendments eliminated the NMFS's part of the administration and enforcement, which caused concern among many environmentalists because APHIS officials are not as experienced in working with marine mammals as are personnel of the NMFS. The Humane Society of the United States appealed for reinstatement of the NMFS as a joint authority, but APHIS was delegated sole authority. Zoos and aquariums supported APHIS control, as that agency made it easier for these institutions to capture and transport marine mammals. APHIS required only that public facilities that already owned any marine mammals send notification to APHIS after acquiring additional mammals, whereas previously the institutions had been required to obtain permits before acquiring such animals.

Other changes brought about by the amendments eased regulations on scientists and researchers, who no longer are required to obtain permits to conduct studies of marine mammals, unless their work has the potential to harm the animals. Many ecologists reacted positively to the 1994 amendments to the MMPA because the amendments emphasized the importance of maintaining healthy ecosystems, particularly in the waters off the northwestern and northeastern coasts of the United States, where the seal and sea lion populations had been declining at an alarming rate.

The MMPA made illegal many of the human activities that threatened marine mammal populations in the past. Sea otters had been overhunted for their skins, but government protection has allowed their populations to recover around Prince William Sound and off the California coast. Whales are protected by the MMPA and by the ban on whaling instituted by the International Whaling Commission. However, the natural renewable resources on which whales feed may be endangered, as some countries are harvesting large quantities of krill, which is the mainstay of many whales' diets and an important link in the marine food chain.

As human populations increase worldwide and become more industrialized, demands on the oceans as a food source and a place to dump chemicals also increase, as does noise pollution from sonic testing and boat traffic. Continued publicity and pressure from environmental advocacy groups such as Greenpeace and the World Wide Fund for Nature have led wildlife managers, fishers, animal rights supporters, and scientists to work together under the terms of the MMPA as amended in 1994.

Dale F. Burnside with Aubyn C. Burnside

FURTHER READING

Curnutt, Jordan. *Animals and the Law: A Sourcebook.* Santa Barbara, Calif.: ABC-CLIO, 2001.

Ray, G. Carleton, and Jerry McCormick-Ray. *Coastal-Marine Conservation: Science and Policy.* Malden, Mass.: Blackwell, 2004.

SEE ALSO: Dolphin-safe tuna; International whaling ban; International Whaling Commission; Ocean dumping; Ocean pollution; Polar bears; Seal hunting; Whaling.

Marsh, George Perkins

CATEGORIES: Activism and advocacy; ecology and ecosystems

IDENTIFICATION: American statesman, diplomat, and author

BORN: March 15, 1801; Woodstock, Vermont

DIED: July 23, 1882; Vallombrosa, Italy

SIGNIFICANCE: Marsh's widely read book *Man and Nature: Or, Physical Geography as Modified by Human Action* (1864), a treatise on environmental history, became one foundation for the conservation and environmental movements of the twentieth century.

George Perkins Marsh's childhood home in Vermont was located at the base of a mountain along the Quechee River, which originally flowed all summer. However, because of logging and sheep grazing,

Statesman and author George Perkins Marsh. (Library of Congress)

the river flow changed to flooding in the spring and little or no flow in the late summer. The problem of the river's flow was explained to the young Marsh by his father, an eminent lawyer, and the situation eventually became a major influence on Marsh's thinking. In the short term, however, he followed the example of his father and grandfather and trained for the law. He then entered the Vermont legislature and later represented his state in the U.S. Congress.

In 1850 Marsh became the U.S. minister to Turkey. While serving there, he traveled around the eastern Mediterranean region and became impressed with the evidence of civilization's impacts on the land. He realized that deforestation and grazing by goats were important causes of desertification in arid regions. Traveling to France, he saw both the severe erosion that followed deforestation in the mountains and the value of reforestation for restoring the land.

After five years abroad, Marsh returned to the United States, and the governor of Vermont appointed him state fish commissioner. He published *Report on the Artificial Propagation of Fish* (1857), which explained the impacts of logging, livestock, farming, and industry on fish streams. In 1861 President Abraham Lincoln appointed Marsh minister to Italy, in which position he remained for the rest of his life. At his wife's urging, Marsh had begun writing what would become the book *Man and Nature: Or, Physical Geography as Modified by Human Action* in 1860, and he completed the manuscript in Italy in 1863. The book was published in 1864 to such success that it was soon reprinted, and Marsh produced a revised edition in 1874. Italian editions were published in 1869 and 1872.

Man and Nature argues urgently that humankind must learn to manage both natural and domesticated resources wisely. Marsh believed that science should provide the guidance for such management. To him the most relevant discipline was geography; today his subject might be called applied ecology. *Man and Nature* includes chapters on wildlife, water, sand, and the side effects of engineering projects. Its longest chapter is on the ecology of forests and the consequences of deforestation. Marsh's book helped to convince the U.S. government to establish a forest policy and then the Forest Service. It also exerted influence on prominent scientists in Europe, and it became one foundation for the conservation and environmental movements of the twentieth century.

Frank N. Egerton

FURTHER READING

Elder, John. *Pilgrimage to Vallombrosa: From Vermont to Italy in the Footsteps of George Perkins Marsh*. Charlottesville: University of Virginia Press, 2006.

Lowenthal, David. *George Perkins Marsh: Prophet of Conservation*. Seattle: University of Washington Press, 2000.

SEE ALSO: Conservation policy; Ecology in history; Leopold, Aldo; Logging and clear-cutting; Overgrazing of livestock; Sustainable forestry.

Marshall, Robert

CATEGORIES: Activism and advocacy; forests and plants; preservation and wilderness issues

IDENTIFICATION: American forester and plant physiologist

BORN: January 2, 1901; New York, New York

DIED: November 11, 1939; on a train en route from New York to Washington, D.C.

SIGNIFICANCE: Marshall influenced both government policy and public opinion through his numerous writings on the need for wilderness conservation and through his participation in the Wilderness Society, an organization he cofounded.

Robert Marshall was born in 1901. He was the son of Florence Lowenstein Marshall and constitutional lawyer and conservationist Louis Marshall, who had been a delegate at the New York State Constitutional Convention of 1894, which placed in the state constitution the famous provision that the New York Forest Preserve shall be "kept forever wild." Marshall's extensive utilization of the family library introduced him to books and topographical surveys of the Adirondack Mountains. At the age of fourteen, he, along with his brother George and a guide, ascended a high Adirondack peak, thus cementing a lifelong love affair with wilderness exploration and celebration.

Marshall's higher education at three universities (bachelor of science in forestry at Syracuse, master's degree in forestry at Harvard, and Ph.D. in plant pathology at The Johns Hopkins University) served him well as he developed a literary and resource management career. During his early professional work in the U.S. Forest Service, he had the opportunity to observe large, unbroken wilderness conditions, which served as the catalyst and foundation for his first major article on extensive natural landscapes. Titled "The Problem of the Wilderness" and appearing in the February, 1930, issue of *The Scientific Monthly*, it was a clarion call for setting aside and protecting large tracts of land in their natural and, to the extent possible, primeval condition. The article elucidated four salient wilderness themes: its great beauty and wildness, with integrated aesthetic, mental, and physical values; the rapid disappearance of wilderness; the need for human beings to look beyond commodity value of resources in wilderness as the sole arbitrator of its value; and the urgency to act for wilderness preservation.

In 1931 Marshall settled in Washington, D.C., and immediately devoted his efforts to writing assignments. Collaborating with the U.S. Forest Service on a book titled *A National Plan for American Forestry* (1932), he contributed sections on national parks, wilderness, and recreation. One year later he published *The People's Forests*, in which he articulated the importance of conserving water, soil, and forests. Again he upheld the need to preserve forested areas through arguments related to human aesthetic needs, arguing that such preservation is of pivotal importance to contemporary society.

In 1933 Marshall was appointed director of forestry at the Office of Indian Affairs, where he helped develop sixteen wilderness areas on Indian reservations. Two years later, he was a leader of eight people who founded the Wilderness Society. In 1937 he became chief of the new Forest Service Division of Recreation and Lands and immediately began moving official Forest Service policy toward supporting wilderness. He also drafted new administrative regulations relating to a classification system for wilderness and wild areas. Approval for these regulations came just months before Marshall's untimely death in 1939 at age thirty-eight. In 1964, as the Wilderness Act became law, the twentieth area to be named to the National Wilderness Preservation System was the Bob Marshall Wilderness Area in the Flathead and Lewis and Clark national forests.

Charles Mortensen

Further Reading

Herron, John P. *Science and the Social Good: Nature, Culture, and Community, 1865-1965.* New York: Oxford University Press, 2010.

Marshall, Robert. *Alaska Wilderness: Exploring the Central Brooks Range.* Edited by George Marshall. 3d ed. Berkeley: University of California Press, 2005.
_____. *The People's Forests.* 1933. Reprint. Iowa City: University of Iowa Press, 2002.

See also: Conservation; Forest and range policy; Forest management; Forest Service, U.S.; Leopold, Aldo; Muir, John; Wilderness Society.

Massachusetts v. Environmental Protection Agency

Categories: Treaties, laws, and court cases; weather and climate
The Case: U.S. Supreme Court ruling concerning the regulation of greenhouse gases
Date: Decided on April 2, 2007
Significance: The Supreme Court's decision in the case of *Massachusetts v. Environmental Protection Agency* substantially enhanced the opportunity of states to challenge the decisions of federal agencies in court, but it did not quickly produce an EPA regulation governing greenhouse gas emissions.

In 1999 a group of environmental organizations petitioned the U.S. Environmental Protection Agency (EPA) to exercise its authority under the Clean Air Act to make a rule regulating greenhouse gases from vehicle emissions. In 2001 the EPA requested public comment, and in 2003 the agency denied the petition, giving three reasons: First, the EPA lacked the authority under the Clean Air Act to issue carbon dioxide emission standards; second, scientific evidence had not conclusively established causal links between human activity, an increase in greenhouse gases, and a rise in global air temperature; and third, EPA regulation of greenhouse gases would conflict with the president's climate change policy and hamper his ability to negotiate with other nations to reduce greenhouse gases. Massachusetts, joined by state and local governments and environmental organizations, appealed the decision. After the U.S. Court of Appeals denied review in 2005, the state appealed to the U.S. Supreme Court.

The Supreme Court first addressed the issue of the state's standing to invoke the Court's jurisdiction. Justice John P. Stevens, writing for the Court, stated that the Court relaxed its standing test for two reasons: First, the dispute involved the interpretation of a federal statute, the Clean Air Act, intended to protect Massachusetts; and second, the party seeking review was a sovereign state that owned the territory alleged to be harmed, an interest that was entitled to particular generosity. Applying its three-part standing test, the Court granted Massachusetts standing to challenge the EPA's refusal to grant the state's rule-making petition, because the state had demonstrated sufficient injury by alleging that it had lost a significant portion of its coastline, that the EPA's refusal to regulate auto emissions (though such emissions are only one source of greenhouse gases) contributed to this injury, and that the EPA's use of its Clean Air Act authority to regulate carbon dioxide from new motor vehicles would be likely to redress the injury by slowing global warming.

The Court then addressed the two issues that had been appealed. First, it held that the EPA had authority under the Clean Air Act to regulate greenhouse gas emissions from new motor vehicles because the statute's broad definition of "air pollutant" includes all airborne pollutants. Second, it held that the EPA's exercise of its authority had to be grounded in the Clean Air Act, which granted the agency discretion to determine if an air pollutant causes or contributes to the endangerment of public health or welfare, not the discretion to make policy arguments that it would be unwise to regulate at the present time. Since the EPA provided no reasoned explanation for its denial of the rule-making petition, its action violated the Clean Air Act. The EPA was mandated to provide a reasoned explanation for its decision grounded in the statute.

The Supreme Court's decision substantially enhanced the opportunity of states to challenge federal agency decisions in court, but it did not quickly produce an EPA regulation governing greenhouse gas emissions. In May, 2007, President George W. Bush directed the EPA to write new regulations, but by the end of 2008, none had been issued. In April, 2009, the EPA under President Barack Obama's administration concluded its scientific review and announced a proposed finding that greenhouse gases, including carbon dioxide, endanger public health and welfare. Then, on December 7, 2009, the opening day of the United Nations Climate Change Conference in Copenhagen, Denmark, the EPA finalized its endangerment finding. This decision began the agency's process of writing a new rule regarding the emissions from motor vehicles, power plants, and factories that contribute to global warming.

William Crawford Green

FURTHER READING

Ferrey, Steven. "Air Quality Regulation." In *Environmental Law: Examples and Explanations*. 5th ed. New York: Aspen, 2010.

Heinzerling, Lisa. "The Role of Science in *Massachusetts v. EPA*." *Emory Law Journal* 58 (2008): 411-422.

Sugar, Michael. "*Massachusetts v. Environmental Protection Agency*." *Harvard Environmental Law Review* 31 (2007): 531-544.

SEE ALSO: Air pollution; Air-pollution policy; Automobile emissions; Carbon dioxide; Clean Air Act and amendments; Environmental law, U.S.; Environmental Protection Agency; Global warming; Greenhouse gases.

Mather, Stephen T.

CATEGORIES: Activism and advocacy; preservation and wilderness issues

IDENTIFICATION: American conservationist

BORN: July 4, 1867; San Francisco, California

DIED: January 22, 1930; Brookline, Massachusetts

SIGNIFICANCE: As the first director of the U.S. National Park Service, Mather personified the national parks movement during the early decades of the twentieth century.

A descendant of one of America's early Puritan families, Stephen T. Mather was born and raised in California, where he came to love the mountains and forests of the western United States. He joined John Muir's Sierra Club in 1904. He was educated at the University of California at Berkeley, where he made many lifelong contacts. After graduation he went into the borax business, in which he accumulated considerable wealth.

Mather, a Republican, followed Theodore Roosevelt, the most famous conservationist in the United States, into the Progressive Party in 1912. However, Mather had the ability to transcend partisan politics. Several earlier attempts had been made to bring order to the small number of national parks that had already been established, and in 1913 President

Stephen T. Mather, right, with U.S. secretary of the interior Albert B. Fall at Glacier Point, Yosemite Valley, in 1921. (National Park Service Historic Photograph Collection)

Woodrow Wilson's secretary of the interior, Franklin K. Lane, decided to upgrade the parks by appointing Adolph C. Miller, a fellow University of California graduate, as assistant secretary. After Miller was reassigned elsewhere, Lane challenged Mather to take over the position. He did so in 1915, with another Berkeley graduate, Horace Albright, as his chief assistant. In 1916 the Park Service Act became law, and Mather became the National Park Service's first and arguably most significant director.

As director of the service, Mather faced a series of interrelated issues: The parks required both financial and political support from Congress, which in turn depended on public opinion. Energetic, charismatic, and brilliant at public relations, Mather generated popular support for the national park system through various avenues, including national publications such as the *Saturday Evening Post* and *National Geographic*

Magazine—more than one thousand articles about the parks appeared from 1917 to 1919. He organized conferences, brought politicians and writers to the parks, and used his own funds in support of the park system; he personally paid the salary of Robert Sterling Yard, who became the parks' first publicist. In order to provide better access to the parks, Mather strongly supported at least some development; he worked with railroads and automobile associations in building roads and providing other public amenities in some parks. Through the years he had his differences with politicians, bureaucrats, and others, but he worked successfully with both Republicans and Democrats.

Mather retired in 1929 because of ill health and was succeeded by Albright, but he left the National Park Service as a major icon in America's consciousness. Under his tenure the park system increased from thirteen parks, eighteen national monuments, a total of 1.9 million hectares (4.8 million acres) of area, and 334,000 annual visitors to twenty national parks, thirty-two national monuments, 3.3 million hectares (8.3 million acres), and three million visitors each year. However, development engendered criticism, and Mather was accused of being too prodevelopment. Despite the criticism, he had no doubt about his responsibility as director; he stated, "Our job in the Park Service is to keep the national parks as close to what God made them as possible." No director since Mather's time has had as much impact on the U.S. National Park Service, and no one since has been able to generate as much public and political support for the park system.

Eugene Larson

FURTHER READING

Gonzalez, George A. "The Political Economy of the National Park System." In *Corporate Power and the Environment: The Political Economy of U.S. Environmental Policy.* Lanham, Md.: Rowman & Littlefield, 2001.

Heacox, Kim. *An American Idea: The Making of the National Parks.* Washington, D.C.: National Geographic Society, 2001.

Nash, Roderick. *Wilderness and the American Mind.* 4th ed. New Haven, Conn.: Yale University Press, 2001.

SEE ALSO: Kings Canyon and Sequoia national parks; Muir, John; National Park Service, U.S.; National parks; Yellowstone National Park; Yosemite Valley.

Medfly spraying

CATEGORY: Pollutants and toxins

DEFINITION: Aerial application of pesticides to eliminate the Mediterranean fruit fly, an agricultural threat

SIGNIFICANCE: Efforts to eradicate Medfly infestations through the spraying of pesticides have met with criticism by many environmentalists because of concerns about the possible harmful effects of the pesticides on other insects, wildlife, livestock, aquaculture, water supplies, and human health.

The small, two-winged Mediterranean fruit fly, or Medfly (*Ceratitis capitata*), belongs to a group of insects commonly called fruit flies. Considered a major agricultural pest around the world, the Medfly is a threat to more than 250 vegetables and fruits, including peaches, cherries, avocados, pears, and citrus fruits. A longtime inhabitant of Hawaii and the Tropics, the Medfly has repeatedly tried to establish itself in the continental United States since its first unsuccessful foray into Florida in 1929. Attempts to control the fly with pesticides have so far proven successful but controversial.

The incompatibility of the Medfly with humans stems from the insect's fondness for domestic crops. Typically, the female fly lays eggs—as many as several hundred at a time—in the fleshy parts of fruits or vegetables. When the eggs hatch, the larvae tunnel through the fruit, making it unfit for human consumption. When the damaged fruit falls to the ground, the larvae exit and burrow into the ground until they mature into flies and start the cycle over again.

Because of the Medfly's capacity for causing widespread crop destruction, governments around the world have imposed various quarantines, embargoes, and postharvest treatment requirements on any fruits and vegetables that originate in areas known to be infested with the insect. Various eradication programs were used against the Medfly in the United States during the twentieth century. Early efforts in Florida included the use of a compound of arsenate and copper carbonate that was applied with handheld equipment. Another control method was the removal of infested fruit trees.

The most common form of eradication practiced is the aerial spraying of the pesticide malathion mixed with a bait, such as syrup. A poison also employed to control mosquitoes, malathion is used in weaker amounts in the war against the Medfly. Although all aerial spraying attempts in the United States must receive prior approval from the Environmental Protection Agency (EPA), the use of malathion has generated controversy in California and Florida. Opponents commonly complain about the pesticide's possible harmful effects on other insects, wildlife, livestock, aquaculture, water supplies, and human health—especially that of children. In response to these concerns, the EPA now requires various federal and state agencies involved in eradication programs to seek more environmentally friendly methods to battle the Medfly.

One alternative to the spraying of malathion involves the use of domestically raised flies that have been sterilized through radiation; these flies are released to mate unsuccessfully with wild Medflies, thus reducing the regenerative capacity of the population. Another method that has been used involves a pesticide made from a mixture of two dyes, phloxine B and uranine, which are commonly used to tint drugs and cosmetics. Once ingested by the

A Mediterranean fruit fly is about two-thirds the size of a common housefly. The Medfly is a threat to more than 250 kinds of vegetable and fruit crops. (AP/Wide World Photos)

Medfly, the dye particles absorb light, which in turn produces oxidizing agents that destroy cell tissues. As a result, most flies die within twenty-four hours. The dyes quickly lose their potency and become nontoxic. Preliminary tests of this method in Hawaii indicated that the use of these dyes may be more effective and safer for the environment than malathion.

John M. Dunn

FURTHER READING

Gullan, P. J., and P. S. Cranston. "Pest Management." In *The Insects: An Outline of Entomology.* 4th ed. Hoboken, N.J.: John Wiley & Sons, 2010.

Hamilton, Denis, and Stephen Crossley, eds. *Pesticide Residues in Food and Drinking Water: Human Exposure and Risks.* Hoboken, N.J.: John Wiley & Sons, 2004.

SEE ALSO: Agricultural chemicals; Biopesticides; Integrated pest management; Malathion; Pesticides and herbicides.

Mediterranean Blue Plan

CATEGORY: Water and water pollution
IDENTIFICATION: A multinational effort to curb pollution in the Mediterranean Sea
DATE: Initiated in 1980

SIGNIFICANCE: Action plans developed by researchers working with Blue Plan Regional Activity Centers have helped to reduce pollution in the Mediterranean, despite some difficulties posed by the need for international cooperation.

In 1980, under the auspices of the United Nations Environment Programme (UNEP), the nations bordering the Mediterranean Sea signed an agreement setting forth ways in which they would cooperate to reduce pollution of their common sea. The agreement, which soon came to be known as the Blue Plan (or Plan Bleu) for its efforts to clean the Mediterranean's waters, represented the culmination of several years of international efforts. In 1975, for example, UNEP had provided more than $7 million to the Mediterranean Action Plan (MAP), an earlier program designed to help Mediterranean countries fight pollution. Early in 1979, however, UNEP informed signatories to the 1975 agreement that it would cut back future financial support for MAP. UNEP thus called a February, 1979, conference in Geneva, Switzerland, to prepare a new approach to budgetary demands for immediate environmental remedies and to map out a strategy for protecting the ecology of the Mediterranean basin. In Geneva, a program was drafted identifying twenty-three environmental protection projects demanding immediate attention. A budget of $6.5 million was established, one-half to

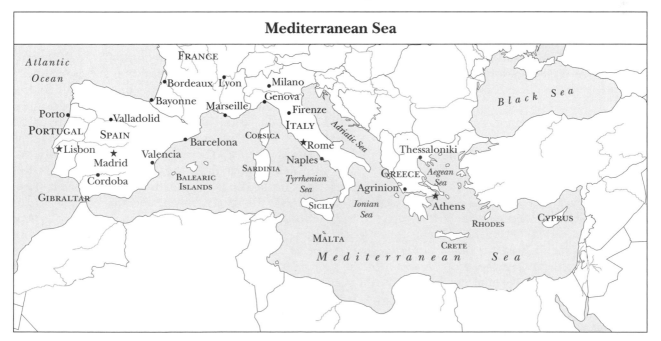

Mediterranean Sea

come from the participating countries, one-fourth from UNEP, and the remainder from contributions of services and staff time by environmental organizations.

Staffs of international researchers formed Blue Plan Regional Activity Centers, which provide information to signatory governments to help them plan future economic development in such a way as to prevent a repetition of the environmental damages that had been done to the sea and its coastline in earlier decades. The role of UNEP is to facilitate communication among these centers and to sponsor international meetings to share findings and propose solutions on a regular basis. Blue Plan researchers have focused their attention on food production, industry, energy use, tourism, and transport. In the early years of the plan, for example, jointly sponsored research suggested that some ecologically harmful industries, such as mining and metallurgical processing and petrochemical production, were overproducing in Mediterranean areas; in such cases, plan officials suggested ecologically preferable and economically logical adjustments.

The attainment of such goals, however, has sometimes been complicated by political and economic factors. For example, efforts to streamline supply and production of coal and steel on a geographic basis were hindered by long-standing tensions between Turkey and Greece. Similarly, Tunisia and Algeria have resisted energy market cooperation with each other, although Tunisia needs the natural gas and petroleum that neighboring Algeria produces; Tunisia thus has continued to pursue, at substantial economic and ecological cost, its own limited petroleum production. Despite such setbacks, Blue Plan efforts have had some effect, and pollution levels in the Mediterranean have dropped.

Alexander Scott

FURTHER READING

Blondel, Jacques, et al. *The Mediterranean Region: Biological Diversity Through Time and Space.* 2d ed. New York: Oxford University Press, 2010.

Skjærset, Jon Birger. "The Effectiveness of the Mediterranean Action Plan." In *Environmental Regime Effectiveness: Confronting Theory with Evidence,* by Edward L. Miles et al. Cambridge, Mass.: MIT Press, 2002.

Wainwright, John, and John B. Thornes. *Environmental Issues in the Mediterranean: Processes and Perspectives from the Past and Present.* New York: Routledge, 2004.

SEE ALSO: Ocean pollution; United Nations Environment Programme; Water pollution; Water-pollution policy.

Mendes, Chico

CATEGORIES: Activism and advocacy; forests and plants; ecology and ecosystems

IDENTIFICATION: Brazilian rubber tapper and trade union leader

BORN: December 15, 1944; Acre, Brazil

DIED: December 22, 1988; Xapuri, Acre, Brazil

SIGNIFICANCE: Mendes spent his entire life working against the forces of environmental destruction in the Amazon forest in order to sustain a way of life for his fellow rubber tappers and other indigenous peoples of western Brazil. He earned international recognition as a defender of the Amazon ecosystem.

Francisco "Chico" Mendes Filho's early years were like those of all rubber tappers in Brazil. His family lived on a *seringal* (rubber estate) owned by a master. Tappers received artificially low prices for latex, and they had to purchase all their tools and foodstuffs from the estate store. Although education of tapper children was prohibited, Mendes secretly learned to read and write from a fellow tapper.

In the 1960's the Brazilian military government embarked on a massive effort to open the rain forests in the Amazon region to development. Roads were built, and large cattle ranches were developed to make cheap meat available to the frustrated working class. The area also became a dumping ground for the surplus population in the east. By 1980 deforested land in the Amazon totaled 123,590 square kilometers (47,718 square miles), and this change threatened the way of life of the region's indigenous peoples.

In 1971 Mendes began organizing rural rubber tappers; his work culminated in the formation of a union in Xapuri in the northern state of Acre in 1977. Unions then began to spring up in other towns in Acre as well. Their main tactic in challenging developers and ranchers was the *empate*, a standoff between

Trade union leader and ecologist Chico Mendes. (AFP/Getty Images)

tappers and laborers who were clearing the forest during the dry season. From the mid-1970's to 1988, the Acre rubber tappers staged more than forty major *empates*, persuading the laborers to lay down their chain saws and go home. More than 810,000 hectares (2 million acres) of forest were saved during this period.

Mendes's leadership was recognized in 1985 when the first National Rubber Tappers Congress was held in Brasília. Soon after, Mendes gained international recognition through the efforts of a maverick Brazilian agronomist and a British journalist. Although his first concern was sustaining his people rather than promoting an environmental agenda, Mendes eventually realized that environmental concerns would give his cause attention and shape world opinion in the rubber tappers' favor. In 1987 he traveled to Miami, Florida, to speak with the Inter-American Development Bank, which was financing road building in the Amazon. Mendes also traveled to Washington,

D.C., to testify before a U.S. Senate committee that was considering providing funds for the bank.

Mendes's crowning achievement came in October, 1988, when the Brazilian government declared a 24,686-hectare (61,000-acre) tract of rubber tapper territory near Xapuri an extractive reserve. An extractive reserve—a concept originated by Mendes—is an area designated solely for sustainable use. The local ranchers were strongly opposed to the establishment of such reserves, however, and Mendes was assassinated in December, 1988, by a rancher's hired gunman.

Ruth Bamberger

FURTHER READING

Revkin, Andrew. *The Burning Season: The Murder of Chico Mendes and the Fight for the Amazon Rain Forest.* 1990. Reprint. Washington, D.C.: Island Press, 2004.

Rodrigues, Gomercindo. *Walking the Forest with Chico Mendes: Struggle for Justice in the Amazon.* Edited and translated by Linda Rabben. Austin: University of Texas Press, 2007.

SEE ALSO: Amazon River basin; Deforestation; Greenhouse effect; Rain forests.

Mercury

CATEGORY: Pollutants and toxins

DEFINITION: Silver-white metal element that is liquid at room temperature

SIGNIFICANCE: Mercury is highly toxic, and exposure to mercury compounds can cause acute episodes of poisoning and chronic health effects. Mercury is found in the environment in elemental and combined forms; people are exposed to its various forms in workplaces, in homes, and through mercury-laden foods.

Mercury, often called quicksilver, is a relatively rare element, making up only about 3 parts per billion of the earth's crust. Mercury occurs in its elemental form as a liquid, but it is more commonly found combined with other elements to form various minerals. The most common mercury-containing mineral is cinnabar (mercury sulfide), which is found mainly in Spain, Algeria, and China.

Mercury and its inorganic compounds have been used in producing caustic soda, dry-cell batteries, scientific measuring devices, dental amalgams, and mercury vapor lamps. Workers in factories making such products may be exposed to relatively high levels of mercury vapors and compounds. Dental assistants may also be exposed to relatively high levels of mercury vapor. Mercury is also released into the environment by the waste incineration of these products and may eventually accumulate in food sources such as grains and fish. For a long time, mercury was used as an additive to latex paints to inhibit the growth of bacteria and fungi, but this use was eliminated in the United States in 1991. Mercury compounds have also been used in agriculture as fungicides; workers who handle these compounds may experience negative health effects. Other sources of mercury in the environment are mercury mines and smelters, which may release considerable amounts of mercury into surrounding regions.

Some bacteria can convert inorganic mercury into organic compounds, such as methylmercury. Methylmercury bioaccumulates in biotic systems and biomagnifies until the highest members of the aquatic food chain have high levels of mercury in their fatty tissues. People who eat significant amounts of mercury-rich fish, such as tuna, pike, swordfish, and shark, may therefore suffer adverse health effects from mercury exposure.

Since elemental mercury is a liquid at room temperature and easily volatilizes (passes off in vapor), it can be inhaled and cause damage to the central nervous system, lungs, and kidneys. Mercury is a strong neurotoxin that at low levels of exposure can cause a series of effects such as memory loss, tremors, and excitability; higher levels may cause more severe mental problems and even death. Inhalation of high concentrations of mercury can lead to hallucinations, delirium, and even suicide in some cases. It is not clear whether elemental mercury causes cancer in humans; the U.S. Environmental Protection Agency (EPA) considers elemental mercury to be "not classifiable as to human carcinogenicity."

The inorganic forms of mercury are not easily absorbed by the body and pose relatively small health risks. Oral ingestion of inorganic mercury can cause nausea, stomach pains, and vomiting. The EPA states that the acute lethal dose of most inorganic mercury compounds for a normal adult is 1 to 4 grams (0.035 to 0.141 ounces), or 14 to 57 milligrams (0.0005 to 0.002 ounces) per kilogram (2.2 pounds) for a person weighing 70 kilograms (approximately 154 pounds). The EPA classifies inorganic mercury as a possible human carcinogen.

Organic mercury compounds, particularly methylmercury, are absorbed easily and constitute the greatest health risk for most people. Methylmercury exposure is greatest from the consumption of fish and fish products. Low-level poisoning by methylmercury may cause vision problems, paresthesia (the sensation of prickly skin), speech difficulties, shyness, and a general malaise. Acute exposure can cause numerous problems of the central nervous system, including blindness, deafness, and loss of consciousness. The EPA estimates that the minimum lethal dose of methylmercury is 20 to 60 milligrams (0.0007 to 0.002 ounces) per kilogram for a person weighing 70 kilograms. Lower doses are lethal for more sensitive people, particularly children. Methylmercury has been shown to have significant developmental effects on babies born to women who ingested

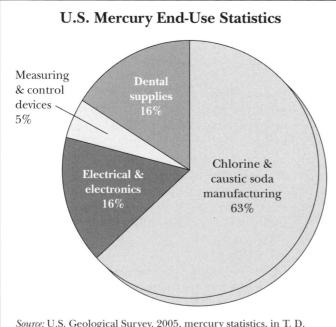

U.S. Mercury End-Use Statistics

Measuring & control devices 5%

Dental supplies 16%

Electrical & electronics 16%

Chlorine & caustic soda manufacturing 63%

Source: U.S. Geological Survey, 2005, mercury statistics, in T. D. Kelly and G. R. Matos, comps., *Historical Statistics for Mineral and Material Commodities in the United States,* U.S. Geological Survey Data Series 140.

high levels during their pregnancy. The infants from such pregnancies may exhibit developmental disabilities, visual problems (including blindness in some), and cerebral palsy. The EPA classifies methylmercury as a possible human carcinogen.

Mercury has caused serious environmental and human health problems when released into ecological systems and when used inappropriately. One of the most infamous incidents is the mercury poisoning that took place at Minamata Bay in Japan in the 1950's and early 1960's. Inorganic mercury waste was released into the bay, and bacteria converted the inorganic mercury to methylmercury, which was then consumed by marine organisms and passed up the food chain, eventually resulting in human mercury poisoning. At least 43 people died of "Minamata disease" from eating fish and other marine organisms that contained high levels of mercury. Even more people were killed in Iraq by the inappropriate used of methylmercury fungicide. In 1972 more than 460 people died from mercury poisoning after eating bread made from wheat that had been treated with the fungicide. Many of the animals and plants within the area were also poisoned by the methylmercury.

Jay R. Yett

FURTHER READING

Bakir, F., et al. "Methylmercury Poisoning in Iraq." *Science* 181 (July, 1973): 230-241.

Eisler, Ronald. *Mercury Hazards to Living Organisms.* Boca Raton, Fla.: CRC Press, 2006.

Montgomery, Carla W. "Water Pollution." In *Environmental Geology.* 9th ed. New York: McGraw-Hill, 2010.

Timbrell, John. "Environmental Pollutants." In *Introduction to Toxicology.* 3d ed. New York: Taylor & Francis, 2002.

SEE ALSO: Biomagnification; Food chains; Heavy metals; Minamata Bay mercury poisoning; Oak Ridge, Tennessee, mercury releases.

Methane

CATEGORIES: Pollutants and toxins; atmosphere and air pollution; weather and climate

DEFINITION: Colorless, odorless gas that is the principal component of natural gas

SIGNIFICANCE: Methane in the form of natural gas represents an abundant source of energy, and methane combustion is cleaner than petroleum or coal combustion. Methane is a powerful greenhouse gas, however; scientists are thus faced with the challenge of optimizing energy production using methane while minimizing unwanted methane emissions.

Methane is produced when bacteria digest organic matter under anaerobic (without air) conditions, creating natural gas. Natural gas contains 50-90 percent methane. Most natural gas is found with coal and petroleum deposits buried deep underground and is a product of the decomposition of ancient swamps and bogs.

The sources of methane emissions are both anthropogenic (human-influenced) and natural. Anthropogenic sources include fossil-fuel production, livestock raising (enteric fermentation in the stomachs of animals such as cattle and pigs produces approximately 37 percent of all human-induced methane), rice cultivation, biomass burning, and waste management (sewage treatment and landfills). The Intergovernmental Panel on Climate Change estimates that more than 60 percent of global methane emissions are related to such activities. Natural sources of methane emissions include wetlands, which provide habitat conducive to bacteria that produce methane during their decomposition of organic material; the digestive processes of termites (the second-largest natural source of methane emissions); and oceans, where methane emissions come from anaerobic digestion by marine zooplankton and fish and by methanogenesis in marine sediments.

Methane is also stored as a hydrate (methane hydrate, a crystalline solid consisting of gas molecules surrounded by a cage of water molecules) in immense amounts in marine sediments and in the frozen Arctic tundra. The worldwide amount of carbon stored as methane hydrate is estimated to total twice the amount of carbon found in all known fossil fuels on earth. Methane can be released from the hydrates by increases in temperature and other factors.

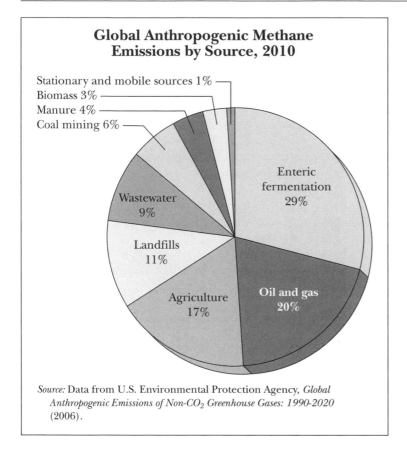

Global Anthropogenic Methane Emissions by Source, 2010

- Stationary and mobile sources 1%
- Biomass 3%
- Manure 4%
- Coal mining 6%
- Wastewater 9%
- Landfills 11%
- Agriculture 17%
- Oil and gas 20%
- Enteric fermentation 29%

Source: Data from U.S. Environmental Protection Agency, *Global Anthropogenic Emissions of Non-CO₂ Greenhouse Gases: 1990-2020* (2006).

Methane is used in industrial and chemical processes and can be transported as a refrigerated liquid (liquefied natural gas, or LNG). It is useful as a fuel for cooking, for powering motor vehicles, and for heating homes and commercial buildings. Many factory furnaces burn methane, and it is also used to generate electricity. Compressed natural gas is thought to be the cleanest-burning form of fossil fuel available, since the simplicity of the methane molecule results in lower emissions of various pollutants than are produced by other fossil fuels. Methane is also a key raw material for producing solvents (such as methanol) and other organic chemicals.

METHANE REMOVAL FROM THE ATMOSPHERE

According to the U.S. Environmental Protection Agency (EPA), released methane can remain in the earth's atmosphere for nine to fifteen years. Once emitted, methane can be removed from the atmosphere by a variety of processes, frequently called sinks. The dominant sink is oxidation by photochemically produced hydroxyl radicals (OH). The majority of methane molecules react with OH to form the methyl group CH_3 and water in the tropospheric layer of the atmosphere, with smaller amounts of methane destroyed in the stratosphere. These two OH reactions account for almost 90 percent of methane removal. Two smaller sinks are microbial uptake of methane in soils and methane's reaction with chlorine atoms in the marine boundary layer.

The balance between methane emissions and methane removal processes ultimately determines atmospheric methane concentrations. The methane remaining after methane removal processes can absorb terrestrial infrared radiation that would otherwise escape to space. This property can contribute to the warming of the atmosphere, which is why methane is considered to be a greenhouse gas. It is more than twenty times more effective in trapping heat in the atmosphere than carbon dioxide (CO_2) over a period of one hundred years.

RESEARCH DIRECTIONS

Methane research is proceeding in two major directions: Some focuses on energy generation, seeking ways to make bioconversion of wastes to methane more economically attractive as an alternative fuel, and some focuses on the environment, seeking ways to limit the release of methane into the atmosphere because of its properties as a greenhouse gas. A shared goal of these two areas of research is to find ways in which much of the methane released into the atmosphere could be harnessed for energy production. Methane has advantages over petroleum and coal as a fuel because it burns more cleanly than do those fossil fuels.

In the energy sector, operators of coal mines are looking for ways to isolate the methane produced as a result of mining activities, instead of venting it into the atmosphere. The EPA estimates that up to 40 percent of the methane that migrates to the atmosphere can be used for heat and power generation, injection into pipeline systems for transport, methanol production, or on-site applications such as coal drying. Landfill gas-to-energy projects, which collect the methane that forms in landfills, offer another promising way to reduce atmospheric release and provide inexpensive energy.

The U.S. Global Change Research Program has identified as a priority research activity the development of global monitoring sites to measure atmospheric methane levels. The Carbon Cycle Greenhouse Gases Group of the National Oceanic and Atmospheric Administration's Climate Monitoring and Diagnostics Laboratory also makes ongoing atmospheric methane measurements from land and sea surface sites and aircraft and continuous measurements from baseline observatories and towers. Measurement records from international laboratories are integrated and extended to produce a globally consistent cooperative data product called GLOBALVIEW.

Bernard Jacobson

FURTHER READING

Buell, Phyllis, and James Girard. *Chemistry Fundamentals: An Environmental Perspective.* 2d ed. Sudbury, Mass.: Jones and Bartlett, 2003.

Demirbas, Ayhan. *Methane Gas Hydrate.* London: Springer, 2010.

Khalil, M. A. K., ed. *Atmospheric Methane: Its Role in the Global Environment.* New York: Springer, 2000.

National Academy of Sciences. *Climate Change Science: An Analysis of Some Key Questions.* Washington, D.C.: National Academies Press, 2001.

_____. *Methane Generation from Human, Animal, and Agricultural Wastes.* Washington, D.C.: National Academies Press, 2001.

Soliva, Carla Riccarda, Junichi Takahashi, and Michael Kreuzer, eds. *Greenhouse Gases and Animal Agriculture.* Boston: Elsevier, 2006.

SEE ALSO: Bioremediation; Cattle; Dams and reservoirs; Landfills; Resource recovery; Synthetic fuels; Waste management.

Methyl parathion

CATEGORIES: Pollutants and toxins; agriculture and food

DEFINITION: Neurotoxin used on agricultural crops to kill pests, especially boll weevils, mites, and mosquito larvae

SIGNIFICANCE: Because methyl parathion is a highly toxic compound, its use in the United States is subject by law to many restrictions.

The molecular formula for methyl parathion is $C_8H_{10}NO_5PS$, and the chemical name is O,O-dimethyl (4-nitrophenyl) phosphorothioate. Also known as parathion-methyl and metafos, and commonly called cotton poison and roach milk, methyl parathion has been marketed under such trade names as Metacide and Penncap-M and packaged in dust, liquid, and emulsifiable concentrate forms. The U.S. Environmental Protection Agency (EPA) classifies methyl parathion as a restricted-use pesticide (RUP) that can be sold and utilized solely by certified applicators. Methyl parathion packaging is distinguished with the word "danger" to alert users of potential harmful contact. The EPA forbids agricultural workers from being in a field treated with methyl parathion for forty-eight hours after application of the pesticide.

Methyl parathion is an organophosphate that inhibits the enzyme cholinesterase, which is crucial to healthy functioning of the nervous system. If absorbed through the skin, inhaled, or ingested, the pesticide blisters tissues, retards reproductivity, and can cause death. Methyl parathion is toxic to aquatic organisms such as fish, shrimp, and crabs and harmfully affects carnivores that consume tainted organisms. Contaminated game animals pose a danger to hunters who eat the animals' flesh.

By killing algae-eating insects and crustaceans, methyl parathion has the unintended side effect of causing algae to reproduce rapidly; the algae then use oxygen that other water inhabitants need. In addition, methyl parathion poisons small mammals such as rats and rabbits, is harmful to birds and to beneficial insects such as honeybees, and has been reported to injure alfalfa and sorghum.

Methyl parathion usually requires several months to break down in soil, but it does not leave dangerous residues or kill soil microorganisms. The speed of degradation increases with higher temperatures, with exposure to sunlight, and in flooded soil. While small amounts of methyl parathion do not contaminate groundwater, spills require longer to degrade. The pesticide degrades more quickly in flowing water, in fresh water, and in water bodies with sediments. In plants, methyl parathion is almost completely metabolized within a week.

Although it is considered safe only for outdoor use, methyl parathion is effective against cockroaches and rats; because of this effectiveness, unlicensed exterminators often spray the chemical inside homes. In 1996 the EPA evacuated residents of one thousand contaminated homes in Mississippi and spent millions of dollars to rebuild the structures. To prevent misuses of the pesticide, manufacturers, in agreement with the EPA, instituted a computerized system to track the locations of containers of methyl parathion.

Methyl parathion poses combustion hazards above 100 degrees Celsius (212 degrees Fahrenheit) and releases toxic fumes, including dimethyl sulfide, sulfur dioxide, and carbon monoxide. Concerns about adverse toxic effects of pesticide products have resulted in studies of methyl parathion's potential carcinogenic tendencies. In 1992 producers of methyl parathion stopped supporting its use on such agricultural crops as strawberries and tobacco and for mosquito control in forests. In 1999, the EPA eliminated use of the pesticide on crops particularly likely to form large parts of children's diets; these included apples, peaches, pears, grapes, nectarines, cherries, plums, carrots, certain peas, certain beans, and tomatoes.

Elizabeth D. Schafer

FURTHER READING

Friis, Robert H. "Pesticides and Other Organic Chemicals." In *Essentials of Environmental Health.* Sudbury, Mass.: Jones and Bartlett, 2007.
Hilgenkamp, Kathryn. "Pests and Pesticides." In *Environmental Health: Ecological Perspectives.* Sudbury, Mass.: Jones and Bartlett, 2006.

SEE ALSO: Agricultural chemicals; Pesticides and herbicides; Runoff, agricultural; Water pollution.

Middle East

CATEGORIES: Places; ecology and ecosystems; resources and resource management

SIGNIFICANCE: Because the Middle East is mostly arid or semiarid in terms of annual rainfall, access to fresh water is a very important issue in the region. An additional feature of the Middle East that has bearing on the relationship between its environment and the world economy is that a significant portion of the world's supplies of fossil fuels—60 percent of known reserves of petroleum and 40 percent of known reserves of natural gas—are located in the Middle East.

The geographical region of the world known as the Middle East encompasses countries and climatic zones of several continents, including parts of western Asia and the north and northeast coasts of Africa. Its marine boundaries are shared by the Mediterranean, Black, Caspian, and Red seas, the Persian Gulf, and the Indian Ocean. The sovereign polities that make up the Middle East include but are not limited to Turkey, Syria, Lebanon, Israel, Jordan, Saudi Arabia, Egypt, Sudan, Somalia, Libya, Iraq, Iran, Azerbaijan, Turkmenistan, Yemen, Oman, Qatar, the United Arab Emirates, Bahrain, Afghanistan, and Pakistan.

DESERTIFICATION

Desertification has increased in a number of places in the Middle East as a result of the combination of climatic change and poor human management of resources, including deforestation (which may be done in ignorance, out of desperation, or even deliberately owing to political instability, as in the burning of pine forests in the Levant as acts of terrorism and war) and wind erosion of soil caused by overgrazing of livestock. Based on satellite date collected between 1979 and 2005, some climate change in the region has been verified as natural, especially where possible expansion of the Tropics by 1 degree of latitude over twenty-five years—about 129 kilometers (80 miles)—resulted in increasing droughts and widespread fires and desertification owing to jet stream movements toward both poles, making places such as the northern Middle East susceptible to even dryer conditions.

It is unknown how much of the increased desertification is human-induced or the result of natural climatic variation. It is well known that the planting of pine forests in Israel from the 1960's onward not only resulted in several hundred million pine trees in large forested tracts, mostly Aleppo pine (*Pinus halepensis*) and Turkish pine (*Pinus brutia*), but also increased rainfall from orographic precipitation where normally rising windborne water vapor from Mediterranean evaporation was cooled by the forests. Precipitation from mitigated cooler land temperature in the hill country rose from 25 to 30 centimeters (10-12 inches) per year during the early 1960's to approximately 76 centimeters (30 inches) per year during the 1990's as condensation and dew point caused the rising moisture-laden air to release its water in the Levant rather than lose it to evaporation over the heated land. This type of reforestation could be a boon to

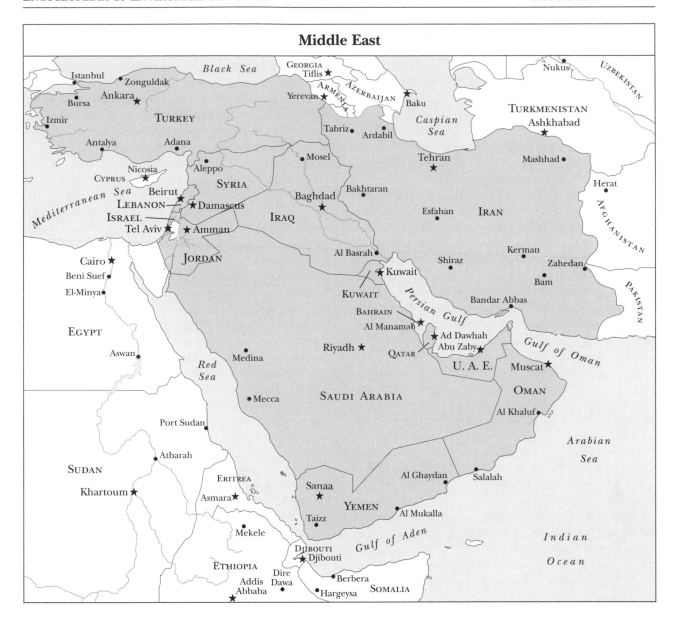

Middle East

marginally arid land in the coastal Middle East, reducing aridity and possibly reversing desertification.

WATER RIGHTS

Because of the region's aridity, many states of the Middle East are dependent on imported water. Some, including Egypt, Iraq, Kuwait, Bahrain, Egypt, Syria, and Israel, are dependent on external suppliers for more than 50 percent of their water, and many of the states in the Middle East import more than 25 percent of all the water they use.

Disputes over water rights in the Middle East are often acrimonious. Some watersheds, such as that of the Jordan River, are shared by multiple stakeholders whose diplomatic relationships have always been strained even without considerations of who owns the water rights or disputes over how to partition water supplies—whether by population, rainfall, commercial agricultural needs, or other political constraints.

Since 1990 war in the region has also had impacts on the hydrology of the Tigris-Euphrates basin. In Iraq alone in 1990, the Gulf War resulted in great damage to the water-supply system, and this was compounded in 2003 during the Iraq War, when the Shatt al Arab waterways near Basra in the south of Iraq were a major military target. Iraqi hydrology was subsequently destabilized by continued insurgencies and regional strife between Shia and Sunni factions and terrorism. From 1990 onward in Iraq eight major dams were destroyed or their watersheds polluted, and more than twenty-eight municipal water management facilities were destroyed. The hydrological infrastructure of Iraq was badly damaged between 2003 and 2008, and more than 70 percent of the population was without adequate water supplies for agriculture, drinking, and personal hygiene.

In Iran, before 1968 more than 50 percent of the nation's water still derived from ancient *qanat* subterranean aqueducts or ancient surface canals; by 1987, however, considerable population growth in large cities such as Tehran made this ancient system woefully inadequate, and large dams had to be constructed. Only 10 percent of Iran—mostly its western Zagros montane region and the Caspian plain fed by the Elburz montane region—receives adequate rainfall for sustainable agriculture, and the central Dasht-e-Lūt desert receives almost no precipitation. New or revitalized dams are planned or under construction in Iran in the southern Zagros watershed in Khūzestān, and these dams could have considerable environmental impact. Elsewhere, war between Israel and Lebanon's Hezbollah impaired water delivery in Lebanon; bombings and counterstrikes resulted in drastic damage to hydrology systems, including the destruction of more than forty water-supply distribution points, pipelines, and pumping stations.

The statement "He who controls water controls life" has been attributed to Gudea of Lagash, a ruler in southern Mesopotamia (the area that now comprises Iraq and parts of Syria, Turkey, and Iran) around 2100 B.C.E. The control of water is increasingly vital for the arid Middle East, where various countries fight over marginal water sources and major diplomatic impasses characterize attempts by the United Nations to mediate the problems. Water rights are expected to become more problematic in the twenty-first century if projected global warming further reduces water supplies in the region.

SOLAR POWER

Much of the energy used in the Middle East depends on the petroleum that is abundant in the region, but because this fossil-fuel resource is finite in supply and is also a significant source of pollution, great incentives exist for Middle Eastern nations to reduce this dependence. Because the Middle East has many cloudless days per year, the region has substantial potential for the development of solar power. In January, 2010, during the launching of a major national initiative for solar power development, Saudi Arabia's petroleum and mineral resources minister, Ali Al-Naimi, stated that the nation plans to export as much solar energy to the world in the future as it has exported oil in the past.

DESALINATION

Because the Middle East is so dependent on imported water, removing salt from seawater has been an important priority in the region for decades. The task of increasing supplies of fresh water through desalination is complicated by two major environmental issues: the high salinity of the water available for desalination and the large amounts of energy needed to operate desalination plants. Because they are surrounded by evaporitic basins where most of the water is lost to vapor before it can become precipitation, the bodies of water adjacent to Middle Eastern nations are among the highest salt-bearing waters in the world, with an average salinity of 41,000 ppm (parts per million) in the Red Sea and 38,000 ppm in the eastern Mediterranean Sea, meaning that salt content is about 4.1 percent and 3.8 percent, respectively. Fossil fuels are often used to generate the electricity needed to run desalination plants, but because the burning of these fuels contributes to pollution, attempts are ongoing to implement the use of wind or solar energy in such plants.

The Middle East produces 75 percent of global water desalination. In 2010 Saudi Arabia—the world's largest producer of desalinated water—was operating thirty desalination plants supplying 70 percent of national drinking-water needs for a population of some 29 million. Israel is also a high-tech leader in desalination, producing a volume of 100 million cubic meters (3.5 billion cubic feet, or 26.4 billion gallons) per year of fresh water. The world's largest desalination distillation operation is the Jebel Ali plant in Dubai, in the United Arab Emirates, which is expected eventually to provide up to 250 million cubic meters (8.8 billion cubic feet, or 66 billion gallons) of water per year from the Persian Gulf.

If the Middle East could implement massive water desalination on a scale that increases early twenty-first century output of fresh water by as much as 1,000 percent across many national boundaries, it is conceivable that agriculture and reforestation could thrive in the region, with a resultant rise in relative humidity and a sustainable hydrology cycle that would considerably change the dominant desert climate. It is important that the Middle East not deplete the fossil-fuel energy sources on which its overall economy depends before such a transformation can be effected.

Patrick Norman Hunt

FURTHER READING

Allan, J. A. *The Middle East Water Question: Hydropolitics and the Global Economy.* London: I. B. Tauris, 2001.

Braverman, Irus. *Planted Flags: Trees, Land, and Law in Israel/Palestine.* New York: Cambridge University Press, 2009.

Fu, Qiang, et al. "Enhanced Mid-latitude Tropospheric Warming in Satellite Measurements." *Science* 312 (May 26, 2006): 1179.

Ghafour, P. K. Abdul . "Solar Energy Initiative Launched." *Arab News* (Jeddah, Saudi Arabia), January 25, 2010.

Issar, Arie S., and Mattanyah Zohar. *Climate Change: Environment and Civilization in the Middle East.* New York: Springer, 2004.

SEE ALSO: Africa; Aswan High Dam; Desalination; Desertification; Gulf War oil burning; Oil crises and oil embargoes; Overgrazing of livestock; Solar energy.

Migratory Bird Act

CATEGORIES: Treaties, laws, and court cases; animals and endangered species

THE LAW: U.S. federal legislation designed to protect migratory birds

DATE: Became effective on March 4, 1913

SIGNIFICANCE: The Weeks-McLean Act was an important first step in U.S. government protection of migratory birds.

The Migratory Bird Act of 1913, also known as the Weeks-McLean Act (for its sponsors, Congressman John W. Weeks of Massachusetts and Senator John P. McLean of Connecticut), was preceded by the Lacey Act of 1900, which was introduced by John Lacey of Iowa in the House of Representatives. Both laws were passed in an effort to protect wildlife, to eliminate the slaughter of wild birds and the interstate market for their parts for the lucrative millinery trade, and to establish a regulatory system of federal, state, and foreign statutes to govern and protect fowl and other wildlife.

It is estimated that during the late nineteenth century nearly 200 million birds were slaughtered for their feathers and carcasses, which were used to decorate women's hats. The calamitous species extinction of the passenger pigeon aroused the attention of conservationists and protectionists nationwide. George Bird Grinnell founded the first Audubon Society in 1886 in an effort to organize opposition to the bird trade. Ornithology and bird-watching were popular activities pursued by professionals and amateurs alike at the time. In 1896 Harriet Hemenway founded the Massachusetts Audubon Society and quickly attracted influential members who championed opposition to the wanton destruction of birds. President Theodore Roosevelt was an avid supporter of the Audubon Society. In 1910 New York State passed the Audubon Plumage Law, banning the sale of the feathers of birds native to the state.

In 1913 the Weeks-McLean Act put birds under federal jurisdiction, regulating the spring hunting season. This act was followed in 1916 by a convention signed by Canada and the United States to establish federal guidelines protecting both game and nongame birds on both sides of the border. Signed into law in the United States as the Migratory Bird Treaty Act of 1918, this treaty was amended in 1937 to include Mexico in a comprehensive plan to protect all migratory birds on the North American continent. The Weeks-McLean Act is considered one of the first environmental laws of the United States.

Victoria M. Breting-García

FURTHER READING

Bean, Michael J. "Historical Background to the Endangered Species Act." In *Endangered Species Act: Law, Policies, and Perspectives,* edited by Donald C. Bauer and William Robert Irvin. 2d ed. Chicago: American Bar Association, 2010.

Burnett, J. Alexander. *A Passion for Wildlife: The History of the Canadian Wildlife Service.* Vancouver: University of British Columbia Press, 2003.

Price, Jennifer. *Flight Maps: Adventures with Nature in Modern America.* New York: Basic Books, 1999.

SEE ALSO: Convention on International Trade in Endangered Species; Darling, Jay; Endangered Species Act; Endangered species and species protection policy; National Audubon Society; Passenger pigeon extinction; Whooping cranes.

Minamata Bay mercury poisoning

CATEGORIES: Disasters; human health and the environment

THE EVENT: Contamination of the local food chain in Japan's Minamata Bay by industrial waste containing mercury

DATES: 1950's-1960's

SIGNIFICANCE: The environmental health tragedy caused by the disposal of industrial waste in Minamata Bay represents one of the first identified cases of a clear cause-and-effect relationship between toxic chemical discharge and severe harm to humans and their environment.

The city of Minamata is located on the southwest coast of Kyūshū, the southernmost of Japan's four main islands, in Kumamoto prefecture. During the 1950's, a local chemical plant owned by Chisso Corporation was engaged in the production of acetaldehyde and vinyl chloride. One of the chemicals used as a catalyst in the production processes was mercury oxide (HgO). The industrial waste from the plant, including mercury, was discharged into Minamata Bay. At the time, such waste disposal was an acceptable practice, and the amount of discharge increased during the early 1950's. It was generally understood that mercury, a heavy metal like lead, cadmium, and arsenic, could be injurious to the health of persons who mishandled, ingested, or inhaled it, but because mercury is dense and quite insoluble in water, it was presumed that it would quickly sink into the sediment at the bottom of the bay, where it would be slowly buried and disappear.

By 1953, unusual medical symptoms began to appear in area residents; most of the symptoms were neurological, such as tremors and impairment of senses. In 1956, the first case of a distinct medical condition was reported. By 1959, the cause of the affliction had been identified and the effects established: The first patients had acute or near-acute levels of mercury poisoning, a condition that became known as Minamata disease.

The residents of the area had not realized that the dense mercury (Hg) on the bay bottom was being acted on by microorganisms such as bacteria and algae and was being converted into methylmercury and dimethyl mercury. Methylmercury, the more injurious and toxic of the two substances, is much less dense than mercury itself and is more soluble. In the bay en-

vironment, it worked its way to the upper sediment, was taken up in the food chain by bottom-dwelling shellfish and fish, and was then consumed by the local people, for whom seafood was a dietary staple. Mercury can also be taken up by fish-eating birds and mammals. Heavy metals tend to accumulate in the bodies of organisms that ingest them, as in the edible flesh of fish. They concentrate up the food chain, as one species retains and accumulates the contaminant from its regular diet.

In natural environmental cycling and processes, elements such as mercury are weathered out of rock to soil and enter the hydrosphere (lakes, rivers, groundwater, and oceans). They might become more concentrated, but organic decay eventually returns them to the soil or water. This cycling can be disrupted when industrial or other human activities introduce anomalously large amounts of the element into the cycle that cannot be accommodated by natural means. Whether an element is toxic to humans or others is a function of concentration; some elements are essential in small amounts but become toxic in large concentrations.

Mercury occurs as a trace element in average crustal rock at a level of 0.1 parts per million (ppm); in seawater, it is typically under 0.05 parts per billion. At Minamata Bay, mercury levels in fish and shellfish were measured up to 50 ppm (25 ppm represents a biomagnification factor of 500,000 over normal seawater). For comparison, the U.S. Food and Drug Administration (FDA) prohibits fish with more than 1 ppm of mercury from being commercially marketed. Residents diagnosed with Minamata disease had mercury levels of 20 to 100 ppm in their livers and kidneys and 3 to 25 ppm in their brains. Minamata Bay sediment levels of mercury were as high as 7,000 ppm.

The medical effects of consuming mercury include progressive damage to the central nervous system and birth defects from prenatal exposure. Symptoms include shakiness (tremors), blurred vision, impaired speech, numbness in limbs, loss of memory and intelligence, nervousness, and, in extreme cases, death. By the time symptoms are apparent, the damage is irreversible. Of the first 34 people diagnosed with Minamata disease, 20 died within six months. By 1960, 43 people had died and 116 were permanently affected. About one-third of the infants born in the area were affected by symptoms. As of 1992, 2,252 people had become ill and been officially diagnosed with the disease, and 1,043 had died. Several thousand more claimed to have the disease but had not been officially diagnosed by the examining board set up by the Japanese government.

The developing environmental health problem of the 1950's at Minamata Bay, however inadvertently it may have started, became a broader and more persistent tragedy because of the lack of corporate or governmental response. Chisso Corporation denied responsibility for the problem, arguing that the dumping of industrial waste was an accepted practice and was legal at the time. Some have speculated that the company may have suspected the cause of the growing problem but concealed it. The national government declined to act on the matter. At the time, there was little governmental consciousness of environmental quality issues, and intervention might have harmed the chemical industry and slowed Japan's post-World War II industrialization and economic recovery. A similar problem occurred at

Niigata in west-central Japan in the 1960's, when mercury waste was dumped into the Agano River. The national government thus did not intervene in the Minamata situation until 1968, and the dumping of mercury into the bay was not stopped until 1971, twelve years after the problem became known.

The National Institute for Minamata Disease was set up in 1978 to assist with medical studies, conduct research, and follow the progress of victims over time. The Japanese government provides medical care for diagnosed victims, along with financial compensation. Active victims' movements and support groups have used the court system to pursue responsibility and liability.

It is estimated that the total discharge of mercury from the Chisso plant was between 70 and 150 tons. Major efforts were implemented to clean up the site and restore the local environment so that it would once again be fit for sea life and human use. From

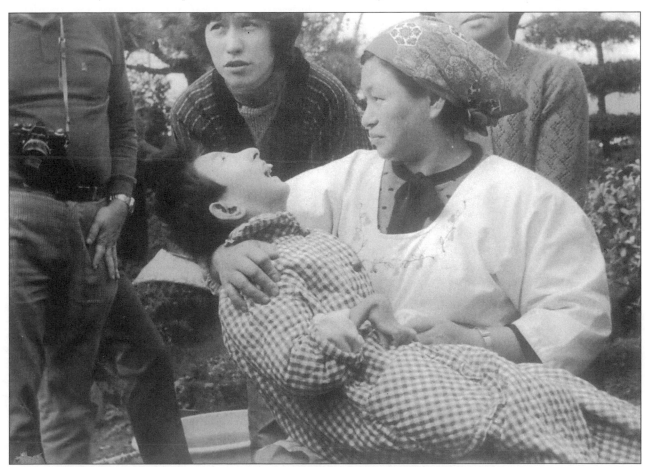

A woman holds a victim of mercury poisoning in Minamata, Japan, in 1973. Physical deformities, such as this young woman's malformed hand, were common among those who suffered so-called Minamata disease. (AP/Wide World Photos)

1977 to 1990 the mercury-contaminated sludge was dredged from 58 hectares (143 acres) of the bay in a project funded by the national and prefectural governments and Chisso Corporation. In 1974 a huge net was placed to isolate 380 hectares (939 acres) of the bay area to prevent fish from swimming in or out. In 1997, Minamata Bay was declared to be free of mercury contamination, and the net was removed.

Robert S. Carmichael

FURTHER READING

D'Itri, Frank M. "Mercury Contamination—What We Have Learned Since Minamata." *Environmental Monitoring and Assessment* 19, nos. 1-3 (1991): 165-182.

Eisler, Ronald. *Mercury Hazards to Living Organisms.* Boca Raton, Fla.: CRC Press, 2006.

George, Timothy S. *Minamata: Pollution and the Struggle for Democracy in Postwar Japan.* Cambridge, Mass.: Harvard University Asia Center, 2001.

Montgomery, Carla W. "Water Pollution." In *Environmental Geology.* 9th ed. New York: McGraw-Hill, 2010.

SEE ALSO: Biomagnification; Environmental health; Environmental illnesses; Food chains; Hazardous waste; Heavy metals; Mercury; Oak Ridge, Tennessee, mercury releases; Reclamation Act; Water pollution.

Mine reclamation

CATEGORY: Land and land use

DEFINITION: Process of returning disturbed land areas to stable and productive uses after minerals have been removed through mining

SIGNIFICANCE: The reclamation of lands no longer used for mining is important to prevent various environmental problems, such as soil erosion, pollution of groundwater by acid drainage, and the physical dangers posed by abandoned shafts and tunnels.

Failure to reclaim mined land may result in substantial loss of biological productivity of the land surface, significant degradation of nearby water bodies, and hazards to human health and safety. Degradation of land and water is attributable to erosion and sedimentation of soils, acid mine drainage, and damage

to groundwater aquifers. Hazards to human health and safety include open mine shafts, mine fires, subsidence of the ground above underground tunnels, clifflike surfaces known as highwalls, and landslides on steep slopes. Land areas disturbed by mining that have not been reclaimed are referred to as abandoned mine lands.

The amount of land to be reclaimed at any mine is determined by the amount of mineral removed and the type of mining operation. Underground mines require little reclamation except near the tunnel entrance. Surface or strip mines disturb larger areas and volumes of soil and rock than do underground mines, and therefore they require more reclamation. In all forms of surface mining, rock and soil located above and between seams of the mineral are removed to expose the mineral for extraction.

Before soil and rock can be removed, all vegetation covering the land surface to be mined must be removed. Next, topsoil and subsoils are excavated and used in an adjacent area that is being reclaimed, or they are separated from other overburden rock (rock that overlies deposits of minerals to be mined) and stockpiled for later use. The process of segregating and reusing fertile topsoils during mining is critical to the later success of reclamation efforts because a suitable growing medium is essential to the reestablishment of viable plant communities. Reclamation of abandoned mine lands where topsoil was not separated from other overburden is generally more difficult and expensive than reclamation at operating mines.

RECLAMATION ACTIVITIES

Reclamation encompasses three activities: backfilling and grading, replacement of topsoil, and revegetation. Backfilling and grading occur after the mineral has been removed. Overburden is replaced in the mined area to reestablish a stable land surface that is consistent with the surrounding area and compatible with the intended postmining land use. Front-end loaders, heavy trucks, bulldozers, and graders are used to fill and grade the contours of highwalls, overburden piles, and depressions to approximate original slopes. The resulting surfaces must be stable—that is, not prone to landslides or erosion—and should blend in with the surrounding natural topography.

Mining operations sometimes unearth natural materials that are toxic or acid-forming. When exposed

Plantings take hold on land reclaimed from what was formerly a strip mine. (©Shuli Hallak/CORBIS)

to the atmosphere, these materials may alter the acidity of surrounding soil or water bodies, destroying the organisms that live there. Such materials must be isolated from surface water and groundwater, soils, and vegetation so that cannot contaminate the environment. This generally means placing them below the root zone of plants during backfilling and grading. During backfilling and grading, heavy equipment repeatedly crosses the work area, causing compaction of the ground surface. Prior to redistribution of topsoil, it may be necessary to rip up this surface to relieve compaction. This helps prevent slippage of topsoil by creating a roughened surface and aids root penetration by vegetation, thus improving surface stability.

After backfilling and grading are completed, a layer of topsoil is spread over the graded overburden to a depth determined by the intended postmining land use and the amount of topsoil available, often 1.2

meters (4 feet) or more. Topsoil stockpiled for more than two or three years begins to lose nutrients, beneficial bacteria, and fungi that aid in plant establishment, so soil tests are used to determine what soil amendments may be needed. Where nutrients are lacking, they are replenished using fertilizers similar to those used on home lawns and gardens.

Revegetation must occur soon after placement of topsoil to control the effects of wind and water erosion. A fast-growing annual grass or cereal grain cover crop, as well as mulch, may be used to stabilize the soil until the first normal planting season. Shrubs and small trees may also be planted. The goal of revegetation is the establishment of a diverse, permanent vegetative cover of a seasonal variety native to the area, or of a variety that supports the intended postmining land use.

In the eastern United States, where water is plentiful, it may take five years to determine whether a mine reclamation project has been successful; in the semiarid western United States, determination may require ten years. Common uses of reclaimed mined land include cropland agriculture, commercial forestry, recreational areas such as parks, public works (such as airfields, roads, housing developments, and industrial sites), and fish and wildlife conservation.

RECLAMATION LAWS

In the United States, reclamation of land disturbed during the mining of coal has been required since 1977 under national legislation known as the Surface Mining Control and Reclamation Act (SMCRA). Principal responsibility for enforcing this law rests with the U.S. Department of the Interior's Office of Surface Mining Regulation and Enforcement and state regulatory authorities, with programs approved under the statute. A small fund is available for the reclamation of abandoned mine lands, financed by fees on each ton of coal produced by active mining operations.

The requirements of SMCRA apply only to coal mines operating since 1977. Lands disturbed by the mining of gold, silver, nickel, copper, bauxite, limestone, and other minerals and industrial materials are not subject to uniform national standards but may be subject to requirements for reclamation imposed by some states; such state regulations vary greatly, however. The knowledge and technology necessary for successful reclamation of land disturbed by mining are available for almost all ecological systems, except desert and alpine climate conditions that do not favor the rapid plant growth necessary to stabilize reclaimed soil.

Michael S. Hamilton

FURTHER READING

Hossner, Lloyd R., ed. *Reclamation of Surface-Mined Lands.* Boca Raton, FL: CRC Press, 1988.

Otto, James M. "Global Trends in Mine Reclamation and Closure Regulation." In *Mining, Society, and a Sustainable World,* edited by J. P. Richards. New York: Springer, 2009.

Pipkin, Bernard W., et al. "Mineral Resources and Society." In *Geology and the Environment.* 5th ed. Belmont, Calif.: Thomson Brooks/Cole, 2008.

Schor, Horst J., and Donald H. Gray. *Landforming: An Environmental Approach to Hillside Development, Mine Reclamation, and Watershed Restoration.* Hoboken, N.J.: John Wiley & Sons, 2007.

SEE ALSO: Acid mine drainage; Erosion and erosion control; Groundwater pollution; Restoration ecology; Strip and surface mining; Surface Mining Control and Reclamation Act; Water pollution.

Mississippi River

CATEGORIES: Places; ecology and ecosystems; water and water pollution

IDENTIFICATION: Major inland tributary with headwaters at Lake Itasca in northern Minnesota, running south through ten U.S. states to a delta below New Orleans, Louisiana

SIGNIFICANCE: The second-longest river in the United States, the Mississippi serves as the drainage basin for 40 percent of the country, supplies nearly one-fourth of the country's drinking water, and is the principal route for inland waterborne commerce in the nation. Efforts to control its flow to accommodate commercial interests have created significant environmental problems that have adversely affected ecosystems and human communities along its banks.

The Mississippi River flows irregularly for approximately 3,750 kilometers (2,330 miles) through the midsection of the United States. It is fed by more than one hundred tributaries, including the Missouri and Ohio rivers. The Upper Mississippi, the region between the headwaters in Minnesota and the area in Illinois and Missouri where its two major tributaries join, is fairly narrow and shallow, at many points running between high bluffs. By contrast, the Lower Mississippi, shaped by activity during the Pleistocene glacial advance, is wide and deep. It runs through an alluvial floodplain that, before human intervention, frequently flooded in the spring.

The Mississippi River has been home to thousands of life-forms, including several endangered species; hundreds of species of birds and mammals populate its shorelines. It is the most important bird and waterfowl migration route in North America. Its bottomlands are the largest wetlands area and support the largest hardwood forest in the United States. Since the nineteenth century, however, engineering projects undertaken to benefit commercial enterprises have caused serious degradation of the Mississippi as a site for balanced natural ecosystems. Major damage has occurred principally in two forms: pollution introduced directly from commercial enterprises or indirectly by projects designed to control the flow of the river for commercial or recreational purposes, and flooding exacerbated by efforts to channel the river to maintain optimal shipping lanes.

COMMERCIALIZATION AND ITS IMPACTS

Since the nineteenth century the Mississippi River has served as a major route for commerce traveling from America's heartland to the Gulf of Mexico and from there to locations around the world. Numerous commercial enterprises and ports sprang up along the river during the nineteenth century, and farming became a major activity all along the Mississippi; in the lower regions, farms and plantations in the rich bottomlands took away natural habitats for wildlife, weakened the riverbanks through erosion, and diminished natural cover for the soil.

Charged with managing the nation's waterways, the U.S. Army Corps of Engineers began improving navigability on the Mississippi in 1879. Over the years

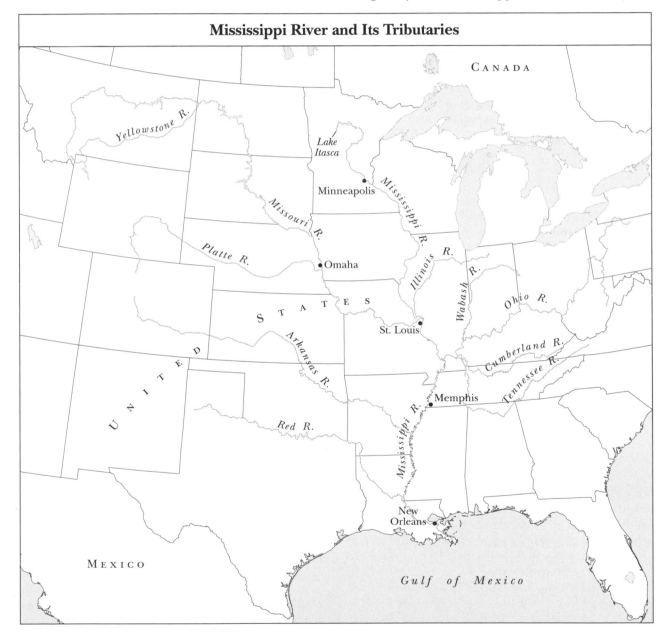

Mississippi River and Its Tributaries

the Corps has undertaken a number of engineering projects to optimize the river's value for commerce. After a disastrous flood in 1927, the Corps initiated several programs aimed at flood control. A series of twenty-seven locks and dams were constructed to create a continuous nine-foot shipping lane from Minneapolis to St. Louis, Missouri, guaranteeing the easy passage of barge traffic. Large earth-and-concrete levees were constructed to prevent flooding below St. Louis. In addition to flood prevention, the levees were designed to increase the force of the water to maintain a deep channel in the river. Nevertheless, for decades the Corps has been forced to conduct systematic dredging, especially near the delta, where silt buildup makes it difficult for larger ships to travel into and out of the Gulf of Mexico.

These initiatives allowed commercial use of the river to increase dramatically; for example, in 2000 more than 83 million tons of cargo moved down the river. Farm crops make up the majority of this cargo, and farmers' ability to use the river to send goods to ports around the country and the world has kept down the need for overland transport. Another twentieth century development was the construction of paper mills, chemical plants, and oil refineries along the river's banks, which proved an ideal location for companies wishing to ship large quantities of their products.

POLLUTION PROBLEMS

The benefits of commercial use of the Mississippi have been balanced, and perhaps outweighed, by damage done to the river's ecosystems. Numerous species of fish and wildlife disappeared from the upper river as pools and backwaters between dams became silted. The high levees below St. Louis caused wetlands adjacent to the river to dry up, eliminating habitat for numerous species. Cities and towns have often dumped sewage into the river, and chemical runoff from farms, particularly phosphorus and nitrogen, has polluted the waterway from its northernmost reaches down to the Gulf. As far north as Minnesota, chemical plants have dumped toxic waste into the Mississippi, polluting it with a variety of toxins, including furan, trichlorobenzene, dichloro-diphenyl-trichloroethane (DDT), trichloroethane (TCA), and polychlorinated biphenyls (PCBs). The problem has become exacerbated farther south, especially from Baton Rouge, Louisiana, to the mouth of the river, where chemical plants and oil refineries are concentrated. The cumulative effects of toxic waste disposal,

oil spillage, and other forms of refuse dumping have materially degraded the use of the river and its banks for fishing, hunting, and trapping and have led to the creation of a hypoxic (depleted of oxygen) area, known as a dead zone, at the mouth of the river extending into the Gulf of Mexico. This area is one of the largest dead zones in the world.

Until the 1960's, many commercial enterprises used the river as a dumping ground for waste, and little was done to protect the natural environment. Since the 1960's, however, environmental groups such as the Izaak Walton League and Greenpeace, as well as organizations sponsoring recreational use of the river, have been increasingly active in demanding that the federal government take measures to protect the Mississippi's ecosystems. Their efforts have sparked modest programs to reduce pollution and reverse the adverse environmental effects of the levees, locks, and dams. A government-sponsored environmental management plan implemented in 1986 is aimed at restoring wetlands. Monitoring by the government and private groups has resulted in a reduction in point source pollution (that is, pollution caused by specific sources, such as refuse dumped by a specific business or city), but nonpoint source pollution, especially runoff from agricultural regions, has been much harder to control.

FLOODING AND FLOOD CONTROL

Traditionally, spring runoff from as far away as the Rocky Mountains to the west and the Appalachians to the east has caused the Mississippi River to rise above its banks and flood adjacent lands. The natural environment historically benefited from this seasonal event, as flooding created rich alluvial soil and wetland habitats. Efforts to control the flow of the Mississippi changed flood patterns significantly, however, as the river was hemmed in by high levees. Waters rushing downstream now do so more rapidly until, in extreme cases, their power creates levee breaks. Debris from the natural landscape and human structures swept away by the current and pollutants being carried downstream have wreaked havoc in flooded areas, and attendant damages to residential and commercial centers built up along the river have often run into the billions of dollars. In turn, the Army Corps of Engineers has been called upon to devise more stringent measures to keep the river within manageable boundaries, further reducing the wetlands environments along the banks and concentrating pollutants within the river's main channel. The Corps has also been active in implementing measures to keep the Mississippi from changing its main channel, a phenomenon common before the nineteenth century. Such changes could cause cities and towns along the river to become landlocked, which would materially affect their commerce.

Efforts to control flooding while maximizing the river's commercial potential have also had negative effects downstream, especially around New Orleans. In 1965 the Corps created the Mississippi River Gulf Outlet (MRGO), a direct channel giving commercial ships a shorter route from the city to the Gulf of Mexico. The channel had little impact on reducing ship traffic, but its presence led to considerable erosion of marshes and wetlands south of the city. The impact was most noticeable when Hurricane Katrina hit in 2005. While the levees along the Mississippi were not breached, the surge from the Gulf traveled up the MRGO and over terrain bereft of natural barriers, causing breaches in flood walls of canals leading off the river into urban areas. Waters inundated New Orleans, creating the worst natural disaster in the history of the United States.

Laurence W. Mazzeno

FURTHER READING

Anfinson, John O. *The River We Have Wrought*. Minneapolis: University of Minnesota Press, 2003.

Fischer, Katherine. *Dreaming the Mississippi*. Columbia: University of Missouri Press, 2006.

Fremling, Calvin R. *Immortal River: The Upper Mississippi in Ancient and Modern Times*. Madison: University of Wisconsin Press, 2005.

Hall, B. C., and C. T. Wood. *Big Muddy: Down the Mississippi Through America's Heartland*. New York: Dutton, 1992.

Hilliard, Sam B., ed. *Man and Environment in the Lower Mississippi Valley*. Baton Rouge: Louisiana State University School of Geoscience, 1978.

Meyer, Gary C. "Preservation and Management of the River's Natural Resources." In *Grand Excursions on the Upper Mississippi River: Places, Landscapes, and Regional Identity After 1854*, edited by Curtis C. Roseman and Elizabeth M. Roseman. Iowa City: University of Iowa Press, 2004.

SEE ALSO: Biodiversity; Commercial fishing; Dredging; Ecosystems; Floods; Watershed management; Wild and Scenic Rivers Act.

Mobro barge incident

CATEGORY: Waste and waste management

THE EVENT: Months-long journey by the barge *Mobro* along the U.S. Atlantic and Gulf coasts in search of a place to unload a cargo of garbage

DATES: March-September, 1987

SIGNIFICANCE: The *Mobro* incident raised Americans' awareness of the serious problem of waste management and helped spark national interest in recycling as a solution to the increasing problem of lack of landfill space.

On March 22, 1987, the *Mobro* left New York loaded with 3,186 tons of garbage from the Long Island town of Islip and New York City. The garbage was to be unloaded in the rural Jones County, North Carolina, town of Morehead City, where it would be converted to methane gas. The project was the brainchild of Lowell Harrelson, an Alabama entrepreneur who envisioned shipping garbage from northern states where landfill space was becoming scarcer to cheaper southern landfills that had adequate space. Jones County manager Larry Meadows saw the plan to accept out-of-state garbage as a way to boost the county's tax base.

When the *Mobro* arrived in Morehead City sooner than expected, North Carolina officials questioned whether the barge's hasty arrival meant that its cargo might contain hazardous waste. North Carolina officials denied the *Mobro* permission to dock, and the barge set forth in search of another port. The news media learned of the *Mobro* and began to report on its journey. Before it finally ended up back where it began in New York, the *Mobro* traveled to Louisiana, Alabama, Mississippi, Florida, New Jersey, the Bahamas, Mexico, and Belize (a distance of some 9,600 kilometers, or 6,000 miles), only to be turned away at every port. Amid intense publicity, the "not in my backyard" syndrome, or NIMBYism, quickly took hold. Mexico tracked the *Mobro* with gunboats, and Belize threatened to dispatch its air force to prevent the barge from approaching its waters.

As a result of media coverage, the barge became a symbol of U.S. waste disposal problems. New York environmentalist Walter Hang spoke of a garbage crisis on television. Other environmentalists referred to a nation buried in trash. Some Environmental Protection Agency (EPA) officials took note of the *Mobro* incident and publicly acknowledged a national garbage crisis. By September, 1988, EPA administrator J. Winston Porter asked each state and municipality to formulate a recycling plan that would reduce disposal needs by 25 percent. In an EPA report titled "Solid Waste Dilemma," Porter implied that the country was running out of space to bury garbage. Others in the EPA disputed Porter's claim and maintained that the United States had an average of twenty-one years of remaining landfill capacity.

The crisis ended after the *Mobro* returned to New York. Lawsuits filed in New York State Supreme Court by the borough of Queens and the New York Public Interest Group (NYPIRG) failed to prevent incineration and burial of the garbage. Queens officials feared a health hazard, while NYPIRG contended that the garbage contained high levels of cadmium and lead. All three thousand bales of trash were finally burned in September, 1987, at Brooklyn's Southwest Incinerator, and the resulting 500 tons of ash were buried at Islip's Blydenburgh landfill.

Michele Zebich-Knos

FURTHER READING

Lamar, Jacob V., Jr. "Don't Be a Litterbarge: No One Wants the Wretched Refuse of New York's Teeming Shore." *Time*, May 4, 1987, 26.

Melosi, Martin V. *Garbage in the Cities: Refuse, Reform, and the Environment*. Rev. ed. Pittsburgh: University of Pittsburgh Press, 2005.

Miller, Benjamin. *Fat of the Land: The Garbage of New York—The Last Two Hundred Years*. New York: Four Walls Eight Windows, 2000.

SEE ALSO: Incineration of waste products; Landfills; NIMBYism; Recycling; Solid waste management policy; Waste management.

Molina, Mario

CATEGORY: Weather and climate
IDENTIFICATION: Mexican chemist
BORN: March 19, 1943; Mexico City, Mexico
SIGNIFICANCE: Molina's pioneering work concerning the formation and catalytic decomposition of ozone in the stratosphere led to greater scientific attention to the issue of climate change.

The ozone that is produced naturally in the earth's stratosphere provides a filter to restrict the

Chemist Mario Molina. (©Brooks Kraft/Sygma/CORBIS)

amount of harmful ultraviolet radiation that strikes the surface of the planet. It was the work of Mario Molina, Paul Crutzen, and Frank Sherwood Rowland that first drew attention to the catalytic role played by chlorofluorocarbons (CFCs) in the decomposition of ozone. CFCs were long used in aerosol spray cans, electronic parts cleaning, and refrigeration systems. Their production was banned in 1987 by an international agreement called the Montreal Protocol.

Molina was born in Mexico City in 1943 to Roberto Molina Pasquel and Leonor Henríquez de Molina. He obtained a degree in chemical engineering in 1965 from the Universidad Nacional Autónoma de México. He earned a master's degree at the University of Freiburg in Germany in 1967 for work done on the study of rates of polymerization reactions. Molina returned to his undergraduate institution as an assistant professor for one year and then began working on a doctorate at the Berkeley campus of the University of

ENCYCLOPEDIA OF ENVIRONMENTAL ISSUES

California. He earned his Ph.D. in 1972 in physical chemistry and remained at Berkeley as a postdoctoral student.

In 1973 Molina accepted a postdoctoral appointment with Rowland at the University of California at Irvine. At this time Molina began to look at the question of what happens to chemicals and pollutants in the troposphere and stratosphere, where they are subject to high-intensity ultraviolet radiation. A 1974 article by Molina in the journal *Nature* began a series of papers pointing out the connection between ozone destruction and CFCs. Molina's continuing research led to the prediction of the ozone hole that was later discovered over Antarctica in 1985. Molina then began researching the interface between the atmosphere and the biosphere; he hoped his investigations would lead to an understanding of global climate change processes.

Molina has held appointments with the Jet Propulsion Laboratory at the California Institute of Technology as a senior research scientist and with the Department of Earth, Atmospheric, and Planetary Sciences and the Department of Chemistry at the Massachusetts Institute of Technology (MIT) as a professor. In 2004, he accepted positions in the Department of Chemistry and Biochemistry at the University of California, San Diego, and at the Center for Atmospheric Sciences at the Scripps Institution of Oceanography.

Molina's work has been recognized with numerous honors, including the American Chemical Society Esselen Award (1987), the American Association for the Advancement of Science Newcomb-Cleveland Prize (1988), the National Aeronautics and Space Administration Medal for Exceptional Scientific Advancement (1989), and the United Nations Environment Programme's Global 500 Award (1989). In 1995, Molina shared the Nobel Prize in Chemistry with Crutzen and Rowland for his pioneering work concerning the formation and catalytic decomposition of ozone in the stratosphere.

Kenneth H. Brown

FURTHER READING

Hill, Marquita K. "Stratospheric Ozone Depletion." In *Understanding Environmental Pollution*. 3d ed. New York: Cambridge University Press, 2010.

Joesten, Melvin D., John L. Hogg, and Mary E. Castellion. "Chlorofluorocarbons and the Ozone Layer." In *The World of Chemistry: Essentials*. 4th ed. Belmont, Calif.: Thomson Brooks/Cole, 2007.

Miller, G. Tyler, Jr., and Scott Spoolman. "Climate Change and Ozone Depletion." In *Environmental Science: Problems, Concepts, and Solutions*. 13th ed. Belmont, Calif.: Brooks/Cole, 2010.

SEE ALSO: Chlorofluorocarbons; Freon; Montreal Protocol; Ozone layer; Rowland, Frank Sherwood.

Monkeywrenching

CATEGORIES: Activism and advocacy; philosophy and ethics

DEFINITION: Direct-action tactics used by radical environmentalists to disrupt activities they believe degrade the environment

SIGNIFICANCE: The activities known as monkeywrenching are the source of considerable controversy. Many environmentalists see such radical tactics as damaging to their work to gain the support of the public for government policies and regulations aimed at preserving and protecting the environment.

The term "monkeywrenching" was first used in 1904 to refer to the sabotaging of factory machinery by throwing a monkey wrench, or a spanner with a movable jaw, into the works. Such acts were preceded in early nineteenth century England by the actions of Luddites and machine breakers who protested the mechanization of workplaces during the Industrial Revolution. The term was appropriated later by environmentalist and author Edward Abbey in his 1975 novel *The Monkey Wrench Gang*, which details the exploits of three men and one woman who take the law into their own hands to defend the wilderness from excavation. They perform acts of sabotage, or ecotage, on road-building equipment and entertain visions of blowing up Arizona's Glen Canyon Dam.

The radical environmental group Earth First!, founded in 1980, has used civil disobedience and monkeywrenching tactics to defend wilderness areas from developers. The general idea behind monkeywrenching is to create a stir, delay or halt projects, and gain publicity for the cause. Dave Foreman, one of the founders of Earth First!, published *Ecodefense: A Field Guide to Monkeywrenching* in 1985 and *Confessions of an Eco-Warrior* in 1991. He credits Abbey's book as a major motivation and inspiration. In 1990 Earth First!

promoted the Redwood Summer, a ten-week campaign to slow the logging of redwoods. Among the tactics used during the campaign was tree spiking, in which large, long nails are driven into trees to dissuade loggers. Because the nails can potentially shatter chain saws and thus hurt loggers, Foreman, in *Ecodefense*, advised monkeywrenchers to warn loggers when an area has been spiked. Incidents have occurred, however, in which no notice of spiking was given.

Another tactic used by monkeywrenchers is tree sitting, in which activists physically occupy trees to prevent their being cut down. In 1998 a California woman lived for many months in a tree to prevent logging of a particular grove. Earth First! activists have also been known to knock down billboards and to destroy heavy equipment used for land clearing and development. The self-proclaimed "navy of Earth First!'s army" is the Sea Shepherd Conservation Society, directed by Paul Watson, who published *Ocean Warrior* in 1994. This group uses direct action to prevent whaling ships and others from killing and capturing marine mammals.

Monkeywrenchers tend to be determined and not easily reformed by the experience of incarceration; many activists have been detained multiple times. The ideological viewpoint of those supporting monkeywrenching is strongly preservationist. Although monkeywrenchers are often influenced by the writings of such early preservationists as John Muir and Henry David Thoreau, they are far more militant in their activities.

The theory behind monkeywrenching stems from a field of thought known as deep ecology, the core tenet of which is biocentrism, a belief that the human species is just one member of a biological community in which all species have equal standing. This view places humans on the same level as every other living thing. Followers of this philosophy feel that human conduct should proceed from an understanding that all forms of life, no matter how big or small, have an equal right to exist. Therefore, they claim to fight for organisms that cannot defend themselves against industrial intrusion.

Oliver B. Pollak and Aaron S. Pollak

FURTHER READING

Foreman, Dave, and Bill Haywood, eds. *Ecodefense: A Field Guide to Monkeywrenching*. 3d ed. Chico, Calif.: Abbzug Press, 2002.

Scarce, Rik. *Eco-Warriors: Understanding the Radical Environmental Movement*. Updated ed. Walnut Creek, Calif.: Left Coast Press, 2006.

SEE ALSO: Abbey, Edward; Biocentrism; Deep ecology; Earth First!; Ecotage; Ecoterrorism; Foreman, Dave.

Mono Lake

CATEGORIES: Places; ecology and ecosystems; water and water pollution

IDENTIFICATION: Natural saline lake located in east-central California

SIGNIFICANCE: The diversion of stream water from Mono Lake for use by the city of Los Angeles led to lowered water levels and higher-than-normal salinity in the lake, which had serious impacts on the lake's delicate ecosystem.

Mono Lake, which covers about 150 square kilometers (58 square miles), receives most of its water from underground water flow and from streams that drain from the Sierra Nevada to the west. The lake has no outlet, so evaporation is the main natural cause of water loss. Since 1941, the city of Los Angeles has diverted water from the streams that drain into Mono Lake into the Los Angeles Aqueduct to supply the city with water. This loss of stream water into the lake resulted in a drop in the lake's water level of more than 15 meters (49 feet), with a consequent increase in dissolved constituents in the lake.

By the early twenty-first century, Mono Lake's water contained about three times the concentration of dissolved salts found in seawater. The main kinds of dissolved ions (charged particles) found in the lake are sodium, chloride, and carbonate-bicarbonate. Lesser amounts of potassium, magnesium, and sulfate ions are also present. In addition, the waters have a very low hydrogen ion concentration (low acidity) and high hydroxide ion concentration compared with most natural waters. Nitrate, phosphate, and ammonium ion concentrations are also fairly high compared with most natural waters, especially in the lower portions of the lake. These ions help to stimulate the growth of phytoplankton in the lake waters.

The fairly high concentrations of calcium and carbonate-bicarbonate ions in some places in Mono Lake have caused chemical precipitation of tufa tow-

Mono Lake's tufa towers, composed of calcium carbonate, were formed underwater when freshwater springs mixed with minerals in the lake water; they became visible above the surface as the lake water receded after water began to be diverted to Los Angeles. (AP/Wide World Photos)

ers (composed of calcium carbonate) underwater, often helped by the action of algae. As the lake levels have dropped, the tufa towers have become exposed around the edges of the lake in certain places. The towers provide nesting places for some owls, falcons, and small mammals.

Plants and Animals

Mono Lake is too salty for fish and most water birds. In the spring, however, the lake explodes with millions of tons of algae and other small plants. Two larger organisms feed on the algae: brine shrimp and small alkali flies. Brine shrimp are small crustaceans, often red in color and each only a little more than 1 centimeter (0.4 inch) long, with small appendages that they use to move algae into their mouths. They are found in Mono Lake, mostly in the upper, oxygen-rich waters, only from spring to fall; it has been estimated that some seven trillion shrimp may inhabit the lake at their seasonal peak. Alkali flies occur in swarms along the shoreline of the lake. They feed on the algae and lay eggs in algal mats underwater.

A few species of birds have adapted to being able to use Mono Lake at least during some parts of the year, especially when there are large quantities of brine shrimp and alkali flies to eat. The California gull is one such bird; these gulls are abundant in the summer, when they form nests on some of the small islands in the lake. The gulls migrate toward the ocean during the winter. The eared grebe is another bird that lives in the lake in the summer and fall, consuming vast amounts of brine shrimp before flying south in the winter to the Gulf of California. The red-necked phalarope and Wilson's phalarope stop over at Mono Lake, where they consume large quantities of alkali flies, before flying more than 4,800 kilometers (about 3,000 miles) south in the winter to places in South America. The American avocet, a long-legged wading bird, stops at the lake in the spring when flying to the north and in the fall when flying to the south. Many kinds of ducks also stop by the lake, although the waters are too salty for them to spend much time there. Before Los Angeles diverted the freshwater streams draining into Mono Lake, many more kinds and numbers of ducks used to stop at nearby streams and visit the less salty waters in the lake near the streams.

ENVIRONMENTAL PROBLEMS

The diversion of stream flow from Mono Lake to the Los Angeles Aqueduct that began in the 1940's resulted over time in a large drop in the average volume of water in the lake and thus lowered the lake level. The salinity of the waters eventually doubled in concentration, reducing the production of algae and thus the populations of brine shrimp that fed on the algae. The lake's recreational uses were also affected, as boat docks and beaches were left without water near them as the water level dropped. Up to five-sixths of the average flow of stream water that had originally gone to Mono Lake was being diverted to Los Angeles, and in years with low rainfall, some streams simply dried up. The fish populations, vegetation, and wetlands that depended on these streams were severely damaged.

Beginning in 1979, environmentalists brought lawsuits intended to force the city of Los Angeles to stop diverting water from Mono Lake. After many court decisions and appeals, in the mid-1990's Los Angeles was required to reduce its diversion of Mono Lake waters so that the original level and salinity of the lake could eventually be restored.

Robert L. Cullers

FURTHER READING

Carl, David, and Don Banta. *Mono Lake Basin, California.* Mount Pleasant, S.C.: Arcadia, 2008.

Flaherty, Dennis, and Mark A. Schlenz. *Mono Lake: Mirror of Imagination.* Santa Barbara, Calif.: Companion Press, 2007.

Hart, John. *Storm over Mono: The Mono Lake Battle and the California Water Future.* Berkeley: University of California Press, 1996.

Mono Basin Ecosystem Study Committee. *The Mono Basin Ecosystem.* Washington, D.C.: National Academy Press, 1987.

SEE ALSO: Ecosystems; Endangered species and species protection policy; Environmentalism; Habitat destruction; Los Angeles Aqueduct; Public trust doctrine; Water rights.

Monoculture

CATEGORY: Agriculture and food
DEFINITION: Agricultural practice of growing only one crop over a large section of land
SIGNIFICANCE: Almost all industrialized countries utilize monoculture as the major form of crop production. Monoculture is the most efficient means, in terms of both yield and cost, to produce food and fiber for large numbers of people, but some environmental concerns are associated with the practice. For example, monoculture increases the reliance on agricultural chemicals and decreases the genetic diversity of crop plants.

The continuing increase in the number of humans inhabiting the earth has placed greater and greater demands on agricultural systems to produce sufficient food and fiber to feed and clothe the population. In order to meet these demands, mechanized agriculture is utilized in countries where there is sufficient money and land to support it. Mechanized agriculture depends on the availability of large sections of relatively level land in order to facilitate the efficient operation of the machinery. In addition, mechanized agriculture relies on monoculture, the practice of planting the same crop on vast areas of land to increase the efficiency of planting, cultivating, and harvesting. Mechanized agriculture is practiced in most of Europe, many of the republics that once composed the Soviet Union, much of South America, almost all of North America, and a few other developed countries.

HISTORICAL PERSPECTIVE

Throughout most of human history, farms and ranches were owned by families who primarily practiced sustenance agriculture. Each farm or ranch produced a variety of crops sufficient to feed the resident family as well as a small excess that was sold for cash or bartered for other goods or services. Because almost all agricultural activities required human or animal labor, large families were desirable. Until the nineteenth century, agricultural tools such as plows were made of wood; however, the arrival of the Industrial Revolution changed the agriculture industry just as it did almost all other industries. Eli Whitney invented the cotton gin, a machine that could separate cotton fibers from the seeds of the plant, in 1793. Cyrus

McCormick invented the mechanical reaper in 1833, and John Lane and John Deere began the commercial manufacture of the steel plow in 1837. Inventions such as these led the way for the development of many different types of agricultural machinery, which resulted in the mechanization of most farms and ranches. Most agricultural enterprises in the United States were mechanized by the early part of the twentieth century.

The Industrial Revolution also produced a significant change in the social nature of the United States. The nation went from being predominantly an agrarian society to being predominantly an urban society, as many people who had been involved in agricultural production left the farms to go to the cities to work in factories.

As the population continued to grow, the selection and production of higher-yielding crops became increasingly important, and the Green Revolution of the twentieth century helped to make this possible. Biological scientists supplied basic information that allowed agricultural scientists to develop new, higher-yielding varieties of numerous crops, particularly the seed grains that supply most of the calories necessary for maintenance of the world's population. These higher-yielding crop varieties resulted in tremendous increases in the world's food supply. The utilization of these new crop varieties, along with the increase in mechanization, also led to an increased reliance on monoculture.

Advantages

The major advantage to monoculture is increased efficiency, in terms of both crop yield and economics. Limited numbers of people are available for work as farm laborers, and this limited supply of manual labor resulted in the development of machinery that can be used to plant, cultivate, and harvest different crops. In many instances, different kinds of machines are necessary for different crops, so if farmers grow many different crops on numerous small plots of land, they need to change the machinery being used each time they handle a different crop. This requires large amounts of time and is expensive, in both labor and equipment costs. Farm machinery is very expensive, and few farmers can afford the equipment necessary to handle more than one or two different crops. If farmers intersperse different crops in one large field, they must also expend costly fuel for their machinery while they skip one part of the field to get to another.

Many of these problems have been reduced through the use of monoculture. The farmer can buy fewer pieces of equipment, the equipment does not have to be changed out as often, and large tracts can be harvested at one time without the farmer's having to skip part of the field.

The use of mechanized monoculture tremendously increased the amount of agricultural produce available throughout the world during the twentieth century. Not only did the number of hectares under cultivation increase owing to the increased efficiency of machinery, but also the yield per hectare being farmed increased dramatically in many parts of the world. For example, corn yields almost doubled in the United States from 1970 to 2010. This increase came about for a variety of reasons, including improved farming methods, improved irrigation, new varieties of crops, the use of fertilizers and pesticides, and the availability of more efficient machinery, all of which developed in response to the increased reliance on monoculture. In the twenty-first century, the highly mechanized modern agricultural unit requires relatively few employees to produce large amounts of crops for an increasing number of consumers.

Disadvantages

Although the practice of growing only one crop over a vast number of hectares has resulted in much higher yields, it has also caused some environmental problems. One of the problems associated with monoculture is the loss of genetic diversity. Farmers most often plant hybrid varieties of crops that have been developed to give high yields and to produce plants that are equal in size and other characteristics that provide uniformity in order to facilitate mechanized harvesting. When large numbers of farmers plant the same hybrid variety for a number of years, important genetic material that might provide pest resistance or some other important trait could be lost. It has been estimated that two-thirds of all seeds planted in developing countries belong to the same strains. This could prove disastrous in the future if these strains were to become infected with an uncontrollable pest. The use of the same hybrid by large numbers of farmers in a given area also results in pest control problems. Vast fields of the same crop attract large numbers of pests, including weeds, fungi, and insects. Monoculture is thus highly dependent on the use of pesticides to control these pests, and the long-term, large-scale application of pesticides could prove detrimental to other plants, animals, and humans.

The practice of monoculture generally means that the same crop may be planted year after year, with little or no crop rotation. As a rule, agricultural activity decreases the availability of a number of important plant nutrients found in the soil, and the lack of crop rotation can hasten the depletion of certain essential soil nutrients. This means that the farmer will have to apply fertilizer more often to restore soil fertility. Some of the nutrients in fertilizers, particularly nitrogen and phosphorus, can end up in nearby streams and lakes, where they contribute to water pollution.

In addition, the use of monoculture can often mean that large tracts of land remain uncovered between the harvesting of one crop and the planting of the next. When soil is left bare without any covering vegetation, it is subject to wind and water erosion, which results in the removal of the topsoil. Intensive mechanized monoculture can also result in the loss of organic matter from the soil. Soil suitable for growing plants contains organic matter that will remain in the soil if part of the crop is turned back under after it is harvested; however, in mechanized monoculture, the entire plant is often removed, thereby decreasing the organic matter content over time. Polyculture cropping systems (that is, the planting of several diverse crops) avoid some of the problems associated with monoculture, but the trade-off is lower overall agricultural productivity.

D. R. Gossett

FURTHER READING

Enger, Eldon D., and Bradley F. Smith. *Environmental Science: A Study of Interrelationships.* 12th ed. Boston: McGraw-Hill Higher Education, 2010.

Hill, Marquita K. "Pesticides." In *Understanding Environmental Pollution.* 3d ed. New York: Cambridge University Press, 2010.

McBrewster, John, Frederic P. Miller, and Agnes F. Vandome, eds. *Industrial Agriculture: Factory Farming, Livestock, Aquaculture, Agribusiness, Monoculture, Agroecology, Organic Farming, Urban Agriculture.* Phoenix, Ariz.: Alphascript, 2009.

Miller, G. Tyler, Jr., and Scott Spoolman. *Environmental Science: Problems, Concepts, and Solutions.* 13th ed. Belmont, Calif.: Brooks/Cole, 2010.

Wojtkowski, Paul. *Landscape Agroecology.* Boca Raton, Fla.: CRC Press, 2003.

SEE ALSO: Agricultural chemicals; Agricultural revolution; Forest management; Green Revolution; Intensive farming; Organic gardening and farming; Pesticides and herbicides; Sustainable agriculture.

Monongahela River tank collapse

CATEGORIES: Disasters; water and water pollution

THE EVENT: Collapse of an oil storage tank near the Monongahela River in Pennsylvania that resulted in a massive oil spill

DATE: January 2, 1988

SIGNIFICANCE: When a faulty storage tank released millions of gallons of oil near the Monongahela River, creating the worst inland oil spill in U.S. history to that date, the environmental damage included the deaths of thousands of fish and waterfowl.

In the early evening of January 2, 1988, the rupture of a storage tank at the Ashland Oil terminal in Floreffe, Pennsylvania, 40 kilometers (25 miles) southeast of Pittsburgh, released 3.9 million gallons of diesel oil. The oil spilled over a containment dike and flowed across a road into a ravine, and much of the oil eventually found its way into a storm sewer leading to the Monongahela River.

Because of darkness and freezing weather, the extent of the damage was not fully recognized until the next morning, by which time nearly 750,000 gallons of oil had flowed through the storm sewer into the Monongahela. Nevertheless, response efforts from a number of agencies, including the local volunteer fire department, borough police, and the Mt. Pleasant Hazardous Materials Team, began almost immediately. Within hours, a team from the Pennsylvania Emergency Management Agency (PEMA) was en route.

The first concern of those responding was to stop the flow of fuel, but darkness and cold made the operation difficult. Moreover, a strong odor of gasoline indicated an additional gas leak of unknown origin. The mixture of gasoline and diesel fuel presented a dangerous situation made even more serious by the presence of hazardous chemicals at a nearby chemical plant, and the decision was made to evacuate twelve hundred nearby residents. Emergency crews and firefighters worked throughout the night to contain the oil.

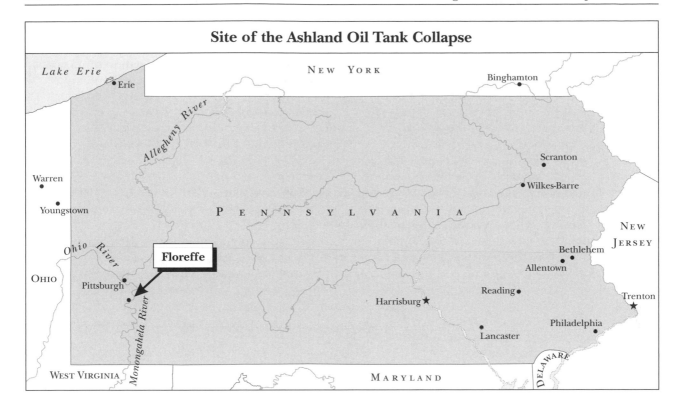

Site of the Ashland Oil Tank Collapse

A coordinator from the U.S. Environmental Protection Agency (EPA) arrived the following morning and discovered the condition of the river to be worse than expected, as the oil had dispersed through the water volume rather than remaining on the surface. Water intakes for communities downstream on the Monongahela and the Ohio River were shut off as a preventive measure.

Containment booms placed downstream had little effect, so deflection booms were used to move the oil to collection areas. Because of the cold temperatures, the oil formed heavy globs that could be picked up from the river edges and bottom. Over the next two months, nearly 205,000 gallons of oil, or 29 percent of the total spilled into the river, were recovered.

The damage caused to wildlife in and near the river was difficult to assess, as many species were hibernating or otherwise inactive in the winter weather. Researchers estimated that the spill killed eleven thousand fish and two thousand waterfowl.

Investigations into the cause of the collapse discovered a small flaw in the steel plates at the base of the storage tank. The tank's forty-year-old steel had been weakened by reassembly, and the temperature was low enough to cause a brittle fracture. When the tank was filled, the resulting stress caused a crack near a weld.

Ashland Oil had not secured a written permit before constructing the tank, and the tank had not been properly tested before it was filled. The company was held liable for all damages, and it paid a federal fine of $2.25 million for violating the Clean Water Act and the Refuse Act; it also paid $11 million in cleanup costs and tens of millions more to other injured parties.

Alexander Scott

FURTHER READING

Fingas, Merv. *The Basics of Oil Spill Cleanup.* 2d ed. Boca Raton, Fla.: CRC Press, 2001.

Lehr, Jay, et al. "Oil Spills and Leaks." In *Handbook of Complex Environmental Remediation Problems.* New York: McGraw-Hill, 2002.

SEE ALSO: BP *Deepwater Horizon* oil spill; Oil spills; PEMEX oil well leak; Santa Barbara oil spill; Water pollution.

Montreal Protocol

CATEGORIES: Treaties, laws, and court cases; atmosphere and air pollution

THE TREATY: International agreement with the specific goal of limiting atmospheric inputs of chlorofluorocarbons for the purpose of protecting the ozone layer

DATE: Opened for signature on September 16, 1987

SIGNIFICANCE: The Montreal Protocol was a political innovation in that it called for a gradual reduction in chlorofluorocarbon production and was created to be flexible enough to respond to new scientific information. The agreement led to reductions in atmospheric concentrations of chemical compounds known to deplete ozone in the earth's atmosphere.

The ozone layer, which is 10 to 20 kilometers (6 to 12 miles) above the earth's surface, screens out most of the sun's ultraviolet radiation. Ultraviolet radiation can lead to mutations and cancer in living things. The nations participating in the Montreal Protocol were motivated to act by four major developments: The accumulation of chlorofluorocarbons (CFCs) in the atmosphere was observed in the early 1970's; CFC decomposition in the atmosphere was demonstrated to cause ozone destruction in 1974; a hole in the ozone layer was discovered over Antarctica in the early 1980's, and scientists found evidence linking the ozone hole to CFCs in 1985; and CFC substitutes were developed by important CFC producers.

On September 16, 1987, the Montreal Protocol on Substances That Deplete the Ozone Layer was signed by 46 nations, including the United States. These nations represented two-thirds of the total global production and consumption of CFCs and other halogenated compounds. The protocol entered into force on January 1, 1989. By 2010, 196 nations had ratified the treaty.

The Montreal Protocol was designed to control the production and consumption of CFCs and other halogenated compounds suspected of causing ozone destruction. The United States and other industrialized countries committed themselves to freezing CFC production immediately at 1986 levels and to reducing total CFCs leading into the twenty-first century. Consumption of the major CFCs was to be frozen at their 1986 levels by mid-1989, reduced to 80 percent of 1986 levels by mid-1993, and reduced to 50 percent of 1986 levels by 1998. Other halogenated compounds were to be frozen at 1986 consumption levels in 1992.

The protocol was amended in 1990 when new scientific evidence suggested that ozone was being depleted above Antarctica more dramatically than previously assumed. Measurements above Antarctica

Milestones Leading to the Montreal Protocol

YEAR	EVENT
1930's	Chlorofluorocarbons (CFCs) are developed by Du Pont as a safe alternative to toxic refrigerants.
1970	James Lovelock's electron capture detector reveals the accumulation of CFCs in the atmosphere.
1974	Mario Molina and Frank Sherwood Rowland show that CFCs degrade through photodecomposition and release ozone-depleting chlorine molecules.
1978	Pressure from environmentalists causes the United States to ban CFCs as an aerosol propellant; however, worldwide use continues to grow.
1981	The United Nations Environment Programme forms the Ozone Group and discusses a global treaty to protect the ozone layer.
1985	Scientists observe a hole in the ozone layer over Antarctica that is linked to stratospheric chlorine.
1986	Major CFC producers advocate international efforts to limit the growth of CFC emissions.
1987	The Montreal Protocol is signed; 46 nations, including the United States, commit to a plan to reduce and eventually eliminate CFC production.
1989	The Montreal Protocol enters into force; signatory nations are required to begin the phaseout of CFCs.
1990	Revised estimates of ozone depletion lead to a call to cease CFC production by 2000.
1993	Scientists detect a measurable reduction in atmospheric CFCs.

showed that ozone concentrations declined to between 50 and 95 percent of 1979 levels during certain times of the year. Improved atmospheric models also suggested that the goals of ozone protection could be met only through more stringent curbs on CFC production. The amended Montreal Protocol called for a total phaseout of specified CFCs, halons, and carbon tetrachlorides by 2000 and methyl chloroform by 1995. It also accelerated the rate at which the phaseout would be conducted for CFCs in general. Periodic amendments to the protocol since 1990 (in 1992, 1995, 1997, 1999, and 2007) have further accelerated the phaseout schedules. The amendments have also added controlled substances to the protocol, including hydrobromofluorocarbons (HBFCs), hydrochlorofluorocarbons (HCFCs), and bromochloromethane. In 2008, the United Nations Environment Programme estimated that if compliance with the protocol continues, the Arctic ozone layer should return to its pre-1980 level by 2050; the Antarctic ozone layer is expected to recover by 2060-2075.

The Montreal Protocol was a political innovation because it called for a gradual reduction in CFC production and allowed for adjustments in the activities of each treaty member that were flexible enough to respond to new scientific information. A total ban on CFCs would have been unworkable; without reasonably inexpensive alternatives, the distribution of temperature-sensitive medical supplies such as blood and food shipments would have been imperiled because CFCs were essential to most refrigerating units. Many workplaces dependent on air-conditioning also would have been adversely affected. In addition, considerable amounts of industrial machinery with productive lifetimes of twenty to thirty years would have become obsolete immediately.

Considerable disagreement still exists regarding the extent of ozone depletion and its effects. Debate continues over whether the economic cost of finding alternatives to CFCs is outweighed by the estimated increase in skin cancer caused by a global reduction in ozone. Scientists have found evidence that ozone depletion can lead to climate change, reduced productivity in Antarctic waters, and decreased reproduction in amphibians worldwide. Important arguments among the initial signers of the Montreal Protocol concerned the level of production cuts that would be required to end the problem of ozone depletion (one reason the protocol was amended in 1990) and the level of support to which developing nations were entitled for complying with the protocol and forgoing the benefits of cheap CFCs (particularly for refrigeration) that developed countries enjoyed.

Mark Coyne

FURTHER READING

Barrett, Scott. "The Montreal Protocol." In *Environment and Statecraft: The Strategy of Environmental Treaty-Making.* New York: Oxford University Press, 2003.

Benedick, Richard. *Ozone Diplomacy: New Directions in Safeguarding the Planet.* Enlarged ed. Cambridge, Mass.: Harvard University Press, 1998.

Kaniaru, Donald, ed. *The Montreal Protocol: Celebrating Twenty Years of Environmental Progress—Ozone Layer and Climate Protection.* London: Cameron May, 2007.

Susskind, Lawrence. *Environmental Diplomacy: Negotiating More Effective Global Agreements.* New York: Oxford University Press, 1994.

Zerefos, Christos, Georgios Contopoulos, and Gregory Skalkeas, eds. *Twenty Years of Ozone Decline: Proceedings of the Symposium for the Twentieth Anniversary of the Montreal Protocol.* Dordrecht, the Netherlands: Springer, 2009.

SEE ALSO: Air pollution; Chlorofluorocarbons; Climate change and human health; Climate change and oceans; Climate change skeptics; Climate models; Freon; Greenhouse effect; Ozone layer.

Mount Tambora eruption

CATEGORIES: Disasters; atmosphere and air pollution; weather and climate

THE EVENT: Massive volcanic eruption in the East Indies that sent such large amounts of ash and sulfur dioxide into the atmosphere that global temperatures were depressed in the following year

DATES: April 5-12, 1815

SIGNIFICANCE: The Mount Tambora eruption is the most dramatic and unequivocal historical example of the widespread effects of volcanic eruptions on world climate. The main lesson of this event, however, is that even an eruption of such magnitude is unlikely to produce long-lasting climatic perturbation in the absence of additional reinforcing factors. Although the year was long remembered as one of great hardship, no severe famine or political upheaval occurred in the affected areas in 1816-1817.

On April 5, 1815, Mount Tambora, an enormous volcano on the island of Sumbawa, abruptly came to life. Over the next several days the eruption intensified, culminating in a cataclysm on April 12 that pulverized and sent into the atmosphere some 100 cubic kilometers (24 cubic miles) of mountaintop, approximately five times that ejected by Krakatoa in 1883. Pyroclastic flows coursed down the flanks of Mount Tambora, incinerating everything in their path and triggering tsunamis that devastated the coasts of Lombok, Bali, and Sulawesi. Violent eruptions continued through July of 1815. The eruptions and tsunamis killed upward of 40,000 people outright, and an estimated 107,000 died of famine and disease on Sumbawa, Lombok, and Bali owing to the destruction of crops by a thick blanket of volcanic ash. Tambora ranks high among purely natural disasters in terms of immediate loss of life.

The Tambora eruption was the most massive and violent volcanic event in the preceding 10,000 years. (The Mount Mazama eruption, which took place around 5500 B.C.E., was of the same magnitude overall but occurred in three distinct phases over a period of 100 years.) The force of the Tambora eruption propelled fine ash and sulfur dioxide more than 25 kilometers (15.5 miles) into the atmosphere, where it was carried around the globe, blocking incident solar radiation. Observers in Europe and North America noted spectacular sunsets and dirty snow in 1815.

Temperatures in the fall and winter of 1815 remained near normal for the decade 1810-1820, which was unusually cold. Climatologists who have examined snow cores suspect that an as-yet-unidentified volcanic eruption somewhere in the Tropics in 1808 started a cooling trend that intensified the effects of Tambora. May, 1815, was unusually cold in both Europe and North America, delaying the planting of crops, and killing frosts occurred in the northeastern United States in June, July, August, and September, destroying sensitive crops as soon as they were planted. In northern and central Europe, low temperatures and high rainfall depressed the yields of cereal grain crops and effectively prevented hay harvesting. In economies entirely dependent on animal power, the failure of feed crops was a major disaster.

Warm temperate and tropical regions were less affected. Difficulties in distributing the food crops produced in these regions highlighted deficiencies in transportation, especially in North America, where people living inland suffered great privation despite adequate food supplies in the southern coastal areas. The experience of 1816—often called the year without a summer—helped stimulate the building of the Erie Canal and accelerated westward migration as people abandoned farms in New England.

People experiencing the nonsummer of 1816 and the bitter winter of 1816-1817 were unaware of the volcanic connection, although Benjamin Franklin had suggested a possible link earlier. Some ascribed the lack of usual summer weather to very conspicuous sunspots, and one Italian scientist confidently predicted that the sun was about to fail. The bizarre weather sparked a number of notable literary works, including Lord Byron's "Darkness" (written in July, 1816) and Mary Wollstonecraft Shelley's *Frankenstein* (first published in 1818). Hard times associated with crop failures and the end of the Napoleonic wars produced riots and civil disturbances in Great Britain and elsewhere, but these had no lasting impact. Most of the high mortality rate seen in this period resulted from typhus epidemics that raged in Ireland and Central Europe.

Martha A. Sherwood

FURTHER READING

Evans, Robert. "Blast from the Past." *Smithsonian* 33 (July, 2002): 52-57.

Grove, Jean N. *Little Ice Ages, Ancient and Modern.* 2d ed. New York: Routledge, 2004.

Stommel, Henry, and Elizabeth Stommel. *Volcano Weather: The Story of 1816, the Year Without a Summer.* Newport, R.I.: Seven Seas Press, 1983.

SEE ALSO: Airborne particulates; Climatology; Krakatoa eruption; Sulfur oxides; Volcanoes and weather.

Mountain gorillas

CATEGORY: Animals and endangered species

DEFINITION: Subspecies of the eastern gorilla found in remote areas of the mountains of Rwanda, the Democratic Republic of the Congo, and Uganda

SIGNIFICANCE: Although the population of mountain gorillas remains under constant threat from loss of habitat and poaching, the active involvement of international conservation organizations and individual scientists has helped to raise both public awareness and funds for the gorillas' protection.

Mountain gorillas in Volcanoes National Park, Rwanda. (©Dreamstime.com)

Gorillas, the largest living primates, are closely related to humans. Males are less than 1.8 meters (6 feet) tall and weigh approximately 136 kilograms (300 pounds), while females can grow up to 1.5 meters (5 feet) in height and weigh about 90 kilograms (200 pounds). Both have a brownish-gray fur that turns gray with age. Three main types of gorilla are found in Africa: the western lowland gorilla (*Gorilla gorilla gorilla*), the eastern lowland gorilla (*Gorilla gorilla graueri*), and the mountain gorilla (*Gorilla gorilla beringei*).

Mountain gorillas, the most endangered of the three, live in the high-altitude (up to 3,700 meters, or 12,000 feet) forests of the Virunga Volcanoes National Park in Rwanda and the Bwindi Reserve in Uganda. They are ground-dwelling animals that walk on their hind limbs and the knuckles of their forelimbs. They eat shrubs and fruit. They live in small groups, each of which consists of a group-leading mature male (also known as the silverback), two to four adult females, and an equal number of younger gorillas. The males

are generally peaceful and become aggressive only when they meet other group leaders. Females produce single offspring after a gestation of almost nine months; they lose about 50 percent of their offspring within the first year of life. The average life span of the mountain gorilla is estimated to be close to thirty-five years.

The end of World War II led to great changes in the colonization of Africa. Under mounting pressure, the governments of countries such as England, France, and Belgium realized that sustaining their pre-World War II colonial attitudes toward Africa was too expensive, both financially and politically. Several of the African colonies' struggles for independence from European colonial leadership resulted in the rise of corrupt leaders whose interests were self-centered rather than patriotic. The poaching of wildlife was occasionally banned, but bribery, greed, and reckless disregard of the law continued to thrive in a few of these nations. Gorillas were in great demand for their palms, skins, and skulls, which sold for high prices.

Mountain gorillas were also indirectly affected by the poaching of other animals, such as antelope, buffaloes, and elephants, which were driven to unusually higher altitudes and grazed at the expense of the mountain gorillas' food supply.

The forests of Uganda, Rwanda, and what was then known as Zaire (now the Democratic Republic of the Congo) were overexploited with no regard for the environment, which led to a dangerous decline in the gorilla population. Individual environmentalists such as Dian Fossey, who eventually died under mysterious circumstances in Africa, ensured that people around the world became aware of the imminent extinction of mountain gorillas. Since 1977 organizations such as the World Wildlife Federation, the African Wildlife Federation, the Gorilla Conservation Program, and the Fauna and Flora Preservation Society have been actively involved in protecting the endangered mountain gorilla. The activities of these organizations include conservation, reforestation, law enforcement, environmental education, and controlled tourism and building development.

Soraya Ghayourmanesh

FURTHER READING

De la Bédoyère, Camilla. *No One Loved Gorillas More: Dian Fossey—Letters from the Mist.* Washington, D.C.: National Geographic Society, 2005.

Eckhart, Gene, and Annette Lanjouw. *Mountain Gorillas: Biology, Conservation, and Coexistence.* Baltimore: The Johns Hopkins University Press, 2008.

Robbins, Martha M., Pascale Sicotte, and Kelly J. Stewart, eds. *Mountain Gorillas: Three Decades of Research at Karisoke.* New York: Cambridge University Press, 2001.

SEE ALSO: Extinctions and species loss; Fossey, Dian; Poaching; Wildlife management.

Muir, John

CATEGORIES: Activism and advocacy; preservation and wilderness issues

IDENTIFICATION: Scottish American naturalist, preservationist, and writer

BORN: April 21, 1838; Dunbar, Scotland

DIED: December 24, 1914; Los Angeles, California

SIGNIFICANCE: Muir, one of America's most notable preservationists and a founder of the Sierra Club, introduced Americans to California's Sierra Nevada and worked hard to protect much of the region's wilderness, including Yosemite, against development.

Born in Scotland, John Muir emigrated to the frontier area of Wisconsin in 1849 with his family. Muir's father was a dominant but negative influence on his life. The elder Muir was deeply religious but viewed the Christian god as one of justice rather than of love. Muir turned away from his father's repressive religion, substituting for it an intense love for nature; for Muir, the divine seemed manifest in the wilderness.

Naturalist and preservationist John Muir. (Library of Congress)

While laboring on the family farm, Muir became an observer of the environment. He was also something of an inventor and worked as a craftsman. After attending the University of Wisconsin, from which he did not graduate, Muir trekked through the southern United States in 1867 hoping to journey on to South America. Instead, he boarded a ship in New York that was bound for California; he arrived in San Francisco in 1868. Muir had no love for cities; he could cross a mountainous wilderness without maps, but he was lost in any urban area. To him, cities were mentally and morally corrupting. He once wrote, "There is not a perfectly sane man in San Francisco."

Muir spent his first summer in California in the Central Valley; in the spring of 1869 he journeyed into the Sierra Nevada as supervisor of a sheep herd. That year was an epiphany for Muir. Yosemite Valley was still almost virginal, and the surrounding cliffs and mountains drew him on as nothing before had done. He stayed in Yosemite the following winter, working in a lumber mill and escaping into the wilderness whenever possible. He wrote of the experience, "I am bewitched, enchanted. . . . I have run wild." With only bread, tea, and a blanket, Muir explored much of the central Sierras.

In 1870 the philosopher and essayist Ralph Waldo Emerson visited Yosemite, and Muir eagerly sought him out. Parlor Transcendentalist Emerson and natural man Muir had much in common, but Muir was unable to convince Emerson to camp under the trees and stars. Muir's attachment to the Sierras was not merely religious, philosophical, or emotional. He was a keen natural scientist. One of his major scientific contributions was his theory that glaciation is the key to explaining the existence of Yosemite Valley and the other canyons and mountains of the Sierras. This idea contravened the accepted doctrine that Yosemite Valley had been formed by a gigantic catastrophe that caused the valley floor to fall several thousand feet. Muir's beliefs were initially dismissed, but in the high country he discovered residual glaciers, validating his claim. In 1879 Muir became one of the first people to explore Alaska's Glacier Bay.

Something of a loner and an eccentric, Muir had

Roosevelt on Muir

In his autobiography, Theodore Roosevelt described his 1903 visit to Yosemite with John Muir:

When I first visited California, it was my good fortune to see the "big trees," the Sequoias, and then to travel down into the Yosemite, with John Muir. Of course of all people in the world he was the one with whom it was best worth while thus to see the Yosemite. He told me that when Emerson came to California he tried to get him to come out and camp with him, for that was the only way in which to see at their best the majesty and charm of the Sierras. But at the time Emerson was getting old and could not go.

John Muir met me with a couple of packers and two mules to carry our tent, bedding, and food for a three days' trip. The first night was clear, and we lay down in the darkening aisles of the great Sequoia grove. The majestic trunks, beautiful in color and in symmetry, rose round us like the pillars of a mightier cathedral than ever was conceived even by the fervor of the Middle Ages. Hermit thrushes sang beautifully in the evening, and again, with a burst of wonderful music, at dawn.

I was interested and a little surprised to find that, unlike John Burroughs, John Muir cared little for birds or bird songs, and knew little about them. The hermit thrushes meant nothing to him, the trees and the flowers and the cliffs everything. The only birds he noticed or cared for were some that were very conspicuous, such as the water-ouzels always particular favorites of mine too. The second night we camped in a snow-storm, on the edge of the cañon walls, under the spreading limbs of a grove of mighty silver fir; and next day we went down into the wonderland of the valley itself. I shall always be glad that I was in the Yosemite with John Muir and in the Yellowstone with John Burroughs.

Source: Theodore Roosevelt, *An Autobiography* (New York: Charles Scribner's Sons, 1913).

few close friends. One of these was Jeanne Carr, whose husband, Ezra, was a professor of geology and chemistry at Wisconsin who later moved to Berkeley, California. Jeanne was very supportive of Muir and his ventures, and it was she who convinced him to write about his experiences, in articles published first in the *Overland Monthly* and later in New York's *Harper's* and *Scribner's* magazines. Muir labored with his writing, but his articles and books are marvelously descriptive, with a visual immediacy that has kept his work in print for many decades. It was through these articles that Muir and his beloved Sierras became known to many Americans.

Muir was always fearful of the impact of civilization on the wilderness. The fragile alpine environments were unsuited to heavy grazing by sheep and cattle, and human visitors to Yosemite and beyond often wished to bring all of their civilized amenities with them. One of Muir's greatest successes was in preserving Yosemite, then under state control and in danger of overdevelopment. With the support of *Century* magazine's Robert Underwood Johnson, Muir mounted a successful campaign that saw Yosemite Valley and much of the surrounding territory become a national park in 1890.

Muir's greatest failure, however, also concerned Yosemite. San Francisco's water system was antiquated and inadequate, and local officials saw a solution in tapping the waters of the Tuolumne River, which ran through Hetch Hetchy Valley; Muir had long praised the beauty of the valley, which, at his urging, had been included within the boundaries of Yosemite park. The issue of damming Hetch Hetchy became a national conflict in the early twentieth century. Some noted conservationists favored it, most notably Gifford Pinchot, chief forester under U.S. president Theodore Roosevelt and, briefly, William Howard Taft. Conservationists such as Pinchot believed that nature and its resources should be made available for human use. Muir, on the other hand, was a preservationist who believed that nature has a right to exist without human interference. Muir and his supporters fought the proposal for years but finally lost in 1913. Soon Hetch Hetchy Valley disappeared in the rising waters when the river was dammed.

Muir died one year later. He left a stirring legacy, not least in the Sierra Club, which he helped found in 1892 and for which he served as president until the end of his life. Many later conservationists and environmentalists have been inspired by Muir's preservationist ethic. No other figure in the environmentalism movement has been so widely honored; California alone boasts many sites named for Muir, including the Muir Woods near San Francisco, Muir Grove in Sequoia National Park, and the Sierras' John Muir Trail.

Eugene Larson

FURTHER READING

Cohen, Michael P. *The History of the Sierra Club, 1892-1970.* San Francisco: Sierra Club Books, 1988.

Fox, Stephen. *John Muir and His Legacy: The American Conservation Movement.* Boston: Little, Brown, 1981.

Miller, Rod. *John Muir: Magnificent Tramp.* New York: Forge, 2005.

Muir, John. *Essential Muir.* Edited by Fred D. White. Berkeley, Calif.: Heyday Books, 2006.

Perrottet, Tony. "John Muir's Yosemite: The Father of the Conservation Movement Found His Calling on a Visit to the California Wilderness." *Smithsonian* 39, no. 4 (July, 2008): 48-55.

Turner, Frederick. *John Muir: Rediscovering America.* 1985. Reprint. Cambridge, Mass.: Perseus, 2000.

SEE ALSO: Hetch Hetchy Dam; National parks; Pinchot, Gifford; Preservation; Roosevelt, Theodore; Sierra Club; Yosemite Valley.

Multiple-use management

CATEGORIES: Resources and resource management; forests and plants; land and land use

DEFINITION: The management of resources, especially forests, for many purposes, including recreation, plant and animal habitats, grazing, groundwater and flood protection, and economic uses such as timber harvesting and mineral extraction

SIGNIFICANCE: The systematic management of resources for multiple uses while providing for conservation and sustainability is the goal of multiple-use management. Because some uses of natural resources are not compatible with each other and some can cause damage to the environment, multiple-use management poses a number of challenges.

The management of natural resources for multiple uses involves trade-offs among the current and future ecological, social, and economic use demands of citizens, governments, and private entities for goods and services. Multiple uses may take place on private and public forestlands, farmland, open lands, and along coastal resources. Forests are most often managed for multiple uses, as they provide not only wood resources but also aesthetic and environmental benefits, including regulation of climate, reduction of air pollution, provision of wildlife habitats, stabilization of soil, and preservation of water resources, some of which are used for public water supplies.

The management of forests, open lands, and coastal resources for diverse purposes such as conservation, recreation, and commerce enhances these lands' productivity. Multiple-use management policies allow for various uses of land in tandem with practices that sustain wildlife habitats and protect the environment. Some of the recreational activities that take place on multiple-use lands include hiking, boating, swimming, rafting, camping, hunting, fishing, mountain biking, snowmobiling, and hang gliding. Commercial activities on such lands include logging and grazing, and mineral extraction is a common subsurface use. In the United States these commercial uses generate billions of dollars in revenue each year for federal, state, and local governments. Multiple-use management also takes place in coastal ecosystems, where uses include regulated development along coastlines, commercial fishing, oil and gas extraction, and conservation.

Foresters are aware of the multiple benefits that forests provide, and they manage forest resources based on the capacity of the land to accommodate multiple uses simultaneously without destroying the environmental benefits derived from forestland. Multiple-use forest management practices include sustaining the production of renewable resources such as trees and vegetation while protecting wildlife habitats, allowing recreation, and managing for fire and diseases. Private entities and individuals are also involved in multiple-use management when they engage in sustainable land-use planning that ensures development of compatible uses. Multiple-use management plans may designate areas for exclusive uses, such as timber production; dual uses, such as cultural heritage preservation and recreation; and general uses, where many compatible uses such as watershed protection, recreation, timbering, and wildlife refuges coexist simultaneously.

In the United States the Bureau of Land Management (BLM) and the Forest Service manage many of the federal multiple-use properties. In 1976 Congress passed the Federal Land Policy and Management Act, which mandated the BLM to manage public lands and their varied resources for multiple uses. Some of the other federal laws that promote multiple-use management include the Mineral Leasing Act of 1920, which allows for subsurface mineral extraction and mining on public lands, and the Taylor Grazing Act of 1934, which supports the grazing of cattle as well as wild horses, burros, sheep, and bison on public rangelands.

Carol A. Rolf

FURTHER READING

Bowes, Michael D., and John V. Krutilla. *Multiple-Use Management: The Economics of Public Forestlands.* Washington, D.C.: Resources for the Future, 1989.

Lafortezza, Raffaele, et al., eds. *Patterns and Processes in Forest Landscapes: Multiple Use and Sustainable Management.* New York: Springer, 2008.

McElfish, James, and Rebecca Kihslinger. *Nature-Friendly Land Use Practices at Multiple Scales.* Washington, D.C.: Environmental Law Institute, 2008.

SEE ALSO: Bureau of Land Management, U.S.; Carrying capacity; Forest management; Grazing and grasslands; Habitat destruction; Hunting; Logging and clear-cutting; Range management; Strip and surface mining; Wild horses and burros; Wildlife refuges.

Mumford, Lewis

CATEGORY: Urban environments
IDENTIFICATION: American historian and social philosopher
BORN: October 19, 1895; Flushing, New York
DIED: January 26, 1990; Amenia, New York
SIGNIFICANCE: Although Mumford could not be called an environmental activist in the usual sense, his advocacy of garden cities, his criticism of urban sprawl and the directions taken by modern technology, and his opposition to the use of nuclear energy show his deep-seated concern for the human environment.

Social philosopher and urban critic Lewis Mumford was one of the twentieth century's most original thinkers. His writings, which include some thirty books and countless essays, cover a vast range of disciplines, including sociology, technology, philosophy, history, and literature. He is perhaps best known for his architectural criticism and works on urban history and planning.

Mumford did not believe in the efficacy of political action and consequently did not participate in any environmental organizations or ecological crusades in a formal manner. He did, however, speak eloquently and forcefully on a range of topics that had, as their natural corollary, beneficial consequences for the human environment.

Mumford was a strong advocate of regional planning, which was a response to the overpopulation and decline of cities in the United States. He was concerned about urban noise, congestion, senseless growth, and the way cities sprawled into the surrounding countryside, despoiling nature in the process. He proposed the creation of medium-sized garden cities located near metropolises. These garden cities would be true cities rather than suburbs, with a balance of housing, industry, commerce, and necessary cultural and civic amenities; they would be surrounded by greenbelts made up of parks and agricultural land. The cities would exist in harmony with nature and the surrounding countryside and would relate to the soil, climate, and topography of their regions. Unlike the central cities, the garden cities would be built to human scale and would not be dominated by the automobile, of which Mumford was a severe critic because of its potential to create noise, congestion, and pollution.

Mumford was also a strenuous critic of the course of modern technology, as evidenced in his major two-volume work *The Myth of the Machine*, consisting of

Historian and urban theorist Lewis Mumford. (©Bettmann/ CORBIS)

Technics and Human Development (1967) and *The Pentagon of Power* (1970). He blamed technology for encouraging the blind pursuit of power and productivity, often with dehumanizing and debasing consequences for the human personality. In its quest for profits, Mumford argued, technology ignores human wants and desires while it strives to dominate and conquer nature. Mumford's solution was visionary: He called for a complete reordering of human values—an internal transformation—to curb the disastrous trend. Many critics misunderstood Mumford's arguments and complained that he was behaving like a petulant Old Testament prophet who hated science and scientists.

A final area that stamped Mumford as a friend of the environment was his early opposition to the atomic bomb. He criticized this weapon on moral, political, and environmental grounds. He was among the first to question the peaceful application of atomic energy. He asserted that too little was still known about fissionable material to justify the cavalier risks that were being taken, particularly because no one knew how to dispose of nuclear waste material. Mumford feared for both humans and nature if the use of nuclear energy was not constrained or eliminated.

Throughout his career, Mumford praised the virtues of moderation, self-restraint, and balance in human activity, and he showed respect and love for the countryside and nature. For these reasons, he deserves an honored place among those concerned about the future of the environment.

David C. Lukowitz

FURTHER READING

Guha, Ramachandra. "The Historical Social Ecology of Lewis Mumford." In *How Much Should a Person Consume? Environmentalism in India and the United States.* Berkeley: University of California Press, 2006.

Mazmanian, Daniel A., and Michael E. Kraft, eds. *Toward Sustainable Communities: Transition and Transformations in Environmental Policy.* Cambridge, Mass.: MIT Press, 2009.

Miller, Donald L. *Lewis Mumford: A Life.* 1989. Reprint. New York: Grove Press, 2002.

SEE ALSO: Antinuclear movement; Greenbelts; Heat islands; Open spaces; Planned communities; Urban planning; Urban sprawl.

N

Nader, Ralph

CATEGORIES: Activism and advocacy; human health and the environment

IDENTIFICATION: American political activist and consumer advocate

BORN: February 27, 1934; Winsted, Connecticut

SIGNIFICANCE: As a consumer advocate, Nader has been a champion of the underdog—including the poor, the elderly, and members of minority groups—against corporate and political power structures in the United States.

Ralph Nader's interest in environmental issues began long before the environmentalist movement emerged in the late 1960's. When Nader was a child, his parents instilled in him a respect for the duties of citizenship in an industrial democracy. While a student at Princeton University, Nader fought to prevent the spraying of trees on the campus with the pesticide dichloro-diphenyl-trichloroethane (DDT). At Harvard, where he attended law school, he was an editor of the *Harvard Law Record*, which he attempted to change from a dry academic publication to a forum for ideas on social reform.

Nader took his energy and vision for change to Washington, D.C., in the mid-1960's. He was hired by the U.S. Department of Labor's policy planning division as a consultant on highway safety. This job enabled him to amass extensive information on the automobile industry, which he used in writing *Unsafe at Any Speed* (1965), a blatant indictment of the Detroit automobile industry, specifically General Motors. The publication of this book moved Nader to center stage as a consumer advocate.

Beginning in 1969, Nader founded a number of public interest organizations, including the Center for Study of Responsive Law, the Corporate Accountability Research Group, the Public Interest Research Group, and the Clean Water Action Project. These groups focus on a wide range of consumer issues and give high priority to such environmental concerns as safety standards in the workplace, clean and renewable energy resources, automobile emissions standards, clean water standards, and food safety. Nader

and his organizations were instrumental in securing passage of major environmental legislation during the 1970's. Nader also played a key role in the establishment of the U.S. Environmental Protection Agency, the Occupational Safety and Health Administration, and the Consumer Product Safety Commission.

Even though the consumer rights and environmentalist movements lost some momentum in the 1980's, Nader's efforts for reform did not decline. The organizations he founded drew supporters who feared that President Ronald Reagan's administration would undo earlier legislative victories. Nader adapted his agenda to address new threats to consumers, such as proposed limits on liability for corporate

Consumer advocate Ralph Nader speaks at a news conference after he was named the Green Party's candidate for the U.S. presidency in 1996. (AP/Wide World Photos)

negligence, restrictions on regulatory agencies protecting consumers, and trade agreements threatening workers and weakening environmental standards. In 1996 Nader ran for president of the United States on the Green Party ticket; this move gave him a new forum from which to articulate his environmental agenda and expose corporate and political leaders' priorities on protecting their narrow interests over the public welfare. He ran for president as the Green Party's candidate again in 2000, and he campaigned for the office as an independent in 2004 and 2008.

Nader has coauthored and coedited several books on environmental topics, including *The Menace of Atomic Energy* (1977), with John Abbotts, and *Who's Poisoning America* (1981), with Ronald Brownstein and John Richard. He published the autobiographical *Crashing the Party: Taking on the Corporate Government in an Age of Surrender* in 2002 and the memoir *The Seventeen Traditions* in 2007.

Ruth Bamberger

FURTHER READING

Carter, Neil. "Party Politics and the Environment." In *The Politics of the Environment: Ideas, Activism, Policy.* 2d ed. New York: Cambridge University Press, 2007.

Marcello, Patricia Cronin. *Ralph Nader: A Biography.* Westport, Conn.: Greenwood Press, 2004.

Nader, Ralph, Ronald Brownstein, and John Richard, eds. *Who's Poisoning America: Corporate Polluters and Their Victims in the Chemical Age.* San Francisco: Sierra Club Books, 1981.

SEE ALSO: Alternative energy sources; Automobile emissions; Clean Water Act and amendments; Environmental health; Environmental Protection Agency; Right-to-know legislation.

Naess, Arne

CATEGORY: Philosophy and ethics
IDENTIFICATION: Norwegian philosopher
BORN: January 27, 1912; Oslo, Norway
DIED: January 12, 2009; Oslo, Norway
SIGNIFICANCE: Naess's ideas, in particular his introduction of the concept of deep ecology, have had a great deal of influence on environmental philosophy and activism.

The wide-ranging philosophy and academic thought of Arne Naess were enriched by Naess's extensive experience with mountaineering. His early interests encompassed the philosophy of science, empirical semantics, and the critique of dogmatism. In the late 1960's Naess began to develop an ecological philosophy (or ecosophy) that he called Ecosophy T, which was heavily influenced by the metaphysics of Baruch Spinoza, the social activism of Mohandas Gandhi, and the spiritual insights of Hinduism and Buddhism.

At the core of Ecosophy T is an intuitive identification with all of life and a spontaneous feeling of unqualified value for the flourishing of all things. This intuition involves a wider sense of self that includes all of existence without dissolving individuality. From this comes an inclination toward maximizing biodiversity, pluralism in philosophy and religion, and decentralization in economics and politics. A central but controversial principle that emerges from Ecosophy T is biospherical egalitarianism—that is, the principle that all things have an equal right to live and bloom. Considered by itself as an absolute, this principle leads to crippling ethical paradoxes, but Naess combined it with a recognition of the inevitability of harming other life and the need to develop a hierarchy of obligations.

In 1972 Naess presented a paper titled "The Shallow and the Deep, Long-Range Ecology Movement" in which he described the concept of deep ecology and its radical orientation to the environment. Deep ecology's holism has been criticized as devaluing individuals, but Naess's philosophy does not present a simple monism that places value only on a single whole. Instead, its nondualistic holism includes an affirmation of the reality and value of individuals and relationships.

Central to Naess's method was an insistent questioning of conventional ideas and values, as well as a probing into the root causes of environmental problems. Such inquiry, Naess believed, leads beyond mere policy debates to critical engagement in worldviews, which are rooted in foundational intuitions. While reason cannot establish these intuitions, it can help people deduce general principles and particular beliefs. Similarly, Naess rejected conventional notions of duty and altruism, which are based on an atomistic view of independent selves. Naess saw morality as grounded in the intuitive experience of identification with others and a spontaneous sense of caring for the well-being of all things. Naess was a phenomenologist; he emphasized a direct experience of reality that goes beyond the subject-object duality. One result of this emphasis on the direct intuition of nature is a pervasive sense of joy.

Naess was not concerned simply with articulating his own ecosophy. True to his antidogmatism, he presented deep ecology as an open-ended philosophical and social movement. Along with George Sessions, Naess created the deep ecology platform, a list of eight principles on which radical environmentalists with different philosophies can agree and that they can use as the basis for activism. Naess championed a diversity of worldviews and beliefs while arguing for the need for solidarity in engaging environmental problems.

David Landis Barnhill

FURTHER READING

De Steiguer, J. F. "Arne Naess and the Deep Ecology Movement." In *The Origins of Modern Environmental Thought.* Tucson: University of Arizona Press, 2006.

Katz, Eric, Andrew Light, and David Rothenberg, eds. *Beneath the Surface: Critical Essays in the Philosophy of Deep Ecology.* Cambridge, Mass.: MIT Press, 2000.

SEE ALSO: Biodiversity; Deep ecology; Ecofeminism; Environmental ethics; Social ecology; Speciesism.

Nanotechnology

CATEGORY: Human health and the environment

DEFINITION: Diverse field of science and technology that seeks to understand and control matter at dimensions of 1 to 100 nanometers

SIGNIFICANCE: Nanotechnology receives large-scale public and private investment around the world. The field holds the promise of revolutionizing much of science and technology, but the potential impacts of the products of nanotechnology on the environment, health, and safety are poorly understood. Preliminary research findings suggest that much further study is needed.

Nanotechnology is a relatively new field concerned with a wide range of materials and processes. The increasing speeds and decreasing sizes of personal electronic devices owe much to nanotechnology, which enables the production of lighter and stronger materials that reduce energy usage and prolong the life spans of the devices made with them. Battery and lighting systems have been developed that use nanotechnology to be more fuel-efficient. Nanotechnology has also contributed to the development of improved medical diagnostic devices and drug-delivery systems. Research into the use of nanotechnology to detect pollutants and build more effective water-purification systems is ongoing. The promise of nanotechnology is immense, but the field is new, and much remains uncertain about its overall impacts on the environment and human health.

DEFINITION

The term "nano" is a prefix for one-billionth, or 1×10^{-9}, of a unit. Nanotechnology is concerned with the study, development, and control of materials at nanoscale—that is, in the range of 1 to 100 nanometers (or 1 to 100 billionths of 1 meter). Nanotechnology includes processes and instruments that engineer materials at nanoscale, and nanoparticles have at least one dimension in the nanoscale range.

Many naturally occurring materials have nanoscale dimensions, including proteins, the genetic molecule deoxyribonucleic acid (DNA), and viruses. Volcanoes and forest fires produce nanoparticles, and many soils contain organic and inorganic nanoparticles. Nanoparticles are also produced by human activities, such as cigarette smoking and the burning of fuels in combustion engines. The polymer and plastics industry makes chemical molecules with nanoscale dimensions. Environmental issues associated with nanotechnology thus overlap other areas.

A significant segment of nanotechnology is concerned with newly discovered nanoparticles. These can have completely different electrical, magnetic, or biological properties compared to larger particles of the same substance. For example, gold nanoparticles can be red, blue, or gold, depending on their precise size. Titanium dioxide and zinc oxide are used in sunscreens to block the sun's ultraviolet (UV) rays, but large particles of these substances leave a white coating on the skin; as nanoparticles, the same substances are transparent and more appealing to sunbathers. Questions have arisen, however, about what else might be different about these nanoparticles and what effects they might have when they get into the body or are washed into the environment.

Richard E. Smalley has been called the grandfather of nanotechnology. In 1986 he and others discovered

Kevin Ausman, a scientist from Rice University, displays a bottle of carbon nanotubes, key components in the present and future of nanotechnology. (AP/Wide World Photos)

a completely new form of carbon. They named it buckminsterfullerene, but as the molecules look like soccer balls, they are more commonly called buckyballs. Several sizes and shapes have been identified and collectively are called fullerenes. They are being investigated as possible devices to transport drug molecules to specific tissues and cells.

Many different kinds of nanoparticles have been developed and given names such as dendrimers, nanowires, and quantum dots. Carbon nanotubes are particularly interesting. Identified in 1991, they are highly organized carbon atoms that form sheets that roll into long tubes. Having different forms, they usually are a few nanometers wide and can be millimeters long. They are extremely strong for their weight, can transport other materials, and have unique electrical properties. They may have uses in reinforcing car and airplane bodies and in making comfortable bulletproof clothing; they may also have applications in medicine and in new battery technology. Carbon

nanotubes have been at the center of early debate over the potential environmental impacts of nanotechnology.

PRELIMINARY NANOTOXICOLOGY

Nanoparticles can enter living cells, making them useful as drugs but also raising concerns. Some nanoparticles enter the nuclei of cells, where genetic material is stored. This may have beneficial uses, but it could also lead to genetic damage. Nanoparticles smaller than 35 nanometers can penetrate the blood-brain barrier, which prevents most chemicals from reaching the brain. This property could help deliver drugs for brain disorders, but it might also cause harmful side effects.

The field of science that investigates such concerns is known as nanotoxicology. Some early nanotoxicological studies have provided worrisome results. Nanosilver has antibacterial properties and has been used in special clothing and on surfaces where bacteria might grow. Bulk silver is normally safe, but laboratory experiments have shown that nanosilver might interfere with the human immune system.

Carbon nanotubes account for much of nanoparticle manufacturing, and their production is predicted to increase dramatically. However, a 2009 review of research into the toxicity of carbon nanotubes found that only twenty-one studies had yet been conducted, and all had shown some damage to tissues and animals. None of the research studies had examined the effects of carbon nanotubes on humans.

Hardly any research has examined nanoparticles after they enter the environment from normal wear and tear or when products are discarded. One of the first studies exposed largemouth bass to buckyballs for forty-eight hours. Most of the organs of the fish were unharmed, but their brains showed evidence of oxidative damage. Because buckyballs are highly fat-soluble and thus prone to cause environmental damage, the researchers urged the widespread application of the precautionary principle to avoid the sort of environmental damage seen from earlier chemicals with similar solubility profiles.

NANOBOTS AND GRAY GOO

The vision for nanotechnology originated in a 1959 talk by American physicist Richard Feynman. He predicted the development of very fast and small computers, as well as tiny machines that could circulate through the body. Such hypothetical devices have come to be called nanobots or nanites and are commonly included in science-fiction scenarios. They also come up in discussions about the potential environmental impacts of nanotechnology.

Another early proponent of nanotechnology, K. Eric Drexler, published *Engines of Creation: The Coming Era of Nanotechnology* in 1986. This book had a significant impact on the development and popular understanding of nanotechnology. Drexler's proposal, called molecular manufacturing, involved building nanomachines (or assemblers) that would make things with atomic precision. Drexler's proposal remains scientifically controversial, with critics such as Richard Smalley asserting that the approach is physically impossible.

Others have used the idea of such assemblers to portray nanotechnology as extremely dangerous. Drexler suggested that nanomachines could be programmed to assemble copies of themselves. This would increase production but raises concerns about how to control them. Nanotechnology molecular assemblers have become associated with self-replication, although the ideas are not necessarily linked. Science-fiction authors have invented scenarios in which marauding nanobots wreak environmental havoc.

Drexler eventually distanced himself from the idea of assembler self-replication, but he coined the term "gray goo" to describe a situation in which self-replicating nanomachines get out of control and consume everything around them. In a later edition of his book, Drexler lamented how nanotechnology had become associated with so-called gray goo scenarios. He used the term in only one passage of the book, while repeatedly stressing nanotechnology's potential either to destroy the world or to remake it, curing illness and restoring the environment.

Such grand claims make it more difficult to evaluate the real potential and actual threat of nanotechnology. A distinction must be made between "normal" nanotechnology and "futuristic" nanotechnology. Molecular assemblers and gray goo scenarios belong well into the future, although the steps that current scientists take may determine whether or when these might develop. Normal nanotechnology in the early twenty-first century focuses on recent discoveries about nanoparticles and their properties. Because of the newness of the field, much remains unknown and uncertain, either positive or negative, and practical environmental concerns are being raised.

EXAMINING ENVIRONMENTAL CONCERNS

Given the scientific uncertainty about the impacts of nanoparticles, many scientists and policy makers have called for caution in the study and use of such particles. The European Union has adopted a precautionary approach in its chemical regulatory agency and in its voluntary code of conduct for nanotechnologists. Worldwide, concerted efforts have been undertaken to understand and regulate nanoparticles. The U.S. Environmental Protection Agency launched a major research initiative in 2009 into the health and environmental concerns raised by nanotechnology.

While nanotoxicology is starting to be addressed, many remain concerned about the funding of research into environmental concerns. In the United States, the National Nanotechnology Initiative (NNI) coordinates federal investment in nanotechnology. Between 2005 and 2010, less than 4 percent of the $9 billion invested went directly to the examination of environmental, health, and safety issues. In 2008 the NNI published its strategy for such research, but the National Research Council was critical of the approach, concluding that it overestimated how much research had already been conducted and thereby underestimated the funding necessary to address environmental issues adequately.

The potential benefits of nanotechnology are enormous, but they will be realized only if concerns about possible negative health and environmental impacts are addressed. A gray goo scenario is not necessary for nanoparticles to damage the environment, and much research has yet to be done in this young field before the risks of nanotechnology are fully understood and steps can be taken to ensure that any potential harms are minimized.

Dónal P. O'Mathúna

FURTHER READING

Allhoff, Fritz, et al., eds. *Nanoethics: The Ethical and Social Implications of Nanotechnology.* Hoboken, N.J.: Wiley-Interscience, 2007.

Drexler, K. Eric. *Engines of Creation: The Coming Era of Nanotechnology.* New York: Anchor Books, 1986.

Edwards, Steven A. *The Nanotech Pioneers: Where Are They Taking Us?* Weinheim, Germany: Wiley-VCH, 2006.

O'Mathúna, Dónal P. *Nanoethics: Big Ethical Issues with Small Technology.* London: Continuum, 2009.

Ray, Paresh Chandra, Hongtao Yu, and Peter P. Fu. "Toxicity and Environmental Risks of Nanomaterials: Challenges and Future Needs." *Journal of Environmental Science and Health*, Part C, Environmental Carcinogenesis and Ecotoxicology Reviews 27, no. 1 (2009): 1-35.

SEE ALSO: Environmental ethics; Environmental Protection Agency; European Environment Agency; Polychlorinated biphenyls; Precautionary principle; Risk assessment.

Narmada River

CATEGORIES: Places; ecology and ecosystems

IDENTIFICATION: Fifth-largest river in the Indian subcontinent

SIGNIFICANCE: The Narmada River is symbolic of a global struggle against the environmental impacts of large dam construction. Since the 1980's, the Narmada Valley Development Project has been the subject of protest and opposition by environmentalists and other activists because of the ecological imbalances and human displacement created by the dams the project has built on the river.

Originating from the Narmada Kund lake on the Amarkantak hill in the eastern part of the Indian state of Madhya Pradesh, the Narmada River flows westward for 1,312 kilometers (815 miles) before discharging into the Gulf of Khambhat of the Arabian Sea. Sandwiched between the Vindhya and Satpura mountain ranges, the Narmada basin, which is 98,796 square kilometers (38,145 square miles) in area, is a geologically unique horst-graben formation.

A strong rainfall gradient leads to starkly different vegetation cover in the Narmada basin: from moist evergreen forests in the upper catchment and along the banks of water bodies to dry deciduous forests in the lower catchment. These highly biodiverse forests are home to 76 species of mammals, including the Bengal tiger, leopards, bears, wild boars, and wild dogs, as well as 276 species of birds.

Because a number of protected areas have been established in this region—including the Pachmarhi Biosphere Reserve and Satpura, Bandhavgarh, and Kanha national parks—significant parts of the forests are relatively undisturbed. The region along the Narmada River is also home to several *adivasi* (indigenous) communities, such as the Bhil, Bhilala, Gond, and Korku. They have subsistence economies and distinct cultural and religious practices. The region is also an archaeological treasure site, with fifteen-thousand-year-old prehistoric rock shelter paintings.

Since the 1980's, the Narmada Valley Development Project has built 30 large, 135 medium, and 3,000 minor multipurpose dams on the Narmada River and its tributaries. Because of the river's environmental and social importance, it became the focus of a movement in the 1980's by various environmental action groups and by communities that were directly affected by the land submergence that accompanied the building of

In 1999, protesting villagers, led by activist Medha Patkar, rear, cross the swollen Narmada River while shouting slogans against the government's decision to raise the height of a dam on the river. (AP/Wide World Photos)

dams on the river. During the late 1980's, several of these groups merged under the leadership of the activist Medha Patkar to form the Narmada Bachao Andolan (NBA); the group's name translates as "Save the Narmada Movement." The NBA stands in direct opposition to the Sardar Sarovar Project (SSP), the largest dam on the Narmada River. According to the Indian government, when completed the SSP will irrigate 1.8 million hectares (4.5 million acres) of land, provide drinking water to 40 million people, supply more than 5 billion units of electricity, and create more than 600,000 jobs. Researchers have found that it will also submerge almost 37,000 hectares (91,430 acres) of land and displace more than 200,000 people. In addition, it is a threat to aquatic habitat and will lead to soil salinization and waterlogging.

During the 1990's, the mobilization to save the Narmada River spread to include other hydroelectric projects. Detailed studies of large dams found that the benefits to agriculture had been significantly overestimated by the government. Moreover, since these projects are mostly funded with public moneys, rising capital expenditures and operational losses raised serious concerns. Added to this, the absence of a proper rehabilitation policy for the 21-33 million people displaced by dam building made the construction of large dams unsustainable. Critics argued that the lack of comprehensive environmental impact assessments preceding the building of large dams and other development projects seriously undermined the possibility that preventive and mitigative measures could be taken.

On October 18, 2000, a public interest lawsuit filed by the NBA to stop the construction of the SSP was rejected by the Supreme Court of India. Since that time the NBA has continued to fight for rehabilitation for those displaced by dams in the Narmada Valley region.

Niharika Banerjea

FURTHER READING

Baviskar, Amita. *In the Belly of the River: Tribal Conflicts over Development in the Narmada Valley.* New York: Oxford University Press, 2004.

Wood, John. *The Politics of Water Resource Development in India.* Thousand Oaks, Calif.: Sage, 2007.

SEE ALSO: Aswan High Dam; Dams and reservoirs; Environmental refugees; Franklin Dam opposition; Glen Canyon Dam; Hetch Hetchy Dam; Three Gorges Dam.

National Audubon Society

CATEGORIES: Organizations and agencies; activism and advocacy; animals and endangered species

IDENTIFICATION: American nonprofit organization devoted to the conservation of birds and other wildlife

DATE: Established in 1905

SIGNIFICANCE: The National Audubon Society has been successful in publicizing environmental issues relating to wildlife preservation, in lobbying for and supporting federal and state conservation legislation, and in disseminating greater knowledge of wildlife species and their natural environments through its publications.

The National Audubon Society derives its name from John James Audubon, the noted French American painter who was author and illustrator of *The Birds of America* (1838). The earliest Audubon Society was founded in 1886 by George Bird Grinnell, a pioneering conservationist who grew up with the Audubon family. Though it did not last, becoming defunct in 1888, this society spawned several state Audubon societies, the first of which was the Massachusetts Audubon Society founded by Harriet Hemenway in 1896. Primarily through the efforts of ornithologist William Dutcher, these groups affiliated into the National Association of Audubon Societies in 1905 (with Dutcher as first president), though each state society was, and remains, independent, with its own particular emphasis and program agenda, depending on prevailing local conditions. In 1940, the National Association became the National Audubon Society.

Even before the incorporation, the societies were successful in pushing for legislation to curtail or prohibit the killing of waterfowl for the purpose of using the birds' plumage in clothing. (In honor of that success, the National Audubon Society adopted an image of the great egret as the organization's logo emblem in 1953.) In 1918, the National Association achieved a major victory through its role in the passage of the Migratory Bird Treaty Act.

Although the society's purpose as originally conceived was to raise awareness of, and forestall, the potential extinction of bird species, its mission has expanded over the years to include a more general and overarching protection of the environment as a whole and, in particular, the plant and animal species inhabiting it—without abandoning its pronounced emphasis on birdlife. Wildlife sanctuaries have long been a cornerstone of Audubon Society activity, and the national organization and individual state societies operate many such refuges, including Pelican Island and Corkscrew Swamp in Florida, Paul J. Rainey Sanctuary in Louisiana, Hog Island in Maine, Tennille Creek in Oregon, and the Lillian Annette Rowe Bird Sanctuary in Nebraska.

In 1934 Roger Tory Peterson became the society's educational director and also published *A Field Guide to the Birds*, the first nature field guide. Subsequently, the publication of a series of pocket guides covering all aspects of nature developed into a major National Audubon Society initiative. In addition to such guides and other publications, the society publishes the bimonthly magazine *Audubon*, which began as *Bird-Lore* in 1899. An important element in the society's success and continued influence is its focus on the popular, amateur, and local/community levels, particularly through its sponsorship of significant birding (birdwatching) activities.

Raymond Pierre Hylton

FURTHER READING

Anderson, John M. *Wildlife Sanctuaries and the Audubon Society: Places to Hide and Seek.* Austin: University of Texas Press, 2000.

Graham, Frank, Jr., with Buchheister, Carl W. *The Audubon Ark: A History of the National Audubon Society.* New York: Alfred A. Knopf, 1990.

Obmascik, Mark. *The Big Year: A Tale of Man, Nature, and Fowl Obsession.* New York: Free Press, 2004.

Rhodes, Richard. *John James Audubon: The Making of an American.* New York: Alfred A. Knopf, 2004.

SEE ALSO: Audubon, John James; Carson, Rachel; Dodo bird extinction; Migratory Bird Act; Northern spotted owl; Passenger pigeon extinction; Whooping cranes; Wildlife refuges.

National Environmental Policy Act

CATEGORIES: Treaties, laws, and court cases; ecology and ecosystems

THE LAW: U.S. federal law that established national policies and objectives for the protection and maintenance of the environment

DATE: Enacted on January 1, 1970

SIGNIFICANCE: The National Environmental Policy Act has raised public awareness of the concept of environmental impacts, highlighting the need for governments and citizens to be aware of the unintended consequences of federally supported actions that affect the human environment.

Sometimes referred to as the environmental Magna Carta of the United States, the law officially known as the National Environmental Policy Act of 1969 (NEPA), although it was not enacted until the first day of 1970, was the first U.S. legislation to require a comprehensive and coordinated national environmental policy that embraced public review of environmental impacts associated with the actions of federal agencies. Oversight of NEPA compliance was facilitated through the Council on Environmental Quality (CEQ), which was created through a provision of the act.

In response to widespread public interest in environmental quality during the late 1960's, the U.S. Congress held separate House and Senate committee hearings to identify the best method for legislating a national policy on environmental protection. Among the problems discussed was the tendency of mission-oriented federal agencies involved in development to overlook environmentally preferable alternatives in their decision-making processes. Draft versions of House and Senate environmental bills were later integrated and approved by Congress as NEPA. Signed into law by President Richard Nixon on January 1, 1970, NEPA recognized the profound impact of human activity on the interrelations of all components of the natural environment, particularly the influences of population growth, high-density urbanization, resource exploration, and new and expanding technological advances. Language used in the bill embraced many of the philosophies of the conservation movement of the early twentieth century and later environmentalism of the 1960's.

Although other U.S. environmental statutes provide robust protection for the environment, they focus only on specific categories of resources. In contrast, NEPA serves as an umbrella statute that outlines a set of procedures and embraces the importance of public participation in federal decision making when the quality of the environment is at stake. NEPA does not demand explicit results such as limits on pollution emissions or specific actions to protect endangered species, nor does it serve as a substitute for other federal planning activities or regulatory processes. Rather, the act requires that decisions be made on the basis of thoughtful analysis of the direct, indirect, and cumulative environmental impacts of proposed actions.

NEPA prescribes the completion of a series of steps for all actions involving federal participation that may affect the environment, emphasizing in spirit and intent the importance of public participation in safeguarding the environment. The first step in federal project review is completion of an environmental assessment (EA) with input from local governments, Native American tribal governments, the public, and other federal agencies. An EA documents influences on the environment associated with a proposed federal action, including the types and levels of significant impacts. NEPA further requires that, following completion of the EA, a second document be prepared for all actions going forward. A finding of no significant impact (FONSI) provides documentation in cases where it has been determined that the planned actions will have no significant effect on the quality of the human environment. Federal actions that may significantly alter the quality of the human environment, including possible degradation to threatened or endangered species or their habitats, must be evaluated in greater detail. An environmental impact statement (EIS) involves additional analysis of pertinent social, demographic, economic, and ecological information and consideration of alternative courses of action. NEPA requires that the EIS process be carried out using a framework involving public input through a variety of mechanisms, including individual or group responses to proposed alternatives.

Implementation and Criticisms

Initially, NEPA implementation was difficult within agencies struggling to establish guidelines. Within the act's first two years, federal

agencies had completed more than 3,600 EIS's but were also involved in nearly 150 associated lawsuits. Although the number of EIS's submitted each year fell thereafter, more than 26,000 were written from 1970 to 2000. These reports document actions ranging from the construction of highways to the development of facilities for holding toxic wastes. About 80 percent of EIS's have been produced by a small number of federal agencies, including the Forest Service, the Bureau of Land Management, the Department of Housing and Urban Development, the Federal Highway Administration, and the Army Corps of Engineers. There is no formal tracking process for EAs, but a much large number of such assessments have been prepared for other federal projects and activities.

Critics argue that the EIS process is not cost-effective in terms of human resources or dollars spent and that NEPA guidelines are often inconsistently applied. As an overarching policy, NEPA cannot prevent agencies

National Environmental Policy Act

The National Environmental Policy Act declared its major "purposes" as follows:

To declare a national policy which will encourage productive and enjoyable harmony between man and his environment; to promote efforts which will prevent or eliminate damage to the environment and biosphere and stimulate the health and welfare of man; to enrich the understanding of the ecological systems and natural resources important to the Nation; and to establish a Council on Environmental Quality.

Title I
Congressional Declaration of National Environmental Policy

The Congress, recognizing the profound impact of man's activity on the interrelations of all components of the natural environment, particularly the profound influences of population growth, high-density urbanization, industrial expansion, resource exploitation, and new and expanding technological advances and recognizing further the critical importance of restoring and maintaining environmental quality to the overall welfare and development of man, declares that it is the continuing policy of the Federal Government, in cooperation with State and local governments, and other concerned public and private organizations, to use all practicable means and measures, including financial and technical assistance, in a manner calculated to foster and promote the general welfare, to create and maintain conditions under which man and nature can exist in productive harmony, and fulfill the social, economic, and other requirements of present and future generations of Americans.

from implementing unwise actions or even from concluding that other values are more important than environmental considerations. Despite ambiguity in its language, however, the act has been credited with significantly modifying the actions of both government agencies and private industry by preventing hundreds of activities with potentially severe environmental effects.

NEPA has raised awareness of the concept of environmental impact, highlighting the need for governments and citizens to be aware of the unintended consequences of federally supported actions that affect the human environment. NEPA has also assisted in creating pathways for other federal statutes that consider environmental issues in decision making, such as the 1980 Comprehensive Environmental Response, Compensation, and Liability Act (CERCLA), also known as Superfund. Having withstood the regulatory reform commissions of several presidential administrations, NEPA stands out for its brevity and simplicity. Perhaps the best evidence of the act's success can be seen in the fact that it has been emulated by one-half of the state governments in the United States and by more than eighty national governments throughout the world.

Thomas A. Wikle

FURTHER READING

Buck, Susan J. "Environmentalism in the United States." In *Understanding Environmental Administration and Law.* 3d ed. Washington, D.C.: Island Press, 2006.

Caldwell, Lynton K. *Science and the National Environmental Policy Act: Redirecting Policy Through Procedural Reform.* Tuscaloosa: University of Alabama Press, 1982.

Clark, Ray, and Larry Canter, eds. *Environmental Policy and NEPA: Past, Present, and Future.* Boca Raton, Fla.: CRC Press, 1997.

King, Thomas F. "The Umbrella: The National Environmental Policy Act." In *Cultural Resource Laws and Practice: An Introductory Guide.* 2d ed. Walnut Creek, Calif.: AltaMira Press, 2004.

SEE ALSO: Clean Air Act and amendments; Clean Water Act and amendments; Council on Environmental Quality; Endangered Species Act; Environmental impact assessments and statements; Environmental Protection Agency; National Forest Management Act; Superfund.

National Forest Management Act

CATEGORIES: Treaties, laws, and court cases; forests and plants; resources and resource management

THE LAW: U.S. federal legislation providing for the protection of national forests in regard to logging, cutting rates, and reforestation

DATE: Enacted on October 22, 1976

SIGNIFICANCE: The National Forest Management Act represents an attempt to protect national forestlands from potential ecological damage by setting guidelines for tree maintenance, water issues, and species management. The act provides legal shelter for ancient forests and vulnerable animal groups.

The National Forest Management Act of 1976 (NFMA) was signed into law by President Gerald Ford. An amendment to the Forest and Rangeland Renewable Resources Planning Act of 1974, the legislation was passed largely as the result of controversial logging practices that had removed too many trees in given areas. The two locations that were most problematic were the Bitterroot National Forest in Montana and the Monongahela National Forest in West Virginia. In 1972, Senator Frank Church of Idaho issued a report that resulted in a decline in clear-cutting in national forests and defined specific land categories that could not sustain any foresting. In addition, the Izaak Walton League of America sued to stop clear-cutting in the Monongahela National Forest. The U.S. Court of Appeals for the Fourth Circuit agreed in 1975 that this type of foresting is problematic and banned clear-cutting in national forests.

Developed by foresters, representatives of the timber industry, members of Congress, and officials from the U.S. Department of Agriculture and its Forest Service, the NFMA was written to serve several major purposes, including protection of the national forests and management of these lands so that they can be used for various reasons and offer sustained yields. The act put strict limits on clear-cutting, defined marginal lands and endangered species, and regulated timber management in marginal lands; it also included rules for the evaluation of species diversity and for the determination of tree age in regard to cutting. In addition, the act expanded protections of soil, water, and air resources in national forests, which had been especially affected by clear-cutting. It prohibited any timber harvesting practices that would result in the permanent destruction of these resources in the

Ford on the National Forest Management Act

In his statement on signing the National Forest Management Act on October 22, 1976, President Gerald R. Ford addressed the need to balance many interests in the management of the nation's natural resources.

In America's first century, our forests and their vast resources seemed to our forefathers inexhaustible. By the late 19th century, however, the spirit of expansion and development had led to much abuse of our forest lands. Fires frequently raged out of control over millions of acres, devastating floods were increasing and our wildlife was being depleted.

With wisdom and timeliness, this Nation began to establish Federal forest reserves to protect our forest lands and to guarantee that future generations would enjoy their benefits. . . .

From its inception, the National Forest System was administered not only to protect forest lands but also to restore their productivity. After an early period of basic custodial protection, a philosophy evolved to manage the National Forests in such a way that they provided a variety of uses and benefits for present and future generations. . . .

While the National Forest Management Act of 1976 evolved from a timber management controversy, the act goes far beyond a simple remedy of the court's decision. Basically, the act expands and refines the forest resource assessment and planning requirements of the Forest and Rangeland Renewable Resources Planning Act of 1974—one of the first acts I signed upon taking office. This act reaffirms and further defines the concept of multiple-use, sustained-yield management and outlines policies and procedures for land management planning in the National Forest System. Emphasis throughout the act is on a balanced consideration of all resources in the land management process.

Of equal importance, this act guarantees the public full opportunity to participate in national forest land and resource planning. Finally, it recognizes the importance of scientific research and cooperation with State and local governments and private landowners in achieving wise use and management of the Nation's forest resources.

In my consideration of this legislation, a statement made in 1907 by Gifford Pinchot, the first Chief Forester of the Forest Service, was brought to my attention. Mr. Pinchot said:

There are many great interests in the National forests which sometimes conflict a little. They must all be fit into one another so that the machine runs smoothly as a whole. It is often necessary for one man to give way a little here, another a little there. But by giving way a little at the present, they both profit by it a great deal in the end.

---

form of blocked watercourses, water temperature modifications, accumulations of sediment, or damage to fish habitats.

The NFMA requires the U.S. secretary of agriculture to create plans for the resource management of each portion of the national forest system and to appraise the condition of all national forests every fifteen years. The main criticism of the NFMA has been that it lacks sufficient specific detail, and thus interpretations vary regarding the approaches to management planning that are required.

Theresa L. Stowell

FURTHER READING

Blackman, Ted. "Forest Service Comes Down with a Case of the 'Warm Fuzzies.'" *Wood Technology* 127, no. 1 (2000): 20.

Ellis, Donald J., and Jo Ellen Force. *National Forest Planning and the National Forest Management Act of 1976: An Annotated Bibliography, 1976-1986.* Bethesda, Md.: Society of American Foresters, 1988.

Hays, Samuel P. *Wars in the Woods: The Rise of Ecological Forestry in America.* Pittsburgh: University of Pittsburgh Press, 2007.

Lawrence, Nathaniel S. W. "A Forest of Objections: The Effort to Drop NEPA Review for National Forest Management Act Plans." *Environmental Law Reporter* 39, no. 7 (2009): 10651-10655.

SEE ALSO: Balance of nature; Deforestation; Extinctions and species loss; Forest and range policy; Forest management; Forest Service, U.S.

National forests

CATEGORIES: Forests and plants; resources and resource management

DEFINITION: Forestlands owned and managed by a national government

SIGNIFICANCE: The U.S. national forest system was one of the first and most successful resource management and conservation programs initiated by the federal government. Similar programs have been developed in most of the forested nations of the world to manage the forests' economic and ecological functions.

National forests are distinctly different from national parks. National parks are aesthetically pleasing, culturally or historically significant, or ecologically important tracts of land protected in perpetuity for the public benefit. National forests are publicly owned resource management areas operated to maximize the cash value of timber and forest products, benefit the domestic economy, and ensure perpetual productivity of the land.

In the United States, the nineteenth century saw increased efforts to preserve timber resources at the same time that unrestrained national growth encouraged increased timber harvesting. In 1827 U.S. president John Quincy Adams expressed interest in developing a plan for sustainable forestry to ensure the availability of masts for ships. The American Association for the Advancement of Science discussed the need for sustained-yield forestry during the 1860's and in 1873 asked Congress to preserve and manage the nation's forests.

The first congressional actions to preserve forests focused on areas of significant natural beauty, such as Yellowstone National Park and Yosemite Valley. The Division of Forestry was established within the U.S. Department of Agriculture in 1876 to promote harvestable timber development. Protections for harvestable stands of timber were first approved in the Creative Act of 1891, which authorized the president of the United States to withdraw public lands previously open to preemption and homesteading rights and to establish forest reserves.

President Benjamin Harrison established America's first national forest, now known as Shoshone National Forest, in Wyoming in 1891. Confusion arose over the process for including land within the na-

tional forest system, however, and Congress suspended the law while it debated the purpose of the reserves, eventually authorizing selective cutting and marketing of timber from the existing reserves. Controversy concerning the vesting of the forest resources in the Department of the Interior or the Department of Agriculture ended in 1905 when the forest reserve system was established. President Theodore Roosevelt vested responsibility for national forest resources in the Bureau of Forestry of the Department of Agriculture under Gifford Pinchot, chief of the Division of Forestry. The reserve system became the national forest system in 1907.

A RESOURCE WITH MANY USES

From its inception, the U.S. national forest system was intended to serve multiple purposes. The purposes of the original legislation were to improve water flows and furnish a continuous supply of timber. This evolved into the triple purposes of resource protection (especially fire protection), "wise use" of timber resources, and multiple use of system lands. The forest service lands are protected, harvested, and open to multiple other public uses. The Multiple Use-Sustained Yield Act of 1960 defined "multiple use" as a combination of outdoor recreation, fish and wildlife management, and timber production intended to meet the needs of the American people but not necessarily giving the maximum dollar value return.

The National Forest Management Act of 1976 increased citizen involvement in decisions concerning timber harvesting, resource conservation, and multiple uses. The 1976 act also limited the technique of timber harvesting by clear-cutting, a practice condemned by most environmentalists. In addition, the act limited logging on fragile lands and encouraged actions to maintain the diversity of plants and animals and to conserve plants, animals, soils, and watersheds in the forests. However, the act also emphasized the importance of multiple uses such as mining, oil and gas exploration, grazing, farming, hunting, recreation, and logging. The conflicts among conservation, harvesting, and multiple uses have continued to plague forest policy decision makers. Emerging forest policies in the early twenty-first century include sustained-yield forestry; substitution of alternative harvestable crops for harvestable timber, including crops such as nuts, fruits, gums, extracts, syrups, tars, and oils; and ecosystem maintenance.

The federal government owns and manages approximately 266 million hectares (657 million acres) of public land, about 29 percent of the land area of the United States. Almost 30 percent of the federal land area is part of the national forest system. In 2010, the national forest system comprised roughly 78 million hectares (193 million acres) of land in 155 national forests and 20 national grasslands located in forty-four states, Puerto Rico, and the Virgin Islands. Two-thirds of U.S. national forests are located in the West, the Southeast, and Alaska. Twenty-three eastern states share about 50 forests. Roughly one out of every six hectares in the national forest system is part of the National Wilderness Preservation System (NWPS), and forest system lands represent about 33 percent of the NWPS.

OTHER NATIONAL FOREST PROGRAMS

Around the world, 143 countries have forest policy statements, and 156 countries have specific forest laws. Almost 75 percent of the world's forests are covered by national forest programs, and about 80 percent are under public ownership.

According to 2010 statistics from the United Nations Food and Agriculture Organization, the world's total forest area is just over 4 billion hectares (9.9 billion acres), or approximately 31 percent of the globe's total land area. Forests in the Russian Federation, Brazil, Canada, the United States, and China account for more than half of the world's forested area. During the 1990's global forests declined at a rate of about 16 million hectares (39.5 million acres) per

Recreational Uses of U.S. National Forests, 2007

Activity	Thousands of Visitors
Viewing natural features and visiting historic sites	103,700
Nature center activities, nature study, and relaxing	91,600
Backpacking, hiking/walking, horseback riding, and bicycling	84,000
Viewing wildlife	71,800
Camping, resort use, picnicking	45,300
Driving for pleasure	42,400
Fishing and hunting	40,800
Downhill skiiing and cross-country skiing	33,300
Snowmobiling, motorized water activity, motorized trail activity, and other motorized activity	20,000

Source: National Visitor Use Monitoring Results, National Summary Report. United States Department of Agriculture, Forest Service, 2009.

year. The decline slowed to 13 million hectares (32 million acres) per year between 2000 and 2010. In addition, during this later period natural forest expansion in combination with tree-planting programs added 7 million hectares (17.3 million acres) of new forest annually.

Between 1990 and 2010, the world forest area designated primarily for wood and nonwood forest products decreased, while the area designated for multiple uses increased. By 2010 about 30 percent of the world's forests were designated for productive purposes, 24 percent for multiple uses, and 8 percent for protective functions such as soil and water conservation, avalanche control, sand dune stabilization, and desertification control.

The global area of forested lands in legally protected areas such as national parks and wilderness areas increased between 1990 and 2010 by more than 94 million hectares (232 million acres). In 2010 legally protected areas accounted for about 13 percent of the world's total forest area. Environmental activism, ecotourism, and debt-for-nature swaps encourage the continuing development of national forest systems and other conservation measures in most nations of the world.

Gordon Neal Diem
Updated by Karen N. Kähler

FURTHER READING

Hayes, Tanya, and Elinor Ostrom. "Conserving the World's Forests: Are Protected Areas the Only Way?" *Indiana Law Review* 38, no. 3 (2005): 595-618.

Hays, Samuel P. *The American People and the National Forests: The First Century of the U.S. Forest Service.* Pittsburgh: University of Pittsburgh Press, 2009.

_____. *Wars in the Woods: The Rise of Ecological Forestry in America.* Pittsburgh: University of Pittsburgh Press, 2007.

Hirt, Paul W. *A Conspiracy of Optimism: Management of the National Forests Since World War Two.* Lincoln: University of Nebraska Press, 1994.

United Nations Food and Agriculture Organization. *Global Forest Resources Assessment, 2010.* Rome: Author, 2010.

SEE ALSO: Conservation Reserve Program; Forest and range policy; Forest management; Forest Service, U.S.; Logging and clear-cutting; National Forest Management Act; Renewable resources; Sustainable forestry; Wilderness Act.

National Oceanic and Atmospheric Administration

CATEGORIES: Organizations and agencies; atmosphere and air pollution

IDENTIFICATION: U.S. federal agency that conducts scientific investigations concerning the conditions of the world's oceans and atmosphere

DATE: Established on October 3, 1970

SIGNIFICANCE: The research and monitoring conducted by the scientists of the National Oceanic and Atmospheric Administration and its many divisions have contributed to a better understanding of the earth's environment.

The National Oceanic and Atmospheric Administration (NOAA), which is part of the U.S. Department of Commerce, is the governmental agency charged with monitoring and conducting research concerning the oceans and the atmosphere. Among the general public perhaps the most widely known division within the NOAA is the National Weather Service, which provides storm warnings, weather reports, and observations of the weather to the public on a daily basis. NOAA's functions, however, include much more than weather-related matters; a variety of divisions within the agency provide advice and scientific information to state and local agencies to aid them in environmental management and policy decisions. In addition to the Weather Service, these are the National Marine Fisheries Service, the National Ocean Service, the Office of Oceanic and Atmospheric Research, the Office of Program Planning and Integration, and the National Environmental Satellite, Data, and Information Service.

The National Ocean Service is dedicated to investigating fresh- and saltwater resources and maintaining their economic and ecological viability. The Office of Oceanic and Atmospheric Research provides support for NOAA's research activities, some of which include investigations into severe storms, El Niño events, and deep-sea thermal vents. The Office of Program Planning and Integration assists NOAA personnel with management and performance issues. The National Environmental Satellite, Data, and Information Service manages data on the earth's ecosystems gathered by the Polar Operational Environmental Satellite and the Geostationary Operational Environmental Satellite, both of which gather a wide range of environmental information.

When NOAA was formed in 1970, it combined three of the oldest agencies in the federal government. The oldest of these, the U.S. Coast and Geodetic Survey, dated back to 1807; established during the presidency of Thomas Jefferson, it was the first agency dedicated to scientific geographic investigation of the coasts and surrounding waters of the United States. The National Weather Service dated back to 1870 and the days of the Army Signal Corps. Ultimately this agency became the National Weather Bureau and then, under NOAA, the National Weather Service. The Commission of Fish and Fisheries, which became the National Marine Fisheries Service, had its beginnings in 1871; it is the nation's oldest agency dedicated to the conservation of food fish.

Under NOAA significant research has been accomplished that has contributed to a better understanding of the earth's environment. For example, NOAA's establishment of systems for monitoring sea temperatures in the South Pacific helped establish reliable El Niño climate predictions. The National Weather Service's programs for tracking severe storms have saved untold numbers of lives. NOAA's coastal monitoring of ocean pollution and its fisheries programs have contributed to environmental policies that support economic growth and sustainability. NOAA's use of environmental sensing satellite platforms as well as other technologies, such as the Global Positioning System for navigation and the development of modern electronic maritime charts, has also contributed to safer conditions for seafarers.

M. Marian Mustoe

FURTHER READING

Artiola, Janick F., Ian L. Pepper, and Mark Brusseau, eds. *Environmental Monitoring and Characterization.* Burlington, Mass.: Elsevier Academic, 2004.

National Research Council. *Environmental Data Management at NOAA: Archiving, Stewardship, and Access.* Washington, D.C.: National Academies Press, 2007.

SEE ALSO: Climate change and oceans; Climate models; Climatology; Cloud seeding; Coastal Zone Management Act; Intergovernmental Panel on Climate Change; Ocean currents; United Nations Framework Convention on Climate Change; U.S. Climate Action Partnership.

National Park Service, U.S.

CATEGORIES: Organizations and agencies; preservation and wilderness issues
IDENTIFICATION: Agency within the U.S. Department of the Interior responsible for managing national parks, monuments, historic sites, battlefields, recreational areas, and heritage areas
DATE: Established on August 25, 1916
SIGNIFICANCE: The U.S. National Park Service is a world leader in the preservation and protection of natural and cultural resources and has served as a model for the management of national park systems worldwide.

The U.S. Congress established the National Park Service in 1916 to protect natural and historic resources for the enjoyment of future generations, making sites that would otherwise have been lost to development or neglect accessible to the general public. Units within the national park system range from parks covering many thousands of hectares, such as Grand Canyon National Park, to sites consisting of only one building, such as Thaddeus Kosciuszko National Memorial in Philadelphia. Although the designation "national park" has a specific legal meaning, for convenience employees of the National Park Service often refer to all management units within the system as "parks" regardless of size.

A number of federal preserves were designated prior to creation of the National Park Service, such as Hot Springs in Arkansas and Yosemite in California, but management of those sites was split among several different cabinet departments, including the War Department, which detailed troops to patrol Yellowstone and Yosemite. Advocates for wilderness preservation and a uniform national parks policy, such as John Muir, argued that oversight of national preserves should be the responsibility of one agency devoted solely to that task. In 1906, under President Theodore Roosevelt, the U.S. Forest Service had been established within the Department of Agriculture to manage the nation's forest reserves for wise use; promoters of a park service believed a similar agency should be established to manage the nation's parks and monuments.

At the time the Park Service was established, the system consisted primarily of sites noted for their spectacular natural beauty and unique features, such as Yellowstone and Yosemite. However, the 1916 legislation stated the mission of the National Park Service would be "to conserve the scenery and the natural and historic objects and the wild life therein and to provide for the enjoyment of the same in such manner and by such means as will leave them unimpaired for the enjoyment of future generations." This allowed the national park system to include sites important for their historic associations, such as Valley Forge in Pennsylvania, or their archaeological value, such as Mesa Verde in Colorado.

National Park System Statistics, 2000-2007

	2000	2005	2007
Expenditures	$1.833 billion	$2.451 billion	$2.412 billion
Revenue from operations	$234 million	$286 million	$346 million
Recreational visitors	285,900,000	273,500,000	275,600,000
Overnight stays	15,400,000	13,500,000	13,800,000
Park system lands (acres)	78,153,000	79,048,000	78,845,000

Source: U.S. Census Bureau, Statistical Abstract of the United States, 2009, 2009.
Note: Includes visitor data for national parks, monuments, recreation areas, seashores, and miscellaneous other areas.

RESPONSIBILITIES

In the twenty-first century, the U.S. National Park Service, globally recognized as a leader in the preservation of both natural and historic resources, is responsible for managing sites and cultural artifacts in all fifty states, the District of Columbia, Puerto Rico, the U.S. Virgin Islands, Guam, the Northern Marianas, and American Samoa. Sites managed by the Park Service are categorized as national parks, national preserves, national scenic rivers, national recreational areas, national monuments, national historical parks, national historic sites, and national parkways. In addition, the Park Service manages some national trails, administers the National Register of Historic Places and National Historic Landmark programs, and plays an advisory role in the management of National Heritage Areas.

For administrative purposes, the National Park Service is divided into seven regions: Northeast, National Capital, Southeast, Midwest, Intermountain, Pacific West, and Alaska. Each regional office is staffed by experts in cultural and natural resource management, such as wildlife biologists, to provide specialized support to the individual parks within the region. The director of the National Park Service is appointed by the president, but most other employees are members of the civil service.

DESIGNATION OF SITES

National monuments can be created by presidential executive order, but congressional action is required to create sites in all the other categories under the Park Service. The executive branch's power to protect unique natural and cultural resources dates back to President Theodore Roosevelt and the 1906 Antiquities Act, legislation that was passed to allow the president to provide immediate protection to threatened resources. The fact that a site has been designated as a unit of the national park system, whether through congressional action or by executive order, does not guarantee permanent protection of that site. Congress can remove any unit in the system by passing legislation to do so. This rarely happens; in general, once a site becomes part of the national park system, it is more likely that the site will be enhanced and expanded rather than the reverse.

The history of the Park Service and the expansion of the national park system are reflections of U.S. history overall and of changes in thinking among Americans regarding what is worthy of preservation. Since establishment of the National Park Service in 1916, numerous sites throughout the United States have been proposed as additions to the system. When a site is suggested, Congress first authorizes a study to determine whether the proposed unit meets the criteria for its intended designation—for example, whether a historic site's significance is truly of national importance. Creation of a national park can be an intensely political process, with vigorous lobbying for and against the proposal. In addition, the vision of what a national park should be has changed over time and continues to evolve, so that sites that may have been rejected in the past may eventually come to be seen as appropriate.

In addition, partnerships between the Park Service and nonfederal agencies and organizations are becoming more common, as is seen in parks such as the Lewis and Clark National and State Historical Parks in Washington and Oregon and Keweenaw National Historical Park in Michigan. Both parks include some sites owned and managed directly by the National

Park Service and some sites owned and managed by local government agencies or nonprofit organizations.

Nancy Farm Männikkö

FURTHER READING

Albright, Horace M., and Marian Albright Schenk. *Creating the National Park Service: The Missing Years.* Norman: University of Oklahoma Press, 1999.

Duncan, Dayton, and Ken Burns. *The National Parks: America's Best Idea—An Illustrated History.* New York: Alfred A. Knopf, 2009.

Farabee, Charles R. *National Park Ranger: An American Icon.* Lanham, Md.: Roberts Rinehart, 2003.

Heacox, Kim. *An American Idea: The Making of the National Parks.* Washington, D.C.: National Geographic Society, 2001.

SEE ALSO: Department of the Interior, U.S.; Everglades; Grand Canyon; Hetch Hetchy Dam; Kings Canyon and Sequoia national parks; Mather, Stephen T.; National parks; Wilderness areas; Yellowstone National Park; Yosemite Valley.

National parks

CATEGORY: Preservation and wilderness issues

DEFINITION: Areas of scenic, historic, or other value that are set aside by federal governments for the preservation of animals and wildlife and for human recreation

SIGNIFICANCE: National park systems throughout the world provide protection for wildlife and plant life, preserve spaces for outdoor recreation, and educate people about the importance of protecting ecosystems.

National parks are places where preservation for future generations must be balanced with present-day enjoyment. In the United States this balance has proved difficult to achieve, but nevertheless more than 390 park units—including national parks, historic sites, historical parks, memorials, memorial parks, battlefields, battlefield parks, battlefield sites, lakeshores, seashores, monuments, parkways, scenic trails, scenic rivers, scenic riverways, rivers, capital parks, and recreation areas—have been established across the United States since the U.S. Congress made

Yellowstone the world's first national park in 1872 (the politicians were willing to protect and preserve the geologic wonders of Yellowstone primarily because they were convinced that the lands were economically useless). Each of these park units was established to preserve and protect geologic wonders, spectacular scenery, wildlife, or a particular aspect of American history or culture.

Other countries have also established national parks, frequently using the United States as a model. Starting in the final decades of the twentieth century, the United Nations worked with countries to protect these areas. In 1972, on the occasion of the one hundredth anniversary of the founding of Yellowstone National Park, the United Nations Educational, Scientific, and Cultural Organization (UNESCO) formed the World Heritage Committee. By 2009, 186 countries had ratified the World Heritage Convention. The committee, at its first meeting in 1977, formulated the World Heritage List, which names cultural and natural sites considered to be of "outstanding universal value"; approximately thirty new sites have been added to the list each year. By the end of the first decade of the twenty-first century, the World Heritage List named almost nine hundred cultural and natural sites. The list contains more cultural than natural sites, and not all the sites are parks, but many parks have received monetary support and advice from the United Nations to help ameliorate environmental problems, which abound in the United States and the rest of the world; the causes of these problems include the difficulties posed by the poaching of wildlife, exploitation of mineral deposits, and finding a balance between use and preservation.

NATIONAL COORDINATING OFFICES

Initially, national park operations in the United States were complicated because no single central federal office existed to coordinate activities. After much controversy, Congress passed the National Parks Act, or the Organic Act, of 1916. The act established a central authority called the National Park Service and stated its responsibilities, which include conserving and providing for the enjoyment of the scenery, natural and historic objects, and wildlife in the parks while leaving them unimpaired for the enjoyment of future generations.

This was not the first national office of national parks. In 1911 the Canadian parliament had passed the Dominion Forest Reserves and Parks Act, which

"Fixing" the National Parks

National Park Service director Conrad Wirth presented his plan to "fix" the park system. Wirth's proposal, "Mission 66: Special Presentation to President Eisenhower and the Cabinet" (January 27, 1956), reads, in part:

As you know, these are the areas where the Nation preserves its irreplaceable treasures in lands, scenery—and its historic sites—to be used for the benefit and enjoyment of the people, and passed on unimpaired to future generations. The areas of the National Park System are among the most important vacation lands of the American people. Today, people flock to the parks in such numbers that it is increasingly difficult for them to get the benefits which parks ought to provide, or for us to preserve these benefits for Americans of tomorrow.

This is why the Department of the Interior and the National Park Service have surveyed the parks and their problems, and propose to embark upon Mission 66—a program designed to place the national parks in condition to serve America and Americans, today and in the future.

The problem of today is simply that the parks are being loved to death. They are neither equipped nor staffed to protect their irreplaceable resources, nor to take care of their increasing millions of visitors.

Where else do so many Americans under the most pleasant circumstances come face to face with their Government? Where else but on historic ground can they better renew the idealism that prompted the patriots to their deeds of valor? Where else but in the great out-of-doors as God made it can we better recapture the spirit and something of the qualities of the pioneers? Pride in their Government, love of the land, and faith in the American Tradition—these are the real products of our national parks.

provided for the administration of forest reserves and dominion parks, and allowed dominion parks to be established from forest reserves. Thus the Dominion Parks Service, created as a new branch in the Canadian Department of the Interior, became the first distinct bureau of national parks in the world. John Bernard Harkin served as commissioner from the service's inception in 1911 to 1936. In addition to working to separate the administration of the parks from that of the forests, Harkin emphasized resource preservation.

The numbers of national parks, as well as the numbers of visitors to the parks, continued to grow in both

the United States and Canada. After World War II, as the automobile became ubiquitous, visits to the national parks increased rapidly. The facilities in place at the parks were old, however, and staffing was minimal. In 1956 the director of the U.S. National Park Service, Conrad L. Wirth, decided that instead of asking Congress for annual appropriations, he would package the national parks' needs into a program called Mission 66. This one-billion-dollar, ten-year restoration program was designed to end in 1966, the fiftieth anniversary of the National Park Service. Canada was also influenced by Mission 66, and the Parks Policy of 1964, under the leadership of John I. Nicol, director of the Canadian National and Historic Parks Branch, emphasized the importance of protecting natural resources in the parks.

PROBLEMS OF OVERUSE

The numbers of visitors to national parks in both countries increased steadily from the 1960's onward. With larger crowds came overuse, which created its own serious problems, including congestion, pollution, and—in some cases—destruction. Some popular parks banned automobiles from their roads, replacing them with shuttle buses. The park services of both Canada and the United States are continually faced with the problem of encouraging use while protecting valuable national resources and leaving them unimpaired for future generations.

In the United States, most of the service businesses in the national parks, such as restaurants, hotels, and souvenir shops, have traditionally been operated by private concessionaires. The operators usually sign long-term contracts under which small percentages of their profits are returned to the federal government. In the late twentieth century, as public funding for the parks declined and the numbers of visitors continued to increase, the National Park Service intensified its ties with private agencies to provide public services at national parks and other sites administered by the service. Some proposed projects became controversial and were never implemented, such as a giant theater that was to be built at Gettysburg Battlefield.

The National Park Service recognizes the importance of cooperation with the communities immediately adjacent to the parks, and partnerships and resulting plans have been developed around several national parks. Several of Canada's national parks, such as Banff and Jasper, actually contain towns, and the animals that live in the parks roam freely among automobiles, homes, and stores.

MANAGEMENT OF PLANTS AND WILDLIFE

Both the Canadian and the U.S. national park services must balance use with preservation of wildlife and plants. In general, the aim is to allow native species to flourish within the parks while protecting the parks from exotic, or nonnative, species. However, conflicts frequently develop regarding the remedial approaches that should be taken when exotic species are found in the parks. Some parks, such as the Shenandoah and Great Smoky Mountains national parks, have been invaded by exotic insects that have destroyed whole forested areas and the component parts of the ecosystem.

One of the most controversial decisions in recent U.S. National Park Service history was the decision to reintroduce the gray wolf into the Yellowstone ecosystem in 1995. Wolves in the area had been exterminated by rangers in 1915 because they were seen as a menace to other native animals such as elk, deer, and mountain sheep. After many years of debate, the National Park Service was given permission and funds (approximately $6.7 million) to reintroduce wolves to the Yellowstone ecosystem, which includes the park and neighboring parts of Montana and Idaho. Many park visitors, upon being surveyed, had indicated that they wanted the wolves brought back. Local livestock owners, however, were concerned about the safety of their animals. To balance the competing needs—to serve the desires of park visitors, to restore the natural ecology of the park, to address the economic concerns of the livestock owners, and to support efforts to improve the wolf population—the Park Service developed plans to protect neighboring livestock before it

Mammoth Hot Springs in Yellowstone National Park, the world's first national park. (Courtesy, PDPhoto.org)

undertook the careful reintroduction of wolves. In January, 1995, the first set of gray wolves from Hinton, Alberta, Canada, were brought to Yellowstone. The introduction was deemed successful, and the ecology of the park returned to a more natural state as the wolf population increased and the wolves became significant predators of elk, moose, and deer.

One problem area related to wildlife that affects almost all national parks involves the dangers that arise from human-animal interaction. It is illegal for visitors to feed any wildlife in U.S. and Canadian national parks, but some visitors break the law. Human food is not part of the animals' natural diet, and some foods can cause digestive problems in animals and even endanger their lives; in addition, animals that come to associate humans with a ready food source can be dangerous to humans. When wild animals enter campgrounds and motor vehicles in search of something to eat, they occasionally harm humans. In extreme cases, to prevent animals from hurting people, park rangers may have to transport, or even kill, animals that have become too aggressive in searching for humans' food.

ECONOMIC ACTIVITIES AND ENVIRONMENTAL PROBLEMS

The economic activities taking place outside park boundaries can cause environmental problems inside the parks. In some cases, neighboring mining or logging operations have contaminated waterways or hastened soil erosion. Sometimes compromises are reached that can reduce the impacts of such problems. For example, a planned gold mining operation outside Yellowstone National Park was moved when the U.S. government agreed to exchange other federal lands for the original intended site.

In other cases, distant power plants, factories, or even urban areas as a whole contribute to air pollution within park boundaries. In many parks in the United States, the combination of human-caused pollution and natural geology and meteorology has led to diminished views, more rapid weathering of natural wonders, and harm to wildlife. In South Africa, the industrial district of Saldanha Bay has contributed to air and water pollution in the West Coast National Park.

National parks in parts of Africa and Central America face serious problems of wildlife poaching because of inadequate park administration and law enforcement. In areas such as Benin, disease-carrying livestock sometimes invade parks, resulting in wildlife deaths. Parts of national parks in Côte d'Ivoire and Senegal have been cultivated by farmers, but the governments of both countries have developed successful resettlement programs, thus lessening the impact on the parks.

Individual countries often look for help from the World Heritage Committee in establishing and maintaining national parks. The committee keeps a list of sites known as World Heritage in Danger, with the intent of bringing to worldwide attention the "conditions which threaten the very characteristics for which a property was inscribed on the World Heritage List." In 2010, thirty-one sites were on the World Heritage in Danger list. Sometimes, successful intervention results in a site being removed from the list, as occurred in 1998 with Plitvice Lakes National Park in Croatia. The park had been overdeveloped and overused, but after it was added to the World Heritage in Danger list in 1992, its underground water supply was protected and a new road was built to decrease truck traffic through the park.

Margaret F. Boorstein

FURTHER READING

Allin, Craig W., ed. *International Handbook of National Parks and Nature Reserves.* New York: Greenwood Press, 1990.

Grusin, Richard. *Culture, Technology, and the Creation of America's National Parks.* New York: Cambridge University Press, 2004.

Heacox, Kim. *An American Idea: The Making of the National Parks.* Washington, D.C.: National Geographic Society, 2001.

Ridenour, James. *The National Parks Compromised: Pork Barrel Politics and America's Treasures.* Merrillville, Ind.: ICS Books, 1994.

Runte, Alfred. *National Parks: The American Experience.* 4th ed. Lanham, Md.: Taylor Trade, 2010.

Sellars, Richard. *Preserving Nature in the National Parks: A History.* New ed. New Haven, Conn.: Yale University Press, 2009.

SEE ALSO: Grand Canyon; Kings Canyon and Sequoia national parks; National Park Service, U.S.; Nature preservation policy; Preservation; Serengeti National Park; World Heritage Convention; Yellowstone National Park; Yosemite Valley.

National Trails System Act

CATEGORIES: Treaties, laws, and court cases; preservation and wilderness issues

THE LAW: U.S. federal law authorizing a national network of scenic, historic, and recreational trails protected from intrusive development

DATE: Enacted on October 2, 1968

SIGNIFICANCE: Hundreds of thousands of people use the system of protected trails created by the National Trails System Act every year to gain access to scenic, historic, and culturally important sites across the United States.

When the Appalachian Trail, a hiking trail running from Georgia to Maine, was developed during the 1920's and 1930's, it was intended to provide a wilderness experience for hikers from the cities. By the 1960's, however, that wilderness was threatened by commercial and residential development, with its accompanying roads and utilities. Supporters of the Appalachian Trail lobbied Congress to protect the trail, and in 1965 President Lyndon B. Johnson issued his "Special Message to the Congress on Conservation and Restoration of Natural Beauty," in which he called for a nationwide system of trails that would be largely maintained by volunteers. The next year, Secretary of the Interior Stewart Udall directed the Bureau of Outdoor Recreation to study the issue, and the result was an influential report titled *Trails for America*, which outlined plans for a system of trails throughout the country.

The National Trails System Act, as enacted in 1968, defined and created three different types of trails as described in *Trails for America*: national scenic trails, national recreation trails, and connecting and side trails. National scenic trails are trails and protected corridors that are more than 161 kilometers (100 miles) long, linking points of extraordinary scenery. Motorized vehicles are forbidden on these trails, to protect the wilderness experience, and only rudimentary development, such as campsites or shelters, is permitted. Only Congress can create national scenic trails, and the responsibility for maintaining them falls to the Department of the Interior. Congress has created eight such trails; these include the Appalachian National Scenic Trail and the Pacific Crest National Scenic Trail, both established in 1968 with the signing of the act. No new national scenic trails have been designated since 1983.

In 1978, Congress amended the National Trails System Act to add national historic trails, which mark historically important routes of travel. The Iditarod Trail, the Lewis and Clark Trail, the Mormon Pioneer Trail, and the Oregon Trail were designated as the first in this group; by 2009 the list had grown to include nineteen trails.

National recreation trails are generally smaller, more locally controlled, and more open to a variety of uses, including the riding of motorized vehicles. Designation of these trails, which number more than eight hundred, is not subject to congressional approval, and the trails do not receive federal funding. Side and connecting trails are trails that connect to other trails in the National Trails System; although such trails are provided for in the original act, only two have been established.

National Trails System Act

The National Trails System Act of 1968 was landmark legislation that provided expanded outdoor recreation opportunities at the same time that it ensured protection of the nation's historic trails system. The act, in part, states:

In order to provide for the ever-increasing outdoor recreation needs of an expanding population and in order to promote the preservation of, public access to, travel within, and enjoyment and appreciation of the open-air, outdoor areas and historic resources of the Nation, trails should be established (i) primarily, near the urban areas of the Nation, and (ii) secondarily, within scenic areas and along historic travel routes of the Nation which are often more remotely located.

The purpose of this Act is to provide the means for attaining these objectives by instituting a national system of recreation, scenic and historic trails; by designating the Appalachian Trail and the Pacific Crest Trail as the initial components of that system; and by prescribing the methods by which, and standards according to which, additional components may be added to the system.

The Congress recognizes the valuable contributions that volunteers and private, nonprofit trail groups have made to the development and maintenance of the Nation's trails. In recognition of these contributions, it is further the purpose of this Act to encourage and assist volunteer citizen involvement in the planning, development, maintenance, and management, where appropriate, of trails.

The National Trails System has proved to be a great success, attracting thousands of visitors each year to places of scenic and historic importance. Many of the trails are still maintained by volunteers, as President Johnson originally envisioned. Each year, on the first Saturday in June, thousands of organizations and businesses observe National Trails Day with hikes, horseback rides, mountain bike rides, trail-grooming events, and educational workshops.

Cynthia A. Bily

FURTHER READING

Baldwin, Pamela. *Federal Land Management Agencies.* Hauppauge, N.Y.: Nova, 2005.
Dilsaver, Lary M., ed. *America's National Park System: The Critical Documents.* Lanham, Md.: Rowman & Littlefield, 1997.
Johnson, Sandra L. "The National Trails System: An Overview." In *National Forests: Current Issues and Perspectives,* edited by Ross W. Gorte. Hauppauge, N.Y.: Nova, 2003.

SEE ALSO: Back-to-the-land movement; Environmental law, U.S.; John Muir Trail; Land-use policy; Nature preservation policy; Wilderness areas.

National Wildlife Refuge System Administration Act

CATEGORIES: Treaties, laws, and court cases; animals and endangered species; ecology and ecosystems

THE LAW: U.S. federal legislation coordinating the administration of hundreds of national wildlife refuges

DATE: Enacted on October 15, 1966

SIGNIFICANCE: After nearly seventy years of setting aside vast areas of land for wildlife preservation, the U.S. government acted to provide a system for managing those refuges by placing them under the administration of the Department of the Interior. In addition, the legislation enumerated specific guidelines for the use and development of those lands that would be in keeping with the larger mission of protecting their rich biodiversity.

By 1966 more than six decades had passed since President Theodore Roosevelt placed Florida's remote Pelican Island under federal protection to provide a refuge from excessive trapping for brown pelicans, herons, and egrets that nested on the island. During that period, the U.S. government had come to direct a network of more than 28.3 million hectares (70 million acres) of refuges protecting more than 500 species of birds, 150 species of mammals, 200 species of reptiles, and 150 kinds of fish, a network far larger than the higher-profile national park system. Under the protocol articulated by Roosevelt, these refuges were conceived not merely as vehicles for protecting threatened or endangered species but also as sanctuaries where the diversity of a wide range of biomes—prairies, wetlands, marshes, swamps, tidal basins, and taiga and tundra forests—could be preserved for future generations.

Given the loosely controlled nature of this refuge system (over the years, presidents and the Congress had designated additional lands as protected) and its far-ranging network of refuges (more than five hundred) under the direction of hundreds of bureaucrats and scientists, Congress decided to act in 1966 to define a clear mission for the system by passing legislation that would place the administration of the refuges under the Department of the Interior through the U.S. Fish and Wildlife Service. The legislation further specified a mission for the refuges: to maintain the integrity of each biome by prioritizing the conservation of fish, birds, animals, and plants from a variety of threats from both nature (such as disease, insect infestations, and food and water source depletion) and human activities (such as fires, pollution, poaching, increasing demands for recreational sites, and urban development).

In addition, the act provided that any decision concerning the use and development of these lands must be approved by the secretary of the interior to ensure that such development does not violate the mission of the refuge system. The act also provided guidelines for specific recreational uses of the public lands: controlled hunting and fishing as a way to manage game populations, photography, educational uses for students of all levels, and environmental education and scientific research. Finally, the act defined a protocol for the acquisition of new lands, either through donation or through government funding.

The National Wildlife Refuge System Administration Act did not satisfy everyone; under pressure from vested interests, Congress did not include farming, grazing, or oil and gas drilling among the activities restricted in wildlife refuges, for example (these activities were included in the 1997 National Wildlife Refuge System Improvement Act). The act accomplished a great deal, however, in creating the world's largest system of public lands designated as sanctuaries for plant life and wildlife.

Joseph Dewey

FURTHER READING

Bean, Michael J., and Melanie J. Rowland. *The Evolution of National Wildlife Law.* Westport, Conn.: Greenwood Press, 1997.

Brinkley, Douglas. *The Wilderness Warrior: Theodore Roosevelt and the Crusade for America.* New York: Harper, 2009.

Butcher, Russell D. *America's National Wildlife Refuges: A Complete Guide.* 2d ed. Lanham, Md.: Taylor Trade, 2008.

Clark, Jeanne. *America's Wildlife Refuges: Lands of Promise.* Portland, Oreg.: Graphic Arts Center Publishing, 2003.

Fischman, Robert L. *The National Wildlife Refuges: Coordinating a Conservation System Through Law.* Washington, D.C.: Island Press, 2003.

SEE ALSO: Back-to-the-land movement; Biodiversity; Biomes; Department of the Interior, U.S.; Environmental law, U.S.; Fish and Wildlife Service, U.S.; Great Swamp National Wildlife Refuge; Nature reserves; North America; Poaching; Roosevelt, Theodore; Wildlife management; Wildlife refuges.

Natural burial

CATEGORY: Philosophy and ethics

DEFINITION: Interment of unembalmed human remains using only materials that, along with the body, will decay naturally

SIGNIFICANCE: Advocates of the practice of natural burial assert that it offers an alternative to the environmentally wasteful commercial funeral practices common in industrialized nations.

Natural burial is a relatively new funeral practice in industrialized nations, such as the United States and Canada, but it is actually a return to traditional methods of interring the dead. In a natural burial, the deceased person is not chemically embalmed, the materials used for interment (such as caskets made from wood or paper) are biodegradable, and interment takes place in a cemetery where the burial site may be left unmarked and the vegetation managed as wildlife habitat. Cemeteries devoted to natural burials thus resemble natural meadows with wildflowers, in contrast to the typical mown turf grass of traditional cemeteries.

Advocates for natural burial emphasize that the practice both promotes a more realistic understanding of death and is better for the environment than most commercial funeral practices. Because the body is not embalmed, funeral home workers are exposed to fewer toxic chemicals, and no artificial chemicals are released into the soil as the body inevitably decays. Some advocates have described natural burial as their last act of recycling. Caskets used for natural burials are made from materials that will rot with the body, such as wood or paper. In some cases, no casket is used at all; the body is simply wrapped in a shroud made of natural fibers and placed directly into the grave.

Proponents of natural burial note that the practice encourages friends and family members of the deceased to prepare the body for interment rather than rely on a commercial funeral home. The body is generally bathed, wrapped in a shroud or dressed in clothing made from fibers that will biodegrade (such as cotton or linen), and then placed in a casket for the funeral service. Advocates of the practice assert that preparing the body in this way offers friends and relatives the opportunity to perform one last, loving service for their loved one; the funeral preparation becomes both a celebration of the decedent's life and a part of the grieving process.

Until the twentieth century, most people in the United States experienced death and funeral practices as intimate events taking place primarily in the home. Bodies were not embalmed, and they were buried in wooden caskets placed directly into the ground. There was a general understanding that when a person died, the body would decay, just as do the remains of all once-living things. Over time, however, concerns about sanitation and disease control arose, along with political lobbying by morticians and others in the funeral industry for laws that would discourage home funerals. These developments led to the increasing commercialization and depersonalization of funeral practices. Embalming, for example, was first promoted as a way to control disease; only later did the emphasis on the unnatural preservation of the body become a selling point.

U.S. state and local laws vary widely regarding natural burial, as do individual cemetery regulations. For example, many traditional cemeteries require the use of a concrete vault, which is sometimes referred to as a grave liner, to hold the casket, making a true natural burial impossible. Cemeteries devoted to natural burial remain relatively rare, although their numbers have increased since the late twentieth century.

Some advocates of natural burial consider cremation followed by the scattering of the ashes to be a form of the practice. Critics of cremation point out that it is energy-intensive and thus not as carbon-neutral as interment in a natural burial cemetery, although it is less harmful to the environment than traditional interment with embalming, metal casket, and grave liner.

Nancy Farm Männikkö

FURTHER READING

Butz, Bob. *Going Out Green: One Man's Adventure Planning His Natural Burial.* Traverse City, Mich.: Spirituality & Health Books, 2009.

Harris, Mark. *Grave Matters: A Journey Through the Modern Funeral Industry to a Natural Way of Burial.* New York: Charles Scribner's Sons, 2007.

SEE ALSO: Carbon cycle; Composting; Ecological footprint; Recycling.

Natural capital

CATEGORIES: Philosophy and ethics; resources and resource management

DEFINITION: The stock of natural materials that produces a flow of goods and services while being replenished through its being a part of an ecosystem

SIGNIFICANCE: Natural capital must be considered from economic, ecological, and ethical perspectives because natural capital partially makes up and completely depends on whole ecosystems. From an ecological perspective, natural capital is not interchangeable with human-made capital. Therefore, because humans are dependent on both ecosystems and natural capital, they must take care in how continued economic development affects stocks of natural capital.

In economics, "capital" refers to a stock of materials that produces a flow of goods or services into the future. As a stock, capital is both a quantity of material and a quantity with an estimable value. Traditionally, "capital" refers to goods that are human-made. However, when functionally defined as a stock that produces a flow of goods or services, it can be extended to include human, social, and natural capital, among others.

Examples of natural capital are stocks of oaks and salmon that provide ongoing yields of trees and fish. The sustainable flow of goods from the trees (lumber) or fish (food) is natural income. Other examples of natural income, and thus examples of natural capital, include the recycling of waste materials by fungi and erosion control by a bank of trees.

During the 1980's the idea of sustainable development rose to prominence in environmental economics. The introduction of the concept of natural capital followed from the recognition of the need for economists to balance the seemingly contradictory goals of economics (maximizing production of goods and services) and environmental conservation (limiting the degradation of the environment). It was in this context that the economist David Pearce introduced the concept of natural capital in 1988. Pearce categorized natural capital in relation to the idea of sustainable development, which he took to be the holding of stocks of natural capital constant while the economy pursues what it takes to be most appropriate.

Natural capital's relationship to sustainable development can perhaps be best understood through consideration of its interchangeability with other kinds of capital. With traditional economics' focus on the monetary value of capital, and its employment of the idea of "weak sustainability," which requires only a guarantee of a nondeclining stock of capital, natural capital can be depleted as long as it replaced by other sources of capital. However, it is questionable whether natural capital should be viewed as interchangeable with other kinds of capital.

Ecological economists emphasize the uniquely life-sustaining aspects of natural capital. They assert that because of these aspects, natural capital is not interchangeable with human-made capital, and therefore it should not be valued simply monetarily. Human beings should recognize the need for strong sustainability—that is, the preservation of natural capital stocks above the level required to maintain the integrity of ecosystems.

The nature of natural capital calls for economic, environmental, and ethical consideration. These considerations follow from the valuation perspectives one should take on natural capital. Viewed simply as capital, it has monetary value. Viewed ecologically, natural capital has biophysical value relative to the life-sustaining aspects of the ecosystems of which it is a part. As such, natural capital raises a number of ethical concerns regarding the relationship between human survival and stocks of natural capital, and also the value of nonhuman life and ecosystems.

George Wrisley

FURTHER READING

Åkerman, Maria. "What Does 'Natural Capital' Do? The Role of Metaphor in Economic Understanding of the Environment." *Environmental Values* 12 (2003): 431-448.

Aronson, James, Suzanne J. Milton, and James N. Blignaut, eds. *Restoring Natural Capital: Science, Business, and Practice.* Washington, D.C.: Island Press, 2007.

Prugh, Thomas, with Robert Costanza et al. *Natural Capital and Human Economic Survival.* 2d ed. Boca Raton, Fla.: CRC Press, 1999.

SEE ALSO: Benefit-cost analysis; Contingent valuation; Ecological economics; Environmental economics; Environmental ethics; Free market environmentalism.

Natural Resources Conservation Service

CATEGORIES: Organizations and agencies; agriculture and food; resources and resource management

IDENTIFICATION: Federal agency under the U.S. Department of Agriculture that is responsible for coordinating the conservation of soil and water

DATE: Established on April 27, 1935

SIGNIFICANCE: By partnering with nearly three thousand local conservation districts, as well as with conservation organizations and state governments, the Natural Resources Conservation Service helps to protect soil, water, and wetlands from erosion, contamination, and overuse. The agency's scientists and engineers have done much of the research underlying modern understanding of watersheds and soil.

The Natural Resources Conservation Service (NRCS) began during the 1920's with soil scientist Hugh Hammond Bennett, sometimes called the father of soil conservation. Bennett, who worked for the U.S. Department of Agriculture, recognized the threat posed by soil erosion and coauthored a 1928 government report on the problem titled *Soil Erosion: A National Menace.* He also published articles in popular magazines in which he proposed conservation measures to preserve topsoil. In 1933, Bennett became the first director of a new agency under the Department of the Interior, the Soil Erosion Service, and began an experimental project with farmers in the Coon Creek watershed in Wisconsin to put into practice particular farming methods that best suited the land and watershed in each area. The project demonstrated that intelligent land use could prevent erosion and support clean air and water, as well as safeguard plants and animals.

The Dust Bowl of the 1930's helped Bennett convince the U.S. Congress to take further action, and in 1935 the Soil Conservation Act was passed; this legislation created the Soil Conservation Service (SCS) as a part of the Department of Agriculture, with Bennett at its head. The SCS expanded quickly under President Franklin D. Roosevelt's New Deal and soon was conducting more than one hundred demonstration projects with local farmers, as well as running more than 450 Civilian Conservation Corps (CCC) work camps and employing thousands of Works Progress Administration (WPA) workers. Conservation districts were established, and farmers and local government officials worked with the SCS to tailor plans for each district.

The work of the SCS continued to expand under new laws, including the Flood Control Act of 1944, the Watershed Protection and Flood Prevention Act of 1954, the Clean Water Act of 1972, the Rural Development Act of 1972, and the Food Security Act of 1985. During the 1980's, the SCS collaborated with the U.S. Army Corps of Engineers and the U.S. Fish and Wildlife Service to develop mapping techniques and conventions to identify wetlands on private land. The agency also conducts periodic inventories of soil erosion, cropland, water conservation needs, and natural resources. In 1994, the SCS was reorganized and its name was changed to the Natural Resources Conservation Service, to better reflect the scope of the agency's responsibilities. The NRCS continued to work with watersheds and soil, with new emphasis on

A member of the Natural Resources Conservation Service discusses a laser leveling project in Yuma, Arizona, with local officials. (USDA/NRCS)

the damage caused to water resources by chemical runoff from large farms and on the impacts on water quality of the large meat and dairy operations known as concentrated animal feeding operations, or CAFOs. In the twenty-first century, the NRCS began working with organic farmers to develop targeted conservation practices for organic agricultural operations.

Cynthia A. Bily

FURTHER READING

Blanco, Humberto, and Rattan Lal. *Principles of Soil Conservation and Management.* New York: Springer, 2008.

Conrad, Daniel. "Implementation of Conservation Title Provisions at the State Level." In *Soil and Water Conservation Policies and Programs: Successes and Failures,* edited by Ted L. Napier, Silvana M. Napier, and Jiri Tvrdon. Boca Raton, Fla.: CRC Press, 2000.

National Science Teachers Association. *Dig In! Hands-On Soil Investigations.* Arlington, Va.: NSTA Press, 2001.

Sampson, R. Neil. *For Love of the Land: A History of the National Association of Conservation Districts.* Tucson, Ariz.: Wheatmark, 2009.

SEE ALSO: Agricultural chemicals; Clean Water Act and amendments; Conservation policy; Conservation Reserve Program; Dust Bowl; Environmental law, U.S.; Erosion and erosion control; Flood Control Act; Intensive farming; Organic gardening and farming; Runoff, agricultural; Soil conservation; Soil contamination; Sustainable agriculture; Watershed management.

Natural Resources Defense Council

CATEGORIES: Organizations and agencies; activism and advocacy

IDENTIFICATION: American nonprofit organization dedicated to the protection of natural resources

DATE: Established in 1970

SIGNIFICANCE: The main goals of the Natural Resources Defense Council are to protect the environment from urban sprawl, reduce pollution, prevent habitat destruction, promote actions to mitigate global warming, and increase the use of renewable energy. The organization has fought successfully to protect natural resources and is considered one of the most influential environmental organizations in the United States.

In the 1960's, when the environmental movement was starting to gain traction in the United States, a group of law students and attorneys under the leadership of John Adams created the Natural Resources Defense Council (NRDC). A nonprofit and nonpartisan environmental organization based in New York City, it was founded with a $400,000 grant from the Ford Foundation. Its stated mission is to safeguard the earth, its people, plants, and animals, and the natural systems on which all life depends. The primary strategy of NRDC is to lobby the U.S. Congress and government agencies to promote the implementation of public policies targeting the sound and sustainable use of natural resources. With offices in Washington, D.C., Los Angeles, Chicago, New York, San Francisco, and Beijing, China, and membership above 1.3 million, NRDC is considered one of the most powerful environmental groups in the United States.

NRDC works on a broad range of environmental issues through a number of different programs. Its air and energy program focuses on issues such as air pollution, global warming, and the development of energy-efficient, renewable energy sources. Its waters and oceans program promotes the protection of fish populations, water quality, wetlands, and oceans. Its health program addresses such environmental problems as the contamination of drinking-water supplies by harmful chemicals, while its urban program focuses on environmental justice and the particular problems seen in urban environments, such as poor air and water quality. NRDC's land program strives to promote improvement in such areas as private and national forest management. Its nuclear program keeps track of new developments regarding nuclear weapons and nuclear power, and its international program addresses worldwide environmental issues such as the protection of rain forests and biodiversity, the preservation of wildlife and natural habitats, and the protection of marine life and oceans. NRDC also carries out public education programs and sponsors scientific research projects.

NRDC is a member of the Nuclear Weapons Complex Consolidation Policy Group, which aims to convince lawmakers around the world of the need to reduce nuclear weapons and to explore sustainable sources of energy other than nuclear power. Thus NRDC's position on nuclear energy is that nuclear power is not the solution to weaning the United States from its dependence on foreign oil, because nuclear power generation represents a danger to humans and the environment.

Lakhdar Boukerrou

FURTHER READING

Bevington, Douglas. *The Rebirth of Environmentalism: Grassroots Activism from the Spotted Owl to the Polar Bear.* Washington, D.C.: Island Press, 2009.

Corbett, Julia B. *Communicating Nature: How We Create and Understand Environmental Messages.* Washington, D.C.: Island Press, 2006.

Miller, G. Tyler, Jr., and Scott Spoolman. "Politics, Environment, and Sustainability." In *Living in the Environment: Principles, Connections, and Solutions.* 16th ed. Belmont, Calif.: Brooks/Cole, 2009.

SEE ALSO: Natural Resources Conservation Service; Renewable resources; Resource Conservation and Recovery Act; Resource depletion; Resource recovery; World Resources Institute.

Nature Conservancy

CATEGORIES: Organizations and agencies; preservation and wilderness issues

IDENTIFICATION: American nonprofit organization that works to preserve threatened ecosystems and the plants and animals that inhabit them

DATE: Founded in 1951

SIGNIFICANCE: By purchasing land and by helping other landowners to develop conservation plans, the Nature Conservancy has been successful in protecting unspoiled areas from development and pollution.

The Nature Conservancy, based in Arlington, Virginia, traces its origins to a committee of the Ecological Society of America, which was founded in 1915. In 1946, the committee became the Ecologists' Union, which changed its name to the Nature Conservancy in 1950 and was incorporated as a nonprofit organization in 1951. The group, which included research scientists and conservation advocates, began acquiring land through donation and purchase and collaborating with other agencies to manage public lands to sustain biodiversity.

During the 1960's the Nature Conservancy helped create a new tool for encouraging conservation, the conservation easement, which enables private landowners to receive tax benefits for conserving their land and restricting development while retaining ownership. By 2010 the Nature Conservancy had more than one million members and branches in all fifty U.S. states, as well as in more than thirty other countries. It owned more than 405,000 hectares (1 million acres) of land in the United States and approximately the same amount total in other countries, and was involved with other landowners in hundreds of projects addressing the conservation of water, forests, and marine habitats, as well as issues related to climate change.

The Nature Conservancy partners with corporations, government agencies, and other environmental groups, as well as with private landowners. Most of its funding comes from private donations, although some of its experimental projects are supported by government grants. The organization has helped to create and preserve several important areas throughout the United States, including the Golden Gate National Recreation Area in California, the Tallgrass Prairie Preserve in Oklahoma, the Great Sand Dunes

The Nature Conservancy and the U.S. National Park Service have undertaken a joint effort at Channel Islands National Park in California to protect the tiny island fox, a species found nowhere else in the world. (NPS)

National Park and Preserve in Colorado, and Glacial Ridge National Wildlife Refuge in Minnesota. It also conducts programs in other countries and has helped organize debt-for-nature swaps, under which developing nations have some of their international debt canceled in exchange for conserving tracts of ecologically valuable land. In addition to working internationally on an increasingly large scale, the Nature Conservancy encourages its many members to participate in efforts such as the Plant a Billion Trees Campaign, which plants one tree in Brazil for every dollar donated, and its Adopt an Acre program, which allows donors to designate particular habitats to be protected through their donations.

The Nature Conservancy is the wealthiest environmental group in the world, with more than one billion dollars in assets, and its large size and tremendous influence have created controversy. Critics have charged that some of the organization's acquisitions and logging, drilling, and mining projects, while intended to be models of sustainable development, have ignored the rights of indigenous peoples or threatened endangered species. The group's advocacy of integrated

fire management has also been controversial. In addition, the Nature Conservancy has been criticized for acquiring private lands through charitable gifts and then reselling them against the wishes of donors, often at large profit to the U.S. federal government.

Cynthia A. Bily

FURTHER READING

Birchard, Bill. *Nature's Keepers: The Remarkable Story of How the Nature Conservancy Became the Largest Environmental Organization in the World.* New York: John Wiley & Sons, 2005.

Brewer, Richard. *Conservancy: The Land Trust Movement in America.* Lebanon, N.H.: University Press of New England, 2004.

SEE ALSO: Conservation; Conservation easements; Controlled burning; Debt-for-nature swaps; Endangered species and species protection policy; Indigenous peoples and nature preservation; Nature preservation policy; Nature reserves; Sustainable development; Wildlife refuges.

Nature preservation policy

CATEGORY: Preservation and wilderness issues
DEFINITION: Decisions made and regulations put in place by any level of government to undertake the protection of natural resources
SIGNIFICANCE: Governments generally promote their policies concerning nature preservation through the passage of legislation. In the United States, changes in nature preservation policy over time have reflected the changes that have taken place in the attitudes of the public toward the need for government protection for the natural environment.

The Industrial Revolution, which diminished traditional agriculture while encouraging urbanization and technology, began straining the relationship between humanity and natural resources during the early nineteenth century. As the technological advances being made in Europe rapidly spread west, environmental damage and natural resource depletion escalated in the United States, inspiring the American conservation movement. The conservation and environmental movements have continued to exert tremendous influence on policy making.

American artist George Catlin first proposed setting aside land for wildlife and Native Americans during the nineteenth century, and in 1864 geographer George Perkins Marsh published *Man and Nature: Or, Physical Geography as Modified by Human Action*, the first influential book to address the human impact on nature. The Homestead Act of 1862 greatly encouraged expansion in the western United States by giving more than 405 million hectares (1 billion acres) of land to settlers, a policy that often resulted in barren landscapes. Destructive logging methods were employed, land was rapidly cleared for agriculture, and large-scale fires raged. Some western grasslands experienced such excessive grazing that many regions had not recovered their full productivity by the end of the twentieth century.

As farmers' journals described "wearing out" several homesteads during westward journeys, naturalist Henry David Thoreau and essayist Ralph Waldo Emerson countered with writings that fueled increasing support for nature conservation. The public expressed considerable outrage regarding the near elimination of several wildlife species that previously had existed in massive numbers, such as bison, deer, elk, and beaver. This led to legislation that created the world's first national park in 1872 at Yellowstone, Wyoming, followed by an 1873 petition to Congress by the American Association for the Advancement of Science to curtail the inefficient use of natural resources such as water, soil, forests, and minerals.

The 1891 Forest Reserve Act began the establishment of natural forests, and the 1900 Lacey Act initiated wildlife protection by regulating commercial hunting. The 1894 Buffalo Protection Act provided recognition that a previously abundant natural resource could rapidly become an endangered species. As naturalist John Muir championed numerous wilderness preservation projects and became the first president of the Sierra Club in 1892, federal legislation in the United States began classifying natural resources as renewable or nonrenewable. Renewable resources can be regenerated and even improve under proper management but can be depleted or completely eliminated if misused. Nonrenewable resources are present only in fixed amounts and will not regenerate regardless of human efforts. Examples of renewable resources include plants, animals, soils, and inland waters; nonrenewable resources include minerals and fossil fuels. The founding of private conservation organizations such as the American Forestry Association in 1875, the American Ornithologists' Union in 1883, the Boone and Crockett Club in 1887, and the New York Zoological Society in 1895 increased public influence on conservation legislation at all levels of government.

THEODORE ROOSEVELT AND FRANKLIN D. ROOSEVELT

President Theodore Roosevelt initiated habitat protection for wildlife in 1903 when he set aside Pelican Island in Florida's Indian River as a federal bird sanctuary. Through such initiatives as the 1908 White House Governors' Conference on Conservation, Roosevelt's politics and personality helped establish more than fifty wildlife refuges, five national parks, and eighteen national monuments, and increased the area of national forests by more than 69.7 million hectares (150 million acres). Roosevelt's policies required that certain public lands be held in trust for the "good of the country" and separated many public domain regions from commercial interests.

During and immediately following Theodore Roosevelt's presidency, Forest Service chief Gifford Pinchot and Interior Secretary James Garfield implemented more unified policies governing natural resource planning that relied on scientific principles, leading to development of the discipline of conservation biology. Many nonrenewable resources were then protected from exploitation by private industry by the 1920 Mineral Leasing Act.

Political debates and administration changes during the Great Depression of the 1930's shelved more environmental legislation until President Franklin D. Roosevelt signed the Taylor Grazing Act in 1934, whereby all public domain lands would be managed as part of the public trust. The dry Dust Bowl years of the Great Plains states during the early 1930's severely depleted migratory bird populations, motivating a renewed surge of public conservation activity and passage of the 1934 Duck Stamp Act, which tacked a conservation fee for the acquisition of wetlands onto waterfowl hunting licenses.

In 1933 the Soil Erosion Service (later called the Soil Conservation Service), the Civilian Conservation Corps, and the Tennessee Valley Authority were established to provide water and soil conservation assistance to landowners as farmland in the Midwest continued to deteriorate under improper agricultural practices. Franklin Roosevelt's Civilian Conservation Corps provided unemployed Americans with more than two million jobs planting trees and building irrigation systems and dams. More federal involvement was initiated after dust clouds from the dry soil of midwestern farmland blew east all the way to Washington, D.C. In 1940 the U.S. Congress enacted the Bald Eagle Protection Act to protect the national bird.

Post-World War II Conservation

Advances in technology and economic development in the turbulent era following World War II, combined with the postwar baby boom, put additional stressors on environmental resources. President Harry Truman began a national program for water-pollution control, with later legislation requiring states to set and enforce standards for natural rivers. In attempts to reduce insect-borne disease and increase food production, dichloro-diphenyl-trichloroethane (DDT) and other synthetic pesticides were developed and had considerable initial success, causing the near-complete disappearance of malaria and the production of bumper crops.

The 1947 Forest Pest Control Act provided for the detection and chemical destruction of insects that carried diseases harmful to humans, but the numerous new and powerful experimental substances being invented caused other severe environmental problems, which in many cases caused more damage than those that the pesticides were created to prevent. Grassroots public outcry stimulated several federal restrictions on chemicals such as DDT, with many citizens alerted to these dangers by former U.S. Fish and Wildlife Service biologist Rachel Carson's 1962 book *Silent Spring*. Carson is credited with warning mainstream America about the health and environmental hazards posed by pesticides and other toxic chemicals. Her work stimulated further writings that described human threats to the environment, including *The Population Bomb* (1968), by Paul R. Ehrlich, and *The Limits to Growth* (1972), by Donella H. Meadows, Dennis L. Meadows, Jørgen Randers, and William W. Behrens III.

All forms of environmental pollution greatly increased during the 1950's and 1960's. Television beamed graphic examples of environmental problems into public view, notably the mercury poisoning at Minamata Bay, Japan; killer smog episodes in London, England, and Los Angeles, California; and the 1967 *Torrey Canyon* oil spill in the English Channel. As the prices of land and water rights skyrocketed, the Land and Water Conservation Fund, which was set up by federal legislation during the 1960's to increase outdoor recreation space in the United States, generated revenues from offshore drilling leases. Several catastrophic environmental events occurred in 1969, including toxic waste fires on the Cuyahoga River in Cleveland, Ohio, and a coastal oil spill near Santa Barbara, California. Public pressure regarding these and other concerns led to passage of the National Environmental Policy Act of 1969 (NEPA), which became law on January 1, 1970. During the development of this precedent-setting act, Congress discovered that more than eighty governmental units had activities directly affecting the environment, but no government policies were in place to coordinate and review such activities.

Private individuals and organizations such as the Sierra Club, the Nature Conservancy, the Wilderness Society, the National Wildlife Federation, and the National Audubon Society began lobbying for more laws to establish nature preservation areas for both renewable and nonrenewable natural resources. Two highly visible social programs that influenced public opinion were conducted by the Nature Conservancy: Oklahoma's Tallgrass Prairie National Preserve "Adopt a

Bison" program and Montana's Pine Butte Swamp Preserve, where dinosaur fossils were discovered in 1978. The Endangered Species Preservation Act of 1966 and the Endangered Species Conservation Act of 1969 did not directly protect any species, but they led to later legislation that did.

THE 1970'S AND 1980'S

Following unanimous passage of NEPA by Congress over President Richard Nixon's objection, the 1970's saw the passage and often complicated enforcement of several laws regulating nature preservation. The Environmental Protection Agency (EPA) was established in 1970, followed later that year by passage of significant amendments to the 1963 Clean Air Act. Important pollution-control measures were then implemented by the 1972 Water Pollution Control Act, the 1973 Endangered Species Act, the 1976 Toxic Substances Control Act, the 1976 National Forest Management Act, and the 1977 Clean Air Act amendments. The Endangered Species Act is considered the most effective and wide-reaching act ever passed by Congress to protect natural ecosystems.

Key legislation supporting nature preservation that was passed during the 1980's included the 1980 National Acid Precipitation Act, the 1980 Alaska National Interest Lands Conservation Act (which enabled the size of the refuge and park systems to double), and the 1987 amendments to the 1972 Clean Water Act. Surveys conducted during the late 1980's revealed that more than 60 percent of wildlife areas in the United States were permitting activities that were harmful to wildlife, with the most destructive practices, such as military activities and drilling, not falling under legal jurisdiction of the U.S. Fish and Wildlife Service. This era also saw increased public interest in nature preservation following events such as the 1986 Chernobyl nuclear power plant catastrophe and the 1989 *Exxon Valdez* oil spill, as well as controversies concerning acid rain, tropical deforestation, the harvesting of old-growth timber, and the discovery of a trend toward global warming.

Private corporations that had previously sacrificed important wildlife habitat began to realize that the environment was an important issue to American consumers. In response to pressure from consumers, employees, and stockholders, many businesses implemented stewardship programs designed to protect natural resources and allow public enjoyment of their underdeveloped lands.

THE 1990'S AND BEYOND

Many nature preservation goals first proposed during the Industrial Revolution finally began to be realized with the systematic creation and maintenance of healthy forests; the prevention of timber depletion and siltation of streams; the provision of food, cover, and protection for wildlife; and the establishment of places where human beings could escape growing urbanization by taking part in outdoor recreation. Water reservoirs could now control flooding, provide clean water for humans and livestock, keep the soil fertile for agriculture, provide irrigation, and generate power. Water treatment plants were now effective in keeping rivers clean by processing wastes from urban sewage, while fish hatcheries provided supplemental stocks to natural and human-made reservoirs, streams, and lakes. Scenic easements along riverbanks aided antipollution efforts and reduced erosion, while green spaces required by city zoning regulations held soil and became available for community use while maintaining the environment's natural beauty. Mass interurban transportation systems moved people efficiently, and footpaths and bicycle trails offered outdoor recreation and cultural opportunities. Continual management of resources was more successful in keeping delicate ecosystems in balance, with ongoing environmental efforts including the seeding of wildlife foods, controlled burning to destroy unwanted vegetation, and the closing of wildlife habitats during mating and birthing seasons.

Conservationists-turned-environmentalists greatly influenced nature preservation policies during the 1990's as President George H. W. Bush passed legislation in 1990 that amended the 1970 Clean Air Act to focus more on reducing acid rain and emissions from fossil fuels and nitrogen oxide. However, an activist citizens' commission formed in 1992 by the Defenders of Wildlife found that the United States was "falling far short" of meeting the urgent needs of nature preservation. The public response to this information helped lead to passage of the 1997 National Wildlife Refuge System Improvement Act, which shifted the priorities of nature preservation systems toward the formation of multiple-use environments. This key legislation redefined the mission statement regarding conservation of habitats for fish, wildlife, and plants; designated priority public uses such as hunting, fishing, wildlife observation and photography, and environmental education and interpretation; and required that "environmental health" be maintained on public lands. The principle of multiple use, however, continues to allow mining, drilling, grazing, logging, and motorized recreation, as well as military training such as bombing, tank, and troop exercises, on lands designated for nature preservation.

INTERNATIONAL EFFORTS

Cooperative international nature preservation efforts began with the 1918 Migratory Bird Treaty signed by the United States and Great Britain (for Canada), and later Mexico. The International Union for Conservation of Nature and Natural Resources, founded in 1948, represented the interests of 116 countries toward protecting endangered and threatened "living resources." The United Nations Conference on the Human Environment, hosted by Sweden in 1972, was instrumental in establishing nature preservation as an international concern. Utilizing concepts from this conference, the U.S. Congress passed the 1983 International Environmental Protection Act, which included landmark legislation incorporating wildlife and plant conservation and biological diversity as objectives when the United States provides assistance to developing countries.

The 1973 Convention on International Trade in Endangered Species (CITES) involved the cooperation of more than one hundred nations to regulate the import and export of natural resources. Earth Day, first held on April 22, 1970, as a campus-based event encompassing an estimated twenty million people across the United States, began to be combined with other concerns, and annual observances of Earth Day have continued to demonstrate massive support for conservation issues. The international Greenpeace Foundation was formed in 1971 with the aim of applying pressure to governments and organizations to stop such practices as the testing of nuclear weapons and the dumping of radioactive and toxic wastes.

Sentiments favoring the preservation of nature began taking political form in Europe during the early 1980's, notably with the formation of the Green Party in Germany. International collaboration on environmental preservation issues that influenced later legislation included the 1987 Montreal Protocol to protect the ozone layer, the 1992 United Nations Conference on Environment and Development (also called the Earth Summit) in Brazil, and the 1994 United Nations Population Conference in Egypt. The 1992 Earth Summit in Rio de Janeiro, Brazil, which was the largest international meeting ever held (representatives of 178 nations attended), emphasized an approach to nature conservation that focused on sustainable growth and utilitarian solutions. A paper that resulted from the Earth Summit titled "World Scientists' Warning to Humanity" warned that if current consumption rates continued, the earth's resources would be reduced to the point where the world would be "unable to sustain life in the manner we now know."

This monumental event was followed by another U.N. summit in 1997 in New York City, attended by representatives of more than fifty nations, during which the progress made in the intervening years was reviewed. Many analysts noted that although progress had been made, the local and global issues surrounding nature preservation and environmental problems were continuing to grow in complexity, and many kinds of environmental damage may be irreversible. Although the attendees did agree to take further action on issues of nature preservation, they made few concrete commitments. The five-year review process continued with the 2002 World Summit on Sustainable Development, which was held in Johannesburg, South Africa; the United States did not participate in this conference, which also produced few concrete commitments.

Daniel G. Graetzer

FURTHER READING

Chiras, Daniel D., and John P. Reganold. *Natural Resource Conservation: Management for a Sustainable Future.* 10th ed. Upper Saddle River, N.J.: Benjamin Cummings/Pearson, 2010.

De Steiguer, J. E. *The Origins of Modern Environmental Thought.* Tucson: University of Arizona Press, 2006.

Dowie, Mark. *Losing Ground: American Environmentalism at the Close of the Twentieth Century.* Cambridge, Mass.: MIT Press, 1995.

Kline, Benjamin. *First Along the River: A Brief History of the United States Environmental Movement.* 3d ed. Lanham, Md.: Rowman & Littlefield, 2007.

O'Neill, John. *Ecology, Policy, and Politics: Human Well-Being and the Natural World.* New York: Routledge, 1993.

SEE ALSO: Alaska National Interest Lands Conservation Act; Biodiversity; Earth Summit; Ecosystems; Endangered species and species protection policy; National Environmental Policy Act; Nature reserves; Preservation.

Nature reserves

CATEGORY: Preservation and wilderness issues

DEFINITION: Areas set aside for the protection of animal and plant species, as well as unique geologic formations and other landscape features, in their natural state

SIGNIFICANCE: Protected natural areas serve as repositories of biodiversity. Their preservation helps protect complex ecological, hydrological, and climatological cycles, as well as distinctive landscapes and cultural and historical resources. However, the protection of these sites from development and the exploitation of the natural resources they contain often leads to conflict and controversy.

The nature reserve concept is often traced back to 1872, when the U.S. Congress established Yellowstone as the world's first national park. In fact, the basic idea is much older. In medieval England, the New Forest was a royal preserve. It had special protected status so that the king and his nobles could enjoy fine hunting in a land where farming was rapidly encroaching upon woodlands. Despite the hunting and poaching detailed in history and legend, the New Forest remained a viable natural area for centuries.

As the vast western regions of North America were explored, spectacular places such Banff in Canada (1885) and Yosemite in California (1890) received government protection. The number of protected areas grew over the following decades. By the late twentieth century, thousands of natural sites and parks had been designated for protection at state and federal levels in North America. These range from pristine wilderness to multiple-use areas such as national forests, where some commercial logging is allowed.

The movement to set aside areas for the preservation of nature developed differently on other continents, but by the late twentieth century almost every nation in the world had established nature reserves. By 1980 more than 3,000 protected sites existed worldwide; three decades later, the total exceeded 100,000 sites that cumulatively represented more than 12 percent of the planet's land surface. International treaties give special recognition to three types of reserves: Biosphere reserves, designated by the United Nations Educational, Scientific, and Cultural Organization (UNESCO), shield strictly protected areas with zones of limited, sustainable activity around them; World Heritage Sites, also designated by UNESCO, contain extraordinary natural, cultural, or historical features; and Ramsar sites are wetlands systems protected under the Convention on Wetlands of International Importance, also known as the Ramsar Convention. Most nature reserves are created and managed by governments, although some, often referred to as nature conservancies, are run under private auspices.

Nature reserves serve as reservoirs of native species and genetic diversity; as viable ecosystems that support biological, hydrological, and climatic cycles even outside their bounds; and as places for people to observe and enjoy the natural world. These goals are not innately incompatible. In managing reserves, however, decisions may have to be made that favor one aspect over the others.

ISSUES IN NATURE RESERVE MANAGEMENT

Few people oppose nature reserves in the abstract, but when it comes to real-life choices, many conflicts arise, and clashes between biological or scenic integrity and human economic needs are common. In developed countries, disagreements often take the form of duels between industries seeking to use specific natural resources and government bureaus and conservation groups trying to prevent such activities. Examples include disputes over oil drilling or pipeline construction in remote natural areas and clashes over lumbering and mining in national forests. Such struggles become more heated when unique species are threatened, as in the controversy in the Pacific Northwest regarding logging in the forest habitat of the northern spotted owl.

Developing nations face similar problems and more. By restricting natural areas, poorer countries often deprive their own citizens of traditional food sources and trade items. Banning the sale of natural products may encourage poaching, as the banned products may bring even higher prices on the black market. A well-known example is the illegal trade in African elephant ivory and rhinoceros horns.

Tensions also occur between individuals and groups who value the preservation of pristine wilderness and those who demand access to natural areas for recreational use. Tourists and campers bring cash and public support to parks, but they can disturb delicate natural balances. Automobile traffic brings pollution and noise problems inside the boundaries of nature reserves. In spite of strong warnings about behavior, visitors may start forest fires or introduce alien organisms to an area. Advocates of public access argue, however, that because such lands are public trusts, their reasonable use should not be denied to the citizens who support them.

Other issues arise concerning the management of nature reserves. Fires, epidemics, and unusual weather patterns may threaten nature's balances. When disasters occur, should humans intervene? A consistent environmental ethic might say no—nature will eventually restore itself. However, entire species may be lost before a natural balance reasserts itself. Many such losses to biodiversity are at least partly or indirectly caused by human actions. Science does not provide clear answers. Furthermore, measures taken to prevent or relieve such crises do not always work and may have unanticipated consequences of their own. A number of technical issues related to nature reserves also require further consideration. These include examination of the efficacy of small reserves in protecting specific species, the role of reserves in stabilizing the mix of gases in the earth's atmosphere and slowing global climate change, and the tipping point in the recovery of a species or a damaged ecosystem.

Several promising ideas have enriched the reserve movement. The foremost is international cooperation and joint action. UNESCO's World Heritage Sites program is a major achievement of this approach. Another is the establishment or recognition of natural areas sponsored by nongovernmental groups. Two different but equally positive examples in the United States are the nature conservancy movement and the increasing recognition of the importance of Native American sacred sites that began in the late twentieth century. Yet another concept that has helped reserve efforts to flourish, particularly in the developing world, is ecotourism. Ecotourist dollars represent an economic incentive for countries to preserve their unspoiled wilderness areas.

Emily Alward
Updated by Karen N. Kähler

FURTHER READING

Allin, Craig W., ed. *International Handbook of National Parks and Nature Reserves.* New York: Greenwood Press, 1990.

Carey, Christine, Nigel Dudley, and Sue Stolton. *Squandering Paradise? The Importance and Vulnerability of the World's Protected Areas.* Gland, Switzerland: World Wide Fund for Nature International, 2000.

Chape, Stuart, Mark D. Spalding, and Martin D. Jenkins, eds. *The World's Protected Areas: Status, Values, and Prospects in the Twenty-first Century.* Berkeley: University of California Press, 2008.

Duncan, Dayton, and Ken Burns. *The National Parks: America's Best Idea—An Illustrated History.* New York: Alfred A. Knopf, 2009.

Ghimire, K. B., and Michel P. Pimbert, eds. *Social Change and Conservation.* Sterling, Va.: Earthscan, 2009.

Lockwood, Michael, Graeme Worboys, and Ashish Kothari, eds. *Managing Protected Areas: A Global Guide.* Sterling, Va.: Earthscan, 2006.

Riley, Laura, and William Riley. *Nature's Strongholds: The World's Great Wildlife Reserves.* Princeton, N.J.: Princeton University Press, 2005.

SEE ALSO: Biodiversity; Ecosystems; Ecotourism; Indigenous peoples and nature preservation; International Union for Conservation of Nature; National forests; National parks; Wetlands; Wilderness areas; Wildlife refuges.

Nature writing

CATEGORIES: Activism and advocacy; philosophy and ethics

DEFINITION: Nonfiction writing about the natural environment

SIGNIFICANCE: Since the nineteenth century, nature writers have been highly influential contributors to the environmental movement, often raising public awareness of such issues as the dangers to human health of pollution, the negative impacts of the loss of wilderness and of animal species, and the potential harms of various human activities for the long-term health of the planet.

Nature writing celebrates wilderness while simultaneously discouraging environmental exploitation. The field of nature writing includes environmental writing, environmental journalism, and ecocriticism (or ecological criticism), an interdisciplinary study of the environment and literature combining all the sciences to develop solutions for environmental problems.

Since the first Earth Day in 1970, nature writing has increased dramatically in popularity. Nature writing is heavily based on scientific information, research, and facts about the natural world, but the style in which it is often presented provides a unique and broad perspective that reaches a diverse audience. Nature writers frequently write in the first person, a feature that contributes to the emotional and inspirational tone of their work.

Personal observations and philosophical reflections on nature and the environment make up a large portion of nature writing. Among the many types of nature writing, the most basic form is writing that simply conveys information about the natural world, such as through a field guide or other factual or natural history reporting. Natural history essays include important facts about nature and the environment but also incorporate literary elements and meaning or interpretation. Another form of nature writing is based on the personal experiences of the author in nature and tends to be very emotional and personal. Some forms of nature writing include philosophical interpretations of nature and tend to be abstract and scholarly in tone.

ROLE IN ENVIRONMENTAL AWARENESS

For centuries, writers have explored connections between human beings and the natural world. Early nature writers focused on the environment as secondary to humanity's needs and human progress, but by the middle of the nineteenth century this began to change. Nature writers such as Henry David Thoreau and Ralph Waldo Emerson began to reinterpret the significance of nature and the relationship of humans to nature, initiating a movement that led to a dramatic evolution in environmental thought and ethics. The viewpoint shifted toward one in which the environment was recognized as more than simply a natural resource.

The writings of Thoreau, for example, increased public awareness that the natural environment had considerably more value to human beings than simply the natural resources it offered for exploitation; Thoreau emphasized that nature and the environment could be sources of spiritual truth and support. Together with other prominent nature writers such as John Muir and Aldo Leopold, Thoreau, Emerson, and others served to educate the public about nature and the environment.

PROMINENT NATURE WRITERS

The beginning of modern nature writing can be traced to natural history works popular in the second half of the eighteenth century through the nineteenth century, including writings by Gilbert White, William Bartram, John James Audubon, and Charles Darwin. Other explorers, collectors, and naturalists also contributed to these collections. Thoreau, who wrote many volumes on natural history and the environment, is often considered the father of American nature writing, although his writings were preceded by those of John Bartram and his son William. Many environmental historians consider Thoreau's *Walden: Or, Life in the Woods*, his 1854 memoir, as the beginning of a critical movement in environmental thinking and writing. In subsequent years, this movement flourished and included contributions by John Burroughs, Ralph Waldo Emerson, John Muir, Aldo Leopold, Rachel Carson, and Edward Abbey. Emerson was considered a nature essayist; his writings in the late nineteenth century inspired the works of John Muir, who wrote about his personal experiences in nature, particularly in California's Sierra Nevada.

Rachel Carson's 1962 book *Silent Spring* was critical in raising environmental awareness and fueling public concern about pollution. The book documented the environmental impacts of the uncontrolled spraying of the insecticide dichloro-diphenyl-trichloroethane (DDT) and questioned the rationale that led to the release of substantial amounts of chemicals into the environment when their potential ecological and health impacts were unknown. *Silent Spring* fueled the modern environmental movement by exposing the dangers of indiscriminately used pesticides and fertilizers. It catalyzed a new way of looking at the use of chemicals, industrial practices, and pollution and how they affect the environment and human health, and raised questions about the long-standing belief that scientific progress is, without exception, good for humanity. Although *Silent Spring* was challenged by the pesticide industry, the public reaction to Carson's book eventually led to the establishment of a presidential commission charged with studying the effects of pesticides; the use of DDT was banned in the United States in 1972.

In 1989 Bill McKibben, a prolific nature writer, environmental commentator, and historian, also altered public perceptions with his book *The End of Nature*. By addressing such environmental issues as pollutants, acid rain, the greenhouse effect, and depletion of the ozone layer, McKibben raised awareness of the impacts of human activities on the earth's atmosphere and climate.

C. J. Walsh

FURTHER READING

Buell, Lawrence. *The Future of Environmental Criticism: Environmental Crisis and Literary Imagination.* Malden, Mass.: Blackwell, 2005.

Finch, Robert, and John Elder, eds. *Nature Writing: The Tradition in English.* New York: W. W. Norton, 2002.

Lyon, Thomas J. *This Incomparable Land: A Guide to American Nature Writing.* Rev. ed. Minneapolis: Milkweed Editions, 2001.

McKibben, Bill, ed. *American Earth: Environmental Writing Since Thoreau.* Des Moines, Iowa: Library of America, 2008.

SEE ALSO: Abbey, Edward; Audubon, John James; Burroughs, John; Carson, Rachel; Environmental ethics; Gore, Al; Leopold, Aldo; Muir, John; *Silent Spring*; Thoreau, Henry David.

Neutron bombs

CATEGORY: Nuclear power and radiation

DEFINITION: Nuclear devices that, compared with standard nuclear weapons, are designed to put more energy into radiation and less into blast, heat, and fallout

SIGNIFICANCE: Development of the neutron bomb, an enhanced radiation weapon, was controversial. Opponents argued that the bomb's ability to destroy enemy tanks with less environmental destruction than would be caused by conventional nuclear weapons might actually increase the likelihood that the bomb would be used.

D uring the Cold War, Warsaw Pact armaments and soldiers outnumbered those of the North Atlantic Treaty Organization (NATO) in Europe by more than two to one. Had the Warsaw Pact nations launched a massive tank invasion into Germany, the NATO plan was to halt the invasion with nuclear weapons if necessary. A major flaw in this plan was that Germany might not be any better off if its land were pum-

---

Reagan's Defense of the Neutron Bomb

In a question-and-answer session on August 13, 1981, President Ronald Reagan fielded reporters' questions on a range of national and international topics, including the neutron bomb. The president claimed that the bomb is more humane and more effective than traditional nuclear weapons and is a necessary defense against the Soviet Union's arsenal:

REPORTER: Mr. President, I wonder about the neutron bomb perhaps changing nuclear doctrine. If the Soviets attacked with tanks, might we become the first to use—would we engage in first strike with that weapon?

REAGAN: Well, this is something that seems to be overlooked in all the propaganda that's now being uttered about this weapon, and that is that the present tactical battlefield weapons stationed in Europe are nuclear weapons, far more destructive, far longer in rendering areas uninhabitable because of radioactivity, than the neutron weapon. So, those tactical nuclear weapons are there on both sides already, and this, we think, is a more moderate but more effective version.

You also have to remember that those who are crying the loudest, the Soviet Union, and many of those who under the name of pacifism in Western Europe, who are opposing things like this and opposing the theater nuclear forces and so forth, maybe some are sincere—I'm sure they are—but I think others are really carrying the propaganda ball for the Soviet Union, because there's no mention made of 200 SS-20's, strategic nuclear weapons of medium range, that are aimed at the cities of all of Europe today, and that are not being considered in any of the talk of reduction of theater forces, East and West—just as in SALT II the Soviet Union called our aging B-52 a strategic weapon but did not call their Backfire, modern bomber, a strategic weapon.

So, let's remember the SS-20's before we start worrying too much about what we're thinking about. But remember also that our present 8-inch guns and our present Lance missiles over there are tactical nuclear weapons.

---

meled by NATO nuclear weapons instead of Warsaw Pact tanks and planes. The enhanced radiation weapon known as the neutron bomb was proposed as a way to stop Warsaw Pact tanks without devastating Germany.

Since tanks are resistant to damage by blast and thermal radiation, conventional nuclear warheads with explosive yields of 10 or more kilotons would be required to destroy them in large numbers. A 10-kiloton-yield conventional nuclear warhead could destroy or incapacitate tank crews within a 1.5-square-kilometer (0.6-square-mile) area but would destroy or damage buildings in an area of nearly 5 square kilometers (2 square miles). A 1-kiloton-yield neutron bomb could deliver a prompt neutron dose that would incapacitate any tank crew within the same 1.5-square-kilometer area, but it would destroy or moderately damage urban structures only within a 1-square-kilometer (0.4-square-mile) area. Fallout would be only one-half that of the larger weapon, and within a few hours radiation levels would be low enough for people to pass safely through the area.

Clearly, neutron bombs should be able to halt an invasion with less environmental damage than would be caused by conventional nuclear weapons. Advocates of developing these weapons argued that since the use of neutron bombs would be less costly to Germany's cities and countryside, the Warsaw Pact nations should believe that NATO would be more likely to use them, and this belief would deter a Warsaw Pact attack.

Opponents argued that the Warsaw Pact forces could reduce the effectiveness of neutron bombs by spacing their tanks farther apart and adding neutron shielding. Opponents further noted that the environmental advantages of the neutron bomb made it more likely that NATO would consider using it, and they feared that once nuclear weapons were used in a conflict, the side that was losing would resort to larger nuclear weapons. The confrontation could quickly escalate to full-scale nuclear war, possibly the ultimate human-made environmental disaster.

Beginning in 1981, the United States equipped 40 howitzer projectiles and 350 Lance missiles with neutron bombs. These warheads had yields of 1 kiloton or less. All were withdrawn from service and dismantled at the end of the Cold War. The Soviet Union, France, and China tested enhanced radiation weapons and may have produced them, but by the end of the twentieth century no nation was known to have deployed such weapons.

Charles W. Rogers

FURTHER READING

Muller, Richard A. "Nuclear Madness." In *Physics for Future Presidents: The Science Behind the Headlines.* New York: W. W. Norton, 2008.

Wittner, Lawrence S. *Confronting the Bomb: A Short History of the World Nuclear Disarmament Movement.* Stanford, Calif.: Stanford University Press, 2009.

SEE ALSO: Antinuclear movement; Limited Test Ban Treaty; Nuclear testing; Nuclear weapons production; Radioactive pollution and fallout; Union of Concerned Scientists.

New urbanism

CATEGORIES: Urban environments; land and land use

DEFINITION: Urban design movement that emphasizes sustainable growth

SIGNIFICANCE: The growing movement of new urbanism aims to reduce the negative environmental impacts of urban and suburban development while improving the quality of life for city residents.

The movement known as new urbanism commenced during the 1980's in North America in response to urban sprawl. It had become clear to some developers, urban planners, and architects that outward city development and automobile dependence lead to the destruction of wildlife habitats, air pollution, water scarcity and deteriorating water quality, loss of farmland, high infrastructure costs, racially and ethnically segregated neighborhoods, and the isolation of human beings. In contrast, new urbanism is based on sustainable planning principles that promote compact urban forms and greenfield projects in sparsely populated suburbs and inner-city areas.

New urbanism began with Seaside, Florida, a community created in the early 1980's by developer Robert Davis and architects Andrés Duany and Elizabeth Plater-Zyberk. Seaside was designed to resemble a small, pre-World War II town, with walkable communities, a mix of public and private constructions, and civic spaces. In 1993 Duany and Plater-Zyberk helped to found the Congress for the New Urbanism, a nonprofit organization devoted to promoting the conservation of natural environments; the rebuilding of neighborhoods and regions into mixed residential, business, and retail developments; and the rediscovery of old communities—such as Boston's Back Bay, downtown Charleston, South Carolina, and Philadelphia's Germantown—where social life centers on a courthouse square, public commons, plaza, train station, or main street.

The core principle of new urbanism, community building within a sustainable natural environment, has unfolded on several scales and in several different contexts. New urbanism has come to encompass such varying elements and goals as a regional vision of non-automobile-centric walkable neighborhoods linked to mass-transit lines, towns with distinct centers and edges, compact neighborhoods that preserve farmland and environmentally sensitive areas, mixed land-use planning, public gathering places, infill projects in inner-city neighborhoods, reconstruction of suburban strips, and redevelopment of downtown areas. In the United States many designers, developers, and planners have adopted new urbanist principles in their creation of master plans and design codes. These principles are most frequently implemented in the development of parks and community gardens. The U.S. Department of Housing and Urban Development has also selectively incorporated new urbanism principles in the development of public housing projects.

Proponents of new urbanism argue that lower per-person land consumption and reduction in automobile dependence lead to lower environmental impacts by reducing air pollution and the burning of finite fuels, encouraging local production, and preserving wetlands, woodlands, and animal habitats. Further, high land density and mixed land usage serve to protect hydrologically sensitive areas by preventing land fragmentation and reducing impervious land surfaces. Studies demonstrate that the successful integration of watershed protection techniques in new urbanist developments can reduce environmental degradation in comparison with traditional suburban developments.

Evaluations of new urbanist projects suggest that many have not yet achieved comprehensive environmental benefits. The absence of employment clusters linked to public transit, the existence of cultural norms that favor auto usage over walking or use of public transit, and the relatively low priority given to watershed-based zoning present major challenges. Some environmentalists have criticized new urbanists as prioritizing architectural characteristics over truly environmentally beneficial strategies. They have encouraged planners, developers, and designers to focus on how communities work as opposed to how communities look. New urbanist proposals that incorporate apartments and condominiums and mixed-use designs also face legal challenges from residents of subdivisions who are opposed to high-density development.

Despite such obstacles, new urbanism continues to be a popular urban reform movement. Some new urbanist planners have joined with environmental groups to spread awareness about and gain support for sustainable urban growth, and private developers and public officials are becoming increasingly aware of the value of applying new urbanist principles.

Niharika Banerjea

FURTHER READING

Grant, Jill. *Planning the Good Community: New Urbanism in Theory and Practice.* New York: Routledge, 2006.

Haas, Tigran, ed. *New Urbanism and Beyond: Designing Cities for the Future.* New York: Rizzolli, 2008.

Platt, Rutherford H., ed. *The Humane Metropolis: People and Nature in the Twenty-first-Century City.* Amherst: University of Massachusetts Press, 2006.

Talen, Emily. *New Urbanism and American Planning: The Conflict of Cultures.* New York: Routledge, 2005.

SEE ALSO: Community gardens; Greenbelts; Land-use planning; Open spaces; Planned communities; Urban ecology; Urban parks; Urban planning; Urban sprawl.

Nile River

CATEGORIES: Places; water and water pollution; ecology and ecosystems

IDENTIFICATION: Major river flowing northward from Central Africa into the eastern Mediterranean Sea

SIGNIFICANCE: The Nile River is of inestimable value to the populations of the otherwise dry countries of northeast Africa, but the control of its waters is in dispute and its environmental health is deteriorating.

At 6,820 kilometers (4,238 miles), the Nile is the longest river in the world. It flows through several bioregions and, along with its branches and tributaries, drains one-tenth of the African continent, an area of approximately 3.1 million square kilometers (1.2 million square miles) that is home to some 150 million people.

The branch of the river known as the White Nile rises in Burundi but loses almost half its volume through evaporation and transpiration in the swamps of the Sudd in southern Sudan. Thus the Blue Nile, which is fed by the summer monsoon rains of the Ethiopian highlands and which joins the White Nile at Khartoum, is largely responsible for the annual floods on which the populations downriver depend. The ancient Egyptians dug canals and erected earthworks on the lower reaches of the river in order to irrigate their fields, a practice continued by succeeding generations. During the first half of the nineteenth century,

Turkish viceroy Muḥammad ʿAlī Pasha was responsible for creating a more effective series of diversion dams and canals.

While the Nile's floods are predictable, their size is not. Egypt's British occupiers completed a dam at Aswan in 1902 in the hope of controlling excessive flooding and mitigating the effects of drought, but the dam's reservoir submerged the first cataract of the Nile. Four more dams on the Nile and its tributaries in Sudan followed between 1925 and 1966. The need for hydroelectricity led to the construction of a dam at Owen Falls in Uganda in 1954, which caused the submersion of Ripon Falls near Lake Victoria. Environmental trade-offs continued with the completion of the Aswan High Dam in 1970. The dam brought with it a number of undeniable benefits, including flood control, the generation of hydroelectricity, and the provision of a constant supply of water for irrigation. However, a number of grave consequences ensued as well, including the spread of waterborne diseases and the erosion and salinization of the downriver floodplain. In the twenty-first century, rising sea levels caused by global warming threaten to speed the erosion of the Nile Delta.

Fluctuating weather patterns have compounded the river's environmental problems. Greater-than-average precipitation from 1961 through 1964, blamed on the El Niño weather pattern, destroyed riparian communities and filled the reservoir behind the Aswan High Dam to dangerous levels. Subsequent droughts created massive famine in Ethiopia and exacerbated international rivalries throughout the region. In 1978 Egypt and Sudan began work on the Jonglei Canal, which was designed to allow more of the White Nile's waters to bypass the stagnant Sudd, but construction was halted in 1984 by the Second Sudanese Civil War. In 2008 the two countries agreed to begin work on the project again. Other international cooperative ventures include the Nile Basin Initiative, created in 1999 by nine of the ten countries sharing the basin with the aim of safeguarding its waters and sharing them equitably.

As the population of the Nile basin grows, shortages of fresh water threaten to become increasingly common. By some measurements of per-capita requirements, several countries in the southern basin are already water-deficient, and Egypt—where the Nile's waters have grown heavily polluted—and Ethiopia are expected to become water-deficient in 2025.

Grove Koger

FURTHER READING

Collins, Robert O. *The Nile*. New Haven, Conn.: Yale University Press, 2002.

Hamza, W. "The Nile Estuary." In *Estuaries*, edited by Peter J. Wangersky. New York: Springer, 2006.

Park, Chris S. *The Environment: Principles and Applications*. 2d ed. New York: Routledge, 2001.

SEE ALSO: Africa; Aswan High Dam; Bioregionalism; Dams and reservoirs; Hydroelectricity; Soil conservation; Soil salinization.

NIMBYism

CATEGORY: Philosophy and ethics

DEFINITION: Opposition by community residents to the siting of unwanted development in their local area

SIGNIFICANCE: Although the "not in my backyard" stance on the siting of waste management and similar facilities was once viewed as entirely negative and selfish, NIMBYism has come to be understood as valid in many cases, leading to the appropriate challenging of government decisions and the improvement of planning for facility safety.

The unregulated disposal of solid waste over the years has resulted in environmental problems such as groundwater contamination. Hazardous waste disposal practices have had even more significant impacts on the environment, as well as adverse effects on human health. The enactment of increasingly strict legislation has been successful in decreasing the amounts of hazardous wastes produced and in stimulating the development of better disposal practices, but many existing waste-handling facilities have closed or have been scheduled to close because of their inability to comply with the newer regulations. Consequently, pressing demand has arisen for new facilities that meet the upgraded standards. The proposed sitings of these facilities frequently generate the response known as NIMBY, or "not in my backyard," among the residents of the communities that would be affected.

The controversy develops as a manifestation of the "collective goods" problem. In other words, it is difficult to provide goods that benefit everyone and for which the costs are equally shared. Therefore, the system fosters a situation that provides an incentive to be

A group of Wilmer, Texas, officials and residents block a road in an attempt to stop a truck from unloading a cargo of lead-contaminated soil near their community. (AP/Wide World Photos)

a "free rider." A perception of inequity develops when decisions about site facilities are made without any apparent input from the local community. The basis for NIMBYism, thus, is the desire for basic democratic rights; people who will be affected by change want to be allowed to participate in the decision-making process. Equity issues may further accentuate a negative public response. For example, the sites selected for waste-handling and similar facilities are frequently located in low-income, minority neighborhoods. Therefore, economic and social aspects of the siting decisions often become important considerations. Local community members then raise the question, If the facility will provide for the public good, why does it have to be here as opposed to some other place?

These are all valid concerns that must be considered when decision makers and community members are attempting to reach consensus on equitable siting of solid waste facilities. Engaging the public's participation in a legitimate forum on the health risks, as well as economic and social implications, of a local waste-handling site is a vital part of gaining community support. If advocates of the siting can demonstrate the safety of the proposed project and can show how the quality of the facility can enhance property values and not reflect negatively on the community, residents are more apt to be receptive.

For some time, NIMBYism had a generally negative connotation. It was often characterized as an emotional, almost irrational, selfish response that interfered with the larger public good. Over time, however, the NIMBY responses of community members to some proposed projects began to be perceived in a more neutral, if not a positive, light. The challenging of government decisions is sometimes effective in drawing public attention to valid environmental issues, as well as in forcing consideration of alternatives for more effective and equitable siting of waste management facilities.

Steven B. McBride

FURTHER READING

Rosenbaum, Walter A. *Environmental Politics and Policy.* 7th ed. Washington, D.C.: CQ Press, 2008.

Thomsett, Michael C. *NIMBYism: Navigating the Politics of Local Opposition.* Arlington, Va.: CenterLine, 2004.

SEE ALSO: Environmental ethics; Environmental justice and environmental racism; Hazardous waste; Landfills; Waste management.

Nitrogen oxides

CATEGORIES: Pollutants and toxins; atmosphere and air pollution

DEFINITION: Compounds consisting of nitrogen and oxygen

SIGNIFICANCE: Nitrogen oxides are extremely important environmental pollutants that are closely linked to the energy and agricultural sectors of the global economy. They are causally associated with global climatic change, ozone depletion, acidic precipitation, eutrophication, and photochemical smog formation.

Nitrogen oxides are naturally occurring, biologically active compounds that are produced through a variety of biotic and abiotic processes. Nitrogen-fixing organisms, including mutualists associated with leguminous plants, capture atmospheric nitrogen and produce compounds that are readily converted to nitrites and nitrates by nitrifying bacteria. The biological decomposition of organic matter also leads to the formation of nitrogen oxides. Intense heating causes atmospheric nitrogen and oxygen to react, and lightning is an important naturally occurring, abiotic source of nitrogen oxides in the earth's atmosphere.

The use of fossil fuels and nitrogen-based fertilizers and the cultivation of leguminous plants have substantially increased nitrogen oxide concentrations. Some fossil fuels (such as coal) contain significant amounts of nitrogen, which is oxidized during combustion. Fossil-fuel combustion also produces the heat required to oxidize atmospheric nitrogen.

Nitrous oxide is a powerful greenhouse gas with a warming potential three hundred times greater than that of carbon dioxide. Agricultural practices are primarily responsible for increasing nitrous oxide concentrations in the earth's atmosphere, but combus-

U.S. Nitrous Oxide Emissions by Source, 1990-2008

SOURCE CATEGORY	1990	1995	2000	2005	2006	2007	2008
Agricultural soil management	203.5	205.9	210.1	215.8	211.2	211.0	215.9
Mobile combustion	43.9	54.0	53.2	36.9	33.6	30.3	26.1
Nitric acid production	18.9	21.0	20.7	17.6	17.2	20.5	19.0
Manure management	14.4	15.5	16.7	16.6	17.3	17.3	17.1
Stationary combustion	12.8	13.3	14.5	14.7	14.5	14.6	14.2
Adipic acid production	15.8	17.6	5.5	5.0	4.3	3.7	2.0
Wastewater treatment	3.7	4.0	4.5	4.7	4.8	4.9	4.9
Product uses	4.4	4.6	4.9	4.4	4.4	4.4	4.4
Forestland remaining forestland	2.7	3.7	12.1	8.4	18.0	16.7	10.1
Composting	0.4	0.8	1.4	1.7	1.8	1.8	1.8
Settlements remaining settlements	1.0	1.2	1.1	1.5	1.5	1.6	1.6
Field burning of agricultural residues	0.4	0.4	0.5	0.5	0.5	0.5	0.5
Incineration of waste	0.5	0.5	0.4	0.4	0.4	0.4	0.4
Wetlands remaining wetlands	+	+	+	+	+	+	+
International bunker fuels	1.1	0.9	0.9	1.0	1.2	1.2	1.2
Total for United States	322.3	342.5	345.5	328.3	329.5	327.7	318.2

Source: U.S. Environmental Protection Agency, *Inventory of U.S. Greenhouse Gas Emissions and Sinks, 1990-2008.* Figures are given in teragrams (1 million metric tons) in carbon dioxide equivalents.

tion is also important. The Intergovernmental Panel on Climate Change has identified nitrous oxide as an important driver of human-caused global climate change, and nitrous oxide emissions are internationally regulated under the Kyoto Protocol. Human activities have increased the atmospheric nitrous oxide concentration approximately 18 percent since the dawn of the Industrial Revolution.

Nitrous oxide is also responsible for ozone thinning. Research published in 2009 indicated that nitrous oxide was at that time the most important ozone-depleting compound released by human activities. In the stratosphere, nitrous oxide is converted to nitric oxide, which catalyzes ozone-destroying reactions.

Nitrogen oxides react with moisture in the earth's atmosphere to produce nitric acid, a primary component of acid rain. The most significant environmental damage attributable to nitric acid formation and acidic precipitation is linked to the combustion of fossil fuels and generally occurs downwind of large cities, power plants, and industrial centers.

Nitrogen oxides are important fertilizers, and their production fosters eutrophication, or the overenrichment of bodies of water with nutrients. Nitrogen-limited estuaries and coastal ecosystems may be particularly sensitive to nitrogen inputs, and atmospheric nitrogen loading is associated with harmful algal blooms and a host of other ecological impacts, including declines in species diversity.

The photodissociation of nitrogen oxides leads to the formation of ozone, and the release of nitrogen oxides from automobiles and stationary sources contributes to the formation of photochemical smog. Exposure to abnormally high ambient nitrogen dioxide (NO_2) concentrations may worsen asthma symptoms, and some individuals may run an increased risk of respiratory infection, heart failure, or complications during pregnancy. Average ambient NO_2 concentrations have decreased substantially in the United States since 1980.

Fuel-burning appliances, such as gas stoves, furnaces, fireplaces, and space heaters, produce NO_2, and indoor NO_2 exposure is a recognized public health concern. Research indicates that indoor NO_2 can produce respiratory symptoms among children with asthma, even at concentrations well below the U.S. Environmental Protection Agency's ambient air standard of 53 parts per billion.

Brian G. Wolff

FURTHER READING

Jacobson, Mark Z. *Atmospheric Pollution: History, Science, and Regulation.* New York: Cambridge University Press, 2002.

Vallero, Daniel. *Fundamentals of Air Pollution.* 4th ed. Boston: Elsevier, 2008.

SEE ALSO: Acid deposition and acid rain; Air pollution; Air Quality Index; Airborne particulates; Automobile emissions; Catalytic converters; Cultural eutrophication; Eutrophication; Global warming; Greenhouse effect; Greenhouse gases; Ozone layer; Smog.

Noise pollution

CATEGORY: Pollutants and toxins

DEFINITION: Harmful or annoying sounds in an environment

SIGNIFICANCE: The problem of noise pollution is particularly acute because noise increases with population density; thus a disproportionately large sector of the human population experiences the adverse effects of exposure to noise. The control of noise requires scientific, social, and sometimes political actions.

Music, speech, and noise are the three basic categories of sound. Noise is simply defined as any unwanted sound. The degree to which a sound is unwanted is, however, a psychological question; the results of exposure to noise may range from moderate annoyance to hearing loss from high volume levels. Furthermore, the interpretation of what constitutes noise is subjective; both music and conversation may be regarded as noise in some places, such as in an office or a library.

NOISE IN THE ENVIRONMENT

With few exceptions, technological advances from the mid-twentieth century onward have resulted in a steady increase in the amount of unwanted sound. Examples include jet airplanes, automobiles, and ventilation fans. It was thought at one time that human beings should accept and tolerate the noise that went along with the benefits of many industrial advances, but problems associated with exposure to noise have manifested themselves often enough that there is now serious concern about noise pollution in the environment. Noise can be generated in a great variety of ways, but only a few prominent sources of noise emission are part of the daily lives of people in industrialized nations. Noise in the environment can be greatly reduced if these sources of noise pollution can be controlled.

Noise from airplanes poses a major problem in urban areas. Airplanes produce noise through the efflux of jet engines and the high-pitched whine of engine fans. Since 1969 the Federal Aviation Administration (FAA) has legislated acceptable noise levels for commercial airplanes in the United States. Over the years, remarkable engineering innovations have been made in the reduction of jet noise to satisfy the relatively stringent FAA requirements. In the meantime, however, air travel has become increasingly popular. Airlines in the United States have now captured more than 80 percent of all intercity passenger traffic, and the percentage is still on the rise. It is anticipated that the number of takeoffs and landings near major cities will continue to grow throughout the early decades of the twenty-first century. Furthermore, as land prices rise, residential dwellings are encroaching on noise buffer zones near airports in greater numbers. The control of aircraft noise will certainly continue to be pressing in the future.

The deafening din of high-powered trucks and motorcycles is familiar to nearly everyone. Social surveys in cities consistently rank road traffic noise as one of the primary sources of annoyance. Cities can help to control traffic noise by rerouting heavy traffic and smoothing the flow of traffic so that vehicles avoid unnecessary starts, stops, and acceleration. Requirements that motor vehicles be maintained properly can also reduce traffic noise, as can the construction of sound-barrier walls near highways. Because most of the noise from a vehicle traveling at low speed is radiated from the exhaust system, a good muffler is very effective in controlling the acoustical emission. Given the recognition of noise pollution as a serious problem, the U.S. Environmental Protection Agency (EPA) has made several recommendations for reducing noise from motor vehicles.

Airplanes are a major source of noise pollution in urban areas. (AP/Wide World Photos)

INDOOR NOISE

A high proportion of the U.S. workforce is employed in interior environments in which the workers are subject to long periods of exposure to noise. Indoor noise also affects people living in apartments and in houses of relatively light construction. When indoor noise pollution results from sounds produced outside indoor spaces, lining the walls and ceilings of the spaces with acoustic panels and other sound-absorbing materials can be helpful in reducing the noise. In addition, the operation of most home appliances and factory machinery produces noise. Noise from washing machines, drills, and air-conditioning systems arises from friction, unbalanced rotating parts, and air turbulence created by fans. Such noise can be substantially reduced through proper lubrication, the balancing of rotating parts, and the installment of acoustic insulation. In an effort to encourage the design of quieter mechanical products, the American National Standards Institute has published guidelines for manufacturers to use in rating noise emission in their products. Adherence to the guidelines has been less than uniform, but perhaps this will change if consumers and employers express a willingness to pay higher prices for quieter appliances and machinery.

Other sources of noise pollution include some that are not even detectable by the human ear. Ultrasonic and infrasonic noises possess frequencies above and below the audible range, respectively. Such noises are emitted, for instance, through the background hum of high-voltage transmission cables. Although not technically heard, ultrasonic and infrasonic noises affect people in ways similar to audible noise.

ADVERSE EFFECTS

The adverse physiological and psychological effects of noise on people have been the subject of considerable study. Researchers have found that exposure to noise interferes with people's ability to work and to sleep and also infringes on their enjoyment of recreation. Noise pollution has been associated with fatigue, loss of appetite, indigestion, irritation, and headaches.

High-intensity noise has been shown to have adverse cumulative effects on the human hearing mechanism that may produce temporary or permanent deafness. In fact, noise-induced hearing loss has been identified as a major health hazard by the U.S. Department of Labor's Occupational Safety and Health Administration. Noise pollution decreases worker efficiency and increases worker error rates.

Fai Ma

FURTHER READING

Berg, Richard E., and David G. Stork. *The Physics of Sound.* 3d ed. Upper Saddle River, N.J.: Prentice Hall, 2005.

Chiras, Daniel D. "Air Pollution and Noise: Living and Working in a Healthy Environment." In *Environmental Science.* 8th ed. Sudbury, Mass.: Jones and Bartlett, 2010.

Kotzen, Benz, and Colin English. *Environmental Noise Barriers: A Guide to Their Acoustic and Visual Design.* 2d ed. New York: Taylor & Francis, 2009.

Nadakavukaren, Anne. "Noise Pollution." In *Our Global Environment: A Health Perspective.* 6th ed. Long Grove, Ill.: Waveland Press, 2006.

Rossing, Thomas D., F. Richard Moore, and Paul A. Wheeler. *The Science of Sound.* 3d ed. San Francisco: Addison-Wesley, 2002.

SEE ALSO: Environmental health; Environmental illnesses; Environmental Protection Agency.

North America

CATEGORIES: Places; ecology and ecosystems; land and land use

SIGNIFICANCE: As the location of three heavily populated countries with intense industrial and agricultural activity, North America is a significant contributor to such environmental problems as air and water pollution, habitat destruction, and species loss. The environment management policies of the continent's three nations thus play an important role in the well-being of ecosystems on a global scale.

Although Canada, the United States, and Mexico all function under independent and different governmental systems, the three countries are involved in many collaborative efforts to protect their shared environment. With the increasing awareness of the fact that the political, economic, and social decisions of all countries have environmental impacts far beyond their own borders, all three of the countries of North America have increased their attention to environmental concerns and are committed to a cooperative approach to addressing environmental problems.

Cooperation between Canada and the United States in regard to environmental issues dates back to 1909, when the two countries signed the Boundary Waters Treaty and established the International Joint Commission to manage the bodies of water shared by the two nations. More than thirty agreements regarding the environment have since been signed by Canada and the United States, including the Great Lakes Water Quality Agreement (1972), which protects the ecosystem of the Great Lakes basin, and the Air Quality Agreement (1991), which is concerned with the harmful emissions that cause acid rain. The U.S. Environmental Protection Agency and Environment Canada have implemented and oversee a number of agreements dealing with problems of transboundary movement of hazardous substances ranging from toxic chemicals to solid waste.

Just as the United States and Canada's first cooperative environmental efforts addressed problems along a shared border, so did those of the United States and Mexico. The first agreement between the United States and Mexico concerning the environment was the Mexican Water Treaty of 1944, which concerned the use of the waters of rivers shared by the two nations. In August, 1983, the two countries signed the Agreement on Cooperation for the Protection and Improvement of the Environment in the Border Area, known as the La Paz Agreement, which was intended to reduce risks to public health along the border and sought to restore the area to its natural state. In November, 1993, Mexico and the United States established the North American Development Bank (NADB) and the Border Environment Cooperation Commission (BECC). BECC works with local communities in areas affecting public health, such as solid waste disposal and the treatment of drinking water, and the NADB finances BECC projects. In 2006 the two organizations were combined into a single entity.

With the signing of the North American Free Trade Agreement (NAFTA) in 1992 (the agreement went into effect in January, 1994) by all three nations of North America, concerns arose about the potential impacts of increased industrial and agricultural activity on the environment of Mexico and consequently on North America's environment. As a result, the three countries also signed the North American Agreement on Environmental Cooperation, which entered into effect on the same date as NAFTA. Under this agreement, the Commission for Environmental Cooperation was established to ensure that the new trade agreement among Canada and the United States, two highly developed countries, and Mexico, a country still in the process of economic development, would not have harmful impacts on the environment. The commission not only addresses environmental problems but also works to ensure stronger enforcement of existing environmental laws. Many environmental concerns are shared by all three countries, including air and water pollution, greenhouse gas emissions, land destruction and deforestation, loss of habitat, and loss of land and marine biodiversity.

AIR POLLUTION AND GREENHOUSE GAS EMISSIONS

With its densely populated metropolitan areas and its intensive manufacturing and agricultural activities, North America has serious problems with air pollution. Among the most significant contributors to air pollution across the continent are emissions from the transportation sector. The large numbers of motor vehicles used daily in the major metropolitan areas throughout North America emit large quantities of carbon monoxide, nitrogen oxide, particulate matter, and carbon dioxide into the atmosphere, resulting in ground-level ozone and smog, which are hazardous to human health and to the health of other animals and plant life, including crops and trees. Although primarily produced in metropolitan areas and in areas of heavy industrial production and at power plant locations, ground-level ozone is also a threat to the environment in rural areas, as it is carried into these areas by wind. These emissions, which are primarily greenhouse gases, have also been linked to global warming and climate change.

The petroleum industry and the generation of power by the burning of fossil fuels also contribute heavily to air pollution in North America. Increasing demand for electricity has resulted in increased burning of fossil fuels across the continent. In the late twentieth century, the preference of many North Americans for sport utility vehicles (SUVs) and pickup trucks rather than more fuel-efficient vehicles contributed to increases in the use of fossil fuels. Canada's development of the oil sands in the Athabasca basin in the province of Alberta, a process that requires extraction by open-pit or strip mining, led to an increase in greenhouse gas emissions in that region. In spite of this increase in the petroleum sector, Canada saw a decrease in its overall greenhouse gas emissions in 2008, owing in large part to a sluggish economy.

In 2008 Mexico experienced a 25 percent increase in greenhouse gas emissions, primarily because of problems of flaring and leaks in the production of natural gas and petroleum. The Mexican government has proposed to accomplish its goal of reducing emissions by controlling flaring and leaks in the oil industry and by moving toward more efficient cars and power plants.

With its dependence on motor vehicle transportation and on fossil fuels as the major source of energy generation, the United States emits large amounts of the greenhouse gases that contribute to global warming. While some economic sectors in the United States (particularly those involved with solar energy and electric cars) have done considerable research to find means of reducing emissions, others, such as the coal industry, have been slow to develop and implement measures to protect the environment. The urgency of the need to reduce greenhouse gas emissions remains a topic of debate in the United States.

WATER POLLUTION AND LOSS OF BIODIVERSITY

Water pollution is an important concern in North America. Although both the United States and Canada provide their residents, in general, with better water quality than is found in much of the rest of the world, in rural areas throughout the continent, polluted water from contaminated wells and rivers, lack of proper sewage disposal, and improper disposal of waste from animal confinement systems still can be a serious hazard to human health. Agriculture also contributes to water pollution through irrigation runoff from crop fields, which contains residues of pesticides and fertilizers. Polluted water poses a threat to both land and aquatic species. Toxic chemicals in water affect entire ecosystems as they destroy vegetation and poison fish as well as the animals that eat the fish.

Acid rain remains a serious problem for Canada owing to the transboundary movement of pollutants from coal- and oil-fired power plants, motor vehicles, and smelters in the United States. Canada's soils and lakes lack the alkalinity necessary to neutralize the sulfur, mercury, and other toxic substances present in acid rain, which ruins the soil for crop production, destroys vegetation in the areas affected, and renders lakes and rivers poisonous to fish, other wildlife, and human beings. In addition, acid rain causes stratospheric ozone depletion and makes lakes and rivers more susceptible to ultraviolet radiation.

LAND DESTRUCTION AND DEFORESTATION

Urban sprawl, industrial development, agricultural practices, and other human activities all contribute to land destruction and deforestation. Throughout North America, a long-term trend has been a population shift from rural areas toward the cities, where economic opportunities offer the prospect of a better quality of life. This migration over time has resulted in shifts in land use. As cities expand, land that was prairie, forest, or prime cropland disappears with the building of housing developments, roads, and paved shopping and entertainment centers. Such changes significantly affect the environment through increases in greenhouse gas emissions and the pollution of both water and land as the result of human activities. Further, the loss of natural areas, particularly the loss of forests, contributes to global warming and climate change, because trees are natural cleansers of carbon dioxide from the air.

Although urban sprawl significantly affects the environment as it increases pollutants, agriculture has far greater impacts on the environment. The negative effects are particularly great when land that is not naturally suitable for agriculture is used for that purpose; such land requires intensive irrigation and fertilization to be productive as cropland, and both practices are harmful to the environment. Throughout North America, the overexploitation of land for increased food and biofuel crop production and the overgrazing of grasslands by cattle have caused serious damage to the environment.

The tropical rain forests of south and southeast Mexico have been drastically reduced in area as land has been cleared and converted to pastureland for cattle; this change in land use has resulted in severe soil erosion. In the north and northwestern parts of the country, intensive irrigation has caused not only soil erosion but also desertification, owing to the saline content of the water used.

In the United States, especially in the western states, agriculture uses more water than does any other economic sector. Modern intensive farming methods also cause considerable damage to waterways and to groundwater, contaminating them with pesticides and fertilizers through irrigation runoff. The expansion of farming activities has led to deforestation as woodlands have been cleared and converted to cropland. The major threats to the environment from land use in Canada come from the exploitation of oil sands, which threatens certain forest areas, and from urban expansion in the Great Lakes basin.

LOSS OF HABITAT AND SPECIES ENDANGERMENT

Throughout North America, the natural habitats of many species have been destroyed as forests, grasslands, and wetlands have been converted to other uses, including industrial development and agriculture. This destruction of habitats has resulted in the endangerment of many species native to the continent, such as the peregrine falcon, the condor, and the wolf.

(continued on page 882)

Habitats and Selected Vertebrates of North America

Baleen Whale

Polar Bear

Caribou

Kodiak Bear

Seal

Moose

Musk Ox

TUNDRA

Walrus

Grizzly Bear

Mule Deer

Arctic Wolf

Walrus

Timber Wolf

Wolverine

Hudson Bay

Mountain Goat

ROCKY MOUNTAINS

Red Fox

Lynx

White-Tailed Deer

Steller's Sea Lion

Badger

Beaver

BOREAL FOREST

Mountain Lion

Bison

Porcupine

Seal

Elk

COASTAL REGION

Bighorn Sheep

GRASSLANDS

Gray Squirrel

Sea Otter

Sea Otter

Pronghorn Antelope

Groundhog

California Sea Lion

Prairie Dog

HARDWOOD FOREST

SOUTHWESTERN DESERT

Cottontail Rabbit

Rattlesnake

Atlantic Ocean

Coyote

Opossum

Armadillo

Pelican

Iguana

Spoonbill

Peccary

Gulf of Mexico

Alligator

Parrot

NEOTROPICAL FOREST

Spider Monkey

Macaw

Pacific Ocean

The activities of North Americans have also endangered marine life. One of the most crucial areas of environmental importance affecting the three North American countries involves the extraction of the oil from deposits found beneath the seas off their shores. The process of drilling for the oil can pollute ocean waters, whether through relatively small leaks or through disastrous spills such as the one that resulted from an explosion on BP's *Deepwater Horizon* drilling platform in the Gulf of Mexico in April, 2010. The pollution caused by oil spills endangers fish, shellfish, seabirds, and marine mammals.

Shawncey Webb

FURTHER READING

Beck, Gregor Gilpin, and Bruce Litteljohn, eds. *Voices for the Watershed: Environmental Issues in the Great Lakes-St. Lawrence Drainage Basin.* Montreal: McGill-Queen's University Press, 2000.

Earle, Sylvia A. *The World Is Blue: How Our Fate and the Ocean's Are One.* Washington, D.C.: National Geographic Society, 2009.

Gallagher, Kevin. *Free Trade and the Environment: Mexico, NAFTA, and Beyond.* Stanford, Calif.: Stanford University Press, 2009.

Hillstrom, Kevin, and Laurie Collier Hillstrom. *North America: A Continental Overview of Environmental Issues.* Santa Barbara, Calif.: ABC-CLIO, 2003.

Lombard, Emmett. *International Management of the Environment: Pollution Control in North America.* Westport, Conn.: Praeger, 1999.

López-Hoffman, Laura, et al., eds. *Conservation of Shared Environments: Learning from the United States and Mexico.* Phoenix: University of Arizona Press, 2009.

SEE ALSO: Acid deposition and acid rain; Agricultural chemicals; Deforestation; Desertification; Erosion and erosion control; Grazing and grasslands; Habitat destruction; Land pollution; Oil shale and tar sands; Rain forests; Urban sprawl.

North American Free Trade Agreement

CATEGORIES: Treaties, laws, and court cases; atmosphere and air pollution; water and water pollution

THE TREATY: International agreement providing for the removal of trade barriers and reduction of many important legal and financial restrictions among the United States, Canada, and Mexico

DATE: Signed on December 17, 1992

SIGNIFICANCE: The North American Free Trade Agreement is considered by its supporters to mark new directions in international relations by opening the borders of three very different countries in ways that are intended to enhance not only economic but political and cultural relations as well. Although results have fallen short of original expectations, provisions of the North American Agreement on Environmental Cooperation, a side treaty that went into effect at the same time, are intended to raise awareness of environmental protection issues in all three countries.

The North American Free Trade Agreement (NAFTA), an unparalleled trade accord among the United States, Canada, and Mexico, was signed on December 17, 1992. Once ratified by the legislative bodies of the three nations, the agreement went into effect on January 1, 1994. Although the United States and Canada had established open bilateral trading terms in 1988 (the Canada-United States Free Trade Agreement), NAFTA not only aimed at expanding the earlier agreement, particularly in terms of liberalization of conditions for cross-border private investment, but it also sought to integrate terms of trade and investment between two highly developed national economies and a third emerging, or developing, country. An important aspect of Mexico's role in NAFTA was an expectation that the steps toward liberalization of the Mexican economy begun in 1985 (that is, privatization of traditionally state-run companies and increased emphasis on market-oriented economic activity) would continue at a regular pace.

A first and major aim of NAFTA was to eliminate tariff-based trade "barriers" that historically had shielded domestically produced goods, whether agricultural or industrial, from competition from lower-priced foreign imported goods. High tariffs automatically raise prices for imports. An equally important goal was to eliminate, as much as possible, individual national laws hampering the free flow of labor and capital across the signatories' borders.

Beyond measurable economic results, the "spirit" of NAFTA aimed at improving political relations among the three signatories, especially between the United States and Mexico. A very high priority, for example, was (and continues to be) the need for political cooperation in the war against commerce in illicit drugs—increasingly a military as well as a political necessity. Decades-old concerns over the movement of illegal Mexican immigrants across the border into the United States stood to be reexamined in light of NAFTA's commitment to free trade not only in goods but also in cross-border labor arrangements. Another area dependent on mid- to long-term cooperation across national borders involves proposed programs, largely through shared technology, for ecological sustainability.

Numerous research reports appeared before NAFTA went into effect and continued to appear as analysts tried to estimate the relative attractions and disadvantages for one or another of the signatories that might result from the application of the agreement. In fact, during the first decade and a half of NAFTA's operation, certain patterns, some expected and welcomed, others quite controversial, began to take form. Parties that had opposed the agreement seemed convinced that their negative position had been justified. One argument, for example, was aimed at cross-border investment patterns. Opponents of NAFTA argued that, as U.S. private investment in Mexico increased, levels of investment in the United States itself would be decreased proportionately. Although statistics from the mid-1990's showed some shift in the value of U.S. capital going to Mexico, the overall weight of the movement—not overwhelmingly great—had to be compared with capital going to other, non-NAFTA countries that continue to attract very high levels of U.S. private investment.

On the positive side, supporters of NAFTA could argue that gradual but continuous growth of the Mexican economy would increase its capacity to import a wide variety of goods from its northern neighbors, espe-

cially from the United States. The movement of imports into Mexico would obviously be enhanced by reduced Mexican tariff rates—historically as much as three times the tariff rates of the United States. Estimates of rising levels of U.S. exports to Mexico seemed convincing, rising from about twelve billion dollars in 1986 to more than forty billion dollars in 1993.

Perhaps the strongest arguments against trends facilitated by NAFTA have had to do with potentially controversial effects on labor conditions, particularly on both sides of the Mexico-U.S. border. Critics claim that, as more and more companies opt to move their manufacturing activities to locations in Mexico, where labor costs are considerably lower, U.S. factories are losing orders even to the point of having to close down. Another major concern is symbolized by the Spanish term *maquiladora*, which refers to a fac-

President Bill Clinton speaks to small business owners and operators on November 15, 1993, urging them to encourage their U.S. senators to ratify the North American Free Trade Agreement. (AP/Wide World Photos)

tory set up in Mexico with the specific aim of importing (duty-free under NAFTA's terms) machinery and parts needed for assembly of a wide range of goods (ranging from clothing to automobiles) to be exported (again duty-free) for sale on the U.S. market. Many *maquiladoras* have been criticized for exploitative labor practices (substandard wages and working conditions, tenuous job security, and so on) that, according to critics, have been ignored by those profiting north of the border.

Support for goals set by NAFTA's founders during the 1990's continued to be voiced by Mexico's political leadership through the first decade of the new century. In an interview aired on American television in March, 2010, for example, President Felipe Calderón stressed his continued belief that a complementarity exists between Mexico's labor-intensive economy and the capital-intensive economies of its northern neighbors, especially the United States. Whether his words reflected a realistic appraisal or hopeful idealism, he concluded that the two nations "need each other."

This shared need is especially clear in matters touching on environmental protection. Although initially there were expectations that NAFTA would create positive conditions for ecological improvements in all three member states, such hopes gradually fell short of realities. On one hand, optimists predicted that hoped-for increases in per-capita income in Mexico as a result of higher levels of capital investment, accompanied by an expected shift away from traditional pollution-intensive industries south of the U.S. border (such as cement and base metals production), would help reduce heavy levels of pollution in Mexico. On the other hand, pessimists claimed that laxity in Mexico's regulation of pollution might even attract heavy polluters from the more developed NAFTA members to the north. The outcomes predicted by either side have not materialized in statistically provable terms.

Tremendously high levels of pollution in and around Mexico City aside (as this problem is beyond the purview of NAFTA technical observers), Mexico's record in dealing with soil erosion, water pollution, and urban solid waste pollution remains controversial. Although NAFTA has from the outset been "armed" with an institution specifically designated to coordinate environmental protection efforts (the Commission for Environmental Cooperation, or CEC, which was created by the North American Agreement on Environmental Cooperation, a side treaty of NAFTA), only one-third of the CEC budget (about nine million dollars in the late 1990's) goes to Mexico. In fact, although Mexico did emerge from the major slump that hit its economy in the 1980's, by 2010 expenditures for environmental improvements (including vital inspections of pollution-intensive industries) had never reached target levels laid down by the CEC.

Many environmental protection frustrations identified in the first decades of NAFTA's existence remained and even grew in the first decade of the twenty-first century. Severe budgetary problems that originally seemed to be solely a Mexican dilemma, for example, affected both Canada and the United States when near collapse of the world financial system struck in 2008-2009. This development raised fears of necessary reductions in funding for environmental projects.

Byron Cannon

FURTHER READING

Cameron, Maxwell A., and Brian W. Tomlin. *The Making of NAFTA: How the Deal Was Done.* Ithaca, N.Y.: Cornell University Press, 2000.

Gallagher, Kevin. *Free Trade and the Environment: Mexico, NAFTA, and Beyond.* Stanford, Calif.: Stanford University Press, 2004.

Grinspun, Ricardo, and Yasmine Shamsie, eds. *Whose Canada? Continental Integration, Fortress North America, and the Corporate Agenda.* Montreal: McGill-Queens University Press, 2007.

Hansen, Patricia Isela. "The Interplay Between Trade and the Environment Within the NAFTA Framework." In *Environment, Human Rights, and International Trade,* edited by Francesco Francioni. Portland, Oreg.: Hart, 2001.

Hufbauer, Gary Clyde, and Jeffrey J. Schott. *NAFTA Revisited: Achievements and Challenges.* Washington, D.C.: Institute for International Economics, 2005.

McPhail, Brenda M., ed. *NAFTA Now! The Changing Political Economy of North America.* Lanham, Md.: University Press of America, 1995.

SEE ALSO: Globalization; North America; Sustainable development; Trans-Alaska Pipeline.

North Sea radioactive cargo sinking

CATEGORIES: Disasters; nuclear power and radiation
THE EVENT: Sinking of a ship carrying a cargo of hundreds of tons of radioactive material
DATE: August 25, 1984
SIGNIFICANCE: Although no radioactive material was released into the environment from the cargo carried by the *Mont Louis* when it sank in the North Sea, the incident brought international attention to the dangers of transporting such material.

Uranium hexafluoride is a radioactive material created in the process of refining uranium. Naturally occurring uranium consists of two isotopes: uranium 235 and uranium 238. Only uranium 235 can be split apart in a nuclear reactor to produce energy. In the process used to separate the two isotopes, the uranium is converted to uranium hexafluoride and passed through a gaseous diffusion facility.

On the night of August 25, 1984, the small French ship *Mont Louis* was running eastward in the North Sea, parallel to the coast of Belgium. It carried a cargo of thirty containers holding 225 tons of uranium hexafluoride. At the same time the *Mont Louis* was passing, the West German vessel *Olau Britannia* departed from Flushing, Holland, bound for Sheerness, England. In this situation, the *Olau Britannia* had the right-of-way, but there was heavy fog, and the sailor who should have been the *Mont Louis* lookout had been assigned other duties. The *Olau Britannia* struck the *Mont Louis* in the starboard, or right, side. The two ships remained locked together until they were pulled apart by salvage tugs, at which point the *Mont Louis* capsized and sank in 14 meters (45 feet) of water. Although five members of the ship's crew fell into the sea, they were quickly rescued, and no one on the *Olau Britannia* was seriously injured. Some of the fuel aboard the *Mont Louis* leaked into the sea and fouled a Belgian beach near Ostend.

The uranium hexafluoride on the *Mont Louis* had been created in France, and it had been en route to a gaseous diffusion facility in Riga, Latvia. Most of the uranium hexafluoride had been made from freshly mined uranium, but a small amount had been made from used reactor fuel. While uranium itself is mildly radioactive, uranium hexafluoride made from used reactor fuel is contaminated with highly radioactive materials produced in the reactor.

The *Mont Louis* lay with its starboard side above the water and the containers of uranium hexafluoride still in its hold. Smit Tak International, a salvage company, was hired to recover the uranium hexafluoride and then raise the sunken ship. The salvage workers cut a large hole in the exposed side of the ship and used a floating crane to lift out the containers of uranium hexafluoride. Throughout this process, which took weeks, radioactivity detectors were used, but no significant radioactivity was found. The last container of uranium hexafluoride was recovered on October 4, 1984. The wrecked ship was raised on September 29, 1985. The estimated cost of the salvage operation was $4.6 million.

Rules published by the International Maritime Organization state that ships larger than a certain size are requested, but not required, to notify authorities in those nations whose coasts they pass if they carry dangerous cargo. Because the *Mont Louis* was under the size limit, its owners were not obligated to follow this recommendation. The crew had received no special training in handling radioactive cargo; they may not even have known that the ship carried such cargo. It was fortunate that no uranium hexafluoride was released into the environment in this incident. If a release had occurred, it would have posed a danger to personnel near the scene but not to people onshore.

Edwin G. Wiggins

FURTHER READING
Angler, Natalle. "A Shipwreck Sends a Warning." *Time*, September 10, 1984, 33.
Business Week. "An Accident at Sea Raises Nuclear Alarms." September 10, 1984, 58.
Makhijani, Arjun, Howard Hu, and Katherine Yih, eds. *Nuclear Wastelands: A Global Guide to Nuclear Weapons Production and Its Health and Environmental Effects*. 1995. Reprint. Cambridge, Mass.: MIT Press, 2000.

SEE ALSO: Nuclear accidents; Nuclear and radioactive waste; Nuclear regulatory policy; Ocean pollution.

Northern spotted owl

CATEGORY: Animals and endangered species

IDENTIFICATION: Medium-sized, dark brown owl native to the coniferous forests of southwestern Canada and the northwestern United States

SIGNIFICANCE: Destruction of northern spotted owl habitats by timber companies has led to intense conflict between loggers trying to earn a living and environmentalists striving to save this threatened species and the old-growth forests it inhabits. A 1995 U.S. Supreme Court decision regarding the owl upheld the government's authority to conserve habitat as part of its efforts to protect a threatened or endangered species.

The northern spotted owl (*Strix occidentalis caurina*) favors a habitat of old-growth trees for foraging, roosting, and nesting. The survival of the species has been threatened by the commercial harvesting of old-growth timber in California, Oregon, Washington, and British Columbia. During the 1880's the owl's habitat totaled about 6.1 million hectares (15 million acres); one century later it had dwindled to fewer than 2 million hectares (5 million acres), almost entirely on federal lands.

In 1984 four conservation groups appealed the U.S. Forest Service's regional plans for timber harvesting within the owl's range. Four years later, federal regulators proposed setting aside up to 890 hectares (2,200 acres) of suitable habitat for each nesting pair of owls. While timber companies argued that such a policy would lead to dire economic consequences in the region, environmentalists insisted that more habitat was needed. In 1990 the northern spotted owl was listed as a threatened species under the 1973 Endangered Species Act. The next year, environmental groups sued the U.S. Department of the Interior for failing to protect the owl, and a federal district court in Seattle, Washington, ordered the cessation of logging operations in the region's old-growth federal forests. In 1992 the U.S. Fish and Wildlife Service (FWS) designated critical habitat for the owl on almost 2.8 million hectares (6.9 million acres) of federal lands in the Pacific Northwest. During that year's presidential campaign, candidate Bill Clinton promised that if elected he would convene a summit to consider the spotted owl controversy.

At the resulting Forest Summit in 1993, President Clinton and members of his cabinet met in Portland, Oregon, with representatives from environmental groups, the timber industry, and the salmon fishery industry. Thousands of timber workers gathered in Portland to demonstrate their concerns about the recent loss of between thirteen thousand and twenty thousand jobs in the industry. Clinton subsequently announced a compromise plan that allowed the logging industry to cut about 1.2 billion board feet of timber each year, down from about 5 billion board feet each year during the preceding decade. The administration estimated that only six thousand jobs would be lost as a result of the logging cutback. To compensate for the loss, the president asked Congress to create the Northwest Economic Adjustment Fund, which would provide $1.2 billion in aid over the next five years. The district court of Seattle upheld the legality of the Clinton plan.

The timber industry went to court to challenge the Clinton administration's broad interpretation of the

A northern spotted owl in a tree in the Deschutes National Forest in Oregon. (USFWS)

Endangered Species Act. At issue was whether the government might restrict activities, such as logging, that only indirectly limit the habitat of the northern spotted owl. In *Babbitt v. Sweet Home Chapter of Communities for a Great Oregon* (1995), the U.S. Supreme Court ruled that the government has the discretionary authority to protect any habitat required by an endangered species.

No affected group was entirely happy with the Clinton compromise. In 1997 the U.S. Forest Service released the results of a study that found the spotted owl population was dropping by 4.5 percent per year and that the rate was accelerating. While environmentalists worried about the owl's fate, the timber industry claimed that the settlement would lead to a total of about eighty-five thousand job losses. Numerous lumber mills were forced to close, and many workers either moved elsewhere or retrained for other jobs, which often paid minimum wage. In 1996 the timber industry of Linn County, Oregon, generated only $2.1 million, about 17 percent of what it had generated eight years earlier.

In early 2003 the FWS made an agreement with logging interests to conduct a status review of the northern spotted owl and consider revisions to its critical habitat. The FWS concluded in 2004 that the owl should remain listed as a threatened species. Spotted owl populations were found to be in decline in Canada, Washington, and northern Oregon, despite habitat protection and restoration; in southern Oregon and California, populations were generally stable or experiencing only slight declines or increases.

The FWS status review also led to revision of the owl's critical habitat—a reduction of approximately 647,500 hectares (1.6 million acres), effective in September, 2008. A 2008 FWS plan for the owl's recovery proposed a network of managed habitat blocks on federal land. The FWS noted that securing habitat alone will not be sufficient to effect recovery, citing a larger, more aggressive competing species, the barred owl (*Strix varia*), that threatens to displace the northern spotted owl. The barred owl, an eastern North America native, expanded its range westward during the twentieth century. Some interbreeding and hybridization of the spotted and barred owl species have been noted.

Environmental groups lashed out at the FWS cutback in habitat area. They accused George W. Bush's presidential administration of political interference with the recovery plan and challenged the plan in court. In 2009, under the administration of President Barack Obama, the plan and revised habitat designation were remanded for reevaluation.

Thomas T. Lewis
Updated by Karen N. Kähler

FURTHER READING

Bevington, Douglas. *The Rebirth of Environmentalism: Grassroots Activism from the Spotted Owl to the Polar Bear.* Washington, D.C.: Island Press, 2009.

Courtney, Steven P., et al. *Scientific Evaluation of the Status of the Northern Spotted Owl.* Portland, Oreg.: Sustainable Ecosystems Institute, 2004.

Johnsgard, Paul A. *North American Owls: Biology and Natural History.* Washington, D.C.: Smithsonian Institution Press, 2002.

Keiter, Robert B. "Ecology Triumphant? Spotted Owls and Ecosystem Management." In *Keeping Faith with Nature: Ecosystems, Democracy, and America's Public Lands.* New Haven, Conn.: Yale University Press, 2003.

Stout, Benjamin B. *The Northern Spotted Owl: An Oregon View, 1975-2002.* Victoria, B.C.: Trafford, 2006.

U.S. Fish and Wildlife Service. *Final Recovery Plan for the Northern Spotted Owl (Strix Occidentalis Caurina).* Portland, Oreg.: Author, 2008.

SEE ALSO: Endangered Species Act; Endangered species and species protection policy; Forest Summit; Logging and clear-cutting; Old-growth forests.

Nuclear accidents

CATEGORY: Nuclear power and radiation

DEFINITION: Unplanned events involving the release of radioactive materials into the environment

SIGNIFICANCE: Public concern about reactor safety and the risk of accidents remains an important reason the expansion of nuclear power has stalled in the United States and many other nations. Although the industry's safety record has generally been good, safety lapses and near accidents are sometimes reported, and the public has become increasingly aware that a severe accident could have widespread consequences.

The accident risk in nuclear power plants lies in the possibility that a malfunction caused by equipment failure, operator error, or external events will disrupt the flow of vital cooling water to the intensely hot, dangerously radioactive reactor core, a situation that could result in the fuel melting. Possible problematic external events include natural disasters such as the massive earthquake and subsequent tsunami that damaged reactors at Japan's Fukushima 1 nuclear power plant in March, 2011. In the unlikely event that both the huge steel vessel holding the fuel and the massive concrete containment structure surrounding the reactor are breached, large quantities of dangerous radioactive materials could be dispersed to the environment.

The U.S. Nuclear Regulatory Commission (NRC) is the federal agency that sets the safety requirements that plant owners must meet in the construction and operation of nuclear power reactors in the United States. NRC regulations and technical specifications cover a wide range of areas, including design requirements, safety systems, equipment quality, record keeping, and operator training.

INCIDENTS IN THE UNITED STATES

The worst nuclear accident to have occurred in the United States took place in 1979 at the Three Mile Island (TMI) Unit 2 plant in Pennsylvania, resulting in more than one-half of the fuel melting and the total loss of the reactor. Recurring incidents and near accidents over the years have led to some disquiet among the public and among nuclear regulators. Among these have been the 1975 fire at the Brown's Ferry plant in Alabama that burned for seven hours, during which it took plant personnel several hours to shut down the reactor; the 1983 unexpected failure of the Salem reactor in New Jersey to shut down after a safety system was activated; and the 1985 incident at the Davis-Besse plant in Ohio when multiple equipment failures caused a loss of cooling water that could have initiated a core meltdown if plant operators had not responded quickly. At some plants it has been discovered that important safety equipment has been inoperable for years. The NRC has kept a number of U.S. reactors shut down for long periods until safety-related equipment problems or operator lapses could be corrected.

Proponents of nuclear energy argue that, despite recurrent mishaps, the U.S. industry's overall safety record has been excellent. They assert that the events at TMI showed that systems designed to contain an accident worked. They claim that an accident similar to the one that occurred at Chernobyl in 1986 could not occur in the United States because the design of the older, Soviet-designed Chernobyl-style plants is flawed and permits a severe accident to occur far more easily than in U.S.-designed plants. They note that the number of safety-related occurrences per year at U.S. reactors has declined since the TMI accident. Finally, they point to government-sponsored studies that have concluded that there is a very low probability of a severe accident occurring with significant off-site property damage and injuries to the public.

Antinuclear activists point to numerous recurring safety-related incidents and problems at U.S. reactors, including some cases in which serious accidents were narrowly averted. They argue that the TMI incident showed that a serious accident is not as improbable as the industry claims. While critics acknowledge the design superiority of U.S. reactors compared to the Chernobyl plant, they note that an NRC commissioner testified before Congress in 1986 that an accident at a U.S. reactor with off-site releases equal to or worse than what occurred at Chernobyl could occur under conditions regarded as improbable but not impossible. Critics have also attacked the assumptions and methodology of the government's major nuclear accident risk studies. Finally, opponents observe that those same studies indicate that if a low-probability catastrophic accident should nonetheless occur, it could result in tens of thousands of deaths and injuries, tens of billions of dollars in property damage, and widespread, long-lived radioactive contamination.

The Price-Anderson Act limits the liability of nuclear power plant owners and equipment vendors in the event of a severe accident. This legislation was passed in 1957 after private companies made it clear that they would not participate in the development of nuclear energy without liability protection. Critics argue that the nuclear industry's continuing insistence that such protection is required is inconsistent with its denial of the possibility of a catastrophic accident.

INCIDENTS IN OTHER NATIONS

Much less information is generally available about nuclear accidents and incidents with serious safety implications in countries other than the United States because the public disclosure requirements of most nations are less stringent than U.S. requirements. Worrisome accidents and equipment malfunctions are known to have occurred at reactors in many nations, including Japan, India, and the countries of the former Soviet Union. In Europe, grave concern remains about older Soviet-designed reactors, including some

Chernobyl-style plants, still operating in the countries of Eastern Europe and the former Soviet Union. Western nations have provided some funding for safety improvements and have sought eventual permanent closure of many of these plants.

A number of nuclear accidents have occurred at noncommercial facilities, some of which resulted in serious environmental consequences. In 1952 a fuel melt and explosion severely damaged an experimental Canadian reactor at Chalk River, Ontario. In 1957 an explosion at a waste site at the then-secret Chelyabinsk complex near the city of Kyshtym in Russia contaminated hundreds of square miles of the surrounding countryside. Also in 1957, a fire occurred in England at the Windscale reactor (an atypical plant used to produce plutonium for the United Kingdom's weapons program); the surrounding countryside was contaminated, and radioactive fallout drifted into several neighboring countries.

Phillip A. Greenberg

Aerial view of the Chernobyl nuclear power plant after its April, 1986, accident. The destroyed fourth reactor is seen in front of the plant's chimney. (AP/Wide World Photos)

FURTHER READING

Bodansky, David. *Nuclear Energy: Principles, Practices, and Prospects.* 2d ed. New York: Springer, 2004.

Cooper, John R., Keith Randle, and Ranjeet S. Sokhi. *Radioactive Releases in the Environment: Impact and Assessment.* Hoboken, N.J.: John Wiley & Sons, 2003.

Garwin, Richard L., and Georges Charpak. "Safety, Nuclear Accidents, and Industrial Hazards." In *Megawatts and Megatons: The Future of Nuclear Power and Nuclear Weapons.* Chicago: University of Chicago Press, 2001.

Savchenko, V. K. *The Ecology of the Chernobyl Catastrophe.* New York: Informa Healthcare, 1995.

Wolfson, Richard. *Nuclear Choices: A Citizen's Guide to Nuclear Technology.* Rev. ed. Cambridge, Mass.: MIT Press, 1993.

SEE ALSO: Brazilian radioactive powder release; Chalk River nuclear reactor explosion; Chelyabinsk nuclear waste explosion; Chernobyl nuclear accident; North Sea radioactive cargo sinking; Nuclear regulatory policy; Three Mile Island nuclear accident; Windscale radiation release.

Nuclear and radioactive waste

CATEGORIES: Nuclear power and radiation; waste and waste management

DEFINITION: Broad range of radioactive by-products of nuclear materials used for energy generation, weapons production, research, medicine, and industry

SIGNIFICANCE: Nuclear and radioactive wastes can vary widely in the level and duration of their radioactivity. Some wastes may remain radioactive for as little as a few hours, while others can pose a hazard to human health and the environment for hundreds of thousands of years. Proper disposal of these wastes is fraught with technical and sociopolitical challenges.

Categories of radioactive waste include spent nuclear fuel from nuclear energy facilities, high-level waste (HLW) from spent nuclear fuel reprocessing, radioactive waste by-products from nuclear weapons production at defense processing plants, tailings from uranium ore mining and milling, waste generated during the operation of atomic particle accelerators, and low-level waste (LLW), which includes items such as clothing, rags, equipment, tools, and medical instruments that have become contaminated with radioactivity. The radioactivity of a substance decreases naturally over time, according to the substance's half-life (the time necessary for half of its atoms to decay). While some radioactive wastes may be safe to handle within a relatively short time, some of the more highly radioactive wastes remain unsafe to handle for many thousands of years.

Regulatory control of nuclear and radioactive waste in the United States is vested in a number of government agencies. At the federal level, the Nuclear Regulatory Commission (NRC) and the Department of Energy (DOE) are the primary regulators of highly radioactive waste. Other federal agencies, such as the Environmental Protection Agency (EPA), the Department of Transportation, and the Department of Health and Human Services, can also play a role in the regulation of radioactive material. Individual states are responsible for the regulation, management, storage, and disposal of commercial LLW generated within their boundaries.

After the development of nuclear fission, little concern was initially exhibited over how to dispose of nuclear waste safely. Policy makers perceived nuclear waste disposal as a long-term issue that would need to be addressed sometime in the future. However, as the quantities and types of radioactive and nuclear wastes continued to grow, the issue became increasingly contentious and politicized. The "not in my backyard," or NIMBY, syndrome complicated the process of finding and developing suitable disposal sites. The problem of radioactive and nuclear waste disposal became one of the most controversial and environmentally challenging aspects of the nuclear era.

Waste Disposal Challenges

Only four LLW disposal facilities for commercial and federal, non-DOE radioactive waste are licensed to operate in the United States: one located in Clive, Utah, at what was originally a disposal site for uranium mill tailings; one in Richland, Washington, within the Hanford Nuclear Reservation; one in Barnwell, South Carolina; and one in west Texas, near the New Mexico border. The Utah site is licensed only for

A worker checks the radioactivity of drums containing nuclear waste at the Yonggwang Nuclear Power Plant south of Seoul, South Korea. (AP/Wide World Photos)

those wastes with comparatively short half-lives, whereas the Washington, South Carolina, and Texas facilities accept wastes that will remain radioactive for hundreds of years. Because so few of these disposal sites exist and progress in developing new facilities is slow, much of the country's LLW is stored at the locations where it was produced pending the construction of more permanent facilities. The lack of disposal options has served as an incentive for utilities and other generators of LLW to minimize their waste.

While LLW disposal sites are scarce, HLW disposal sites are nonexistent in the United States. Congress designated Yucca Mountain, Nevada, as the nation's sole deep underground repository for spent nuclear fuel and HLW in 1987. Selection of the site proved highly controversial, with opponents maintaining that a number of factors make it unsuitable for the safe storage of nuclear waste. Among these are the area's volcanic origins, its seismic activity, and the potential for groundwater to act as a transport mechanism for any nuclear waste material that might leak from the disposal facility. DOE scientists have countered that such potential dangers to the integrity of a deep underground storage facility at Yucca Mountain are minimal.

Technical, political, and legal issues have slowed development of a repository at Yucca Mountain in the decades since the site was selected, and the facility has yet to receive any waste for storage. Progress came to a standstill in early 2010, when the federal budget for the coming fiscal year cut funding for the Yucca Mountain project. The DOE subsequently filed a motion with the NRC to withdraw the repository's license application. The withdrawal met with legal opposition, notably from South Carolina and Washington, both states with substantial stores of nuclear waste.

Federally generated, defense-related transuranic waste (waste involving elements with atomic numbers greater than that of uranium) has been disposed of since 1999 in the Waste Isolation Pilot Plant (WIPP) near Carlsbad, New Mexico. At this site waste containers are emplaced in disposal rooms carved into thick salt beds approximately 655 meters (about 2,150 feet, or nearly half a mile) below ground surface. Disposal of commercial wastes, HLW, and LLW at the WIPP is prohibited.

Other Considerations

The delay in establishing a permanent national repository for the disposal of highly radioactive waste has prompted policy makers to search for alternative solutions to the problem of how to store nuclear waste safely. Commercial utilities and defense processing plants have been storing spent nuclear fuel on-site (either in cooling ponds or in aboveground canisters). However, because suitable on-site storage space for highly radioactive waste is running out, maintaining the status quo clearly is not an acceptable long-term alternative. One proposal for a more immediate solution involves the establishment of an interim storage facility for highly radioactive waste. This interim facility—also known as a monitored retrievable storage (MRS) facility—would store nuclear waste aboveground until a permanent repository is ready to accept highly radioactive waste. Environmentalists have expressed concern that any MRS facility could become a permanent aboveground storage facility, which clearly would not be as safe as a deep underground repository. Environmental justice advocates have also argued that Native American populations are likely to be unfairly affected by such facilities; in the past, tribes have been unsuccessfully approached to host MRS sites on their lands in exchange for federal grant money.

Opponents of centralized storage facilities have expressed concerns about the potential for an accident during the transport of nuclear waste to either a national permanent repository or an MRS facility. Transportation routes would carry nuclear waste through a number of highly populated areas. Proponents of regional and national storage facilities counter that a lack of centralization exacerbates management and monitoring problems and increases the possibility of an accident. They also point to the successful existing safety record for the transport of radioactive material.

Nuclear reprocessing has been proposed as a means to make the best use of the waste. Unused uranium and plutonium extracted from nuclear waste through reprocessing can be used to fuel nuclear reactors. Commercial reprocessing was formerly practiced in the United States, but it was suspended in 1976 and banned in 1977 because of concerns that creating stockpiles of uranium and plutonium would increase nuclear weapons proliferation. Reprocessing was a key element of President George W. Bush's Global Nuclear Energy Partnership, proposed in 2006. However, critics cited the dangers of nuclear proliferation and terrorism, the high cost of reprocessing technology, and the fact that reprocessing generates a high volume of LLW as reasons not to pursue commercial reprocessing. In 2009, under the admistration of President Barack Obama, the DOE announced that it was dropping nuclear reprocessing from its agenda.

Kenneth A. Rogers and Donna L. Rogers
Updated by Karen N. Kähler

FURTHER READING

Bayliss, Colin R., and Kevin K. Langley. *Nuclear Decommissioning, Waste Management, and Environmental Site Remediation.* Amsterdam: Butterworth-Heinemann, 2003.

Gerrard, Michael. *Whose Backyard, Whose Risk: Fear and Fairness in Toxic and Nuclear Waste Siting.* Cambridge, Mass.: MIT Press, 1994.

Holt, Mark. *Nuclear Waste Disposal: Alternatives to Yucca Mountain.* Washington, D.C.: Congressional Research Service, 2009.

Lattefer, Arnold P., ed. *Nuclear Waste Research.* Hauppauge, N.Y.: Nova Science, 2008.

McCutcheon, Chuck. *Nuclear Reactions: The Politics of Opening a Radioactive Waste Disposal Site.* Albuquerque: University of New Mexico Press, 2002.

Macfarlane, Allison, and Rodney C. Ewing, eds. *Uncertainty Underground: Yucca Mountain and the Nation's High-Level Nuclear Waste.* Cambridge, Mass.: MIT Press, 2006.

Pusch, Roland. *Geological Storage of Highly Radioactive Waste: Current Concepts and Plans for Radioactive Waste Disposal.* New York: Springer, 2008.

Risoluti, Piero. *Nuclear Waste: A Technological and Political Challenge.* New York: Springer, 2004.

Rogers, Kenneth A., and Donna L. Rogers. "The Politics of Long-Term Highly Radioactive Waste Disposal." *Midsouth Political Science Review* 2 (1998): 61-71.

Vandenbosch, Robert, and Susanne E. Vandenbosch. *Nuclear Waste Stalemate: Political and Scientific Controversies.* Salt Lake City: University of Utah Press, 2007.

Walker, J. Samuel. *The Road to Yucca Mountain: The Development of Radioactive Waste Policy in the United States.* Berkeley: University of California Press, 2009.

SEE ALSO: Antinuclear movement; Hanford Nuclear Reservation; NIMBYism; Nuclear power; Nuclear Regulatory Commission; Nuclear regulatory policy; Nuclear weapons production; Yucca Mountain nuclear waste repository.

Nuclear fusion

CATEGORIES: Nuclear power and radiation; energy and energy use

DEFINITION: Reaction between light atomic nuclei that produces a heavier nucleus and releases energy

SIGNIFICANCE: Nuclear fusion can produce much more energy for a given weight of fuel than any other available energy technology. Controlled fusion reactors could generate high levels of power with uninterrupted delivery and no greenhouse gas emissions. The fusion process produces far less radioactive material than fission reactors, and the by-products generated are less damaging biologically. The radioactive wastes produced by fusion reactors would cease to be dangerous after reasonably short periods of time.

Deuterium, the primary fuel used for nuclear fusion, is abundant in seawater. When a deuterium atom and a tritium atom, both isotopes of hydrogen, are fused together, helium is formed, with the release of a neutron and 17.6 million electronvolts (MEV) of energy. To undergo fusion, the interacting nuclei need kinetic energies on the order of 0.7 MEV to overcome the electrical repulsion between their positive charges. These energies are available at temperatures around 10 million kelvins (18 million degrees Fahrenheit), which poses a major problem for containment of the fusion fuel. In some experiments, the resulting plasma generated by heat from an electrical discharge is contained in the reactor core with appropriately shaped magnetic fields (magnetic confinement). In other experiments, fusion is initiated through the heating and compressing of pellets of the light nuclei with a high-intensity laser beam (inertial confinement). Methods of achieving fusion at lower temperatures, known as cold fusion, continue to be studied, but none has been found that produces more energy than is consumed in the fusion process.

ENVIRONMENTAL ADVANTAGES AND CONCERNS

The natural product resulting from the fusion of deuterium with tritium is helium, which poses no threat to life and does not contribute to global warming. Although tritium is radioactive, it has a short half-life of twelve years, has a very low amount of decay energy, and does not accumulate in the body. It is cycled out of the human body as water, with a biological half-life of seven to fourteen days. Since fusion requires precisely controlled conditions of temperature, pressure, and magnetic field parameters in order to generate net energy, there is no danger of any catastrophic radioactive accident, as heat generation in a fusion reactor would quickly stop if any of these parameters were disrupted by a reactor malfunction. In addition, since the total amount of fusion fuel in the reactor vessel is very small (a few grams) and the density of the plasma is low, there is no risk of a runaway reaction, because fusion would cease in a few seconds if fuel delivery were stopped.

Because of the generation of high-energy neutrons in the deuterium-tritium reaction, the typical structural materials of fusion reactors, such as stainless steel or titanium, tantalum, and niobium alloys, will become radioactive when bombarded by the neutrons. The half-lives of the resulting radioisotopes are typically less than those generated by nuclear fission. Most of this radioactive material would reside in the fusion reactor core and would be dangerous for about fifty years. Low-level wastes would be dangerous for about one hundred years. Since the choice of materials used in a fusion reactor is quite flexible, low-activation materials, such as vanadium alloys and carbon fiber materials, that do not easily become radioactive can be used.

Most fusion reactor designs use liquid lithium as a coolant and for generating tritium when it is bombarded by neutrons coming from the fusion reaction. Because lithium is highly flammable, a fire could release its tritium contents into the atmosphere, posing a radiation risk. Estimates of the amount of tritium and other radioactive gases that might escape from a typical fusion power plant indicate that they would be diluted to safe levels by the time they reached the perimeter fence of the plant.

Another safety and environmental concern associated with fusion reactors is the potential that the neutrons generated in a fusion reactor could be used to breed plutonium for an atomic bomb. In order for a reactor to be used in this way, however, it would have to be extensively redesigned; thus the plutonium production would be very difficult to conceal. The tritium produced in fusion reactors could be used as a component of hydrogen bombs, but the likelihood is minimal.

DEVELOPMENT OF COMMERCIAL FUSION REACTORS

Despite fusion research that started during the 1950's, a commercial fusion reactor will most likely not be available until at least 2050. Several deuterium-tritium fusion reactors using the tokamak design have been built as test devices, but none of them produce more thermal energy than the electrical energy consumed.

The United States, Japan, Russia, China, South Korea, and the European Union are joint funders of the International Thermonuclear Experimental Reactor (ITER), which is being constructed at Cadarache in southern France. ITER is planned to fuse atomic nuclei at extremely high temperatures inside of a giant electromagnetic ring. In May, 2009, Lawrence Livermore National Laboratory announced the development of a high-energy laser system that can heat hydrogen atoms to temperatures that exist in the cores of stars. Such a system would provide the ability to produce more energy from controlled, inertially confined nuclear fusion than is necessary to produce the reaction.

Alvin K. Benson

FURTHER READING

Braams, C. M., and P. E. Stott. *Nuclear Fusion: Half a Century of Magnetic Confinement Fusion Research.* Philadelphia: Institute of Physics Publishing, 2002.

Bryan, Jeff C. *Introduction to Nuclear Science.* Boca Raton, Fla.: CRC Press, 2009.

McCracken, Garry, and Peter Stott. *Fusion: The Energy of the Universe.* Burlington, Mass.: Elsevier Academic, 2005.

Seife, Charles. *Sun in a Bottle: The Strange History of Fusion and the Science of Wishful Thinking.* New York: Viking Press, 2008.

SEE ALSO: Atomic Energy Commission; Nuclear and radioactive waste; Nuclear power; Nuclear Regulatory Commission; Nuclear testing; Power plants.

Nuclear power

CATEGORIES: Nuclear power and radiation; energy and energy use

DEFINITION: Electricity generated through the harnessing of the heat energy produced by controlled nuclear fission

SIGNIFICANCE: Nuclear power has long been the most controversial source of electricity. High plant construction costs, safety concerns, waste disposal problems, and public mistrust all served to slow the growth of the nuclear power industry during the late twentieth century. In the early twenty-first century, factors such as nuclear power's potential to meet electricity demand while reducing greenhouse gas emissions have increased its appeal.

Many of the environmental impacts of nuclear power plants are common to all large-scale electricity-generating facilities, regardless of their fuel. The most important are land use and related impacts on plants, animals, and ecosystems; nonradioactive water effluent and water quality; thermal pollution of adjacent waters; and social impacts on nearby communities. Unique to nuclear power plants is the hazardous radiation emitted by radioactive materials present in all stages of the nuclear fuel cycle. Such radiation is contained in uranium ore when it is mined and processed, in fabricated uranium reactor fuel, in spent fuel that has been fissioned, in contaminated reactor components, and in low-level radioactive waste—such as contaminated tools, protective clothing, and replaced reactor parts—generated by routine plant operation and maintenance.

Spent reactor fuel is highly radioactive and must be isolated from the environment for tens of thousands of years or more. Spent fuel can be reprocessed to separate out usable uranium and plutonium, but doing so generates large volumes of low-level waste that present disposal challenges of their own. Other challenges in waste handling and disposal arise at the end of a plant's useful life, when the contaminated reactor must be dismantled and the radioactive and nonradioactive components disposed of in a process known as decommissioning.

Under normal operating circumstances, commercial nuclear power plants release negligible radioactive emissions. The principal safety concern is that a severe accident could release large quantities of dangerous radioactive materials into the environment, as happened in 1986 at the Chernobyl nuclear plant in Ukraine. Roughly a quarter century after one of Chernobyl's reactors exploded, radiation levels remain dangerously high within the deteriorating concrete containment structure surrounding the reactor, and levels within the exclusion zone that extends in a 30-kilometer (18.6-mile) radius around the plant site continue to be higher than normal background levels.

EARLY HISTORY

The nuclear power industry arose out of the technology developed during World War II to produce the atomic bomb. Postwar enthusiasm for new technology in general, combined with pressure to demonstrate peaceful uses for expensive and fearsome wartime nuclear technology, led to a strong U.S. government effort beginning during the early 1950's to induce industry to develop nuclear energy. Large government subsidies and preferential treatment assisted the industry from its early days. Between fiscal year (FY) 1948 and FY2007, nuclear power received 53.5 percent of all federal energy research and development funds, totaling $85.01 billion in constant FY2008 dollars. One of the industry's unique subsidies was the passage of the 1957 Price-Anderson Act, which limits the liability of nuclear power plant owners and equipment vendors in the event of a reactor accident.

The basic design of the first U.S. nuclear power plants was adapted from early pressurized water reactor technology developed for submarines and other naval propulsion applications. Roughly two-thirds of the 104 commercial reactors in use in the United States are pressurized water reactors. In this reactor design, light (ordinary) water surrounds the nuclear fuel, which is made up of enriched uranium. The system is pressurized so that the fuel heats the water without boiling it. The resulting heat is used to boil a separate water supply, creating steam. The steam spins a turbine to generate electricity. The choice of this reactor design was largely driven by expediency and political considerations rather than an explicit effort to seek safe or reliable design features. The first large-scale commercial nuclear plant in the United States, which began operating at Shippingport, Pennsylvania, on a demonstration basis in 1957 and continued until 1982, was a conversion of a naval reactor project for which funding had been canceled. Some observers believe this early technology decision was largely responsible for the industry's problems in later years.

Another U.S. nuclear power plant that began operations in 1957, the Vallecitos plant near Pleasanton, California, was a research and development facility that became the first privately owned and operated nuclear power plant to provide significant quantities of electricity to a public utility grid. The Vallecitos facility employed a boiling-water reactor, a type of reactor in which the light water surrounding the enriched uranium fuel is converted directly into steam. The steam is piped to a turbine, which rotates to power an electric generator that produces electricity. The Vallecitos reactor was shut down in 1963. Approximately one-third of the commercial nuclear reactors operating in the United States are boiling-water reactors.

Yet another nuclear reactor technology emerged in the postwar years. Beginning in 1946, the United States developed a series of experimental prototype breeder reactors, a type of nuclear reactor in which the reaction is controlled in such a way that more fuel is produced than consumed. In 1963, the first commercial breeder reactor began low-power test operations at the Enrico Fermi Atomic Power Plant in Michigan. An accident in 1966 resulted in a partial core meltdown and caused reactor and fuel assembly damage that took almost four years to repair. Operations resumed in 1970 and continued until the decision was made in 1972 to decommission the reactor.

LATE TWENTIETH CENTURY DEVELOPMENTS

Government, industry, and public opinion were all largely positive about nuclear energy up through the late 1960's. Two-thirds of the U.S. commercial reactors operating in 2010 were issued construction permits between 1966 and 1973, a time of great optimism about the technology. During this period, the National Environmental Policy Act of 1969 was enacted,

Steam rises from the cooling towers of the nuclear reactors at Plant Vogtle in Waynesboro, Georgia. (AP/Wide World Photos)

which forced prospective reactor owners to address environmental impacts in plant proposals; the environmental movement arose, beginning with the first Earth Day in 1970; and there was a widespread increase in environmental activism on the part of the public. By the mid-1970's, several widely publicized safety hearings, plant incidents, and government studies had begun to focus public and media attention on nuclear plant regulatory and safety lapses, accident risks, and the problem of nuclear waste disposal. These forces combined to create a sizable antinuclear movement in the United States, which was bolstered by the 1979 accident at the Three Mile Island plant in Pennsylvania.

Although proponents of nuclear power frequently blamed licensing interventions by antinuclear activists for numerous plant cost overruns and delays, most analyses concluded that in the majority of cases other factors, such as capital availability and shifting regulatory requirements, were primarily responsible for a slowdown in the development of nuclear power. By the late 1980's a decline in the number of plants under construction and a shift of public concern to the threat of the nuclear arms race had begun to reduce the ranks of the antinuclear power movement.

Development of breeder reactor technology in particular slowed significantly in the United States during the late twentieth century. One reason for the decline was the unique safety challenges presented by breeder reactors. Unlike light-water reactor systems, fast-neutron breeder reactors employ molten sodium as a coolant. Sodium burns when exposed to air and reacts explosively with water. Economic considerations also put breeder reactors at a disadvantage. As long as uranium supplies remained abundant, light-water reactor facilities were the more competitive option. Finally, the quantities of plutonium that could be created in breeder reactors raised concerns that the material could be used for nuclear weapons applications. The threat of nuclear proliferation led to a directive from President Jimmy Carter in 1977 that indefinitely deferred commercial reprocessing of spent nuclear fuel and plutonium recycling. This effective ban severely curbed breeder reactor progress in the United States. Congressional funding continued for a demonstration breeder plant in Oak Ridge, Tennessee, despite Carter's opposition, but in 1983 Congress cut funding for the project, which effectively ended the country's breeder reactor development for the rest of the century.

Between 1987 and 1994 nuclear power proponents won long-sought changes in regulations governing emergency planning requirements, the licensing of new reactors, siting, and reactor design certification. The essence of these changes was to facilitate the process for approving new reactors while sharply reducing opportunities for public participation in the regulatory process. Nuclear opponents adamantly opposed many of these changes, especially a 1992 congressional decision authorizing the U.S. Nuclear Regulatory Commission (NRC)—the federal agency responsible for nuclear power licensing and safety regulation—to forgo a long history of issuing separate licenses for construction and operation, each of which allowed for hearings. Instead, Congress mandated the issuance of a single combined license for both activities that permits few opportunities for safety challenges after construction has been completed.

STATUS AND PROJECTIONS: UNITED STATES

The early twenty-first century has seen increasing receptivity to nuclear power in the United States. Among the factors contributing to this renewed interest are the need to meet the nation's continued growth in demand for electricity, the rising prices of fossil fuels, worries over possible interruptions in oil and gas availability, particularly from Middle Eastern sources, and concerns regarding the impact of the burning of fossil fuels on air quality and global climate. Some environmentalists tout nuclear power as a clean-air, carbon-free technology.

In the United States, nuclear power plants contributed approximately 20 percent of the electricity generated in 2009. In mid-2010 there were 104 licensed, operable reactors at sixty-five sites in thirty-one states. The newest of these received its construction license in 1973 and began its operational life in 1996. As of July, 2010, thirteen license applications were under active NRC review for up to twenty-two new reactors.

The nation's FY2011 budget authorized a total of $54.5 billion for nuclear power facilities, $36 billion of it in new loan authority established by the 2005 Energy Policy Act for DOE projects that cut greenhouse gas emissions. In February, 2010, President Barack Obama announced $8 billion in loan guarantees to go toward the construction of two new reactors at the existing Plant Vogtle nuclear power facility near Augusta, Georgia. These reactors are scheduled to come online in 2016 and 2017. If the NRC grants final approval for construction, these will be the first new

Leading World Producers of Nuclear Power, 2010

RANK	COUNTRY	KILOWATT-HOURS (BILLIONS)	PERCENTAGE OF ELECTRICITY	REACTORS
1	United States	798.7	20.2	104
2	France	391.7	75.2	58
3	Japan	263.1	28.9	55
4	Russia	152.8	17.8	32
5	South Korea	141.1	34.8	20
6	Germany	127.7	26.1	17
7	Canada	85.3	14.8	18
8	Ukraine	77.9	48.6	15
9	China	65.7	1.9	13
10	United Kingdom	62.9	17.9	19
11	Spain	50.6	17.5	8
12	Sweden	50.0	34.7	10
	Totals	2267.5	28.1	369
	World as a whole	2560.0	14.0	441

Source: World Nuclear Organization, November, 2010. Nations are listed in descending order of the amount of electricity they generate with nuclear reactors. The percentage column indicates what part of each nation's electricity consumption is supplied by nuclear power. "Reactors" indicates the number of nuclear reactors operating in November, 2010.

nuclear reactors built in the United States in the twenty-first century.

All U.S. reactors were originally granted forty-year operating licenses. Under NRC regulations adopted in 1992, plant owners may seek twenty-year license extensions. In the case of some reactors, competitive economic pressures or the costs of expensive equipment upgrades make retirement preferable to an extended operating life; reactors may even be permanently shut down before their scheduled license expiration dates because competing power sources or needs for expensive repairs make continued operation economically undesirable. In other cases, equipment upgrades and license renewal prove more cost-effective than construction of a new power facility. (The license renewal process alone, not including equipment upgrades, costs $10 million to $15 million.) As of July, 2010, the NRC had granted license renewals to fifty-nine reactors, and more renewal applications were expected.

In 2006 President George W. Bush proposed the Global Nuclear Energy Partnership (GNEP), a program for reducing the risk of nuclear weapons prolif-

eration while minimizing reactor wastes. Spent fuel reprocessing and advanced breeder technologies were components of GNEP. In 2009, under the Obama administration, the DOE announced that it would not be pursuing domestic commercial reprocessing; however, research and development on proliferation-resistant fuel cycles and waste management would continue.

STATUS AND PROJECTIONS: WORLD

As of November, 2010, 441 nuclear reactors were in operation in twenty-nine countries. The number of reactors in the United States (104), France (58), Japan (55), and the Russian Federation (32) accounted for more than half of the world's total operating reactors. The majority of these reactors were between twenty and thirty-nine years old. In eighteen countries, nuclear power supplied at least 20 percent of total electricity for 2009.

While most of the world's nuclear reactors are U.S.-designed pressurized water reactors and boiling-water reactors, other types are also in operation. Many of these employ graphite or heavy water (water enriched in deuterium) to moderate the nuclear reaction, which allows them to use less expensive natural (nonenriched) uranium as a fuel. The advanced gas-cooled reactors that predominate in the United Kingdom, for example, use graphite as a moderator and carbon dioxide instead of water as a coolant. In the pressurized heavy-water reactors common in Canada, heavy water serves as both coolant and moderator. The light-water graphite reactors of the Russian Federation are similar to boiling-water reactors but employ graphite moderators.

Although several countries established breeder reactor programs in the previous century, few of these programs were thriving as of 2010. The United Kingdom had not constructed a new breeder facility since it shut down its last breeder reactor in 1994. Germany's one fast-breeder reactor permanently ended operations in 1991. France shut down Superphénix, the world's only commercial-sized breeder reactor, in 1998, after a spotty history of malfunction and repairs that kept the plant offline for more than half of its operational life. France's Phénix reactor enjoyed a more successful run, but it came to a close in 2010 with no immediate plans for another breeder reactor. In Japan, after a 1995 sodium coolant leak and fire at the Monju reactor, repairs and public controversy suspended the facility's operations until 2010. The only countries constructing new fast-breeder reactors as of August, 2010, were the Russian Federation and India, both of which were motivated to pursue the technology at least in part by concerns regarding uranium supplies.

In 2010 sixty new reactors were under construction in fifteen nations, although some of them may ultimately not be brought online. Fifty-one of them were pressurized water reactors, four were boiling-water reactors, two were pressurized heavy-water reactors, two were fast-breeder reactors, and one was a light-water graphite reactor. China, the Russian Federation, the Republic of Korea (South Korea), and India accounted for roughly 70 percent of the reactors under construction.

Waste- and Contamination-Related Issues

Spent fuel and high-level radioactive waste can pose threats to human health and the environment for many thousands of years, so they must be properly isolated and secured. Deep subterranean storage appears to be the best solution, but finding a repository site with the right geological characteristics is a technical challenge that is invariably complicated by political controversy. In the United States, Congress designated Yucca Mountain, Nevada, as the nation's sole repository in 1987, and site suitability studies were conducted at Yucca Mountain for nearly two decades. Opponents, including the state of Nevada, charged that the DOE's studies were geared more toward preparing the site for operation than for objectively assessing its suitability. The DOE submitted its license application for the repository in 2008, ten years after the original target date for opening. In early 2010, under the Obama administration, funding for the site was cut, and the DOE withdrew its license application. Lawsuits have been filed to challenge Yucca Mountain's closure, which leaves high-level waste in temporary storage at nuclear facilities around the country.

Some countries—notably France, England, Russia, Japan, and India—reprocess spent fuel from nuclear reactors. Reprocessing strips the waste of uranium and plutonium, which can be used to fuel reactors. While reprocessing results in a small reduction of high-level nuclear wastes, it generates large volumes of low-level nuclear wastes, which also require environmentally responsible disposal. Reprocessing is more costly than the single use and disposal of spent fuel. Also, because it creates stockpiles of plutonium, reprocessing has the potential to contribute to nuclear weapons proliferation and terrorism. Separating plutonium from more highly radioactive spent fuel assemblies makes it easier for it to be stolen. With this danger in mind, President Jimmy Carter issued an executive order in 1977 that indefinitely deferred U.S. reprocessing of spent nuclear reactor fuel.

Concern about nuclear power plant accidents continues to trouble the public. Precise accident probability estimates are impossible to derive, given the complexity of nuclear reactor systems. Industry proponents point to several government-sponsored studies that have concluded that the probability of a plant accident with off-site consequences is extremely low. Critics have countered that these studies were methodologically flawed, omitted important factors, and underestimated the true risks. They also emphasize the catastrophic consequences that could result if a low-probability accident should nonetheless occur.

Arguments Pro and Con

Nuclear proponents cite several points in the technology's favor: a good safety record, improved operating performance since the Three Mile Island accident, studies concluding that the risk of a severe accident is low, and the fact that nuclear power plants emit no significant amounts of common air pollutants or gases that contribute to global warming. Critics of nuclear power point to the long-standing failure of any country to establish a site for the permanent disposal of spent fuel and high-level wastes; flaws, uncertainties, and omissions in accident probability studies; the catastrophic consequences that could result from a severe reactor accident; a large number of safety-related incidents and near accidents; and the high cost of nuclear plants compared to competing electricity-generating technologies and energy-efficiency improvements. In addition, for those countries that do not already have nuclear weapons arsenals, inherent proliferation risks are associated with nuclear power technology and fuels, which, if misused, could provide the expertise, infrastructure, and basic materials for a program to develop nuclear weapons.

High costs remain another obstacle to new plant orders. U.S. reactors completed during the 1990's cost $2 billion to $6 billion each, averaging more than

$3 billion. Although proponents argue that new plant designs could be built more cheaply, the industry's history of large cost overruns discomforts prospective plant owners, capital markets, and state regulators. Many analysts believe that unpredictably high capital costs have been among the most important reason for the industry's decline. The picture has been complicated further by the deregulation of the electric utility industry that followed passage of the Public Utility Regulatory Policies Act in 1978 and the Energy Policy Act in 1992. Uncertainty surrounding the future economic and regulatory climate raises questions about the competitiveness of both new and many existing nuclear reactors.

Public opinion surveys in the United States after the events at Three Mile Island showed a growing majority opposing the construction of new nuclear power plants. Numerous influential policy assessments from the late twentieth century concluded that it was unlikely that new nuclear power plants would be built in the United States unless the key issues of waste disposal, costs, safety, and public acceptance were satisfactorily resolved. Although the situation differs from country to country, some combination of these factors served to hamper the growth of nuclear power in most countries during the late twentieth century. In the United States, additional factors that contributed to nuclear power's decline during the 1980's and 1990's included unexpectedly plentiful supplies of natural gas and slower-than-anticipated growth in electricity demand.

The public's mistrust of nuclear power is generally acknowledged as a roadblock to new plant orders. Risk perception studies have shown that the public's lack of trust is deeply rooted, widely felt, and resistant to change. Surveys assessing public support in the United States for the use of nuclear power reflected a decline between 1994, when 57 percent of those polled were in favor, and 2001, when 46 percent were in favor. Since 2001, however, support has increased, reaching 62 percent in 2010. Concerns about such environmental problems as greenhouse gas emissions and the cost and availability of fossil fuels played a role in this upswing in nuclear power's popularity. NRC officials and others have noted that, regardless of one's explanation of the public's views, public acceptance plays an important role in determining whether new plants will be built in the United States.

Phillip A. Greenberg
Updated by Karen N. Kähler

Further Reading

Bodansky, David. *Nuclear Energy: Principles, Practices, and Prospects.* 2d ed. New York: Springer, 2004.

Garwin, Richard L., and Georges Charpak. *Megawatts and Megatons: The Future of Nuclear Power and Nuclear Weapons.* Chicago: University of Chicago Press, 2001.

Hagen, Ronald E., John R. Moens, and Zdenek D. Nikodem. *Impact of U.S. Nuclear Generation on Greenhouse Gas Emissions.* Washington, D.C.: Energy Information Administration, U.S. Department of Energy, 2001.

Hore-Lacy, Ian. *Nuclear Energy in the Twenty-first Century: The World Nuclear University Primer.* London: World Nuclear University Press, 2006.

Mahaffey, James. *Atomic Awakening: A New Look at the History and Future of Nuclear Power.* New York: Pegasus Books, 2009.

Murray, Raymond LeRoy. *Nuclear Energy: An Introduction to the Concepts, Systems, and Applications of Nuclear Processes.* Burlington, Vt.: Butterworth-Heinemann/Elsevier, 2009.

Wolfson, Richard. *Nuclear Choices: A Citizen's Guide to Nuclear Technology.* Rev. ed. Cambridge, Mass.: MIT Press, 1993.

SEE ALSO: Alternative energy sources; Atomic Energy Commission; Breeder reactors; Department of Energy, U.S.; International Union for Conservation of Nature; Nuclear and radioactive waste; Nuclear Regulatory Commission; Nuclear regulatory policy; Nuclear weapons production; Power plants; Superphénix; Yucca Mountain nuclear waste repository.

Nuclear Regulatory Commission

CATEGORIES: Organizations and agencies; nuclear power and radiation

IDENTIFICATION: U.S. federal agency charged with regulating the nuclear power industry and the civilian use of nuclear materials

DATE: Began operating on January 19, 1975

SIGNIFICANCE: The Nuclear Regulatory Commission serves an important role in preventing environmental pollution through its regulation of all civilian use of nuclear materials in the United States.

When it was decided that the dual responsibilities of promoting nuclear energy and regulating radiation safety in the United States created a conflict of interest for the Atomic Energy Commission (AEC), the AEC was abolished by the Energy Reorganization Act of 1974. This act established the Nuclear Regulatory Commission (NRC) to be responsible for regulating nuclear safety and the Energy Research and Development Administration (ERDA) to handle the promotion of nuclear science (ERDA, along with other agencies, became the Department of Energy in 1977).

The mission of the NRC is to develop policies and rules for civilian use of nuclear reactors and materials. These policies and rules must serve the purposes of protecting people and the environment, encouraging security, and providing for the common defense. The NRC's responsibilities may be divided into three main areas: reactors, materials, and waste. Reactor-related regulations cover both power reactors and reactors used for research or training. Materials-related regulations concern the facilities that produce nuclear fuel and the nuclear materials used in academic, industrial, and medical facilities. Waste-related regulations include those concerning the transportation, storage, and disposal of nuclear waste and the decommissioning of nuclear facilities.

The U.S. Congress has passed many different laws since the NRC's establishment that specify responsibilities of the agency and affect its operations. These include the Uranium Mill Tailings Radiation Control Act of 1978, which concerns protection of the environment from radon emitted by the waste produced by uranium mills; the Nuclear Non-Proliferation Act of 1978, which strengthened the NRC's control of exported nuclear materials and increased international monitoring of nuclear materials and facilities; and the Nuclear Waste Policy Amendments Act of 1985, which made U.S. states responsible for the low-level radioactive waste produced within their borders.

A continuing concern for the NRC is the setting of standards for radiation safety. Scientists have not been able to determine what levels of exposure to radiation can be considered safe; thus controversy accompanies any attempt to establish standards in this area. Another ongoing concern of the NRC is national security. The agency has increasingly emphasized the development of countermeasures that may be taken in the event that terrorists attempt to take over nuclear facilities or attack targets in the United States using stolen nuclear materials.

C. Alton Hassell

FURTHER READING

Rosenbaum, Walter A. *Environmental Politics and Policy.* 7th ed. Washington, D.C.: CQ Press, 2008.

U.S. Nuclear Regulatory Commission. *Radiation Protection and the NRC.* Washington, D.C.: Author, 2006.

SEE ALSO: Atomic Energy Commission; Breeder reactors; Chalk River nuclear reactor explosion; Hanford Nuclear Reservation; International Atomic Energy Agency; Nuclear accidents; Nuclear and radioactive waste; Nuclear power; Nuclear regulatory policy; Radioactive pollution and fallout; Yucca Mountain nuclear waste repository.

Nuclear regulatory policy

CATEGORY: Nuclear power and radiation
DEFINITION: Overall plan of action used by a government in the control and licensing of industries involving radioactive materials
SIGNIFICANCE: Governments must give careful consideration to the public health and safety in formulating policies regarding the use and handling of nuclear materials. The mishandling of radioactive wastes, for example, can have long-term negative environmental impacts.

In the United States, the Nuclear Regulatory Commission (NRC) is the federal agency responsible for licensing nuclear power plants and regulating the civilian nuclear power industry. The federal government has broadly preempted the authority of individual U.S. states to regulate the industry. The NRC's mission is threefold: to protect the public health and safety, to safeguard national security, and to ensure environmental protection.

The NRC licenses plant owners to construct and operate nuclear power reactors and issues a wide variety of technical and procedural rules governing reactor operation. In the United States, electrical utility companies have historically owned and operated nuclear power plants, but restructuring of the utility industry may eventually lead to ownership by other kinds of companies.

The nuclear regulatory system is largely based on self-regulation by plant owners, who are responsible for meeting the NRC's requirements. Utility company personnel maintain and inspect their companies' plants and submit regular reports to the NRC. Plant owners must notify the NRC of important safety lapses and equipment malfunctions. As with many industries, federal regulators set forth specific rules, but federal inspectors only spot-check for compliance. After the 1979 accident at the Three Mile Island (TMI) plant in Pennsylvania, the NRC stationed full-time inspectors at all U.S. nuclear power plants to conduct inspections and monitor compliance with NRC rules. Special additional NRC inspection teams are called in when warranted by serious equipment or operator failures.

Utilities found in violation of the NRC's rules may be subject to fines and penalties. When violations are sufficiently serious to raise questions about the safety of a plant's operation, operators customarily shut down the reactor before being ordered to do so. Although the NRC has the authority to order an operating plant to shut down, it has rarely done so. However, the NRC may withhold permission for a plant to restart—sometimes for years—after a shutdown until the utility can demonstrate it has corrected the problems.

Many other countries have based aspects of their nuclear regulatory systems on the U.S. model, particularly with regard to technical oversight, since roughly 80 percent of the world's nuclear plants utilize U.S. reactor designs. However, organizational structure, staffing levels, practices, and requirements differ from country to country. In many nations, such as France and Canada, the relationship between the regulatory agency and plant licensees is much closer than in the United States. In most developing countries, the public has little access to the regulatory and policy processes. In industrialized nations, a wide range exists with regard to opportunities for public participation and the related ability to seek judicial review. In France and Japan, which operate the world's second- and third-largest nuclear programs, regulatory policies are highly centralized and largely closed to public participation.

STATE REGULATION

U.S. states are largely preempted by federal law from health and safety regulation of the nuclear aspects of nuclear power plants. States retain the authority to regulate other aspects of nuclear plants that are common to all electricity-generating facilities. The most important of these are the ability of state public-service commissions to set utility rates and fi-

nancial returns on utility investments and the broad powers of state agencies to regulate land use, facility siting, and nonnuclear environmental impacts.

States and members of the public have long been able to apply to the NRC to be granted status as "intervenors" in NRC licensing proceedings. This aspect of the regulatory system has been fundamentally adversarial. NRC licensing hearings have often resembled judicial proceedings in which parties enjoy certain rights with regard to reviewing documents, cross-examination, the right of appeal to the courts, and so forth.

ANTINUCLEAR ACTIVISM

Beginning in the early 1970's, antinuclear activism and the number of interventions in construction permit and operating license proceedings increased. Industry proponents often claimed that most interventions were mere delaying tactics intended primarily to lengthen licensing times and drive up costs. Critics of the nuclear power industry argued that most interventions were warranted, and many brought to light safety-related problems that would not otherwise have been identified. Studies by government agencies and others concluded that most—but not all—delays in building nuclear plants resulted from factors other than interventions, such as changes in regulatory requirements and construction problems.

With regard to operating plants, the public has little access to the regulatory process. Although NRC regulations permit members of the public to submit a petition to seek a hearing regarding modification or revocation of a utility's license on the grounds of possible safety or environmental problems, the NRC has granted hearings for only a handful of the hundreds of such petitions received.

U.S. REGULATORY HISTORY

The American effort to develop the atomic bomb during World War II was accompanied by a tight government monopoly on nuclear materials and technology. The 1946 Atomic Energy Act created the civilian Atomic Energy Commission (AEC), which oversaw all aspects of atomic energy, and the powerful congressional Joint Committee on Atomic Energy (JCAE), which was given authority over all nuclear-related legislation and oversight. Amendments in 1954 loosened the government's monopoly in an effort to spur the development of a private nuclear power industry. The AEC was given the additional charges of actively promoting the development of commercial nuclear energy and regulating the new industry. Government and private companies cooperated in what was regarded as a joint effort.

Between 1963 and 1974, forty-four U.S. commercial nuclear power plants began operation. Because of the cooperative nature of the endeavor and the relatively few operating facilities, federal regulation was initially modest. However, reactor safety issues soon began to draw media and public attention, and the AEC was widely perceived as too friendly with the industry it was supposed to regulate. In 1974 the U.S. Congress, responding to criticism that the AEC's dual missions of promotion and regulation were conflicting, replaced the AEC with two new agencies. The task of regulation was given to the newly created NRC. Promotion of nuclear energy was assigned to the Energy Research and Development Administration, which was superseded in 1977 by the Department of Energy. In 1976 the JCAE was dissolved for largely the same reasons as the AEC, and congressional authority was divided among various committees.

Critics, including what was then known as the U.S. General Accounting Office (later the Government Accountability Office), noted that the staff of the new NRC was drawn largely from the AEC and thus retained essentially the same mentality. However, the years following the NRC's creation were also marked by increased regulatory scrutiny. Between 1975 and the mid-1980's, NRC safety research increased; by 1981 the inspection and enforcement staff had grown to become the agency's largest division. The 1979 TMI accident created a new atmosphere that was initially marked by a brief moratorium on licensing for new plants and several independent investigations into the accident's causes. An industry group was formed to try to improve safety performance. The NRC responded to public and congressional safety concerns by setting new requirements for construction and operation of plants and by monitoring industry compliance more closely. However, these actions were counterbalanced by the actions of NRC members appointed during President Ronald Reagan's administration, who began initiatives to reduce federal regulation, an effort from which the NRC was insulated but not immune.

CHANGES IN NRC REGULATIONS

Beginning in the late 1980's, the NRC enacted a number of new regulations long sought by the nuclear industry. The most important changes addressed licensing of new reactors, emergency planning, siting, reactor design certification, and operating license extensions. For all reactors licensed through 1996, the NRC followed a two-step licensing process mandated by Congress in 1957. First, a utility company applied for a permit to begin construction. When the plant was completed, the utility applied for an operating license. In numerous cases, intervenors opposed the granting of operating licenses, usually on grounds of safety-related allegations relating to design, construction, or infeasible emergency planning. In 1992 the industry achieved a longtime goal when Congress approved a 1989 NRC regulation change that implemented one-step licensing for new reactors, permitting the NRC to issue a single combined construction and operating license prior to the onset of construction. The rule sharply limited the ability of intervenors to raise environmental and safety issues after a license has been granted, both during construction and before operation begins. Other rule changes in 1989 narrowed intervenors' procedural rights during licensing hearings.

Following the TMI accident, the NRC adopted regulations requiring the agency to approve state and local emergency response plans before a plant could be granted an operating license to ensure that members of the public could be protected or evacuated if a serious accident occurred. In several celebrated cases (the Diablo Canyon, Seabrook, and Shoreham reactors), states refused to cooperate on emergency planning and opposed the issuance of operating licenses on the grounds that no feasible emergency planning was possible because of siting error. In 1987 the NRC adopted new rules permitting the agency to grant an operating license based on emergency plans drafted by the plant owner alone; in 1988 President Reagan issued an executive order authorizing federal agencies to draft emergency plans. These actions effectively bypassed state opposition.

In 1989 the NRC adopted regulations that permitted utilities to seek approval for reactor sites before any license was granted and to retain the ability to use approved sites for up to forty years. Under these rules, utilities are permitted to seek approval of emergency plans for a site before filing any application to build a plant. In 1989 the NRC also adopted new rules allowing reactor vendors to seek approval for "standardized" reactor designs, which would be certified by the agency for fifteen years. The rule prohibits critics from raising design-related safety and environmental issues when a utility seeks to build a new plant using the certified designs.

All U.S. reactors operating by 1996 were issued forty-year operating licenses, 40 percent of which were scheduled to expire by the year 2015. In 1992 the NRC adopted a rule that permits plant owners to seek license extensions of up to twenty additional years. The net results of these regulation changes, which were adamantly opposed by nuclear critics, have been to facilitate the licensing process for new and existing reactors while significantly restricting public access to the regulatory process.

CRITICISM OF THE NRC

The nuclear industry and antinuclear activists alike have criticized the NRC's regulation, although for different reasons. In the late 1990's the nuclear industry argued that the NRC's requirements unnecessarily raise plant operating costs. The industry also complained that the NRC fails to set objective criteria by which adequate levels of safety and licensees' compliance can be judged and that the agency does not rank the relative safety significance of compliance requirements and adjust regulatory priorities and enforcement emphasis accordingly. Meanwhile, antinuclear activists have asserted that the NRC has generally been allied with the industry and has often failed to maintain an arms-length relationship with its licensees. They also argue that the NRC has retreated from tough regulation, has failed to pursue important safety questions, and has been either unresponsive or obstructive to substantive participation by individuals and citizen groups.

The allegations are not mutually exclusive, and there is evidence to support arguments on both sides. However, it is difficult to deny that there has been a fundamental alignment of interest between the NRC and the nuclear industry. Such relationships between federal regulatory agencies and the industries they regulate are widely acknowledged, although the degree of cooperation or tension between the two sides may fluctuate over time. By the late 1990's, assisted by improved safety performance, momentum had shifted toward lighter regulation under the influence of presidential administrations, appointees, and a Congress more sympathetic to the industry's views. Meanwhile, nuclear critics have continued to argue that easing regulatory vigilance was a mistake in the light of repeated plant shutdowns for safety violations, unex-pected aging of plant equipment, and recurring discoveries of safety-related problems in both individual plants and across the industry.

Many U.S. nuclear energy proponents have advocated the amendment of corresponding aspects of the U.S. licensing and regulatory system to resemble the French and Japanese models, which permit little public participation. The key regulatory changes adopted in the United States from 1987 to 1992 indicate movement in that direction. Nuclear power critics oppose such changes, arguing that opportunities for public participation in the regulatory and policy arenas, along with the availability of judicial review, are fundamental to preserving the democratic process in making national decisions about energy and technology policy.

Phillip A. Greenberg

FURTHER READING

Cooke, Stephanie. *In Mortal Hands: A Cautionary History of the Nuclear Age.* New York: Bloomsbury, 2009.

Duffy, Robert J. *Nuclear Politics in America: A History and Theory of Government Regulation.* Lawrence: University Press of Kansas, 1997.

Mazuzan, George T., and J. Samuel Walker. *Controlling the Atom: The Beginnings of Nuclear Regulation, 1946-1962.* Berkeley: University of California Press, 1984.

Murray, Raymond L. "Laws, Regulations, and Organizations." In *Nuclear Energy: An Introduction to the Concepts, Systems, and Applications of Nuclear Processes.* 6th ed. Burlington, Vt.: Butterworth-Heinemann/Elsevier, 2009.

O'Very, David, Christopher Paine, and Dan Reicher. *Controlling the Atom in the Twenty-first Century.* Boulder, Colo.: Westview Press, 1994.

Rees, Joseph V. *Hostages of Each Other: The Transformation of Nuclear Safety Since Three Mile Island.* Chicago: University of Chicago Press, 1994.

Walker, J. Samuel. *Containing the Atom: Nuclear Regulation in a Changing Environment, 1963-1971.* Berkeley: University of California Press, 1992.

_____. *A Short History of Nuclear Regulation, 1946-1999.* Washington, D.C.: U.S. Nuclear Regulatory Commission, 2000.

SEE ALSO: Antinuclear movement; Nuclear accidents; Nuclear power; Nuclear Regulatory Commission; Union of Concerned Scientists.

Nuclear testing

CATEGORY: Nuclear power and radiation

DEFINITION: Explosion of nuclear devices to evaluate their utility as weapons

SIGNIFICANCE: Since 1945 hundreds of nuclear devices have been exploded in the atmosphere and underground as part of the testing of nuclear weapons systems. Such testing has raised concerns about nuclear fallout and its effects on human health and the environment.

The United States began nuclear testing with the shot code-named Trinity on July 16, 1945, at Alamogordo, New Mexico. The United States has tested 219 nuclear devices in the atmosphere, with the last atmospheric test occurring in 1963. Most of these tests were conducted at the Nevada Test Site, but the most powerful tests were done in the South Pacific at Christmas Island and in the Marshall Islands at Bikini Atoll and Enewetak Atoll.

The former Soviet Union conducted 219 atmospheric tests, the last one in 1962. Nearly all these tests were at the Semipalatinsk Test Site in Kazakhstan or at the Northern Test Site, Novaya Zemlya. The higher-yield weapons were tested at the Northern Test Site, including a 50-megaton monster on October 30, 1961, the largest weapon ever exploded. The last of China's 23 atmospheric tests occurred in 1980. These were conducted at the Lop Nur Test Site in northwest China. Britain conducted 21 atmospheric tests off Christmas Island, near Monte Bello Island off Australia, and at sites in South Australia. Britain ended atmospheric testing in 1965. France, which continued atmospheric testing into 1974, conducted 56 such tests at Reganne in Algeria and at Fangataufa and Moruroa in the South Pacific.

The United States ended underground testing in 1992, the former Soviet Union in 1990, Britain in 1991, France in 1996, and China in 1996. Pakistan and India both conducted underground tests in 1998, and North Korea announced that it had conducted a test in 2009. With some notable exceptions, little radioactivity in the atmosphere has resulted from the underground tests by the United States, Britain, and France. The chief adverse environmental effect of most of these underground tests is that the radioactive subterranean rubble might someday pollute groundwater. The former Soviet Union was less successful in containing the radioactivity of underground tests.

TESTING AT BIKINI

The global effect of atmospheric nuclear tests is judged to be small, since the radiation dose that individuals received from global fallout was less than they received from natural background radiation. In contrast, local fallout from atmospheric tests has had serious consequences. The most notorious case of injury from radioactive fallout occurred with the Bravo shot of March 1, 1954, at Bikini Atoll in the South Pacific. The weapon's yield of about 15 megatons was three to four times what was expected. Twelve hours before the test, the wind direction changed, and it was understood that the fallout would no longer be carried safely out to sea. The decision to proceed despite the wind change shows that scientists vastly underestimated the problems that could result from exposure to radioactive fallout.

The nine-man firing team crouched in a bunker only 32 kilometers (20 miles) from ground zero. As their radiation meters began to show unexpectedly high amounts of fallout, they sealed their bunker. Navy ships that were arriving to pick up the firing team and recover scientific data encountered such heavy fallout 50 kilometers (31 miles) out that they retreated at top speed. The fallout ceased after three hours. The ships were washed and returned to rescue the firing party. Covered with bedsheets sealed with masking tape to keep out fallout particles, the members of the firing party drove 1 kilometer (0.6 miles) to the landing site and were taken to the rescue ship by helicopter.

Radiation doses are measured in grays (Gy). (The old unit for an absorbed dose was the rad; 1 gray equals 100 rads.) Normal background radiation results in a dose of around 0.002 Gy per year. Acute doses are doses received over several days instead of over months or years. Acute doses of 0.25 Gy or less generally result in no obvious injuries. Radiation sickness and some obvious injuries begin to occur with acute doses between 0.5 and 1.0 Gy. An acute dose of 6.0 Gy is generally fatal. The men of the firing party received only 0.005 Gy. Had they been outside their bunker, they would have received more than a fatal dose during the first few hours.

Groups farther from Bikini were evacuated as the magnitude of their peril became clear. It was not until two days after the blast that eighty-two people were taken from Rongelap Atoll, 200 kilometers (124 miles) east of Bikini. They received an estimated 1.0 Gy, and some were already showing signs of radiation sickness. A survey flight failed to detect significant radiation over Rongerik Atoll, 240 kilometers (149 miles) east of Bikini, but sensors on the ground did. Again the magnitude of the problem was unclear, and the twenty-eight men at the weather station on Rongerik were not evacuated before they received an estimated 0.3 to 0.5 Gy. The final evacuation was a group of 158 inhabitants on Utirik Atoll, 500 kilometers (310 miles) from Bikini. They received an estimated 0.15 Gy.

The greatest radiation dose was received by twenty-three Japanese fishermen who unwittingly sailed their vessel, the *Lucky Dragon*, into the fallout cloud only 160 kilometers (100 miles) from Bikini. Not realizing that they should wash the fallout from their ship, their dose continued to increase during their two-week voyage to port. By then, their estimated dose was between 1.3 and 4.5 Gy. The ship's radioman, Aikichi Kuboyama, died seven months later of hepatitis, a complication brought on by his treatment for radiation sickness.

AFTEREFFECTS OF THE BIKINI TESTS

Further weapons tests were conducted at Bikini Atoll in 1956 and 1958. By 1967, most of the radioactivity had been swept away by nature or had decayed, and pressure mounted to return the refugees to their home. Radioactive debris from the tests, some plants, and some topsoil were removed from Bikini. New trees were planted, and new houses were built. The first group was resettled on Bikini in 1969. However, some water wells and some local foods were mildly radioactive. Two radioactive elements, strontium 90 and cesium 137, accumulated in the inhabitants' bodies to worrisome levels. They were evacuated again in 1978.

By the mid-1990's, lush vegetation made Bikini seem like paradise, and the atoll's former inhabitants were again anxious to return. In 1997, however, the International Atomic Energy Agency recommended against resettlement of the island because of the risks of radiation from locally grown food. Bikinians argued that the coconut trees would take potassium from the soil instead of the radioactive cesium; therefore, if the food trees were given a heavy dose of potassium fertilizer, and if a few centimeters of topsoil around the living areas were scraped away, it should be safe for the inhabitants to return. Critics of the plan argued that devastating environmental damage would be caused by scraping away so much topsoil. Arguments concerning resettlement of the native population extended into the twenty-first century. Responding to considerable pressure, the U.S. Congress eventually set aside $275 million for health care and other needs of the Marshall Islanders affected by nuclear tests.

Following World War II, five nuclear tests were carried out at the Bikini and Enewetak Atolls. However, logistic and security problems made it attractive for the U.S. government to seek a test site in the continental United States. The Nevada Test Site was chosen because the government already owned the land, its surroundings were only sparsely populated, and it was close to the Los Alamos weapons laboratory. The best evidence available showed that off-site fallout from devices with yields less than 50 kilotons would be minimal. This estimate proved to be too optimistic, however. Although the contamination was light, monitoring teams occasionally closed sections of state

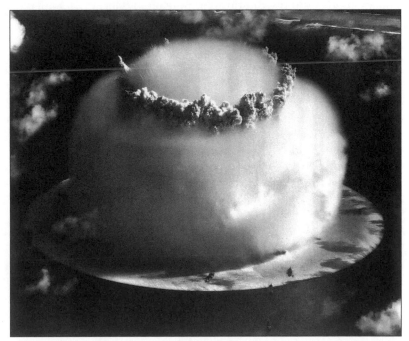

A huge mushroom cloud rises above Bikini Atoll in the Marshall Islands on July 25, 1946, following an atomic test blast conducted by the United States. The dark spots in the foreground are ships that were placed near the blast site to test what a nuclear bomb would do to a fleet of warships. (AP/Wide World Photos)

highways for a few hours or a day and had vehicles washed off. More seriously, on May 19, 1953, fallout near St. George, Utah, amounted to 0.05 Gy for people in the open. A dose that low causes no perceptible damage, but it may result in a slightly increased risk of cancer.

IODINE 131

Years of careful study have shown that iodine 131 is probably the most dangerous element of nuclear fallout. It accounts for 30 percent of the radioactivity produced by a nuclear bomb, but it rapidly decays. Its activity decreases by one-half every eight days, so that after one or two months, it has virtually disappeared. Two factors make iodine 131 dangerous: Cows and goats that graze on plants dusted with fallout concentrate iodine in their milk, and iodine ingested by humans concentrates in their thyroids. This two-step concentration process means that significant doses to the thyroid might result from fallout levels previously considered harmless. U.S. scientists also learned that rainstorms or snowstorms could wash enough fallout from the air to create fallout hot spots as far away from the Nevada Test Site as the East Coast.

Scientists at the National Cancer Institute estimated a per capita thyroid dose in the United States caused by iodine 131 from all the nuclear tests. They accomplished this by using available fallout measurements, the predictions of mathematical models, weather data, and information on patterns of milk consumption. During testing periods, children between the ages of three months and five years received thyroid doses three to seven times the average per capita because they drank more milk and because their thyroids were smaller. The estimated twenty thousand individuals who drank goats' milk were at greater risk, because goats' milk concentrates iodine ten to twenty times more than does cows' milk. Much smaller doses of radioactivity may have come from eating other dairy products or leafy vegetables.

Thyroid cancer is rare, occurring in only 3 or 4 people out of 100,000. Officials at the National Cancer Institute have estimated that nationwide, fallout might have caused 10,000 to 75,000 cases of thyroid abnormalities. It has been pointed out that individuals who were born in the 1950's, drank a lot of milk as children, and lived in areas that received the most fallout would be prudent to seek medical examination of their thyroids.

This fallout was probably allowed to occur because of ignorance of its effects and because of the Cold War mentality. At the time, it was believed that some risk to the public was justified by the needs of the national defense. It also took some years for scientists to collect convincing data that dangerous levels of iodine 131 from fallout could concentrate in the thyroid. Even those doses were no higher than those given by fluoroscopy examinations and some X-ray procedures in the early 1950's.

Scientists at the University of Exeter in the United Kingdom pioneered an ingenious use of radioactive tracers: the estimation of rates of soil erosion and deposition on agricultural land through the measurement of amounts of radioactive cesium at different locations. Tiny amounts of radioactive cesium were deposited worldwide during the years when nuclear weapons were tested in the atmosphere. In undisturbed permanent pastureland or rangeland, the cesium remains near the surface, and scientists need only compare the cesium concentrations from various locations with that of an undisturbed site to determine the amount of erosion.

Charles W. Rogers

FURTHER READING

Cooper, John R., Keith Randle, and Ranjeet S. Sokhi. *Radioactive Releases in the Environment: Impact and Assessment.* Hoboken, N.J.: John Wiley & Sons, 2003.

Dibblin, Jane. *Day of Two Suns: U.S. Nuclear Testing and the Pacific Islanders.* Chicago: New Amsterdam Books, 1990.

Fuller, John G. *The Day We Bombed Utah: America's Most Lethal Secret.* New York: New American Library, 1984.

Hacker, Barton C. *Elements of Controversy: The Atomic Energy Commission and Radiation Safety in Nuclear Weapons Testing, 1947-1974.* Berkeley: University of California Press, 1994.

Hansen, Keith A. "Early Efforts to Limit Nuclear Testing." In *The Comprehensive Nuclear Test Ban Treaty: An Insider's Perspective.* Stanford, Calif.: Stanford University Press, 2006.

Wasserman, Harvey, and Norman Solomon. *Killing Our Own: The Disaster of America's Experience with Atomic Radiation.* New York: Delacorte Press, 1982.

SEE ALSO: Bikini Atoll bombing; Limited Test Ban Treaty; Nuclear weapons production; Radioactive pollution and fallout.

Nuclear weapons production

CATEGORY: Nuclear power and radiation

DEFINITION: Manufacture of explosive devices that are powered by uncontrolled nuclear reactions

SIGNIFICANCE: The manufacture of nuclear weapons produces radioactive waste, which creates disposal problems. Environmental cleanup of the many sites that participated in the production of nuclear weapons in the past continues to require massive amounts of government funding and effort.

In 1942 a research group headed by Enrico Fermi built an atomic pile—literally a pile of graphite blocks interspersed with uranium, uranium oxide, and cadmium control rods. On December 2 they achieved a self-sustaining chain reaction in which neutrons from fissioning uranium nuclei caused a constant number of other uranium nuclei to fission as well. The atomic pile was the direct ancestor of the nuclear reactor. Its successful operation opened the door to the production of plutonium and the plutonium route to nuclear weapons. Since the pile's constituent parts became radioactive, it also began a disposal problem.

The development of the atomic bomb in the United States was a feverish race driven by the fear that Nazi dictator Adolf Hitler would reach the goal first. For the most part, reasonable precautions were taken to protect the scientists and others who worked on the project from dangerous exposure to radiation, but insufficient attention was given to potential environmental problems. The urgency of war dictated that environmental concerns be postponed until more time and resources were available. With the Cold War following on the heels of World War II, this attitude continued for a generation.

The U.S. Department of Energy's Environmental Management Office has the assignment of dealing with the legacy of nuclear weapons production in the United States. The office's responsibilities include restoring the environment where possible, dealing with nuclear waste, shutting down old production facilities, and safeguarding special nuclear materials. Among the tasks to be accomplished is the cleanup of thousands of contaminated areas and buildings, along with 500,000 cubic meters (17.6 million cubic feet) of various levels of radioactive waste. Despite years of planning, permanent disposal sites for high-level nuclear waste remain unavailable, so the worst former production sites can only be stabilized, not fully cleaned up. The estimated cost for this work is $230 billion spent over seventy-five years. For comparison, the estimated cost of repairs after the 1995 earthquake in Kobe, Japan, was $100 billion.

As part of cleanup efforts begun at Hanford Nuclear Reservation in 1989, 2,300 tons of highly radioactive waste was moved from a leak-prone basin to huge underground vaults through the tubes set into the floor of this storage building. Hanford was one of the facilities involved in U.S. nuclear weapons production. (AP/Wide World Photos)

URANIUM

The pollution problems associated with the production of nuclear weapons begin with uranium ore. It is mined, then pulverized and leached to enable extraction of the uranium. After drying, the depleted ore, called tailings, consists of particles the size of sand grains or smaller. More than 242 million tons of uranium tailings are known to exist in the United States and more than 1.9 billion tons worldwide. The two main hazards from these tailings are radon, a gas produced by the decay of radium, and the contamination of groundwater with heavy metals through runoff. Proper control requires that tailings piles be located away from population centers and covered with earth to trap the radon; the earth cover must then be held in place by grasses and other plants. Care must also be taken to prevent runoff from tailings piles from entering the water supply.

Uranium leaves the mill as a compound called yellowcake. For the American atomic weapons program, yellowcake was converted to the gas uranium hexafluoride, and then the isotope uranium 235 was separated from the more common uranium 238 at factories in Oak Ridge, Tennessee. Metal highly enriched in uranium 235 was fabricated into weapons parts at Oak Ridge. Reactor-grade uranium was shipped to the Hanford Nuclear Reservation in the state of Washington, where large nuclear reactors converted some of it into plutonium.

At Hanford, spent reactor fuel, now highly radioactive, was dissolved and its plutonium extracted. The plutonium was shipped to the Rocky Flats plant in Colorado, where it was fabricated into weapons parts. These parts were shipped to the Pantex plant at Amarillo, Texas, where the weapons were assembled. Each plant produced radioactive waste, and the U.S. government continues to deal with these contaminated areas and equipment.

WASTE STORAGE AND CLEANUP

Some 232 million liters (61 million gallons) of nuclear waste liquids and sludges were stored in concrete-encased steel tanks at Hanford. The highly radioactive waste includes chemicals such as nitrates, mercury, and cyanide. An estimated 4 million liters (1 million gallons) have leaked into the ground from the oldest storage tanks. Because it was believed that the pollution would not migrate beyond the large Hanford site, low-level waste, including 1,700 billion liters (449 billion gallons) of processing water, was poured directly onto the ground or into injection wells. The Hanford site represents the largest cleanup of its kind ever undertaken; after government initiation of the project in 1989, the cleanup has continued into the twenty-first century.

At the Weldon Spring, Missouri, site where uranium and thorium were processed, forty-four buildings and several waste disposal pits were contaminated. The buildings have been demolished, and 170,000 cubic meters (6 million cubic feet) of radioactive sludge have been stabilized, enclosed under a thick layer of sand, clay, and rock. In 2002 the cleaned-up site, which features this seven-story high special disposal cell and an interpretive center for visitors, was opened for public inspection. A similar site near Fernald, Ohio, had two hundred structures and 2.4 million cubic meters (85 million cubic feet) of waste. The cleanup of Fernald included shipping some waste elsewhere and storing slightly contaminated soil on-site in a special facility. The project, which turned the site into a nature preserve, was begun in 1990 and was not completed until 2006; the total cost was $4.4 billion. Monitoring of test wells for radioactive contamination continues at the site.

Charles W. Rogers

FURTHER READING

Gerber, Michele Stenehjem. *On the Home Front: The Cold War Legacy of the Hanford Nuclear Site.* 2d ed. Lincoln: University of Nebraska Press, 2002.

Hacker, Barton C. *Elements of Controversy: The Atomic Energy Commission and Radiation Safety in Nuclear Weapons Testing, 1947-1974.* Berkeley: University of California Press, 1994.

Hansen, Keith A. "Early Efforts to Limit Nuclear Testing." In *The Comprehensive Nuclear Test Ban Treaty: An Insider's Perspective.* Stanford, Calif.: Stanford University Press, 2006.

Makhijani, Arjun, Howard Hu, and Katherine Yih, eds. *Nuclear Wastelands: A Global Guide to Nuclear Weapons Production and Its Health and Environmental Effects.* 1995. Reprint. Cambridge, Mass.: MIT Press, 2000.

SEE ALSO: Bikini Atoll bombing; Hanford Nuclear Reservation; Neutron bombs; Nuclear and radioactive waste; Nuclear testing; Nuclear winter; Oak Ridge, Tennessee, mercury releases; Rocky Mountain Arsenal; Yucca Mountain nuclear waste repository.

Nuclear winter

CATEGORY: Nuclear power and radiation

DEFINITION: Severe change in global climatic patterns theorized as the likely result of either a nuclear war or a cosmic impact event

SIGNIFICANCE: Scientists do not all agree regarding the probable severity of the long-term environmental destruction that would result from nuclear warfare, but the concept of nuclear winter has been the subject of widespread discussion.

The nuclear winter theory predicts that a global nuclear war would cause huge amounts of dust and debris to be carried high into the atmosphere, where it could remain for as long as three years. This debris would obstruct incoming sunlight and disrupt photosynthesis, causing the extinction of many plant species. The loss of considerable amounts of plant life would eventually affect the entire food chain. Once the atmosphere cleared, only those species that could survive without sunlight would be left to repopulate the planet.

Discussion of the possibility of nuclear winter began in the early 1980's as a result of the ever-present fear of global nuclear war. Several prominent scientists, notably Carl Sagan of Cornell University, began to examine the possible effects of total nuclear war on the environment. Using computer simulations, they predicted that an enormous amount of dust and particulate matter from the nuclear explosions and resulting fires would be lifted high into the atmosphere. This debris would be quickly distributed around the globe by the earth's prevailing winds and would remain in the atmosphere for a minimum of three months to a maximum of three years. It was predicted that, in the event of a nuclear war, the majority of nuclear impacts would take place in the Northern Hemisphere; the Southern Hemisphere would thus experience a six-month "grace period" before it would feel the full effects of the atmospheric contamination.

The loss of sunlight would affect both plant and animal life. Within a matter of months, the photosynthetic cycle and the natural food chain would be totally disrupted. This would be accompanied by a worldwide drop in temperature, followed by a dramatic increase in the need for heating fuel, which would already be in short supply because of the destruction of energy industries, oil fields, and forests. Many commentators who studied the results of the computer simulations pointed out that the people who died from the exchange of nuclear weapons would be the lucky ones. Those who survived the bombs would have to face starvation and freezing temperatures.

The Persian Gulf War of 1991 offered scientists a firsthand opportunity to study a small-scale test of the nuclear winter concept. During that conflict, huge clouds of black smoke from thousands of burning oil wells entered the atmosphere. From space, astronauts clearly observed distribution patterns as the smoke slowly made its way around the world. Locally, the people of the Middle East experienced effects much like those predicted for a nuclear winter.

Around the time that the concept of nuclear winter was first being discussed, scientists discovered evidence of what was apparently a natural example of the same effect from asteroid impact. A large impact structure was found on the Yucatán Peninsula in Mexico dating back sixty-five million years to the extinction period of the dinosaurs. Scientists quickly employed the scenario of a nuclear winter to dramatize the last days of the dinosaurs. They also pointed out that several other mass extinction events had previously occurred and that such an event is likely to happen in the future.

Paul P. Sipiera

FURTHER READING

Ahrens, C. Donald. "Climate Change." In *Essentials of Meteorology: An Invitation to the Atmosphere.* 5th ed. Belmont, Calif.: Thomson Learning, 2008.

Cotton, William R., and Roger A. Pielke, Sr. "Nuclear Winter." In *Human Impacts on Weather and Climate.* 2d ed. New York: Cambridge University Press, 2007.

Manahan, Stanley E. "Catastrophic Atmospheric Events." In *Environmental Science and Technology: A Sustainable Approach to Green Science and Technology.* 2d ed. Boca Raton, Fla.: CRC Press, 2007.

SEE ALSO: Antinuclear movement; Gulf War oil burning; Nuclear weapons production.

Oak Ridge, Tennessee, mercury releases

CATEGORIES: Disasters; pollutants and toxins

THE EVENT: Contamination of air and water by mercury leaked from a nuclear weapons plant in Oak Ridge, Tennessee

DATES: 1950's; discovered in December, 1981

SIGNIFICANCE: After researchers discovered that mercury from a nuclear weapons plant had contaminated air and water in Oak Ridge, Tennessee, subsequent study of the contamination raised questions about the methods used to determine the risk of toxic pollution to humans.

In December, 1981, Stephen Gough, a biologist employed at Oak Ridge National Laboratory, tested a number of the plants growing near East Fork Poplar Creek and found that they contained a high amount of mercury. Because the creek was near the U.S. Department of Energy's (DOE) Y-12 plant, where nuclear bombs had been built in the 1950's, the possibility was raised that a large amount of mercury had been released into the environment from the plant over the intervening years.

During 1982 the DOE commissioned a study to determine how much mercury had been released. The researchers discovered that an estimated 215,000 kilograms (475,000 pounds) had been released into the creek and that another 907,000 kilograms (2 million pounds) were unaccounted for. The DOE report on the study was made public on May 17, 1983, after a small local newspaper and the Tennessee Department of Health and Environment requested to see the report under the Freedom of Information Act. After a public hearing on June 11, 1983, the DOE was repri-

Oak Ridge National Laboratory was built in 1943 as part of the U.S. effort to build nuclear weapons. (Oak Ridge National Laboratory)

manded by the Science and Technology Committee of the U.S. House of Representatives on November 3, 1983. It was also ordered to assess the risks to human health of the mercury at Oak Ridge and to produce an accurate estimate of how much mercury had been released into the creek and the atmosphere.

The greatest risk to human health from mercury comes from mercury in the methylated form. A problem of this type occurred in Minamata Bay, Japan, in the 1960's, when many people who ate seafood from the methylmercury-contaminated bay developed symptoms of mercury poisoning. The DOE claimed that the mercury at Oak Ridge was not methylated but in elemental form, so the risk was minimal. The DOE also claimed that the mercury was released below the public drinking-water intake and that the mercury traveled downstream before sinking into the earth, where it was trapped by a layer of shale before entering the groundwater. Another danger for humans from mercury is the consumption of contaminated fish; because the fish in Poplar Creek were determined to have twice the level of mercury deemed safe by the U.S. Food and Drug Administration (FDA), fishing in the creek was banned.

A report released by the DOE in 1997 stated that 127,000 kilograms (280,000 pounds) of mercury had entered the East Fork Poplar Creek and that 33,000 kilograms (73,000 pounds) of mercury had entered the atmosphere over several years, with peak years being 1957 and 1958. The study also estimated vegetable and milk production and consumption near the creek during critical years, since the airborne mercury could have contaminated gardens, crops, and grass eaten by cattle. The mercury in the soil was determined to be mercuric sulfide, a stable compound that was not likely to migrate to centers of population.

The DOE's report concluded that small doses of mercury had been ingested by local residents during the decades following 1950 but that such low levels did not pose a health risk. Controversy followed the report, as health experts asserted that individuals react to mercury ingestion in varying ways and that immune disorders and diseases found in people who lived in the Oak Ridge area could be results of the mercury release.

Rose Secrest

FURTHER READING
Eisler, Ronald. *Mercury Hazards to Living Organisms.* Boca Raton, Fla.: CRC Press, 2006.

Johnson, Charles W., and Charles O. Jackson. *City Behind a Fence: Oak Ridge, Tennessee, 1942-1946.* Knoxville: University of Tennessee Press, 1981.
Olwell, Russell B. *At Work in the Atomic City: A Labor and Social History of Oak Ridge, Tennessee.* Knoxville: University of Tennessee Press, 2004.

SEE ALSO: Heavy metals; Mercury; Minamata Bay mercury poisoning.

Ocean currents

CATEGORIES: Weather and climate; water and water pollution

DEFINITION: Continuous surface and deep-level movements of ocean waters in certain directions

SIGNIFICANCE: For environmentalists, the study of ocean currents relates directly to two major concerns: the potential impact of the currents on global climate and weather changes and how the currents can assist in the conservation and sustainable cultivation of food and other marine resources.

Scientists understand a great deal about the significant large-scale movements of all ocean waters, but the specific driving forces are still under investigation. Studies of ocean currents are usually divided between a focus on surface currents, which are largely driven by prevailing winds, and a focus on deep-level currents, which are driven by more complex factors, including saline density, water temperatures, the gravitational pull of the moon, and sea-bottom depths as well as shoreline formations. The impacts of marine life-forms on water circulation, including the movement of microscopic plankton species, are also part of the study of ocean currents.

SURFACE OCEAN CURRENTS

Ocean currents in the upper 400 meters (roughly 1,300 feet) of the ocean depths are largely driven by wind currents, which represents the movement of roughly 10 percent of the ocean water in total. Surface currents and winds have been studied and measured for centuries. While precise modern models are quite complex, in general scientists refer to prevailing winds, which blow from west toward the east in the latitudes between 30 and 60 degrees, and trade winds,

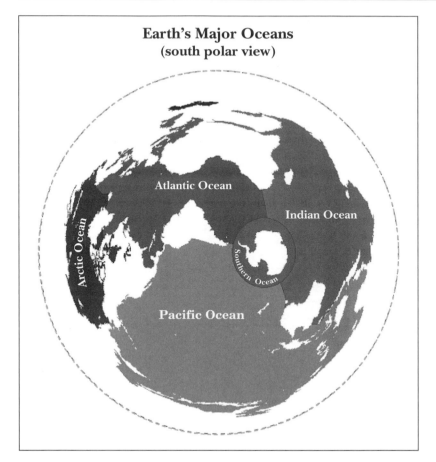

Earth's Major Oceans
(south polar view)

Atlantic Ocean

Indian Ocean

Arctic Ocean

Southern Ocean

Pacific Ocean

does, can draw materials from the bottom of the ocean and mix layers of currents that provide nourishment for the development of microscopic life, starting an important food chain that supports various levels of higher sea life. Many of these rings thus sustain areas historically known for specific species of sea life. Consequently, disruptions in these patterns have potential implications for cultivation of these traditional stocks.

It has become increasingly important for scientists to understand long-term cycles of shifting weather and ocean current patterns. Prevailing winds and currents have been noted to run in regular patterns for thirty to fifty years and then shift for an equal amount of time before returning again to the previous conditions. This has important implications for commercial fisheries, for example, because oceanographers must account not only for the human-made (anthropogenic) dangers of overfishing but also for the apparently natural shifts in current patterns that have impacts on the availability of certain species. These patterns have been found to have positive effects on some species and negative effects on others.

which blow from east to the west closer to the equator (below 30 degrees latitude). These winds create the classic sailing routes, such as the routes between the eastern United States and Europe, where sailing more north assists in a heading toward Europe and sailing more south (toward the equator) assists in a heading toward the eastern United States seaboard.

The study of ocean currents and flows, however, reveals that this general picture is made far more complex by the existence of related currents called meanders, rings, eddies, and gyres. These are relatively smaller patterns of flow within the larger general current pattern. For example, an oceanic gyre or ring is created when winds and limitations of shorelines create circular movement of currents (the circles can be smaller, rings, or involve entire sections of an ocean surface between major continents, such as the South Pacific Gyre, the Indian Ocean Gyre, and the North Atlantic Gyre).

These smaller-scale currents are important for sea life as well as for navigation. Circular rings and eddies, rather like air weather patterns of cyclones or torna-

VERTICAL CURRENTS

In contrast to wind-driven surface currents, which have been known for centuries, below-surface currents were not believed to be significant until relatively recently. In the twentieth and twenty-first centuries, however, deep-ocean currents have received considerable attention. In addition to the horizontal movements in ocean waters are the vertical movements usually referred to as thermohaline circulation. The term "thermohaline" comes from the Greek words for heat and salt, the two major influences on deepwater currents in addition to gravitational pull from the moon and the impacts of coastline formations. Thermohaline circulation is sometimes popularly referred to as the oceanic conveyer belt. The slow movement of these currents can extend into the hundreds of years.

Put simply, dense, and thus heavier, waters are created in the extreme cold conditions of the polar seas. These waters sink, following currents away from the poles and pushing up warmer waters to interact with the cooler atmospheric temperatures. Deep ocean currents affect 75 percent of ocean waters, which have a temperature from 0 to 5 degrees Celsius (32 to 41 degrees Fahrenheit). Deepwater circulation involves the vast majority of ocean waters as opposed to the surface, horizontal movements driven by prevailing winds.

ENVIRONMENTAL IMPACTS

One of the most important implications of the vertical currents of the world's oceans concerns the amount of carbon dioxide in the earth's atmosphere. Colder water absorbs carbon dioxide, whereas warmer water releases it. Another important factor in carbon dioxide distribution is the process known as the biological pump, which moves carbon from the ocean surface to the seafloor. While the presence of carbon dioxide is part of the reason the earth is habitable (the greenhouse effect is one reason the earth is warm enough for human occupation), there is concern that human activity has created abnormal levels of carbon dioxide. The rate at which carbon dioxide is being added to the atmosphere has not been exceeded in the past 420,000 years, and the increase since the mid-nineteenth century has been greater than any observed during the past 20,000 years. Two-thirds of this increase comes from the burning of fossil fuels; the last one-third can be accounted for by deforestation. If global temperatures continue to rise beyond the ability of the ocean currents and the biological pump to mitigate the effects, it is possible that ocean levels will rise enough to inundate low-lying coastlines around the world.

Scientists have also speculated about whether anthropogenic impacts on the atmosphere could actually interfere with the thermohaline circulation. Such interference could potentially have catastrophic impacts on both weather patterns and marine life in and near landmasses around the world. Many European nations, for example, have historically had relatively mild climates in part because of the warming impact of ocean currents on weather patterns. Sudden shifts in these currents could thus result in rapid changes in historic weather patterns. Oceanographic scientists do not all agree, however, regarding the like-lihood of such catastrophic changes. An ongoing debate among scientists concerns the ability to measure long-term cyclical change versus anthropogenic change.

In addition, questions remain concerning impacts on the oceans and the energy sources that drive currents. Nigel Calder, for example, has pointed out that whereas it was once presumed that salinity levels and temperature are primary producers of currents, as recently as 1998 two American oceanographers suggested instead that winds and especially the moon's gravitational pull have a much more significant impact on ocean currents, and particularly on the circulation of ocean waters and their interactions with the land borders. Calder concludes, "If the tidal story is correct, the value of 20th-century computer models of the ocean circulation, based as they were on heat-salt effects, is seriously in question."

Oceanographers are increasingly taking greater numbers of variables into consideration in their analyses of ocean currents and the impacts of weather patterns. Ocean water reacts with sunlight, with various temperatures of air, and with temperature differences between layers of the open sea, and the ways in which changes in ocean currents and temperatures actually take place are not entirely understood. Most scientists agree, however, on the importance of learning more about the potential for change in ocean currents in the short and the long term.

Daniel L. Smith-Christopher

FURTHER READING

Calder, Nigel. "Ocean Currents." In *Magic Universe: The Oxford Guide to Modern Science.* New York: Oxford University Press, 2003.

Longhurst, Alan. *Ecological Geography of the Sea.* 2d ed. Boston: Elsevier, 2007.

Mann, K. H., and J. R. N. Lazier. *Dynamics of Marine Ecosystems: Biological-Physical Interactions in the Oceans.* 3d ed. Malden, Mass.: Blackwell, 2006.

Park, Chris C. "Oceans and Coasts." In *The Environment: Principles and Applications.* 2d ed. New York: Routledge, 2001.

SEE ALSO: Carbon cycle; Carbon dioxide; Climate change and oceans; Coral reefs; Fisheries; Gill nets and drift nets; Global warming; Great Barrier Reef; Greenhouse gases; Ocean pollution; Sea-level changes; Thermohaline circulation.

Ocean dumping

CATEGORIES: Waste and waste management; water and water pollution

DEFINITION: Disposal of waste products in the world's oceans

SIGNIFICANCE: The disposal of sewage, dredging spoil, garbage, chemicals, and other waste into the ocean disrupts individual ecosystems and kills fish and marine life. Although many marine pollution treaties and laws have been written to regulate dumping, such treaties and legislation have proven difficult to enforce.

The ocean is the final stop for much of the garbage generated on land. Sewage processing centers release treated wastewater into the ocean. Waste management officials at overflowing landfills have looked to the ocean as an alternative to burying rubbish and debris. Oil tankers have rinsed and flushed holding compartments in the open ocean. Fishing vessels dump scrap parts and unwanted or spoiled fish into the water. For decades, all these sources of pollution were thought to have little effect on the world's oceans because they are so large. Whether in large or small amounts, however, marine dumping upsets the balance of ocean ecosystems, altering the environments to which fish, plants, and other marine organisms are accustomed. Some survive the changes, but others do not.

Industrial facilities have found the ocean useful for the disposal of waste products as well. Some companies have sealed waste chemicals into metal containers and dumped them "harmlessly" onto the ocean floor. There, the life cycles of crabs, ground fish, and other marine-floor creatures have been disrupted by these alien containers. Over time some of these containers have begun to leak, sometimes releasing radioactive waste or chemicals harmful to the marine environment.

Most of the garbage dumped directly into the oceans has come from ships and boats (although trash continues to make its way from the land into oceans via storm drains, winds, and beach littering). The number of fishing boats, naval vessels, cruise ships, cargo ships, and recreational boats taking to the sea daily is estimated to be in the millions. Many of the larger vessels are like floating towns. With thousands of people on board, these ships can quickly accumulate large amounts of sewage, garbage, and other waste, and space for storing it is often limited. Some ships, such as fishing and naval vessels, remain on the open ocean for weeks or months at a time; it is therefore not surprising that many of these vessels solve their garbage problems by dumping waste overboard. Although the disposal of sewage and wastewater in this way is illegal, many seafaring vessels ignore such laws because they are difficult to enforce.

ANTIDUMPING TREATIES

The environmental threat posed by ocean dumping was apparent by the early 1970's and led to the international Convention on the Prevention of Marine Pollution by Dumping of Wastes and Other Matter, also known as the London Convention, in 1972. One of the earliest global treaties for protection of the marine environment, it entered into force in 1975. The London Convention bans the dumping of certain hazardous materials and requires permits for the dumping of other materials. The 1996 London Protocol, which entered into force in 2006, revises the 1972 treaty to prohibit all dumping except for the following: dredged material; sewage sludge; fish wastes; vessels and platforms; bulky items made up primarily of iron, steel, and concrete; inert inorganic geological material such as mining waste; organic material of natural origin; and carbon dioxide captured from the atmosphere and introduced into the ocean as part of a carbon sequestration system. By 2010 eighty-six nations were parties to the 1972 convention and thirty-seven were parties to the 1996 protocol.

Of the types of garbage that have been dumped, plastics are among the most dangerous to marine creatures. Plastic does not degrade readily and can persist in the ocean for decades, if not centuries. Medical equipment, such as needle syringes and wrappers, is often made of plastic and has been known to wash up on beaches months after being dumped into the deep ocean. Marine birds and mammals often mistake plastic bags and pellets for food. Though an animal feels full after eating plastic, the material offers no nutritional value, and the animal dies. Sea turtles and whales have died after eating plastic bags that became entangled in their intestines. Marine birds and seals have starved to death after their beaks or snouts have become stuck in plastic rings such as those used to join beverage six-packs. Plastic that has broken down into smaller particles may not kill marine life outright, but it presents a dietary source of toxins that becomes increasingly concentrated farther up the food chain. During the late 1980's it was estimated that 7.3 billion kilograms (16 billion pounds) of plastic was being dumped into the ocean annually.

In December, 1988, an annex to the International Convention for the Prevention of Pollution from Ships (known as MARPOL, for "marine pollution") banned the dumping of plastics anywhere in the ocean. Referred to as Annex V of the MARPOL treaty, it also set severe restrictions on other garbage discharges within coastal zones and special areas affected by heavy maritime traffic or low water exchange; required ships to have garbage management plans and to maintain written records of all garbage disposal and incineration operations; and required governments to provide garbage reception facilities at ports. By 2010 the MARPOL treaty had been signed by 150 countries, and Annex V had been signed by 140 countries.

The Fate of Floating Debris

In 1997 oceanographer Charles Moore of the Algalita Marine Research Foundation reported finding a massive area of floating marine trash between California and Hawaii. Dubbed the Great Pacific Garbage Patch, it has proved to be only one of several such areas where wind and ocean currents have trapped and concentrated anthropogenic (human-caused) debris. Although visions of a huge island of floating trash have captured the public imagination, the Pacific Garbage Patch is not actually visible in aerial photographs or satellite images. Rather, it is (as the Ocean Conservancy has described the area) a

A photo provided by the U.S. Army shows the disposal of containers of mustard gas over the side of a barge somewhere in the Atlantic Ocean in 1964. (AP/Wide World Photos)

"chunky soup" of small, broken-down plastic particles intermixed with larger, recognizable objects such as rain boots, food containers, and toothbrushes. Samples have revealed that, in some spots within the Pacific Garbage Patch, plastic is six more times abundant than plankton by weight.

While marine trash can remain caught up in ocean gyres (massive rotating current systems) far from human habitation, other currents carry much of the debris back to shore. A number of conservation groups annually clean this litter from beaches in public-participation events, and many record the amounts and types of materials recovered. The Ocean Conservancy reports that on a single fall day in 2009, volunteers around the globe collected more than ten million individual pieces of trash weighing a total of 3.4 million kilograms (7.4 million pounds) from the world's beaches. By number, the most abundant items recovered that year were cigarettes and cigarette filters, plastic bags, and food wrappers and containers. "Legal and illegal dumping of domestic and industrial garbage, construction materials, and large household appliances" accounted for more than 198,000 of the items found, including cars and car parts, tires, and 55-gallon drums.

Lisa A. Wroble
Updated by Karen N. Kähler

FURTHER READING

Clark, R. B. *Marine Pollution*. 5th ed. New York: Oxford University Press, 2001.

Gorman, Martha. *Environmental Hazards: Marine Pollution*. Santa Barbara, Calif.: ABC-CLIO, 1993.

Hamblin, Jacob Darwin. *Poison in the Well: Radioactive Waste in the Oceans at the Dawn of the Nuclear Age*. New Brunswick, N.J.: Rutgers University Press, 2008.

International Maritime Organization. *Guidelines on the Convention on the Prevention of Marine Pollution by Dumping of Wastes and Other Matter, 1972*. London: Author, 2006.

Laws, Edward A. *Aquatic Pollution: An Introductory Text*. 3d ed. New York: John Wiley & Sons, 2000.

Ocean Conservancy. *Trash Travels: From Our Hands to the Sea, Around the Globe, and Through Time*. Washington, D.C.: Author, 2010.

Ringius, Lasse. *Radioactive Waste Disposal at Sea: Public Ideas, Transnational Policy Entrepreneurs, and Environmental Regimes*. Cambridge, Mass.: MIT Press, 2001.

United Nations Environment Programme. *Marine Litter: A Global Challenge*. Nairobi: Author, 2009.

SEE ALSO: Dredging; London Convention on the Prevention of Marine Pollution; Marine debris; Marine Mammal Protection Act; Ocean currents; Ocean pollution; Seabed disposal; Stormwater management.

Ocean pollution

CATEGORY: Water and water pollution

DEFINITION: Introduction by direct or indirect human action of harmful materials into the marine environment

SIGNIFICANCE: Ocean pollution can harm marine organisms, alter or destroy marine habitats, negatively affect human health, and impede or halt fishing operations. Land-based sources account for approximately 80 percent of the pollutants that enter the marine environment; the remainder come from maritime operations and dumping.

Five major types of pollutants can be found in the oceans. The first type, degradable waste, makes up the greatest volume of discharge and is composed of organic material that will eventually be reduced to stable inorganic compounds such as carbon dioxide, water, and ammonia through bacterial attack. Included in this category are urban wastes, agriculture wastes, food-processing wastes, brewing and distilling wastes, paper pulp mill wastes, chemical industry wastes, and oil spillage.

The second type of pollutant is fertilizer, which has an effect similar to that of organic wastes. Nitrates and phosphorus compounds are carried from agricultural lands by runoff from irrigation and rainfall into rivers and from there into the oceans. Once in the sea, they enhance phytoplankton production, sometimes to the extent that the accumulation of dead plants on the seabed produces anoxic (oxygen-deprived) conditions.

A third kind of pollutant, dissipating waste, is composed mainly of industrial wastes that lose their damaging properties after they enter the sea. Their effects are therefore confined to the immediate area of the discharge. Some examples are heat, acids, alkalis, and cyanide.

A fourth type of pollutant is particulates. These are small particles that may clog the feeding and respiratory structures of marine animals. They can also reduce plant photosynthesis by reducing light penetration; when settled on the ocean bottom, they may smother animals and change the nature of the seabed. Some examples are dredging spoil, powdered ash from coal-fired power stations, china clay waste, colliery (coal-works) waste, and clay from gravel extraction. Larger objects such as containers and plastic sheeting have impacts similar to those of particulates.

The last type of pollutant is conservative waste. Such waste is not subject to bacterial attack and is not dissipated, but it can react with plants and animals. Examples of conservative waste include heavy metals, such as mercury, lead, copper, and zinc; halogenated hydrocarbons, such as polychlorinated biphenyls (PCBs) and dioxins; and radioactive materials.

POLLUTION INPUTS FROM LAND

Pollutants can reach oceans in many ways. According to the United Nations Environment Programme (UNEP), land-based activities are responsible for as much as 80 percent of the pollutant load in coastal waters and deep oceans. Rivers that have been polluted by land runoff are the largest source of harmful substances. River water enters the sea carrying a staggering load of pollutants accumulated along the river's entire length. In agricultural areas, nutrients from fertilizers and livestock wastes, as well as silt from eroding fields, are common pollutants in runoff. Herbicides and pesticides that drain from fields introduce carcinogens and other harmful chemical components into receiving waters. In urban areas, runoff carries an array of pollutants. Typical among these are wastes deposited by motor vehicles, including gasoline, oil, grease, and heavy metals; sediment from construction sites; herbicides and pesticides from lawns and gardens; road salts; viruses and bacteria from inefficient septic systems; and spilled chemicals from industrial sites.

Another type of pollutant input from the land is direct outfall from pipes that empty into oceans. This pollution comes from urban and industrial wastes that are deposited directly into coastal waters. During the nineteenth century, rivers such as the Thames and the Tees in England, the Clyde in Scotland, and the Hudson in New York state became incredibly polluted by this type of input. With increasing pressure to preserve inland sources of drinking water, newer industries that need large quantities of water have sought out coastal locations, which has resulted in increased levels of ocean pollution. Urban areas on the coast also present the problem of municipal wastes and sewage. In many coastal population centers, these wastes

are treated. However, roughly 40 percent of the world's population lives on the coast, often in fertile and productive estuarine and delta areas. This makes it almost impossible to process all wastes before they enter the sea. According to a 2006 UNEP report on the state of the marine environment, in many developing countries more than 80 percent of sewage that enters coastal zones is untreated.

POLLUTION INPUTS AT SEA

On the open water, ships can pollute the ocean in many ways. They often carry toxic substances such as oil, liquefied natural gas, pesticides, and industrial chemicals. Shipwrecks can lead to the release of these chemicals and result in serious damage. Examples include the *Torrey Canyon* (1967), *Amoco Cadiz* (1978), *Exxon Valdez* (1989), and *Sea Empress* (1996) oil spills, which fouled coastlines and caused the deaths of many seabirds and marine animals. Noxious or dangerous materials are frequently carried on deck as a safety precaution and can be lost overboard during storms. During routine operations, ships discharge

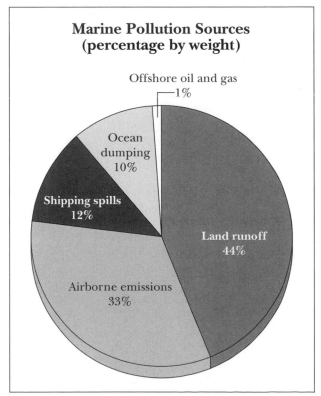

Marine Pollution Sources (percentage by weight)

Offshore oil and gas — 1%
Ocean dumping 10%
Shipping spills 12%
Airborne emissions 33%
Land runoff 44%

Source: Data taken from United Nations Environment Programme, *The State of the Marine Environment.* Oxford, England: Blackwell Scientific Publications, 1991.

oily ballast water, bilgewater, and cargo tank washings (not always legally). They also discard much of their garbage overboard. Marine dumping of any form of plastic, however, has been illegal since 1988.

Another offshore pollutant input comes from activities such as dumping dredging spoil and sewage sludge. Dredging spoil is the waste sediment generated by dredging operations, which may be conducted to keep shipping channels and entrances to ports open, as an underwater mining method, or as a means for removing polluted bottom sediments. While dredged material is sometimes disposed of on land, especially if it is highly toxic and in need of treatment, it is often barged out to sea to be dumped. Dredging spoil from harbors contains large quantities of heavy metals and other contaminants that are transferred to dumping grounds. Even clean spoil, however, can have a negative impact on marine life: Dumping it makes the water column more turbid (cloudy), which can fatally clog the gills of fish and shellfish. Sewage from the treatment of land-based sewage has been dumped at sea in large quantities. This sludge is contaminated with heavy metals, oils and greases, and other substances. Additional offshore industrial activities that can pollute the ocean include oil exploration and extraction, sand and gravel extraction, and mineral extraction. Other harmful wastes that are dumped at sea include ashes from power stations and colliery waste. Until the practice was banned in 1993, containers of low-level radioactive waste were disposed of at deep ocean sites. Conventional munitions, incendiary devices, and chemical weapons were also routinely dumped at sea until the early 1970's.

The last type of input comes from the air. While this type of input is complex and challenging to quantify, scientists believe that it represents a large and significant contribution to pollution. Atmospheric contaminants may exist as gases or aerosols. They are deposited by gas exchange at the sea surface, fall out as particles (dry deposition), or are scavenged from much of the air column by precipitation within clouds (wet deposition). A nontoxic atmospheric gas of increasing concern is carbon dioxide (CO_2). The world's oceans absorb about one-fourth of the CO_2 that human activity adds to the atmosphere annually. The dissolution of CO_2 in seawater, which forms carbonic acid, is creating increasingly acidic oceans. This change in pH is impeding the ability of some marine organisms to form shells and skeletal structures.

IMPACTS

Nutrients from sewage, treated or untreated, are increasing the rate of plant growth in coastal waters, a phenomenon known as cultural eutrophication. Sewage and agricultural runoff are high in nitrogen and phosphorus, which encourage ocean plants to grow. This results in unsightly algal scum on beaches. When the extra algae die and sink to the bottom, the resulting bacterial decomposition uses available dissolved oxygen, causing deoxygenation on the bottom waters. In extreme cases, this can kill fish. Some algal blooms contain toxic substances. Fish and shellfish that eat these algae become unfit for human consumption.

Water-soluble compounds from crude oil and refined products include a variety of compounds toxic to a large number of marine plants and animals. Chemicals released during oil spills from tankers or from offshore drilling operations are toxic to a wide range of plankton, which is at special risk because it lives near the water's surface. Salt marshes and mangrove swamps trap oil from oil spills. When the plants are in bloom, the blossoms become coated with oil and thereafter rarely produce seeds. A 2002 report from World Wildlife Fund Canada estimated that, in the waters of Atlantic Canada alone, roughly 300,000 seabirds die from operational discharges of oil at sea every winter.

Antifouling paints and other coatings that discourage the growth of barnacles and other marine organisms on ships and marine structures can be highly toxic to the ocean environment. Antifouling agents leach from surfaces to which they are applied and persist in the water. Tributylin (TBT), an effective antifouling substance introduced during the 1960's, was subsequently found to cause deformations in oysters and sex changes in whelks. A complete global prohibition on the use of TBT and related compounds for ship antifouling went into effect at the beginning of 2008.

Conservative pollutants are not affected by bacteria; when consumed by animals, they cannot be excreted and thus remain in the body. This process is called bioaccumulation. Animals that eat polluted plants and animals receive an enriched diet of those conservative materials, which they also are unable to excrete. The pollutants can accumulate in their bodies over time, eventually reaching concentrations high enough to cause illness or death. This process, called biomagnification, particularly affects the ocean predators. Bioaccumulation and biomagnification can render seafood unsafe for human consumption.

Discarded plastic items can cause harm to the larger animals of the sea. They can trap mammals, diving birds, or turtles, causing them to drown. Fishing gear lost overboard can continue to drift, snaring fish and entangling seabirds. Marine animals routinely mistake plastic items for food, ingest them, and die. Oceanographers have discovered extensive "garbage patches" far from land where ocean currents have trapped huge masses of plastic particles and intact or semi-intact plastic items. It is not known what effect the increasingly small particles of plastic will have on marine life and human health as they become an inescapable part of the food web.

EFFORTS TO REPAIR THE DAMAGE

In late 1988 a global ban went into effect prohibiting the marine dumping of plastics. However, illegal dumping, uncontrolled runoff, beach littering, and (in the case of lightweight plastic shopping bags) even wind continue to carry plastics into oceans, where they can persist for decades, if not centuries.

A number of international agreements have tackled other aspects of ocean pollution. The Convention on the Prevention of Marine Pollution by Dumping of Wastes and Other Matter, also known as the London Convention, entered into force in 1975; an update, the London Protocol, entered into force in 2006. This treaty regulates ocean dumping. The United Nations Convention on the Law of the Sea, adopted in 1982, addresses economic and environmental issues on the open sea. It was a replacement for related agreements in 1958 and 1960 that were inadequate to handle modern marine pollution problems and other concerns. Agenda 21, approved by governments that attended the Earth Summit held in Rio de Janeiro in 1992, called for all coastal countries to develop integrated coastal zone management plans by the year 2000. Coastal areas remain a high priority because they are most affected by pollution sources on land. Other global efforts to curb ocean pollution include the 1973 International Convention for the Prevention of Pollution from Ships and the 1978 Protocol that modified it. Known as MARPOL 73/78, the treaty and its several annexes address various marine pollution sources and the means for controlling or eliminating them.

Some countries have begun to control pollution in their estuaries and bays, and fish and shellfish have returned to formerly polluted areas. Maritime ship-based activities have considerably less impact on the ocean than in the past, mainly because of the introduction of international treaties limiting the discharge of wastes at sea. In addition, schools, aquariums, nonprofit groups, and governments all work on an ongoing basis to educate the public about the need to prevent ocean pollution.

David F. MacInnes, Jr.
Updated by Karen N. Kähler

FURTHER READING

Allsopp, Michelle, et al. *State of the World's Oceans.* New York: Springer, 2009.

Clark, R. B. *Marine Pollution.* 5th ed. New York: Oxford University Press, 2001.

Field, John G., Gotthilf Hempel, and Colin P. Summerhayes, eds. *Oceans 2020: Science, Trends, and the Challenge of Sustainability.* Washington, D.C.: Island Press, 2002.

Gorman, Martha. *Environmental Hazards: Marine Pollution.* Santa Barbara, Calif.: ABC-CLIO, 1993.

Hofer, Tobias N., ed. *Marine Pollution: New Research.* New York: Nova Science, 2008.

Laws, Edward A. *Aquatic Pollution: An Introductory Text.* 3d ed. New York: John Wiley & Sons, 2000.

Ocean Conservancy. *Trash Travels: From Our Hands to the Sea, Around the Globe, and Through Time.* Washington, D.C.: Author, 2010.

United Nations Environment Programme. *Marine Litter: A Global Challenge.* Nairobi: Author, 2009.

SEE ALSO: Climate change and oceans; Continental shelves; Cultural eutrophication; Dredging; London Convention on the Prevention of Marine Pollution; Marine debris; Ocean dumping; Oil spills; United Nations Convention on the Law of the Sea; Wastewater management.

Odor pollution

CATEGORY: Pollutants and toxins
DEFINITION: Unwanted scents in the environment
SIGNIFICANCE: Human beings are often annoyed by scents they find unpleasant or obnoxious. Although some odors signal the presence of hazardous air pollutants, many are nontoxic; nevertheless, their presence in the environment can have real impacts on quality of life.

Odor complaints are one of the top citizen pollution concerns in many cities in Europe and the

United States; about 10 percent of Americans complain about odor pollution problems. The perception of odor pollution involves a combination of cultural expectations and the ability of individuals to perceive various odors. Cultural expectations play a strong role in reactions to particular odors. For example, people who travel outside their home countries may be exposed to olfactory experiences they consider highly unpleasant, while local residents remain oblivious to the same odors.

Malodorous substances and their sources are often difficult to identify and ameliorate. Because the human nose can detect tiny concentrations of certain chemicals, qualitative and quantitative analyses intended to identify malodorous substances are often inconclusive. Thiols rank high on the list of odor pollution complaints, which is not surprising given the widespread industrial applications of these chemicals and the human ability to smell minute concentrations. Ethanethiol is detectable by smell at concentrations as low as 1 part per 2.8 billion parts of air; much higher concentrations are needed for detection by standard chemical tests.

A variety of chemicals emanating from many sources often contribute to odor problems. Establishing a single cause for an odor is difficult because people who are able to identify a particular odor when it is found alone are often unable to identify it as a component in a complex mixture. Another factor is the concentration of a substance, which may influence the acceptability of the odor. This is particularly true of indole, which is classed as pleasant-smelling in minute concentrations but is overwhelmingly unpleasant in moderate and high concentrations. Because indole is a decomposition product of tryptophan, which is used as a chemical reagent and in the manufacture of perfumes and pharmaceuticals, it is a potential odor pollutant in the vicinity of those industrial settings.

Throughout the United States, new housing subdivisions are encroaching into areas that in the past were reserved for agriculture. In such subdivisions, recently transplanted urban dwellers may complain about a range of odors associated with farming activities, such as mowing hay and manuring fields. Solutions to this type of perceived odor pollution may include working to alter the expectations of new residents and recommending that they close their homes and use air conditioners rather than open their windows for "fresh air."

Within cities, odor pollution is often caused by poor methods of disposal of garbage, including food residues. Large urban areas often develop task forces that combine the resources of a pollution-control office, a sewer department, and refuse collectors to seek out and ameliorate the sources of bad garbage odors. Past investigations of vile odors in San Francisco, California, identified the main culprit to be aging butter discharged into the sewer system by restaurants. The odors emanating from the city's sewer system were eliminated by units similar to those used to clean up toxic spills, which removed the source of the stench and sprayed the area with disinfectant.

Many different industries have been identified as contributing to odor pollution, including food processing, paper manufacturing, electric power generation, and waste disposal. Pollution-control officers frequently find that similar manufacturing facilities have different odor problems. Apparently, differences in effluent gases from the facilities may produce differing odor strengths, resulting in different degrees of annoyance among nearby residents.

The kraft paper industry has received considerable attention for its emissions of highly odorous and unpleasant sulfurous gases. In the manufacture of kraft paper, hydrogen sulfide (H_2S) is the major odor pollutant. However, the presence of nitric oxide (NO) enhances the unpleasant perception of H_2S, which is more readily sensed in an acid gas mixture than in an alkaline gas mixture. Carbonyl sulfide (COS) and sulfur dioxide (SO_2) also affect the perceived odor strength of the effluent gases. A kraft operation that controls the acidity of effluent gases may generate fewer complaints about odor pollution than the industry average. Asphalt plants are another notorious odor source. When these plants are fueled by recycled oil, both the energy source and the product may contribute to odor problems. One technique sometimes employed at these facilities is the use of odor-absorbing products to neutralize emissions of hydrocarbons and sulfur dioxide.

An age-old practice for dealing with unwanted or unpleasant odors is to mask them with neutral or pleasant odors; for example, individuals may apply perfumes to cloak perspiration odor, and stores and offices may use "fresh scent" dispensers in their ventilation systems. Some industries have tried releasing masking odors along with known odor pollutants, with dubious success.

During atmospheric inversion conditions, all air pollutants increase in concentration, including odors.

Transportation-related odors, including diesel and automobile exhaust, mingle with ozone created during photochemical smog. During these photochemical air-pollution episodes, the acrid odors signal a real public health threat.

Anita Baker-Blocker

FURTHER READING

Drobnick, Jim, ed. *The Smell Culture Reader.* New York: Berg, 2006.

Godish, Thad. "Welfare Effects." In *Air Quality.* 4th ed. Boca Raton, Fla.: Lewis, 2004.

Vallero, Daniel. "Effects on Health and Human Welfare." In *Fundamentals of Air Pollution.* 4th ed. Boston: Elsevier, 2008.

SEE ALSO: Air pollution; Landfills; Pulp and paper mills; Sewage treatment and disposal; Smog; Solid waste management policy; Sulfur oxides; Waste management; Waste treatment.

Oil crises and oil embargoes

CATEGORY: Energy and energy use

DEFINITION: Disruptions of oil supply patterns between the Middle East and the industrialized West

SIGNIFICANCE: Oil crises and oil embargoes have typically occurred during times of political tension in the Persian Gulf region. Such episodes have strengthened support in Western nations for greater domestic energy production and for military policies enhancing oil security in the Persian Gulf.

The industrialized West depends on oil as a fuel and chemical feedstock. Given that the majority of the world's long-term oil reserves are located in the Middle East, the West depends on a smooth flow of oil through world markets. On at least three occasions during the last few decades of the twentieth century, political events in the Middle East led to crises in the world oil markets. These crises illustrated the West's vulnerability to oil supply interruptions and price increases.

WORLD OIL MARKET INTERRUPTIONS

In the early years of international oil markets, the United States was the leading producer, accounting for more than one-half of the world's oil production until the 1950's. U.S. oil production, however, peaked at 11.3 million barrels per day in 1970 and began to decline as lower-cost reserves in the Middle East were tapped. Leading oil-producing nations in the Middle East had met in 1960 to form the Organization of Petroleum Exporting Countries (OPEC), but the new organization was largely ignored by the world community. Because oil flowed without significant interruption, there was also little notice of the increasing dependence of the West on Middle Eastern oil into the early 1970's.

The security of oil supply became an immediate world issue on October 6, 1973, with an attack by Egyptian forces against Israeli positions along the Suez Canal. The United States supported Israel in the brief war that followed. In retaliation, King Faisal of Saudi Arabia ordered a 25 percent cut in Saudi oil output and a cutoff of shipments to the United States. The other Arab members of OPEC joined in the embargo and production cutbacks. World oil prices roughly quadrupled, from around three dollars per barrel to twelve dollars. In the United States, emergency conservation measures were implemented, and consumers waited in long lines for limited supplies of motor fuels. The embargo ended about six months later, on March 18, 1974.

The Western economies recovered, and oil prices were relatively stable from 1974 to 1978. Oil demand resumed its growth, and oil again flowed to the West. In 1979, however, the Iranian revolution removed much of Iran's production from world markets at a time when there was little excess capacity. World oil prices again rose sharply, reaching a peak of nearly thirty-five dollars per barrel by 1981. During this incident, unlike 1973-1974, there was no attempt to declare an embargo. OPEC members merely raised prices and continued to ship oil.

The high prices in the aftermath of the Iranian revolution did not persist, as a worldwide surplus of oil drove prices lower during the 1980's. A third crisis, the Persian Gulf War of 1990-1991, elevated prices so temporarily that some analysts referred to the period as "an oil spike," referring to the sharp upward, then downward, movement in oil prices. The upward movement was precipitated by Iraq's invasion of Kuwait in August, 1990. World oil prices reached their highest level in more than eight years, briefly peaking near the forty-dollar-per-barrel mark. Saudi Arabia and other producers began to replace Kuwait's lost output, and prices fell, with a dramatic drop after Operation Desert Storm defeated Iraqi forces in Kuwait in January, 1991. Even the burning of Kuwaiti oil wells by Iraqi forces was unable to keep world prices at their previous high levels.

OIL CRISIS EFFECTS

The early stages of all three crises were characterized by panic buying, with producers and consumers scrambling to assure supplies. Later research showed that the crises affected price more than physical availability. Even during the most severe of the three episodes (1973-1974), the ordered cutbacks amounted to less than 10 percent of world oil supply. Further, after the embargo ended there was evidence that important quantities of supposedly embargoed oil had reached the United States and other Western nations.

During each of the three crises, Western economies entered recessions. Higher oil prices acted as a tax that transferred massive amounts of wealth from consuming nations to producing nations. Later declines in world oil prices had the opposite effect, stimulating the economies of the consuming nations.

The 1973-1974 embargo led to marked changes in energy policy in the United States. During the embargo, President Richard Nixon announced Project Independence, a program designed to enable the United States to achieve self-sufficiency in energy through crash programs along the lines of the Apollo moon missions and the Manhattan Project to develop the first nuclear weapons. Project Independence called for major development of nuclear and coal energy resources, placing a low priority on conservation and environmental concerns. Similar policies continued during the administration of President Gerald R. Ford but lost urgency as the embargo's effects were overcome. The Iranian disruption of 1979-1980 occurred during the term of President Jimmy Carter, adding force to Carter's characterization of energy problems as "the moral equivalent of war." Carter called for conservation and new technology to engineer a transition that he characterized as having the same importance as the transition from wood to coal during the Industrial Revolution. By the time of Operation Desert Storm in 1991, policy concern had shifted to the ability of Western nations to use military force to keep oil flowing. Iraq's initially successful invasion of Kuwait and the implied threat to Saudi Arabia quickly generated support for a military solution by a coalition of nations led by the United States.

KNOWLEDGE GAINED FROM OIL CRISES

Although the oil crises have been extensively studied, their message for the future and the policy lessons to be learned from them are uncertain. One viewpoint holds that oil crises constitute an early warning of resource depletion. The 1973-1974 price increases were initially seen as a signal confirming predictions that the prices of nonrenewable resources would increase as depletion approached. This "depletionist" viewpoint calls for government-enforced conservation of existing resources

---

Major Oil Crises and Oil Embargoes

YEAR	EVENT
1960	Oil-producing nations holding more than 90 percent of known oil reserves form the Organization of Petroleum Exporting Countries (OPEC) to seek higher and more stable crude oil prices.
1970	U.S. oil production peaks. After being the dominant oil producer in the first half of the twentieth century, the United States finds its production declining and begins to import an increasing fraction of its oil.
1970	Libya receives higher prices for its crude oil from Occidental Petroleum.
1973	World oil prices soar after Arab members of OPEC declare a cutback in production and an embargo on shipments bound for the United States and other allies of Israel.
1979	Oil prices double after the fall of the shah of Iran's government removes Iranian capacity from production.
1990	Oil prices briefly reach previous record highs after Iraqi forces invade Kuwait and take over Kuwaiti oil fields.
1991	Operation Desert Storm forces—led by the United States—defeat Iraqi forces, which set numerous oil well fires before retreating.
2003	U.S. occupation of Iraq disrupts Middle East oil markets and raises new concerns about the stability of world oil supplies.

and development of new technologies to postpone or avoid the consequences of oil depletion.

A second interpretation of the oil crises is that instead of reflecting a fundamental problem of depletion, they reflect a problem with distribution; that is, oil resources occur in patterns that are different from their patterns of use. Since large amounts of oil resources are held in the politically unstable Middle East but used in the West, cutoffs and price increases are always possible. The policy problem is viewed as a politically inspired cutoff of oil shipments. There is no threat, in this view, of a calculated economic cutoff of oil because producers would find it more profitable to sell oil at a high price than to cut it off altogether. An embargo would make sense only to demonstrate control of a resource and the will to sustain price increases.

A third interpretation characterizes the oil crises as simple products of monopoly power or conspiracy. During the 1970's oil crises—with their long lines for gasoline—and the rapid 1990 price increases that followed Iraq's invasion of Kuwait, such explanations were common. Early versions of this viewpoint denied resource scarcity as a problem and called for the breakup of oil companies to end the conspiracy. Later, as Middle Eastern oil-producing countries assumed control over operations from Western oil companies, the focus shifted to policy actions to combat the OPEC monopoly.

OPEC POWER

The degree of perceived power held by the OPEC oil cartel has fluctuated considerably since its formation in 1960. OPEC nations held 90 percent of the world's then-known oil reserves. Although this guaranteed the OPEC nations a key role in the future as reserves outside OPEC were depleted, the organization was weak during its first ten years. A hint of OPEC's possible power came in 1970 when a bulldozer accident broke a pipeline carrying oil from the Persian Gulf to the Mediterranean Sea. In the aftermath of the accident, which caused tight oil supplies, Libya insisted on and received higher prices for its crude oil.

During the 1973-1974 embargo OPEC's power was little questioned. The sharp upward movement of oil prices seemed to prove that OPEC had the ability to withhold enough oil to keep prices permanently higher. Further confirmation came with the 1979-1980 incident, when oil prices again doubled. Recessions following the oil crises made the West appear

The 1973 oil embargo imposed by the Organization of Petroleum Exporting Countries (OPEC) led to fuel shortages in the United States. (Library of Congress)

dependent on OPEC's indulgence for its future prosperity.

The collapse of oil prices after the three crises led to a different interpretation. In this view, oil prices were kept artificially low before 1970 by the rapid pumping of oil from Middle Eastern reserves by Western-based companies. The companies were seeking profit with little regard for the long-term future of the resource, since they realized that power would soon shift away from them. After 1970, as the host nations began to realize their power, oil prices recovered from their artificially low levels. The extraordinary oil price increases of 1973-1974, in this view, were from artificially low levels to normal levels. At the time they occurred, they had been seen as increases from normal levels to artificially high levels. Some analysts came to believe that OPEC was simply announcing, rather than causing, higher oil prices.

Energy analysts have observed that oil crises and

embargoes have had a common pattern. First, demand growth catches up with world capacity, leading to generally tight market conditions. Next, a precipitating event occurs in the Middle East, leading to quick price increases and disruptions of ordinary supply channels. If there is no permanent loss of supply, however, the disruption soon ends. At the time of the crisis, policy priority goes to energy security over environmental concerns. With the passage of time, however, energy security loses some of its prominence as a policy goal.

William C. Wood

FURTHER READING

Danielsen, Albert L. *The Evolution of OPEC.* New York: Harcourt Brace Jovanovich, 1982.

Griffin, James M., and Henry B. Steele. *Energy Economics and Policy.* 2d ed. New York: Academic Press, 1986.

Hinrichs, Roger A., and Merlin Kleinbach. *Energy: Its Use and the Environment.* 4th ed. Belmont, Calif.: Thomson Brooks/Cole, 2006.

Parra, Francisco. *Oil Politics: A Modern History of Petroleum.* New York: I. B. Tauris, 2004.

Stobaugh, Robert, and Daniel Yergin, eds. *Energy Future: Report of the Energy Project at the Harvard Business School.* 3d ed. New York: Vintage Books, 1983.

Venn, Fiona. *The Oil Crisis.* New York: Longman, 2002.

SEE ALSO: Alternative fuels; Alternatively fueled vehicles; Fossil fuels; Gulf War oil burning; Oil drilling; Synthetic fuels; Trans-Alaska Pipeline.

Oil drilling

CATEGORIES: Energy and energy use; resources and resource management

DEFINITION: Activities involved in boring through earth and rock to tap petroleum reservoirs

SIGNIFICANCE: Careful management of oil drilling projects is crucial because the processes involved in such projects—including site preparation, equipment setup, drilling, fluid circulation, and waste disposal—all have the potential to degrade the environment.

The process of drilling a hole through the rock layers that overlie a petroleum reservoir is fairly simple, at least in principle. A diamond-tipped drill bit is attached at the base of a length of vertical pipe, and the pipe is rotated. The bit grinds away the rock, producing a hole the width of the drill bit and the length of the attached pipe, or drill string. As the top of the rotating drill string approaches ground level, a new piece of pipe is added, lengthening the drill string and allowing the depth of the hole to increase. The weight of the drill string is supported by a drilling rig, which helps keep the hole straight.

Meanwhile, a fluid known as drilling mud is pumped down the hole to keep the bit cool, help keep the hole from collapsing, carry rock chips (cuttings) up the hole, and prevent overpressurized, subsurface liquid or gas from causing a blowout. Blowouts occur when subsurface zones of high pressure force the fluid in the well to flow out of the hole at a high rate of speed, in some cases blowing the drill string completely out of the hole and destroying the drill rig. Drilling mud is pumped downhole through the interior of the drill string, out through holes in the bit at the bottom of the drill string, then back to the surface between the drill string exterior and the drill hole, a space known as the well annulus. At the surface, the cuttings are collected, and the mud is recirculated downhole. Drilling mud is carefully monitored for the correct viscosity and weight. Chemicals are added, as needed, to modify these characteristics.

At intervals during drilling, the drill string is removed, and a steel liner, or casing, is cemented into place downhole. This supports the sides of the hole and isolates the annulus from the surrounding downhole environment.

If insufficient amounts of oil are found to make the well economically viable, the well is termed a "dry hole." It must be carefully cased from top to bottom and filled with cement so that the hole is sealed permanently, a procedure known as plugging and abandoning the well. If oil is found in sufficient amounts, it is pumped from the well, a process known as primary production. After an oil well's output begins to decline significantly, alternative methods of production may be applied. The term "enhanced oil recovery" refers to these more complicated—and expensive—methods for increasing an oil well's production. The most common method, saltwater injection, involves pumping saltwater brine, produced from the subsurface with the oil, back into adjacent wells, forcing more oil to the surface. Chemicals such as surfactants may also be employed to help mobilize the oil.

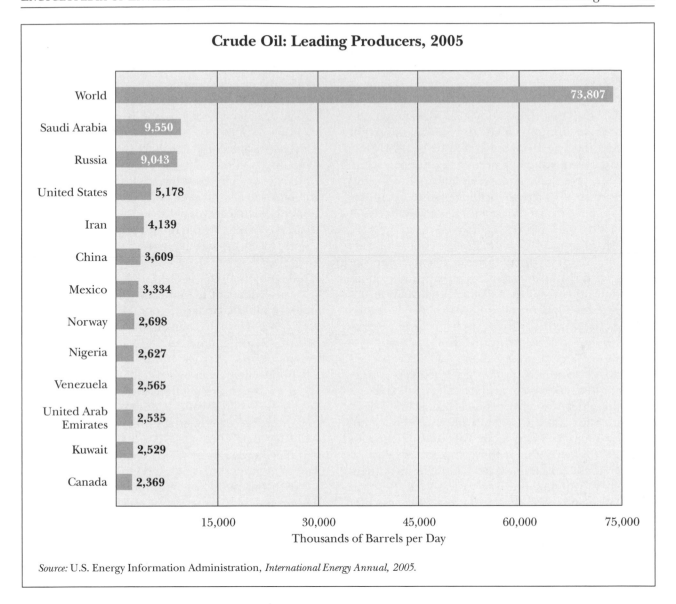

Crude Oil: Leading Producers, 2005

	Thousands of Barrels per Day
World	73,807
Saudi Arabia	9,550
Russia	9,043
United States	5,178
Iran	4,139
China	3,609
Mexico	3,334
Norway	2,698
Nigeria	2,627
Venezuela	2,565
United Arab Emirates	2,535
Kuwait	2,529
Canada	2,369

Source: U.S. Energy Information Administration, *International Energy Annual, 2005.*

ENVIRONMENTAL CONCERNS

The environmental concerns related to oil drilling on land are somewhat different from those related to drilling at sea. In either case, however, the effects of poor operations management can be devastating to the environment, particularly in ecologically sensitive areas. Methods of site access and preparation can have significant impacts on the ecology of land sites, especially in forest preserves, wetlands, and tundra. Soils, as well as surface-water and groundwater supplies, are at risk from oil spills and improper disposal of saltwater, drill cuttings, and drilling mud. Groundwater is also at risk from improper casing, saltwater injection, and well plugging and abandonment. Ground-

water pollution may go undetected for years and can affect relatively large areas of the subsurface.

At sea, much larger areas may be at risk because of down-current transport of pollutants. Environments particularly threatened by marine drilling platforms include coral reefs, oyster banks, mangrove swamps, and tidal estuaries. Disposal of drilling mud and drill cuttings threatens the benthic (seafloor) environment surrounding the platform and the pelagic (open sea) environment down current. Improper operations can release oil and brine into the water column, and faulty casing can create oil and brine seeps on the seafloor.

In the United States, numerous federal laws regulate oil drilling, production, and transport on land and at sea. These include the Federal Water Pollution Control Act of 1972; the Outer Continental Shelf Lands Act of 1953; the Comprehensive Environmental Response, Compensation, and Liability Act (CERCLA) of 1980 (also known as Superfund); the Marine Protection, Research, and Sanctuaries Act of 1972; and the Safe Drinking Water Act of 1974. Most U.S. states have stringent permit requirements, site inspection programs, and accidental spill reporting and response programs for drilling operations. Industry compliance with these programs is generally excellent.

Oil spills from offshore platforms represent only about 1 percent of the petroleum released into the oceans each year. Natural oil seeps contribute about eight times this amount, and their contribution has increased steadily as onshore and offshore oil production has reduced the pressure in hydrocarbon reservoirs. Marine drilling platforms, while perhaps eyesores above water, provide habitats for marine species that otherwise would not be present—these structures might be likened to artificial reefs. Scuba divers and fishing boat captains alike seek out platform substructures for the lush growth of invertebrates encrusting them and the many fish that this growth attracts. When offshore drilling platforms are decommissioned, debates often ensue among environmentalists as to whether the structures should be completely dismantled or allowed to remain intact so that the marine community may benefit.

Clayton D. Harris

FURTHER READING

Boomer, Paul M. *A Primer of Oilwell Drilling.* 7th ed. Austin: Petroleum Extension Service, University of Texas, 2008.

Camp, William G., and Thomas B. Daugherty. "Fossil Fuel Management." In *Managing Our Natural Resources.* 4th ed. Albany, N.Y.: Delmar, 2004.

Mtsiva, V. C., ed. *Oil and Natural Gas: Issues and Policies.* Hauppauge, N.Y.: Nova Science, 2003.

SEE ALSO: BP *Deepwater Horizon* oil spill; *Brent Spar* occupation; Fossil fuels; Ocean pollution; Oil spills; PEMEX oil well leak; Santa Barbara oil spill.

Oil shale and tar sands

CATEGORIES: Energy and energy use; resources and resource management
DEFINITIONS: Oil shale is finely grained sedimentary rock that incorporates a solid hydrocarbon substance in its structure; tar sands are sands that are permeated with a tarlike petroleum that is too thick to flow
SIGNIFICANCE: The conversion of the hydrocarbons in oil shale and tar sands into oil could provide an energy source several times the amount of the earth's current known oil reserves, but the conversion process has a number of environmental costs.

Oil is not found in vast oil-filled underground caverns; rather, it is found in the spaces between mineral grains that make up rock. Permeable rock has more space between grains than does impermeable rock. If the oil is thin enough, it will slowly flow through permeable rock and into the borehole of a well, from which it can be brought to the surface.

The hydrocarbons in oil shale cannot flow because they are in a solid form called kerogen. When some types of kerogen are heated by natural processes (60 to 150 degrees Celsius, or 140 to 300 degrees Fahrenheit) they slowly release crude oil and natural gas. In place of nature's slow process, oil shale may be mined and then burned as a low-grade fuel, or it can be crushed and then heated in a retort to produce crude oil and natural gas. Aboveground retorting is usually done between 300 and 520 degrees Celsius (570 and 970 degrees Fahrenheit). It is also possible to process shale underground (in situ, or in place) by fracturing the rock underground and then using the heat of an underground fire to produce crude oil, which can be pumped to the surface. In place of the fire, steam or other hot fluid can be circulated underground. After retorting, the oil is usually upgraded through a process in which it is passed over a catalyst and treated with hydrogen gas to increase the hydrogen content and remove impurities such as sulfur, nitrogen, iron, and arsenic.

Shale oil is very expensive to produce, although expected increases in production and improvements in the process are projected to bring its cost down significantly over time. The process also has environmental costs. Processing oil shale releases the greenhouse gas carbon dioxide, which has been linked to global warming. Near-surface deposits of oil shale are open-pit mined or strip-mined, both processes that damage surrounding ecosystems; underground mining uses the less efficient room-and-pillar method. In any kind of mining of oil shale, one to five barrels of water are used for each barrel of shale oil produced; in addition, the process produces mountains of crushed rocks that must be disposed of.

The world's use of shale oil peaked in 1980, several years after the 1973 oil crisis. In 2008 Estonia was the top producer at 5,500 barrels per day, followed by Brazil at 3,100 barrels per day. It has been estimated that worldwide deposits of oil shale could produce 2.8 to 3.3 trillion barrels of shale oil. The United States has the largest share, with enough oil shale to produce 1.5 to 2.6 trillion barrels in the Green River Formation (a geologic formation located in parts of Utah, Wyoming, and Colorado). In comparison, the world's conventional oil reserves are estimated at 1.317 trillion barrels.

Different grades of oil shale yield different amounts of oil. About 16 percent of the Green River Formation yields 95 to 380 liters (25 to 100 gallons) of oil per ton of shale, about 33 percent yields 38 to 95 liters (10 to 25 U.S. gallons), and the final 51 percent yields 19 to 38 liters (5 to 10 U.S. gallons) per ton of shale. Unable to compete with cheap conventional oil, oil shale mining ceased in the United States in 1982, but in 2003 an oil shale development program was restarted.

TAR SANDS

Tar sands (also called oil sands) are mixtures of sand, clay, and bitumen, a tarlike form of petroleum. The largest deposits of tar sands are in Canada and

Mining trucks carry oil-laden sand, loaded by the huge shovels in the background, at the Albian Sands tar sands project in Alberta, Canada. (AP/Wide World Photos)

Venezuela, each country having the equivalent of the world's reserves of crude oil. Canada's production of oil from tar sands reached 1.25 million barrels per day in 2009 (at a cost of about $27 per barrel) and provided 47 percent of the nation's oil production. This allowed Canada to become the biggest supplier of petroleum products to the United States. In Canada about two tons of tar sand are processed to produce one barrel of oil. Producing oil in this way has the same environmental problems as producing oil from oil shale: Strip mining degrades ecosystems, and the process releases more carbon dioxide into the atmosphere than does the production of conventional oil, uses large amounts of water, and leaves tons of spent sand that must be disposed of.

Bitumen may be extracted in situ through the injection of steam, hot air, or solvents to allow the bitumen to flow. Alternatively, a deposit may be strip-mined and the materials then hauled to a processing facility, where the bitumen is separated from water, sand, and other waste. Like shale oil, the heavy crude extracted from bitumen must be upgraded through a process in which it is passed over a catalyst and treated with hydrogen gas to increase the hydrogen content and remove impurities such as sulfur, nitrogen, and arsenic. It can then be used as feedstock for conventional refining.

The only significant tar sands in the United States are in eastern Utah. The Utah tar sand deposit is far smaller than Canada's, and the Utah sands are hydrocarbon-wetted rather than water-wetted as the Canadian sands are, so their use would require different processing techniques.

Charles W. Rogers

FURTHER READING

Bartis, James T., et al. *Oil Shale Development in the United States: Prospects and Policy Issues.* Santa Monica, Calif.: RAND Corporation, 2005.

Marsden, William. *Stupid to the Last Drop: How Alberta Is Bringing Environmental Armageddon to Canada (and Doesn't Seem to Care).* Toronto: Vintage Canada, 2008.

Speight, James G. *Enhanced Recovery Methods for Heavy Oil and Tar Sands.* Houston: Gulf, 2009.

SEE ALSO: Acid mine drainage; Carbon dioxide; Oil drilling; Strip and surface mining; Surface Mining Control and Reclamation Act; Synthetic fuels.

Oil spills

CATEGORY: Water and water pollution

DEFINITION: Accidental or intentional discharges of raw or refined petroleum products on land or at sea

SIGNIFICANCE: Oil spills pose both short- and long-term environmental threats, including injuries and deaths among fish, birds, and other wildlife; damage to shoreline recreational areas; and pollution of water supplies.

Oil spills commonly occur in both terrestrial and marine settings. Terrestrial spills affect land areas, including drainage courses and bodies of surface water impounded in lakes and ponds. Subsurface waters (groundwater) are also at risk from leakage of polluted water downward through the vadose zone (zone of aeration) to the water table (zone of saturation). Some crude oil is lost during exploratory drilling, workover operations, and tank storage. Mud pits utilized during rotary drilling usually contain oil recovered from the well during testing or oil derived from an oil-based mud employed in the drilling process. Oil and brine released from wells are sometimes stored in unlined ponds, and these fluids can soak into the ground and kill beneficial microbes in the soil and plant life in the immediate area. Subsurface water can also be contaminated if the oil or brine migrates downward to the groundwater table. Runoff from these areas can enter streams and ponded bodies of water and kill fish and other aquatic animals as well as reduce the natural vegetation.

Terrestrial oil spills also occur during the loading, transportation, and offloading of petroleum from tank trucks and railroad tank cars. Petroleum pipelines, both buried and aboveground, are highly vulnerable to rupture from welding defects, corrosion, earthquakes, and shifting soils. Fire is a constant danger associated with such spills. During the mid-1980's a ruptured pipeline near São Paulo, Brazil, caused the deaths of more than five hundred people and resulted in the destruction by fire of twenty-five hundred homes in the town of Vila Socco. During the 1991 Persian Gulf War, large areas of Kuwait were devastated when Iraqi soldiers damaged oil pipelines and refineries and set fire to hundreds of oil wells. Pools of oil formed near the wells and infiltrated the porous and permeable soil and rock, thereby endangering the water supply of Kuwait City and other areas.

The disposal of petroleum products (diesel fuel, gasoline, kerosene, jet fuel, and used motor oil) is a significant problem worldwide. Some of these products contain carcinogens such as benzene, toluene, and xylene, which require special handling. Many millions of gallons of used oil products are disposed of on land every year, and it has been estimated that more than 6.6 million tons, or about 45 million barrels, of petroleum and petroleum products enter the oceans per year. Of this total, 44 percent is land-derived. These inputs include coastal city contributions from refineries, wastewater outlets, and other sources (13 percent); urban runoff, including storm drains (5 percent); and river runoff (26 percent).

MARINE SPILLS

The waters of oceans and restricted seas can be polluted by oil in several ways. Oil sources include natural seeps along the ocean floor, stream runoff from the land, wastewater drainage outlets from industrial complexes, offshore drilling accidents, deliberate purging of ballast or cargo areas of ships, and accidents involving oil supertankers. The petroleum input into the marine environment from ocean-derived factors has been estimated at about 56 percent. This calculation includes transportation losses during loading and unloading (30 percent), oil spills (5 percent), offshore production losses (1 percent), atmospheric pollution (10 percent), and offshore oil seeps (10 percent).

Supertanker accidents that result in oil spills are major events; causes include shipboard explosions, collisions with other vessels, and grounding on barriers (mostly rocks or coral reefs) because of navigation error, mechanical failure, or inclement weather. Although such oil spills are often quite dramatic, and the damage they do to the environment receives extensive news media coverage, only about 5 percent of the total input of petroleum into the oceans results from oil spills.

Oil slicks are especially problematic in semienclosed bodies of water, such as the Baltic and Mediterranean seas. Because tides are minor in these seas, any oil spilled is not easily eliminated. These places are among the world's most polluted bodies of water.

When an oil spill occurs in the open ocean, the oil floats on the surface of the water because of a density contrast between the two substances. Water has a specific gravity from 1 to 1.2, while oil is lighter, with a specific gravity from 0.7 to 1.0. The lighter components of the spilled petroleum immediately begin to evaporate; also, oil-degrading bacteria in the water begin to feed on the organic deposit and multiply. Warmer water and ambient air temperatures increase the rates of bacterial growth and evaporation. The major part of petroleum dumped into the open ocean evaporates or is reduced by bacteria. With time, the oil turns into an inert, tarlike substance that becomes extremely hard; some such residues have been found with barnacles attached to them.

In areas where the spilled oil is washed ashore and subjected to intense wave action, a gooey emulsion sometimes forms. This foamy mass, which has been described as resembling chocolate mousse, coats everything along the beach, including sand-sized particles, large boulders, aquatic animals and plants, plea-

Time Line of Supertanker Disasters

Year	Vessel Name	Location	Barrels of Oil Lost
1967	Torrey Canyon	Land's End, England	870,000
1976	Argo Merchant	Nantucket, Massachusetts	180,952
1978	Amoco Cadiz	Portsall, France	1,500,000
1983	Castillo de Bellver	Cape Town, South Africa	2,023,810
1989	Exxon Valdez	Prince William Sound, Alaska	261,905
1990	American Trader	Southern California	6,249
1990	Berge Broker	North Atlantic Ocean	95,238
1990	Mega Borg	Gulf of Mexico	75,476
1992	Aegean Sea	Spain	14,821
1993	Braer	Shetland Islands, Scotland	619,048
1996	Sea Empress	Southwest Wales	380,952
1999	New Carissa	Oregon	9,524
1999	Erika	Bay of Biscay, France	83,333
2000	Westchester	Mississippi River, below New Orleans	15,750
2002	Prestige	Spain	1,860
2003	Tasman Spirit	Karachi, Pakistan	3,835
2004	M/V Selendang Ayu	Aleutian Islands, Alaska	9,360
2006	M/T Solar 1	Guimaras Island, Philippines	14,720
2007	Hebei Spirit	South Korea	77,775

sure boats, and human-made shore facilities. The polluted waters can devastate resort beaches, oyster and shrimp beds, fish hatcheries, and prime fishing grounds.

The potential damage resulting from a large oil spill in the ocean includes the loss of substantial numbers of commercial and rough fish, waterfowl, aquatic mammals, shellfish, algae, and plankton. Oil contamination reduces the amount of oxygen in the water column, and this can cause the deaths of large numbers of fish in polluted areas. Millions of fish died as a result of the 1989 *Exxon Valdez* disaster in Alaskan waters, and the Alaskan fishing industry, which is extremely important to the state's economy, was seriously threatened. Although this event is generally considered a major disaster, David McConnell reported in the *Journal of Geological Education* (1999) that, compared with other spills worldwide, the *Exxon Valdez* spill ranked only fifty-third in size.

When seabirds come into contact with spilled oil to the extent that their feathers are soaked, they lose the ability to fly and float, as well as their natural body insulation. Many such birds starve to death, die from exposure, or drown. The *Exxon Valdez* disaster resulted in the deaths of more than 200,000 seabirds. During the Persian Gulf War, more than 20,000 birds perished from oil-related causes. Oil spills also cause the deaths of aquatic mammals, which die from the loss of their food supply, exposure to cold, or poisoning when they ingest the toxic oil. Volunteer workers have saved many birds and mammals at animal centers set up after large spills; workers wash the animals with solvents to remove oil and often hold them until their normal body resilience returns.

Shellfish in bays and marshes can perish in oil-polluted waters from asphyxiation or toxicity. Some effects of oil pollution remain for years. Shellfish in a salt marsh in Massachusetts were quarantined for more than five years after heating oil was spilled

Among the environmental harms of oil spills is the damage done to wildlife such as these sea otters, which died in Prince William Sound off the coast of Alaska as a result of the Exxon Valdez *oil spill.* (AP/Wide World Photos)

nearby in 1969, and traces of petroleum persisted for decades.

CLEANUP PROCEDURES

Oil spills in the open ocean have been reduced or eliminated using several techniques. Igniting the slick, if attempted shortly after a spill, can be effective; however, if the volatile components of the oil have evaporated, it is difficult to start and maintain such fires. In 1967, when the damaged supertanker *Torrey Canyon* began to founder and break up off the coast of England, Great Britain's air force dropped bombs in an effort to ignite the oil that remained in the vessel.

Another technique used to reduce oil slicks is the application of chemicals such as detergents in an attempt to disperse the oil droplets. This procedure is not viable if the temperature of the water is too low for the chemicals to be effective. Floating booms are sometimes used to confine oil streamers or protect inlet areas. These barriers are useful if the waves are not too high. Skimmers may be used to collect floating oil during the early stages of a spill. Plant material (threshed hay, peat moss, or wood shavings) or pulverized rock (chalk or claystone) is sometimes spread over slicks to absorb the oil. Plant material can then be removed and burned or buried; applied rock particles absorb the oil, clump together, and sink to the ocean floor. In many parts of the oceans, oil-eating microbes (mostly bacteria) utilize petroleum as a nutrient. If these microbes are not present or exist only in small numbers in the area, engineered microbes can be introduced. Under ideal temperature and ocean chemistry conditions, the bacteria will rapidly multiply and consume the oil slick. This technique is known as bioremediation.

In nearshore areas, beach sand and boulders can be steamed clean or washed with water or solvents. Pools of oil can be vacuumed into tank trucks and hauled away from the site. The recovered oil can sometimes be processed into useful products. Plant material can also be applied to crude oil along the shoreline. After the 1969 Santa Barbara oil spill, the beaches along the California coast were coated by approximately ten thousand barrels of gooey oil. Numerous volunteers helped spread fresh hay along the beachfront and later recovered the oil-soaked plant material.

Donald F. Reaser

FURTHER READING

Davidson, Jon P., Walter E. Reed, and Paul M. Davis. *Exploring Earth: An Introduction to Physical Geology.* 2d ed. Upper Saddle River, NJ: Prentice Hall, 2002.

Fingas, Merv. *The Basics of Oil Spill Cleanup.* 2d ed. Boca Raton, Fla.: CRC Press, 2001.

Kemp, David D. "Threats to the Availability and Quality of Water." In *Exploring Environmental Issues: An Integrated Approach.* New York: Routledge, 2004.

Lehr, Jay, et al. "Oil Spills and Leaks." In *Handbook of Complex Environmental Remediation Problems.* New York: McGraw-Hill, 2002.

Montgomery, Carla W. *Environmental Geology.* 9th ed. New York: McGraw-Hill, 2010.

National Research Council. *Oil in the Sea III: Inputs, Fates, and Effects.* Washington, D.C.: National Academies Press, 2003.

_____. *Oil Spill Dispersants: Efficacy and Effects.* Washington, D.C.: National Academies Press, 2005.

Ott, Riki. *Not One Drop: Betrayal and Courage in the Wake of the Exxon Valdez Oil Spill.* White River Junction, Vt.: Chelsea Green, 2008.

Walker, Jane. *Oil Spills.* New York: Gloucester Press, 1993.

SEE ALSO: *Amoco Cadiz* oil spill; *Argo Merchant* oil spill; Bioremediation; BP *Deepwater Horizon* oil spill; *Braer* oil spill; *Exxon Valdez* oil spill; Ocean pollution; PEMEX oil well leak; Santa Barbara oil spill; *Sea Empress* oil spill; Tobago oil spill; *Torrey Canyon* oil spill.

Old-growth forests

CATEGORIES: Forests and plants; resources and resource management

DEFINITION: Forests containing many trees that have never been harvested by loggers

SIGNIFICANCE: Persons involved in the timber industry generally view the large trees found in old-growth forests as a renewable source of fine lumber, but environmentalists argue for protection of these trees, asserting that they are part of ancient and unique ecosystems that can never be replaced.

In the 1970's scientists began studying the uncut forests of the Pacific Northwest and the plants and animals that inhabit them. One of the first results of this research was the U.S. Forest Service publication *Ecological Characteristics of Old-Growth Douglas-Fir Forests* (1981). In this report, biologist Jerry Franklin and his colleagues revealed that these forests are not just tangles of dead and dying trees; rather, they constitute unique, thriving ecosystems made up of living and dead trees, mammals, insects, and even fungi.

The lands usually referred to as old-growth forests are located primarily on the western slope of the Cascade Mountains in southeast Alaska, southern British Columbia in Canada, Washington, Oregon, and Northern California. The weather in these regions is wet and mild, ideal for the growth of trees such as Douglas fir, cedar, spruce, and hemlock. Some studies have shown that there is more biomass, including living matter and dead trees, per hectare in these forests than anywhere else on earth. The trees in old-growth forests may be as tall as 90 meters (300 feet) with diameters of 3 meters (10 feet) or more and can live as long as one thousand years. The forest community grows and changes over time in such a forest, not reaching biological climax until the forest consists primarily of hemlock trees, which are able to sprout in the shade of the sun-loving Douglas fir.

One of the most important components of the old-growth forest is the large number of standing dead

A Roosevelt elk grazes in the old-growth forest of Olympic National Park in the state of Washington. (©Natalia Bratslavsky/Dreamstime.com)

trees, or snags, and fallen trees, or logs, on the forest floor and in the streams. The fallen trees rot very slowly, often taking more than two hundred years to disappear completely. During this time they are important for water storage, as wildlife habitat, and as "nurse logs" where new growth can begin. In fact, seedlings of some trees, such as western hemlock and Sitka spruce, have difficulty competing with the mosses on the forest floor and need to sprout on fallen logs.

Another strand in the complex web of the forest consists of micorrhizal fungi, which attach themselves to the roots of the trees and enhance the roots' uptake of water and nutrients. The fruiting bodies of these fungi are eaten by small mammals such as voles, mice, and chipmunks, which then spread the spores of the fungi in their droppings. Numerous species of plants and wildlife appear to be dependent on this ecosystem to survive. The most famous example is the northern spotted owl, whose endangered species status and dependence on the old-growth ecosystem has caused

great political and economic turmoil in the Pacific Northwest.

By the 1970's most of the trees on the timber industry's private lands had been cut. Their replanted forests, known as second growth, would not be ready for harvest for several decades, so the industry became increasingly dependent on public lands for raw materials. Logging of old-growth trees in the national forests of western Oregon and Washington increased from 900 million board feet in 1946 to more than 5 billion board feet in 1986.

Environmentalists claimed that only 10 percent of the region's original forest remained, and they were determined to save what was left. The first step in their campaign was to encourage the use of the evocative term "ancient forest" to counteract the somewhat negative connotations of "old-growth." They then found an effective tool in the northern spotted owl. This small bird was determined to be dependent on the old-growth ecosystem, and its listing under the federal Endangered Species Act in 1990 was a bombshell that caused a decade of scientific, political, and legal conflict.

Under the law, the Forest Service was required to protect enough of the owl's habitat to ensure its survival. An early government report identified 3.1 million hectares (7.7 million acres) of forest to be protected for the bird. Later, the U.S. Fish and Wildlife Service recommended 4.5 million hectares (11 million acres). In 1991 U.S. District Court judge William Dwyer placed an injunction on all logging in spotted owl habitat until a comprehensive plan could be put in place. The timber industry responded with a prediction of tens of thousands of lost jobs and regional economic disaster. In 1993 President Bill Clinton convened the Forest Summit conference in Portland, Oregon, to work out a solution. The Clinton administration's plan, though approved by Judge Dwyer, satisfied neither the industry nor the environmentalists, and protests, lawsuits, and legislative battles continued.

As the twentieth century came to an end, timber harvest levels had been significantly reduced, the Northwest's economy had survived, and environmentalists were promoting additional reasons to value old-growth forests: as habitat for endangered salmon and other fish, as sources for medicinal plants, and simply as repositories of benefits yet to be discovered. The decades-long controversy over the forests of the Northwest had a deep impact on environmental science as well as on U.S. government policy regarding the preservation of natural resources; it also encouraged new interest in other native forests around the world, from Brazil to Malaysia to Russia.

Joseph W. Hinton

FURTHER READING

Dietrich, William. *The Final Forest: The Battle for the Last Great Trees of the Pacific Northwest.* New York: Simon & Schuster, 1992.

Durbin, Kathie. *Tree Huggers: Victory, Defeat, and Renewal in the Northwest Ancient Forest Campaign.* Seattle: Mountaineers Books, 1996.

Keiter, Robert B. "Ecology Triumphant? Spotted Owls and Ecosystem Management." In *Keeping Faith with Nature: Ecosystems, Democracy, and America's Public Lands.* New Haven, Conn.: Yale University Press, 2003.

Kelly, David, and Gary Braasch. *Secrets of the Old Growth Forest.* Salt Lake City: Gibbs-Smith, 1988.

Maser, Chris. *Forest Primeval: The Natural History of an Ancient Forest.* 1989. Reprint. Corvallis: Oregon State University Press, 2001.

Wirth, Christian, Gerd Gleixner, and Martin Heimann, eds. *Old-Growth Forests: Function, Fate, and Value.* New York: Springer, 2009.

SEE ALSO: Ecosystems; Endangered Species Act; Endangered species and species protection policy; Forest Summit; Logging and clear-cutting; Northern spotted owl.

Olmsted, Frederick Law

CATEGORY: Urban environments
IDENTIFICATION: American landscape architect
BORN: April 26, 1822; Hartford, Connecticut
DIED: August 28, 1903; Waverly, Massachusetts
SIGNIFICANCE: Olmsted left a distinct mark on the American environment from New York City to the wilds of California. He synthesized a variety of experiences in his youth and young adulthood to become one of the greatest landscape designers in the history of the United States.

As a child Frederick Law Olmsted traveled widely with his family, becoming familiar with the natural landscapes of various regions. As a young adult he operated an experimental farm and learned how the natural environment interacts with human endeavors to tame it. He later traveled throughout the antebellum American South as a correspondent for *The New York Times*. His observations were compiled and published as *Journeys and Explorations in the Cotton Kingdom* (1861), a discerning analysis of social conditions in the southern slave society.

In 1857 Olmsted and a friend, Calvert Vaux, entered a contest to plan a park on Manhattan Island. Olmsted and Vaux envisioned a park for the people— one that would depart from the grand plans, controlled landscapes, and aristocratic enclaves that characterized European parks. In 1858 Olmsted was appointed chief architect of Central Park, and he devoted his energies to developing a natural environment that would provide a refuge for all of New York City's residents, rich and poor alike.

When the Civil War began, Olmsted set aside his landscaping to organize the U.S. Sanitary Commission; as executive secretary of the commission, he

Landscape architect Frederick Law Olmsted. (Library of Congress)

helped reform the Union Army's medical corps. From there, in 1863, he moved to California to manage the Fremont Mariposa mining enterprise. During his time in the West, he helped with efforts to ensure the preservation of the Yosemite and Mariposa wilderness areas and worked on a draft of what would become the charter for the national park system. By 1875 he was back in the East directing the work of converting Boston's Back Bay from marshes into parkland. In this and other projects—including Prospect Park in Brooklyn, Riverside and Morningside parks in Manhattan, Jackson Park in Chicago, the grounds of the Capitol Building in Washington, D.C., and the grounds of the World's Columbian Exposition of 1893 in Chicago—Olmsted clung to his desire to use and preserve natural environments for all people. At the Biltmore estate, his largest private commission, he designed a managed forest that preserved the natural North Carolina habitat instead of imposing the artifice of a traditional aristocratic estate.

Olmsted also worked as an early city planner and advocate of suburban development. He envisioned suburbs as self-contained communities linked to the city by good roads. He used his ideas in planning a series of parks along parkways in both Boston and Buf-

falo. He also planned every detail of the suburb of Riverside, Illinois, one of the first planned communities in the United States.

Olmsted's legacy is huge. He was the predominant force behind the preservation of Niagara Falls and designed nearly eighty parks and thirteen college campuses (including Stanford University in California). He fathered the American urban park movement and pioneered many of the tenets of American regional planning. In each of his projects, his aim was to move from the merely scenic to a balance between preservation and use of the environment by all people.

Jane Marie Smith

FURTHER READING

Gandy, Matthew. *Concrete and Clay: Reworking Nature in New York City.* Cambridge, Mass.: MIT Press, 2002.

Rybczynski, Witold. *A Clearing in the Distance: Frederick Law Olmsted and America in the Nineteenth Century.* New York: Charles Scribner's Sons, 1999.

Schnadelbach, R. Terry. "Frederick Law Olmsted, 1822-1903." In *Fifty Key Thinkers on the Environment,* edited by Joy A. Palmer. New York: Routledge, 2001.

SEE ALSO: Greenbelts; National parks; Planned communities; Urban parks; Urban planning; Urban sprawl; Yosemite Valley.

Open spaces

CATEGORY: Land and land use

DEFINITION: Areas on developed land that are not occupied by buildings or other structures

SIGNIFICANCE: By requiring the inclusion of open spaces on developed land, local governments seek to enhance residents' quality of life while also using land efficiently and demonstrating sensitivity to the environment.

The inclusion of open spaces is often mandated by municipalities on newly developed tracts of land to encourage more efficient use of land and more flexible development practices that respect and conserve natural resources. Many municipalities require that more than 10 percent of each tract of land be dedicated as open space. When large tracts of land are developed for cluster residential housing, municipalities may require that 40 to 60 percent of these areas be made available as open space. Land designated as open spaces can take the form of lawns, natural areas, recreational areas, croplands or pasturelands, or stormwater management areas.

A lawn is a grassed area with or without trees that may be used by residents for a variety of purposes; such areas are mowed regularly to ensure a neat and tidy appearance. Natural areas are open spaces in which native or natural vegetation has been left undisturbed during construction and land development. Occasionally, natural areas are created after development through the grading and replanting of land that was disturbed during development. Often, natural areas must be maintained by designated authorities. Maintenance tasks in such areas include preventing the proliferation of undesirable plants; clearing litter, dead trees, and brush; and keeping streams freely flowing. Pathways or walking trails may be constructed within natural areas to provide places where local residents may enjoy mild exercise or relax.

Open spaces may also take the form of areas for recreation, such as tennis courts, swimming pools, or ball fields. Safety concerns make these kinds of open spaces expensive to construct and maintain. Most often such recreational facilities are part of public parks. Open spaces in the form of croplands, pasturelands, or lands planted with nursery stock or orchard trees provide visually valuable pastoral settings. Buildings and other structures are usually not permitted in open spaces, but in the case of agricultural areas, flexible development procedures often permit any dwellings or farm-related buildings that were present before development to remain.

Finally, stormwater management features are often incorporated into the open spaces of developed land. Detention or retention basins are integral to the control of stormwater when a tract of land has buildings and paved areas constructed on it. Impervious materials prevent rainwater from percolating into the ground. Unless such measures are taken on developed land, heavy erosion is likely, and neighboring lands may be subjected to flooding.

Anthony J. Nicastro

FURTHER READING

Erickson, Donna. *MetroGreen: Connecting Open Space in North American Cities*. Washington, D.C.: Island Press, 2006.
Schmidt, Stephan. "The Evolving Relationship Between Open Space Preservation and Local Planning Practice." *Journal of Planning History* 7, no. 2 (2008): 91-112.
Van Rooijen, Maurits. "Open Space, Urban Planning, and the Evolution of the Green City." In *Urban Planning in a Changing World: The Twentieth Century Experience*, edited by Robert Freestone. New York: Routledge, 2000.

SEE ALSO: Environmental impact assessments and statements; Greenbelts; Land-use policy; Planned communities; Stormwater management; Urban parks; Urban planning; Wetlands.

Operation Backfire

CATEGORIES: Activism and advocacy; animals and endangered species

IDENTIFICATION: Criminal investigation led by the Federal Bureau of Investigation into a series of destructive actions carried out in the western United States in the 1990's in the name of environmental and animal rights causes

DATES: 2004-2006

SIGNIFICANCE: Operation Backfire is the largest and most highly publicized federal investigation of "ecoterrorism" thus far undertaken in the United States. The operation and subsequent trials fueled national debate over the definition of "terrorism" and the appropriate use of domestic surveillance. Some critics regard Operation Backfire as an attempt to disrupt environmental and animal rights activism.

In 2004 the Portland, Oregon, field office of the Federal Bureau of Investigation (FBI) consolidated seven independent yet related investigations into a single major case that it code-named Operation Backfire. Counting the precursor investigations, Operation Backfire lasted nine years and was assisted by local, state, and federal law-enforcement agencies, including the Bureau of Alcohol, Tobacco, and Firearms. As a result of Operation Backfire's findings, eleven people were initially indicted; by 2010 that number had risen to seventeen.

Those indicted were purported to belong to a Eugene, Oregon-based cell of the Animal Liberation Front (ALF) and the Earth Liberation Front (ELF)—

Types of Ecoterrorism

Category of Ecoterrorism	Percent of Incidents, 1993-2004
Vandalism	77.0
Arson	12.6
Assault and bodily harm	2.0
Bombings	1.1
Other	7.3

decentralized "direct action" animal rights and environmentalist groups considered to be terrorist organizations by the FBI—known as the Family. The charges included arson, conspiracy, use of destructive devices, and destruction of an energy facility. The crimes, which occurred between 1996 and 2001 in California, Colorado, Oregon, Washington, and Wyoming, resulted in approximately $48 million in property damages. Notably included was a highly publicized arson attack on a ski resort in Vail, Colorado, that was carried out to protest a planned expansion that would encroach on habitat of the endangered lynx and resulted in $12 million in damages.

While the indicted parties initially claimed innocence, fifteen of seventeen eventually pled guilty in federal courts. They were sentenced to prison in 2007 for periods ranging from 37 to 188 months; U.S. District Court Judge Ann Aiken imposed extended sentences owing to "terrorism enhancement" penalties. The alleged "mastermind" of the Family, Bill "Avalon" Rodgers, committed suicide in a Flagstaff, Arizona, jail in late 2005 before he could be transferred to Oregon.

Operation Backfire is not without its critics. While John Lewis, a top FBI official, declared in 2005 that "the No. 1 domestic terrorism threat is the ecoterrorism, animal-rights movement," some question the merit and the motive of this designation. In a May, 2005, congressional committee hearing, Senator Frank Lautenberg of New Jersey argued that Americans should not allow themselves to be blinded to more serious terrorist threats that take innocent lives by focusing on the illegal actions of ALF and ELF, because "not a single incident of so-called environmental terrorism has killed anyone." Senator Jim Jeffords of Vermont expressed similar skepticism. In a press release on Operation Backfire, the National Lawyers Guild said that the imposition of "draconian sentences" for property damage offenses without intent

to harm is an unconstitutional and disproportionate punishment. Some have argued that Operation Backfire is part of a broader "Green Scare" (alluding to the twentieth century anticommunist Red Scare), which they allege is an attempt to suppress animal rights and environmental activism by exploiting public fears of terrorism, with the aim of maintaining the political and corporate status quo.

Defending Operation Backfire in 2006, FBI director Robert Mueller stated that "terrorism is terrorism—no matter what the motive." Mueller added that Operation Backfire dealt "a substantial blow" to domestic ecoterrorism and should have a "dramatic impact on persons who contemplate these crimes."

Joel P. MacClellan

FURTHER READING

Bishop, Bill. "Operation Backfire." *Register Guard,* April 15, 2006, A1.

Potter, Will. "The Green Scare." *Vermont Law Review* 33, no. 679 (2009): 671-687.

Scarce, Rik. *Eco-Warriors: Understanding the Radical Environmental Movement.* Updated ed. Walnut Creek, Calif.: Left Coast Press, 2006.

SEE ALSO: Abbey, Edward; Animal rights movement; Ecotage; Ecoterrorism; Monkeywrenching.

Organic gardening and farming

CATEGORY: Agriculture and food

DEFINITION: The growing of crops without the use of synthetic fertilizers or pesticides

SIGNIFICANCE: Interest in organic farming has grown since the late twentieth century for a number of reasons, including concerns about the environmental effects of the chemical pesticides and fertilizers used by conventional large agricultural operations, and particularly about the possibility of negative health effects associated with ingestion of the food produced using such chemicals.

At the beginning of agricultural history, farmers believed that plants "ate the soil" in order to grow. During the nineteenth century, German chemist Justus von Liebig discovered that plants extract nitrogen, phosphorus, and potash from soil. His findings dramatically changed agriculture as farmers found they could grow crops in any type of soil, even sand and water solutions, if they added the right chemicals.

Diversified family farms eventually gave way to huge specialized operations; by the end of the twentieth century, less than 2 percent of the U.S. population was directly involved in crop production. Crop yields were raised with the use of chemicals, but farms and their soil and water were not being cared for as the unique ecosystems they are. The quality of the soil was ignored as chemicals were used to produce high crop yields to feed the 98 percent of the population not involved in farming. Over time, the organic quality of the soil was lost even though the chemicals remained in the soil.

Chemicals used in agriculture were also found to leach into the water supply. In 1988 the Environmental Protection Agency (EPA) found that groundwater in thirty-two U.S. states was contaminated with seventy-four different agricultural chemicals. Aside from the health consequences to humans, leaching causes once friable and fertile soils to turn hard and become nonproductive. Further, the use of chemical insecticides was proving to have toxic effects on both the foods grown and the farmworkers encountering them. About forty-five thousand accidental pesticide poisonings occur in the United States each year; in 1987 the EPA ranked pesticides as the third leading environmental cancer risk.

Organic farming developed as some individuals attempted to return to diversified farming practices that emphasize working with nature to create a renewing, ecologically sound, and sustainable system of agriculture. Organic farmers are committed to crop-growing practices that are free of synthetic chemicals and genetic engineering. Soil must be free of chemicals for at least three years before products grown in it can be certified as organic. The organic certification requirements further state that both plants and livestock must be raised organically, with-

National Standards for Organic Certification

The U.S. Department of Agriculture's National Organic Program published a fact sheet in October, 2002, that explains the national standards for the certification of organic products laid out in the 1990 Organic Foods Production Act:

The national organic standards address the methods, practices, and substances used in producing and handling crops, livestock, and processed agricultural products. The requirements apply to the way the product is created, not to measurable properties of the product itself. Although specific practices and materials used by organic operations may vary, the standards require every aspect of organic production and handling to comply with the provisions of the Organic Foods Production Act (OFPA). Organically produced food cannot be produced using excluded methods, sewage sludge, or ionizing radiation.

CROP STANDARDS

The organic crop production standards say that:

Land will have no prohibited substances applied to it for at least 3 years before the harvest of an organic crop. The use of genetic engineering (included in excluded methods), ionizing radiation and sewage sludge is prohibited. Soil fertility and crop nutrients will be managed through tillage and cultivation practices, crop rotations, and cover crops, supplemented with animal and crop waste materials and allowed synthetic materials.

Preference will be given to the use of organic seeds and other planting stock, but a farmer may use non-organic seeds and planting stock under specified conditions. Crop pests, weeds, and diseases will be controlled primarily through management practices including physical, mechanical, and biological controls. When these practices are not sufficient, a biological, botanical, or synthetic substance approved for use on the National List may be used.

LIVESTOCK STANDARDS

These standards apply to animals used for meat, milk, eggs, and other animal products represented as organically produced. The livestock standards say that:

Animals for slaughter must be raised under organic management from the last third of gestation, or no later than the second day of life for poultry. Producers are required to feed livestock agricultural feed products that are 100 percent organic, but may also provide allowed vitamin and mineral supplements. Producers may convert an entire, distinct dairy herd to organic production by providing 80 percent organically produced feed for 9 months, followed by 3 months of 100 percent organically produced feed. Organically raised animals may not be given hormones to promote growth, or antibiotics for any reason. Preventive management practices, including the use of vaccines, will be used to keep animals healthy. Producers are prohibited from withholding treatment from a sick or injured animal; however, animals treated with a prohibited medication may not be sold as organic. All organically raised animals must have access to the outdoors, including access to pasture for ruminants. They may be temporarily confined only for reasons of health, safety, the animal's stage of production, or to protect soil or water quality.

out the use of chemicals, antibiotics, hormones, or synthetic feed additives. Organic farmers must comply with both organic regulations in their states and the 1990 federal Organic Foods Production Act (OFPA). In the late 1990's the U.S. Department of Agriculture, through the National Organic Standards Board, developed standards and regulations to ensure consistent national standards for organic products.

SOIL FERTILITY

To produce chemical-free crops, organic farmers must rely on organic practices to ensure soil fertility and control unwanted plants and insects. Organic farmers build organic materials in the soil by adding green manure, compost, or animal manure.

"Green manure" is a term used for crops that are grown specifically to introduce organic matter and nutrients into the soil. These crops are raised expressly for the purpose of being plowed under rather than sold to consumers. Green manure crops protect soil against erosion, cycle nutrients from lower levels of the soil into the upper layers, suppress weeds, and keep much-needed nutrients in the soil rather than allowing them to leach out. Legumes are excellent green manure crops because they are able to extract nitrogen from the air and transfer it into the soil, leaving a supply of nitrogen for the next crop. Legumes have nitrogen nodules on their roots; when the legumes are tilled under and decompose, they add more nitrogen to the soil. As plants decay, they also make insoluble plant nutrients—such as carbon dioxide and acetic, butyric, lactic, and other organic acids—available in the soil.

Much of the organic material derived from green manure comes in the form of decaying roots. Alfalfa, one type of legume, sends its roots several feet down into the soil. When the alfalfa plants are turned, the entire root system decomposes into organic material, thus helping to improve water retention and soil quality at the same time. Some examples of legumes used for green manure are sweet clover, ladino clover, alfalfa, and trefoil. Nonlegumes or grasses used for green manure include rye, redtop, and timothy grass.

Further soil fertility can be achieved if green manure is grazed by animals and the manure from those animals is deposited on the soil. Left on their own in a large field, cows, sheep, and horses will choose only their favorite bits of grasses to graze and thus will graze the field unevenly. Organic farmers use porta-

ble fencing to achieve strip rotational grazing, in which animals graze a strip of a field all the way down and then are moved to another strip of the field while the first pasture recovers. This process is repeated until the entire field has been grazed and fertilized. Also, when the grass is grazed, some of the roots die and rot to form a good amount of humus. This humus is the stable organic material that acts as a catalyst for allowing plants to find nutrients.

Compost can work for large or small farmers, but it is particularly useful to small farmers and gardeners who do not have enough land to grow green manure crops to plow under. Composting is a natural soil-building process that began with the first plants that existed on earth and continues as a natural process today. As falling leaves and dead plants, animals, and insects decompose into the soil, they form a rich organic layer. Farmers and gardeners can make compost by alternating layers of carbohydrate-rich plant cuttings and leaves, animal manure, and topsoil in a container or defined area and allowing the components to decay. The rich organic matter that results can then be added to the soil to be cultivated.

Animal manure is used as an organic fertilizer. Rich in nitrogen, the best animal manures come from animals that have a high amount of protein in their diets. For example, beef cattle being fattened for market have a higher level of protein in their solid waste than do dairy cattle that are producing milk for market. The application of manure to fields improves the structure of the soil, raises the organic nitrogen content, and stimulates the growth of soil bacteria and fungi necessary for healthy soil.

CROP ROTATION

Planting the same crop year after year on the same piece of ground results in depleted soil. Crops such as corn, tobacco, and cotton remove nutrients, especially nitrogen, from the soil. Planting legumes every other year can keep soil fertile by adding nitrogen and achieving a balanced nutrient level. Planting a winter cover crop such as rye grass helps to protect the land from erosion, and when the crop is plowed under in the spring, it creates a nutrient-rich soil for the planting of a cash crop. Crop rotation also improves the physical condition of the soil because different crops vary in root depth, are cultivated differently, and respond to either deep or shallow soil preparation.

Traditional agriculture, or monoculture, puts a large source of the same crop in easy proximity to insects that are destructive to that particular crop. Insect offspring can multiply out of proportion when the same crop is grown in a given field year after year. Since insects are drawn to a particular home area, they will not be able to proliferate and thrive if the crop is changed every other year to a crop they do not eat. This is one of the reasons organic farmers rely heavily on crop rotation as one aspect of insect control. Rotating crops can also help control weeds. Some crops and cultivation methods inadvertently allow certain weeds to thrive. Farmers using crop rotation can incorporate successor crops that eradicate those weeds. Some crops, such as potatoes and winter squash, work as "cleaning crops" because of the different styles of cultivation that are used on them.

Organic Insect and Weed Control

Advocates of organic farming methods assert that plants within a balanced ecosystem are resistant to disease and insect infestation and that plants stressed by unfavorable growing situations are much more sus-

ceptible to such problems. The whole premise of organic farming is that farmers should work with nature to help provide healthy, unstressed plants.

Rather than using chemical pesticides, which often create resistant generations of insects and thus the need for newer and stronger chemicals, organic farmers seek natural ways to diminish pest problems. They strive to maintain and replenish soil fertility to produce healthy plants that are resistant to insects. They also try to select plant species that are resistant to insects, weeds, and disease. Crop rotation, as noted above, is another method of keeping insect infestations down. Organic farmers may create barriers to insect pests by planting, in and around the crop field, rows of hedges, trees, or even plants that are not desirable to insects. They may also provide habitats for the insects' natural enemies: birds, beneficial insects, and garden snakes.

The ladybug and the praying mantis are two beneficial insects that can help rid a farm or garden of aphids, mites, mealy bugs, and grasshoppers. Tricho-

A worker walks through rows of greens and lettuce at Rock Star Farms, an organic agricultural operation in Forsyth County, Georgia. (AP/ Wide World Photos)

gramma is a small wasp that will destroy moth eggs, squash borers, cankerworms, cabbage loopers, and corn earworm. Many beneficial insects are available to farmers and gardeners in large quantities through catalog or online retailers. Given that just one ladybug can consume fifty or more aphids per day and can lay up to one thousand fertile eggs, the overall cost of purchasing ladybugs is much lower than the cost of chemical insecticides.

Organic farmers and gardeners also often rely on so-called insecticide crops. Garlic planted near lettuce or peas will deter aphids. Geraniums or marigolds grown close to grapes, cantaloupes, corn, or cucumbers will deter Japanese and cucumber beetles. Herbs such as rosemary, sage, and thyme planted by cabbages will deter white butterfly pests. Potatoes will repel Mexican bean beetles if planted near beans, and tomatoes planted near asparagus will ward off asparagus beetles. Natural insecticides such as red pepper juice can be used for ant control, while a combination of garlic oil and lemon may be used against fleas, mosquito larvae, houseflies, and other insects.

Organic farming relies on the physical control of weeds, especially through the use of cutting (cultivation) or smothering (mulching and hilling). The cultivation method involves the shallow stirring of surface soil to cut off small developing weeds and prevent more from growing. This can be done with tractors, wheel hoes, tillers, or, for small gardens, hand hoes. Mulch is a soil cover that prevents weeds from getting the sunlight they need for growth. Mulching with fully biodegradable materials can help build soil fertility while controlling weeds. Plastic mulches can be used on organic farms as long as they are removed from the fields at the end of the growing or harvest season.

Organic farming offers a safe alternative to the use of synthetic chemicals in the production of food. Growing public awareness of the possible toxic effects of the traditional agricultural reliance on synthetic chemicals is reflected in a growing demand for organically produced foods.

Dion Stewart and Toby Stewart

FURTHER READING

Coleman, Eliot. *The New Organic Grower.* White River Junction, Vt.: Chelsea Green, 1995.

Duram, Leslie A. *Good Growing: Why Organic Farming Works.* Lincoln: University of Nebraska Press, 2005.

Hamilton, Geoff. *Organic Gardening.* Rev. ed. New York: Dorling Kindersley, 2008.

Koepf, Herbert H. *The Biodynamic Farm: Agriculture in the Service of the Earth and Humanity.* Hudson, N.Y.: Anthroposophic Press, 2006.

Kristiansen, Paul, Acram Taji, and John Reganold, eds. *Organic Agriculture: A Global Perspective.* Ithaca, N.Y.: Cornell University Press, 2006.

Lampkin, Nicolas. *Organic Farming.* Rev. ed. 1994. Reprint. Alexandria Bay, N.Y.: Diamond Farm Enterprises, 2002.

Paarlberg, Robert. "Organic and Local Food." In *Food Politics: What Everyone Needs to Know.* New York: Oxford University Press, 2010.

SEE ALSO: Agricultural chemicals; Agricultural revolution; Integrated pest management; Intensive farming; Monoculture; Pesticides and herbicides; Strip farming; Sustainable agriculture.

Osborn, Henry Fairfield, Jr.

CATEGORIES: Activism and advocacy; animals and endangered species; population issues

IDENTIFICATION: American naturalist and conservationist

BORN: January 15, 1887; Princeton, New Jersey

DIED: September 16, 1969; New York, New York

SIGNIFICANCE: Through his work with the New York Zoological Society and his writings, Osborn promoted the preservation of endangered species and their habitats and also raised public awareness of the dangers of human overpopulation.

Henry Fairfield Osborn, Jr., was one of five children of Henry Fairfield Osborn and Lucretia Thatcher Perry. The senior Osborn was a professor of paleontology at Princeton University at the time of Osborn's birth and eventually became the president of the American Museum of Natural History and founder of the New York Zoological Society. The younger Osborn graduated from the Groton School in 1905 and from Princeton University with a bachelor of arts in 1909. He then studied for one year at Cambridge University in England.

After finishing his year in Cambridge, Osborn traveled and held several jobs, including laying track for the railroad in Utah. During World War I he served as a captain in the 351st Field Artillery of the American Expeditionary Force. After returning from the war,

Plundering the "Good Earth"

Henry Fairfield Osborn, Jr., opens Our Plundered Planet *(1948) with the following cautionary words:*

There is beauty in the sounds of the words "good earth." They suggest a picture of the elements and forces of nature working in harmony. The imagination of men through all ages has been fired by the concept of an "earth-symphony." Today we know the concept of poets and philosophers in earlier times is a reality. Nature may be a thing of beauty and is indeed a symphony, but above and below and within its own immutable essences, its distances, its apparent quietness and changelessness it is an active, purposeful, co-ordinated machine. Each part is dependent upon another, all are related to the movement of the whole. Forests, grasslands, soils, water, animal life—without one of these the earth will die—will become dead as the moon. This is provable beyond questioning. Parts of the earth, once living and productive, have thus died at the hand of man. Others are now dying. If we cause more to die, nature will compensate for this in her own way, inexorably, as already she has begun to do.

Osborn worked in the investment banking business on Wall Street. He married Marjorie M. Lamond on September 8, 1914, and together they had three daughters. In 1922, as he continued working in investments, Osborn became a trustee of the New York Zoological Society. The paleontology field trips that he had taken with his father from an early age had convinced him that his real vocation lay in natural science. Osborn left the banking field in 1935 to become the secretary of the New York Zoological Society. He assumed the presidency of the society in 1940 and stayed in that position until 1968, when his health failed. Under his direction the Bronx Zoological Park was greatly improved, and the Marine Aquarium at Coney Island was created. In connection with his fund-raising for these causes, he has been called "the greatest showman since Barnum."

Osborn used his position in the Zoological Society to become a strong advocate of the preservation of endangered species and their habitats. His was also an early voice warning of the dangers of human population growth. He presented his ideas on this topic forcefully in his two books, *Our Plundered Planet* (1948) and *The Limits of the Earth* (1953). He also pushed his ideas about overpopulation as a member

of the Conservation Advisory Committee of the U.S. Department of the Interior and the Planning Committee of the Economic and Social Council of the United Nations. Osborn was often invited to speak at both public and scientific meetings, and he regularly attended national and international conservation conferences. In 1947 Osborn founded the Conservation Foundation; this organization was incorporated into the World Wildlife Fund in 1990. He was presented with the American Design Award in 1948 for his campaign to stop humanity from fighting a losing battle against nature.

Osborn suffered a slight stroke at seventy-nine years of age, which left him with a minor speech impediment. He continued to lecture and work, however, until his death in 1969 at the age of eighty-two.

Kenneth H. Brown

FURTHER READING

Collier, Paul. *The Plundered Planet: Why We Must—and How We Can—Manage Nature for Global Prosperity.* New York: Oxford University Press, 2010.

Shabecoff, Philip. *A Fierce Green Fire: The American Environmental Movement.* Rev. ed. Washington, D.C.: Island Press, 2003.

SEE ALSO: Conservation; International Union for Conservation of Nature; Population growth; Zoos.

O'Shaughnessy Dam. *See* Hetch Hetchy Dam

Ostrom, Elinor

CATEGORY: Resources and resource management
IDENTIFICATION: American political scientist
BORN: August 7, 1933; Los Angeles, California
SIGNIFICANCE: Through her extensive empirical research, Ostrom has shown that it is not inevitable that shared resources will be depleted by overuse.

In 2009 Elinor Ostrom was awarded the Nobel Prize in Economic Sciences "for her analysis of economic governance, especially the commons." The term "the commons" refers to resources that are available to all, such as water, forests, fisheries, and grazing land. It has often been feared that such shared resources are likely to be depleted by overuse (a phenomenon known as the tragedy of the commons), but Ostrom's research has uncovered numerous instances of commons that are well managed by users. Her findings therefore challenge the view that common-pool resources must be regulated by government or privately owned to be properly managed.

After receiving her Ph.D. in political science from the University of California, Los Angeles, in 1965, Ostrom took a teaching position at Indiana University, where she built her career. Through their research into the use of common-pool resources, Ostrom and her colleagues developed a framework, known as institutional analysis and development, to organize the rules governing common-pool resources. Utilizing this framework, Ostrom and others have been able to gain insights into why some common-pool resources are well managed while others are not. Ostrom's research brings order to the complex issues involved in the use of common-pool resources; her findings indicate that users of such resources often come up with better solutions for their management than those imposed by government. Although, as Ostrom has noted, there are no easy, one-size-fits-all answers to the problem of the overexploitation of common-pool resources, her research provides hope and a framework that can be used in the search for solutions to protect these vital resources.

Randall Hannum

SEE ALSO: Aquifers; Fisheries; Grazing and grasslands; Hardin, Garrett; Overgrazing of livestock; Riparian rights; Sustainable forestry; Tragedy of the commons.

Our Common Future

CATEGORIES: Ecology and ecosystems; resources and resource management

IDENTIFICATION: Report of the World Commission on Environment and Development that focused on multilateral global cooperation as the most effective way to address growing environmental crises

DATE: Published in 1987

SIGNIFICANCE: Through *Our Common Future*, the World Commission on Environment and Development gave support to the idea that pressing environmental concerns must be addressed through a concerted international effort that balances the need for economic development with the necessity of protecting and preserving the environment.

By the early 1980's, environmental activists had begun to recognize that individual nations could have little long-term impact in their attempts to address the most significant issues related to the environment, among them air, soil, and water pollution, diminishment of food and water supplies, habitat loss, and species extinction. They came to believe that nations needed to reorient their perceptions of how to go about protecting the earth's resources. Visionary politicians, scientists, and conservationists turned to the United Nations to promote an international effort to address environmental issues.

In 1983, the United Nations established the World Commission on Environment and Development, an international committee of twenty-two members headed by Dr. Gro Harlem Brundtland, former prime minister of Norway (she would be reelected to the office in 1986); Brundtland had long been a passionate advocate for environmentalism. The commission, commonly known as the Brundtland Commission, was to project the course of global environmental development to the year 2000. Brundtland conceived an ambitious agenda: The commission would develop broad recommendations for how humanity might begin to address the damages caused to the planetary ecosystem by unchecked industrial development. The commission dedicated its efforts to defining a radical new vision in which human affairs could be reconciled with natural processes—the inability to seek such reconciliation, the commission warned, would have bleak consequences.

The Need for Sustainable Development

In her foreword to Our Common Future *(1987), Gro Harlem Brundtland notes the interconnections among poverty, inequality, development, and environmental degradation:*

When the terms of reference of our Commission were originally being discussed in 1982, there were those who wanted its considerations to be limited to "environmental issues" only. This would have been a grave mistake. The environment does not exist as a sphere separate from human actions, ambitions, and needs, and attempts to defend it in isolation from human concerns have given the very word "environment" a connotation of naivety in some political circles. The word "development" has also been narrowed by some into a very limited focus, along the lines of "what poor nations should do to become richer," and thus again is automatically dismissed by many in the international arena as being a concern of specialists, of those involved in questions of "development assistance."

But the "environment" is where we all live; and "development" is what we all do in attempting to improve our lot within that abode. The two are inseparable. Further, development issues must be seen as crucial by the political leaders who feel that their countries have reached a plateau towards which other nations must strive. Many of the development paths of the industrialized nations are clearly unsustainable. And the development decisions of these countries, because of their great economic and political power, will have a profound effect upon the ability of all peoples to sustain human progress for generations to come.

Many critical survival issues are related to uneven development, poverty, and population growth. They all place unprecedented pressures on the planet's lands, waters, forests, and other natural resources, not least in the developing countries. The downward spiral of poverty and environmental degradation is a waste of opportunities and of resources. In particular, it is a waste of human resources. These links between poverty, inequality, and environmental degradation formed a major theme in our analysis and recommendations. What is needed now is a new era of economic growth—growth that is forceful and at the same time socially and environmentally sustainable.

Under Brundtland's direction, the commission's landmark report, published in 1987 as *Our Common Future* and commonly referred to as the Brundtland Report, centered on the concept of sustainable development, focusing on how economic development and environmentalism might (and indeed must) work in concert so that the economic needs of the present would not jeopardize future generations. The report endorsed international cooperation among developed and developing nations, with specific concern for the plight of the poor. The commission boldly conceived of the needed campaign to address environmental concerns as holistic and necessarily global; its report emphasized that nations must move beyond centuries-old assumptions about political, religious, and economic boundaries and conceive of the globe as a single organism, its nations as interdependent, and all the crises affecting the global environment as a single interlocked crisis.

In the decades since the report's release, many of the commission's most dire predictions have not come to pass, and efforts to create international cooperation in the face of the urgency of environmental degradation have struggled. Nevertheless, *Our Common Future* has been credited with laying the philosophical foundation for historic international environmental successes of the 1990's, many under the auspices of the United Nations, most notably the 1992 Earth Summit, the Rio Declaration, the Montreal Protocol, the Kyoto Protocol, and Agenda 21.

Joseph Dewey

FURTHER READING
Blewitt, John. *Understanding Sustainable Development.* Sterling, Va.: Earthscan, 2008.
Leman-Stefanovic, Ingrid. *Safeguarding Our Common Future: Rethinking Sustainable Development.* Albany: State University of New York Press, 2000.
Rogers, Peter P., Kazi F. Jalai, and John A. Boyd. *An Introduction to Sustainable Development.* Sterling, Va.: Earthscan, 2007.
Walker, Brian, and David Salt. *Resilience Thinking: Sustaining Ecosystems and People in a Changing World.* Washington, D.C.: Island Press, 2006.
World Commission on Environment and Development. *Our Common Future.* Washington, D.C.: Author, 1987.

SEE ALSO: Agenda 21; Brundtland, Gro Harlem; Brundtland Commission; Earth Summit; *Global 2000*

Report, The; Globalization; Johannesburg Declaration on Sustainable Development; Kyoto Protocol; Sustainable development; United Nations Commission on Sustainable Development; United Nations Environment Programme; World Summit on Sustainable Development.

Overconsumption

CATEGORY: Resources and resource management
DEFINITION: Excessive and unsustainable utilization of resources, goods, and services
SIGNIFICANCE: Human life is impossible to sustain without consumption, but overconsumption often causes multiple external effects (such as air, water, and soil pollution) that have negative impacts on the health of ecosystems and organisms, including those of humans. Depending on the circumstances (culture, geography, social milieu, income, education, and so on), individuals and groups overconsume different kinds of things.

It is useful to distinguish among the reasons for and the agents, objects, and effects of overconsumption. Among the things that are overconsumed in certain parts of the world are natural resources (such as water and timber), energy (such as electricity), and commodities (such as electronic devices and automobiles); services and information may also be overconsumed. Overconsumption of different types can have various environmental effects. The overfishing and overhunting of animals can lead to the extinction of species and thus to the reduction of biodiversity; an example is the extinction of the Caribbean monk seal in the twentieth century as a result of overhunting. The intensive and extensive global usage of gasoline as fuel for combustion engines results in high levels of carbon dioxide emissions, which have been shown to contribute to global warming. Another example of the unsustainable—and intergenerationally unjust—utilization of nonrenewable resources is the consumption of scarce chemical elements such as lithium, tantalum, indium, and hafnium, which are used for semiconductors and liquid crystal displays (LCDs) in televisions, computers, and mobile phones.

REASONS FOR OVERCONSUMPTION
Whereas in the developed world overconsumption often takes the form of the excessive purchasing of consumer goods in affluent social milieus (for example, multiple televisions or vehicles per household), in developing nations overconsumption is more often related to excessive or unsustainable exploitation of raw materials coupled with overpopulation and poverty (such as unsustainable clear-cutting of trees for charcoal or timber export). Such problems raise questions concerning justice or equity within one generation (intragenerational justice) and between generations (intergenerational justice).

Overconsumption is not a modern phenomenon—since Paleolithic times humans have existed who destroyed their environment through excessive utilization of resources and then moved to unspoiled areas. The rate of consumption was intensified and accelerated to a globally unsustainable level by a combination of factors, including the dawn of modern science and technology, the growth of industrialization, the advancement of worldviews and ethics that endorse despiritualization, and the advent of a paradigm of productivity, economic growth, and capitalism. Since the early twentieth century the fields of marketing and advertising have not only intensified but also accelerated the rate of consumption even further by utilizing psychoanalytical, sociological, and cultural strategies. Such strategies play on people's unfulfilled personal desires and attach symbolic values (meta-goods) to commodities to lure potential customers to consume beyond their actual needs.

ATTEMPTS TO REDUCE OVERCONSUMPTION
In addition to causing environmental problems, overconsumption can have significant psychological effects, such as oniomania (compulsive buying disorder), and social side effects, such as social isolation related to excessive usage of electronic media. Environmentalists, educators, and psychotherapists have expressed concerns about marketing strategists' ulterior motives and the hidden agendas of advertisement campaigns. Advertisements can trigger and foster overconsumption by intensifying—or even creating—desires.

Approaches to reducing overconsumption include education about the environmental impacts of consumption habits (for example, through the calculation of ecological footprints, or how much demand exceeds the regenerative capacity of a resource or the biosphere's supply). Other educational strategies focus on analyzing the psychology at work in advertising campaigns and on making consumers aware of social factors that stimulate overconsumption, such as peer pressure, symbolic consumption, and conspicuous consumption. Ecological economists—who locate the economic sector within the world and its ecology—design and suggest political strategies and legal mechanisms that attempt to form development in the sectors of economy, society, and ecology in a sustainable fashion. Campaigns against overconsumption can also take more radical forms that include acts of ecotage (or monkeywrenching) or ecoterrorism.

Roman Meinhold

FURTHER READING

Baudrillard, Jean. *The Consumer Society. Myths and Structures*. 1998. Reprint. Thousand Oaks, Calif.: Sage, 2004.

Meinhold, Roman. "Meta-Goods in Fashion-Myths: Philosophic-Anthropological Implications of Fashion." *Prajna Vihara. Journal of Philosophy and Religion* 8, no. 2 (2007): 1-15.

Penn, Dustin J. "The Evolutionary Roots of Our Environmental Problems: Toward a Darwinian Ecology." *Quarterly Review of Biology* 78, no. 3 (2003): 275-301.

Princen, Thomas. "Consumption and Its Externalities: Where Economy Meets Ecology." *Global Environmental Politics* 1, no. 3 (2001): 11-30.

Schmidtz, David, and Elizabeth Willcott. "Varieties of Overconsumption." *Ethics, Place, and Environment: A Journal of Philosophy and Geography* 9, no. 3 (2006): 351-365.

White, Lynn Townsend, Jr. "The Historical Roots of Our Ecologic Crisis." *Science* 155, no. 3767 (1967): 1203-1207.

SEE ALSO: Ecological footprint; Externalities; Planned obsolescence; Population growth; Recycling; Sustainable development; Waste management; Water use.

Overgrazing of livestock

CATEGORIES: Resources and resource management; agriculture and food

DEFINITION: Excess consumption of natural vegetation by domesticated animals maintained for agricultural purposes

SIGNIFICANCE: Various kinds of environmental damage result when the quantity of grazing livestock exceeds the amount of vegetation land can grow or animals are allowed to graze unrestricted. Overgrazing's negative impacts include desertification, erosion, and loss of biodiversity.

Overgrazing has been a persistent issue since ancient times; historical documents mention insufficient grazing locations for livestock and the poor quality of available fields. The kinds of livestock usually associated with overgrazing include cattle, goats, and sheep, which also damage soil with their hooves while eating vegetation. Human reliance on livestock for food, as well as transportation in many places, intensifies overgrazing problems when effective management is not practiced to protect ranges. Overgrazing reduces the amount of grazing lands available, as areas become degraded and unsuitable for sustaining livestock. The 1930's Dust Bowl in the United States is one of the best-known cases of environmental damage involving overgrazing.

CAUSES

Approximately half of the land on earth is grazed by livestock. As increasing global human population has resulted in greater demand for food, particularly meat, livestock production has accelerated. In 2006, according to a report published by the United Nations Food and Agriculture Organization (FAO), 1.5 billion cattle and 1.75 billion goats and sheep existed worldwide, and some 200 million people earned their income primarily by tending grazing livestock. Expanding herds stress finite grazing resources in diverse ecosystems, including grasslands in tropical and arid regions, overwhelming many ranges. In China's Qinghai Province, according to FAO, farmers graze 5.5 million sheep on lands capable of supplying vegetation for only 3.7 million livestock. Herdsmen expect African grazing lands to feed almost twice the quantity of livestock that available vegetation resources can support. Indian grazing lands supply only 540 million tons of forage instead of the 700 million tons needed to feed livestock.

At the 2007 World Food Summit, researchers focused on the issue of overgrazing. They noted that African, Asian, and Middle Eastern ranges had suffered physical depletion because they had been expected to support too many herds. Lester Brown, president of the Earth Policy Institute, stated that overpopulation of livestock on limited lands had cost livestock producers several billions of dollars because of malnourished animals dying, being unable to reproduce, or providing poor quality meat or milk. FAO acknowledged the loss of grazing lands to urbanization and the cultivation of grain crops for livestock feed, which resulted in remaining pastures often being overcrowded and grazed by more animals than they could sustain.

Property laws differ from country to country, but in many nations public lands are open for anyone to release herds to graze. Damage to the grazing lands occurs because of lack of controls and poor monitoring of livestock movements and access to vegetation. The Bureau of Land Management oversees permits for ranchers in the western United States to graze livestock on almost 65 million hectares (160 million acres), enforcing legislation based on revisions of the Taylor Grazing Act of 1934 and various environmental laws. Some governments offer farmers subsidies to raise livestock, resulting in increased herds and high demand for grazing lands. When the Greek government dispensed such financial incentives without specifying any grazing restrictions, many people who were not farmers bought livestock to qualify for subsidies and then released the animals to roam. Soon, sizable livestock herds, mostly goats, grazed throughout Greece. The number of grazing animals required more vegetation than was available, and many plants, and even trees, were stripped from the land.

ENVIRONMENTAL IMPACTS

By the early twenty-first century, livestock overgrazing had caused serious damage to 20 percent of grazing ranges worldwide, or some 680 million hectares (1.68 billion acres). The damage to the land caused by overgrazing is often irreversible. With the vegetation loss of overgrazing, the land becomes vulnerable to winds, often losing topsoil and becoming eroded. Any

Goats graze in an overgrazed, eroded field in Botswana. (©David Reed/CORBIS)

remaining roots in the soil cease growing because they no longer receive nutrients from aboveground vegetation; thus the soil become less fertile. Loss of vegetation also results in reduced removal of the greenhouse gas carbon dioxide from the atmosphere.

Conservationists estimate that overgrazing in the western United States has harmed 80 percent of riparian resources, especially streams. Without plants near riparian areas, less precipitation soaks into soils. Groundwater amounts decrease and are not restored. Land becomes dry because the soil cannot retain water. Many overgrazed grasslands undergo desertification. It has been estimated that 233,000 hectares (576,000 acres) of Chinese grasslands become desert each year.

Overgrazing causes many native range plants to become endangered or extinct. Weeds and invasive plants, many of them toxic, often displace indigenous grass species when those species are overgrazed and can no longer compete for food, water, and sunlight resources. Plants that are unsuitable as grazing vegetation, such as snakeweed, often appear when overgrazing occurs and thrive because they are hardier than the native plants in the stressed conditions created by overgrazing.

Overgrazing threatens biodiversity by damaging ecosystems. Habitats are altered by the changes in vegetation and water sources that overgrazing causes. When livestock consume all available vegetation, they deprive the native wildlife of shelter from predators, resulting in the death or displacement of both prey species and the predators who feed on them. Birds lose grassy nesting sites. Amphibians, particularly frogs, suffer when overgrazing removes vegetation they need to establish breeding areas in still water. Overgrazing also disrupts the activities of insects and the spawning of fish.

As FAO notes in its 2006 report *Livestock's Long Shadow: Environmental Issues and Options*, the environmentally harmful practice of deforestation is also related to the problem of overgrazing. As grazing ranges have been lost to overgrazing, some people have turned to clearing forestlands—including many millions of hectares of rain forest in South America—to create lands for the feeding of increasing livestock populations.

PREVENTION AND RESTORATION

Conservationists have noted that one practical way to stop overgrazing from occurring is to decrease livestock populations, restricting herds to populations that lands can realistically support with nutritious vegetation. Many governments have developed policies outlining limitations for grazing lands, expecting livestock owners to reduce the sizes of their herds. Education of both government officials and livestock owners has also been shown to be important. A 2005 report on the implementation of the 1994 United Nations Convention to Combat Desertification noted that many African officials and farmers were unaware of the environmental damage caused by overgrazing and the urgency of preventing it.

FAO has noted the importance of herders' managing where and how long their livestock graze, keeping animals from roaming onto vulnerable tracts of land. Also, FAO has recommended that in ranges open to numerous livestock simultaneously, fences blocking livestock should be dismantled so that grazing is not concentrated in some areas, exhausting the vegetation there. Another suggested approach to reducing overgrazing is to lower the demand on grazing land by giving animals alternative sources of food, such as plant stubble remaining in harvested crop fields. In east-central China, this approach has led to livestock eating approximately 500 million tons of wheat, rice, and corn stalks and straw annually.

Attempts to repair overgrazed lands have varied according to locations. Many governments support the reseeding of lands damaged by overgrazing. The Chinese government has proposed cultivating trees to separate overgrazed areas where desertification has occurred from still-viable grasslands. During the early twenty-first century, some Chinese officials encouraged rural residents living on overgrazed lands to relocate to urban communities while efforts were undertaken to renew the overgrazed areas' ecosystems with indigenous plants and animals.

Scientific approaches to combating overgrazing have included the use of satellite imagery to evaluate where overgrazing is most prevalent and identify areas at risk. Data gathered in this way have helped governments decide how to proceed with implementing grazing restrictions and restoration work. In the summer of 2007 researchers activated global positioning systems on livestock in Australia to track where the animals grazed. This surveillance enabled observers to notice the potential for locations to become overgrazed and move livestock before that occurred. Scientists at the International Center for Agricultural Research in the Dry Areas in Syria have designed technology that can help with the replanting of overgrazed land. The device they created distributes grass seeds while forming grooves in the soil to retain the water needed to sustain new plant growth.

Elizabeth D. Schafer

FURTHER READING

Fleishner, Thomas L. "Ecological Costs of Livestock Grazing in Western North America." *Conservation Biology* 8, no. 3 (September, 1994): 629-644.

Ibáñez, Javier, Jaime Martínez, and Susanne Schnabel. "Desertification Due to Overgrazing in a Dynamic Commercial Livestock-Grass-Soil System." *Ecological Modelling* 205, no. 3-4 (2007): 277-288.

Papanastasis, Vasilios P. "Restoration of Degraded Grazing Lands Through Grazing Management: Can It Work?" *Restoration Ecology* 17, no. 4 (July, 2009): 441-445.

Steinfeld, Henning, et al. *Livestock's Long Shadow: Environmental Issues and Options.* Rome: United Nations Food and Agriculture Organization, 2006.

Wuerthner, George, and Mollie Matteson, eds. *Welfare Ranching: The Subsidized Destruction of the American West.* Washington, D.C.: Island Press, 2002.

SEE ALSO: Biodiversity; Bureau of Land Management, U.S.; Cattle; Deforestation; Desertification; Erosion and erosion control; Food and Agriculture Organization; Forest and range policy; Grazing and grasslands; Habitat destruction; Range management; Taylor Grazing Act; United Nations Convention to Combat Desertification.

Ozone layer

CATEGORIES: Atmosphere and air pollution; weather and climate

DEFINITION: Region of the lower stratosphere in which most of the earth's ozone is found

SIGNIFICANCE: The earth's ozone layer protects life on the planet from exposure to dangerous levels of ultraviolet light. Because of the introduction of certain human-made chemicals into the atmosphere, the amount of stratospheric ozone steadily declined during the second half of the twentieth century.

Ozone (O_3) is a molecule made of three atoms of oxygen. It is considered a trace gas because it accounts for only .000007 percent of the earth's atmosphere. Ozone concentrations are measured in terms of Dobson units, which represent the thickness of all the ozone in a column of the atmosphere if it were compressed—on average only 3 millimeters (0.118 inch) thick. Depending on where it is found in the atmosphere, ozone may have either a positive or a negative impact on life. When ozone is near the earth's surface, it is a major air pollutant, a chief constituent of smog, and a greenhouse gas. Fuel combustion and other human activities increase the quantities of ozone in this atmospheric region. Ozone that resides in the stratosphere protects earth's organisms from lethal intensities of solar ultraviolet (UV) radiation. Without this shield of ozone, which is created through naturally occurring processes, life on earth as it is now known would probably cease to exist. Approximately 90 percent of the earth's ozone is found in the stratosphere.

Ozone concentration peaks in the lower stratosphere, between the altitudes of 20 and 25 kilometers (12 and 16 miles). Within this "ozone layer," two sets of chemical reactions, both powered by UV radiation, continuously occur. In one reaction, ozone is produced; in the other, ozone is broken down into oxygen molecules and ions. Slightly more ozone is produced than is destroyed by these reactions, so that stratospheric ozone is constantly maintained by nature. Because UV radiation is a catalyst for both reactions, most of this radiation is used up and prevented from ever reaching earth's surface.

UV radiation is categorized as UVA, UVB, or UVC, depending on wavelength. UVC rays readily kill living cells with which they come into contact. UVB rays, although less energetic than UVC rays, also damage cells. The ozone shield prevents all UVC rays from reaching the surface; however, some UVB radiation does pass through the ozone layer. It is contact with this radiation that causes sunburns, accelerates the natural aging of the skin, and has been shown to increase rates of skin cancer and cataracts. The American Cancer Society estimates that more than one million new cases of skin cancer occur each year in the United States, mainly as a result of UV radiation. Some researchers estimate that every 1 percent decline in ozone concentration causes a 2 percent increase in UV intensity at the earth's surface, resulting in a greater risk of skin cancer, cataracts, and immune deficiencies.

Increased UV radiation also harms plant and animal life. Some studies have suggested that yields from

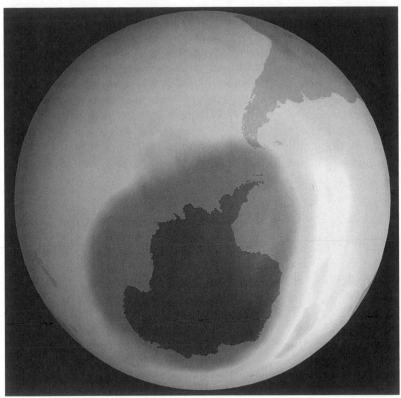

Graphic depicts the Antarctic ozone hole based on data from the Total Ozone Mapping Spectrometer, a satellite instrument that measures ozone values. (UPI/LANDOV)

crops such as corn, wheat, rice, and soybeans drop by 1 percent for each 3 percent decrease in ozone concentration. Increased UV radiation in polar regions impairs and destroys phytoplankton, which makes up the base of the food chain. A decrease in phytoplankton would likely cause population reductions at all levels of the ecosystem. Fewer than 1 in every 100,000 molecules in the atmosphere is ozone, a ratio that both underscores and belies the critical role ozone plays in protecting human health and the global environment.

DISCOVERY OF THE ANTARCTIC OZONE HOLE

Satellites placed in orbit during the late 1970's allowed scientists to observe concentrations of stratospheric ozone. Observations showed that during the Southern Hemisphere's spring (primarily September and October), the ozone layer above Antarctica thinned dramatically, then recovered during November. These findings were initially dismissed as being caused by instrument error; however, the measurements were confirmed by the British Antarctic Survey in 1985. During that spring, the loss of ozone

exceeded 50 percent of normal concentrations. Alarmingly, continued satellite measurements indicated that each year throughout the 1980's and 1990's, the ozone concentration over Antarctica dropped to record lows, and the size of the ozone-depleted area (dubbed the "ozone hole") increased. In 1986 the size of the ozone hole grew larger than the size of the Antarctic continent, and in 1993 the hole was larger than all of North America.

During an intensive Antarctic field program in 1987, extremely high levels of chlorine monoxide (ClO) were found in the stratosphere. This finding was seen by many scientists as evidence that the cause of ozone depletion was chlorofluorocarbons (CFCs), anthropogenic chemicals that make excellent refrigerants, cooling fluids, and cleaning solvents. Since the 1950's, millions of tons of CFCs had been produced in the United States alone. CFC leakage from old refrigerators and automobile air conditioners, combined with a lack of chemical recycling efforts, allowed huge quantities of CFCs to make their way into the atmosphere. Also, one of the CFCs, CFC-11, was used for decades as a propellant in aerosol spray cans until it was banned during the late 1970's.

In the stratosphere, UV radiation breaks down CFCs, causing them to release chlorine (Cl), a gas that readily reacts with ozone. The reaction produces oxygen (O_2) and ClO, which then combine to produce O_2 and Cl. Thus, at the end of these reactions, the chlorine ion is again free to destroy another ozone molecule. It is estimated that a single chlorine ion may reside in the stratosphere for fifty years or longer and destroy hundreds of thousands of ozone molecules.

Most scientists believed the detection of stratospheric ClO was the "smoking gun" that proved that human-made CFCs were the cause of ozone depletion. However, many industrialists and others opposed to what they perceived as overzealous environmental regulations asserted that the chlorine in the atmosphere was a result of natural processes such as volcanic eruptions or sea spray. This theory was finally debunked in 1995 when scientists also found hydrogen fluoride in the stratosphere. Hydrogen fluoride is not produced by any natural process; however, fluoride would be liberated from a CFC molecule when it is broken down by UV radiation. Scientifically, there is no longer any debate that ozone destruction is a direct result of CFCs. Other chlorine-containing compounds have also been implicated in ozone destruction, as have some bromine-containing compounds. While bromine is less abundant in the stratosphere than chlorine, atom for atom it is more effective than chlorine in destroying ozone.

GLOBAL OZONE DEPLETION

Since the initial satellite observations of the 1970's and 1980's, the understanding of the complex science of the formation of the Antarctic ozone hole has greatly improved. Most ozone-destroying compounds are released in the Northern Hemisphere; however, they mix throughout the lower atmosphere in about one year and then mix into the stratosphere in two to five years. A key element that enhances ozone depletion is the presence of polar stratospheric clouds. These clouds form only in the polar regions where temperatures in the stratosphere drop to below −80 degrees Celsius (−112 degrees Fahrenheit). The surfaces of these clouds are sites on which inactive chlorine and bromine are converted into the reactive forms that destroy ozone. The dark Antarctic stratosphere becomes so frigid during winter that these clouds are produced in abundance. As spring begins, the stratosphere is hit with solar UV radiation, accelerating the ozone depletion process. A strong vortex of winds circles the South Pole every winter, keeping the Antarctic stratosphere isolated from neighboring air masses. This allows the ozone-destroying reactions to act on a limited amount of ozone that cannot be replenished by ozone from other latitudes. The result is the appearance of the ozone hole each spring. In late spring the winds weaken, allowing comparatively ozone-rich air from other latitudes to mix into the Antarctic atmosphere.

Ozone concentrations declined globally during the second half of the twentieth century as a result of rising levels of atmospheric chlorine and bromine. It was discovered that an ozone hole also forms over the Arctic in the Northern Hemisphere during the spring, although the hole is smaller in size and higher in ozone concentrations than its Antarctic counterpart. The main reason for the difference between the Arctic and Antarctic is that the winds encircling the Arctic are typically not as strong as those in the Antarctic, thus allowing ozone-rich air to mix constantly into the Arctic. Still, significant ozone depletion occurred throughout the 1990's as Arctic spring concentrations dropped about 30 percent below average levels.

Ozone levels in other parts of the world also declined. Ground stations measured decreasing stratospheric ozone levels at midlatitudes. Satellites indicated a negative trend in ozone concentrations in both hemispheres, with a net decrease of about 3 percent per decade. The depletion increased with latitude and was somewhat larger in the Southern Hemisphere. Over the United States, Europe, and Australia, 4 percent per decade was typical. A new threat to ozone followed the eruption of Mount Pinatubo in the Philippines in 1991, the century's second most violent volcanic eruption and the largest one to occur since ozone monitoring began. Massive amounts of sulfur aerosols were injected into the stratosphere. Volcanic aerosols function like the ice crystals in the Antarctic stratosphere by helping convert chlorine and bromine from their inactive state to a reactive one. During the two years following the eruption, record low levels of annual ozone concentrations were recorded: 9 percent below normal between 30 and 60 degrees north latitude.

As expected, measurements also indicated increased UV radiation reaching the earth's surface. During the 1990's, summer levels of UVB increased 7 percent annually over Canada, while winter levels increased 5 percent per year. At the same time, the incidence of skin cancer grew faster than that of any other form of cancer, with reported cases doubling between 1980 and 1998.

INTERNATIONAL RESPONSE

As evidence grew during the 1980's that CFC production was responsible for ozone depletion, the international community decided to act through a series of treaties and amendments. The multilateral Vienna Convention for the Protection of the Ozone Layer of 1985 outlined the responsibilities of states to prevent human-driven ozone depletion. The Vienna Convention laid the groundwork for the Montreal Protocol on Substances That Deplete the Ozone Layer, which was adopted in 1987. The Montreal Protocol put into place a framework for the phaseout of CFC production in developed countries beginning in 1993. In the following years, as ozone levels kept falling, amendments added more chemicals to the phaseout, accelerated the phaseout timetable, and called for the developing world's participation. By 2010 the Montreal Protocol had been ratified by 196 countries, although not all of them had ratified the subsequent amendments. Under the amended protocol, hydrobromofluorocarbons, CFCs, halons, and carbon tetrachloride have already been phased out; methyl chloroform and methyl bromide have been phased out in developed countries and are scheduled for phaseout in developing countries by 2015; and hydrochlorofluorocarbons are to be phased out by 2020 in developed countries and 2040 in developing countries.

Since the beginning of the twenty-first century, concentrations of ozone-depleting substances in the stratosphere have been on the decline as a result of the Montreal Protocol. These substances have such a long residence time in the atmosphere, however, that even with a ban on production, their concentrations in the stratosphere are not expected to return to pre-1980 levels until 2050 at the earliest. According to the U.S. Environmental Protection Agency (EPA), Antarctic ozone is projected to return to pre-1980 levels between 2060 and 2075. Factors including volcanic eruptions, solar activity, and changes in atmospheric temperature, composition, and air motion could affect the ozone layer's rate of recovery.

Some studies have predicted that maximum ozone depletion will occur between 2000 and 2020, followed by a slow recovery. UV radiation levels at the earth's surface are also expected to increase during the early part of the twenty-first century, bringing with them negative health and ecological effects. The largest ozone hole ever recorded occurred in September, 2006. However, according to an EPA report, over most of the world the ozone layer remained relatively stable between 1998 and 2007.

The story of the ozone layer contains both positive and negative lessons regarding human interaction with the environment. Human production of CFCs and other compounds during the latter half of the twentieth century caused significant, long-term harm to the earth's protective barrier against deadly UV radiation. Human health has been affected, and immeasurable damage has been inflicted on many ecosystems. With the earth in the balance, however, the international community decided to act, and did so forcefully. Without the Montreal Protocol, the fate of the ozone layer and life on earth would have been sealed. It is known that human activity has the power to destroy the global environment, but perhaps human activity may have the power to save it as well.

Craig S. Gilman
Updated by Karen N. Kähler

FURTHER READING

Bakker, Sem H., ed. *Ozone Depletion, Chemistry, and Impacts.* Hauppauge, N.Y.: Nova Science, 2009.

Cagin, Seth, and Philip Dray. *Between Earth and Sky: How CFCs Changed Our World and Endangered the Ozone Layer.* New York: Pantheon Books, 1993.

Joesten, Melvin D., John L. Hogg, and Mary E. Castellion. "Chlorofluorocarbons and the Ozone Layer." In *The World of Chemistry: Essentials.* 4th ed. Belmont, Calif.: Thomson Brooks/Cole, 2007.

Mackenzie, Rob. "Stratospheric Chemistry and Ozone Depletion." In *Atmospheric Science for Environmental Scientists,* edited by C. Nicholas Hewitt and Andrea V. Jackson. New York: Wiley-Blackwell, 2009.

Parson, Edward A. *Protecting the Ozone Layer: Science and Strategy.* New York: Oxford University Press, 2003.

Somerville, Richard C. J. *The Forgiving Air: Understanding Environmental Change.* 2d ed. Boston: American Meteorological Society, 2008.

Turco, Richard P. *Earth Under Siege: From Air Pollution to Global Change.* 2d ed. New York: Oxford University Press, 2002.

United Nations Environment Programme. *Environmental Effects of Ozone Depletion and the Interaction with Climate Change: 2006 Assessment.* Nairobi: Author, 2006.

SEE ALSO: Air-conditioning; Air-pollution policy; Chlorofluorocarbons; Freon; Molina, Mario; Montreal Protocol; Rowland, Frank Sherwood.

Pacific Islands

CATEGORIES: Places; water and water pollution; resources and resource management

IDENTIFICATION: The several thousand islands of Melanesia, Micronesia, and Polynesia scattered throughout the tropical Pacific

SIGNIFICANCE: The environmental challenges in the Pacific Islands region are diverse. They include depletion of nearshore fisheries, pollution of limited freshwater resources, soil degradation, reduction of biodiversity, and waste management problems. These environmental challenges are complicated by such factors as the islands' limited natural resources and their geographic isolation, as well as by the lack of monetary resources to address the problems.

The Pacific Islands region comprises nearly thirty thousand islands scattered over 30 million square kilometers (11.6 million square miles). Approximately one thousand of the islands are inhabited. The region is divided into the three island groups of Melanesia, Micronesia, and Polynesia, which reflect the cultures and ethnic features of the indigenous inhabitants of the islands. The main countries and territories in the region are American Samoa, Cook Islands, Federated States of Micronesia, Fiji Islands, French Polynesia, Guam, Kiribati, Marshall Islands, Nauru, New Caledonia, Niue, Northern Mariana Islands, Palau, Papua New Guinea, Pitcairn Islands, Samoa, Solomon Islands, Tokelau, Tonga, Tuvalu, Vanuatu, and Wallis and Futuna Islands.

The Pacific Islands are also divided into low islands and high islands. The low islands consist mainly of coral reefs and atolls, most of which are only a few meters above sea level; these include the Marshall, Phoenix (Kiribati), Tuamotu (French Polynesia), and Tuvalu groups. The high islands are hilly and some mountainous; these include New Britain, New Caledonia, Papua New Guinea, Fiji, the Marianas, Samoa, the Solomon Islands, and Vanuatu. The total population of the Pacific Islands region in 2008 was about 9.5 million. Population is growing by approximately 2 percent annually and is expected to exceed

15 million by 2030 if this rate continues. The islands with the fastest-growing populations, owing to high birthrates, are the Solomon Islands (2.7 percent) and Vanuatu (2.6 percent). Those islands with decreasing populations, owing to emigration, are Niue (−2.4 percent) and the Northern Mariana Islands (−1.7 percent).

The environmental challenges of the Pacific Islands region are diverse and sundry. They include depletion of nearshore fisheries, pollution of fresh water, soil degradation, urbanization, reduction of biodiversity, damage to nearshore nursery habitats, waste management problems (involving solid, nuclear, and chemical wastes), and stressed natural resources related to tourism. The most serious environmental challenges facing the small island developing states are complicated by traditional approaches to land management, limited natural resources, small and fragile ecosystems, geographic isolation, and poverty (thus a lack of adequate capacity for response). The Pacific Islands region can be divided into three zones, each with its distinguishing environmental problems: the low island states, the small and midsize high islands, and the larger high islands of the western Pacific. All countries in the region share some environmental challenges, however.

LOW ISLAND STATES

The small, low-lying coral islands of the Pacific region (Cook Islands, Kiribati, Tuvalu, Federated States of Micronesia, the Marshall Islands, Niue, Nauru) have extremely limited resources and are economically deprived; thus their capacity to respond to environmental problems is limited. The most serious environmental issues facing most of these countries are the pollution of limited groundwater with sewage and salt, problems with solid waste disposal, lack of land available for agriculture, and rapid population growth. For example, more than 65,000 Marshall Islanders live on 180 square kilometers (69 square miles) of atoll land, which provides each person only 0.3 hectare (0.7 acre) of land. On the other hand, each person has economic sovereignty over 40 square kilometers (15 square miles) of ocean. The population of the Marshall Islands is expected to double by 2025.

The low island states constitute one of the most vulnerable places on earth should sea level accelerate due to global warming. The island group of Tuvalu is often used as an example of this projected problem. However, since instrumentation was installed in 1993 to monitor sea level on Tuvalu's main island, Funafuti, no discernible changes have been seen. Some inundation is evident on the island, but it is the result of erosion, sand mining, and construction projects causing an inflow of seawater. Other factors are also involved, such as excessive use of fresh water for irrigation, a consequence of which is seawater encroachment into vegetable growing pits. Part of the problem is related also to the fact that about one-fourth of the island has been paved over with roads and airport runways. This reduces the infiltration of rainwater into the freshwater layer. When this increased runoff is combined with a high tide, flooding along the coast gives the appearance that the sea level is rising. Perception of trends can also be affected, as increasing population on the islands means people are living on flood-prone land that was previously avoided.

Coral is capable of growing along with sea-level rise, and the atolls are not static. The islands grow as they are replenished by coral that breaks off the reefs and is thrown ashore by storms. In this way atolls are self-maintaining, provided humans do not intervene,

such as by digging coral for use in construction work and building flush toilets that discharge effluent into the sea and affect coral growth.

SMALL AND MIDSIZE HIGH ISLANDS

The people of the small and midsize high islands of the Pacific (Tonga, Samoa, French Polynesia, Palau, Federated States of Micronesia, Guam, American Samoa, and Northern Mariana Islands) are mainly agrarian. The islands have small or no commercial forests and no commercial mineral deposits. The main environmental issues faced by these islands are shortage of land, loss of the surviving native forests (with associated loss in biodiversity), invasion of exotic animal and plant species, decline of coastal fisheries, coral reef degradation, problems with solid waste disposal, and contamination of groundwater and coastal areas by agricultural chemicals and sewage.

Some of the small and midsize high islands are more fortunate than others; Tonga and Samoa, for example, have nearly self-sufficient food supplies and receive high levels of remittances from expatriate island communities living abroad. French Polynesia is a French territory, and Guam, the Northern Mariana Islands, and American Samoa are U.S. territories; all of these nations have high standards of living based on subsidies.

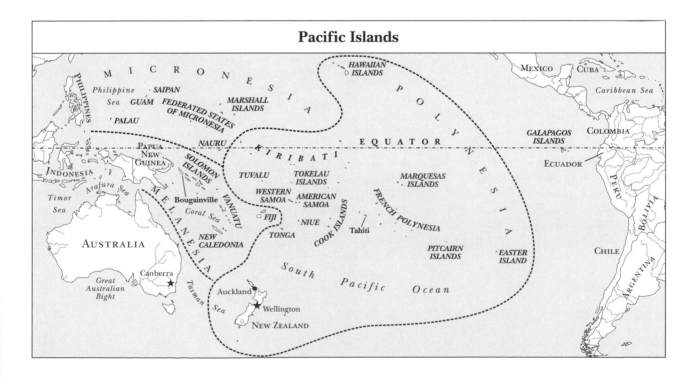

Pacific Islands

LARGER HIGH ISLANDS OF THE WESTERN PACIFIC

The larger high islands of the western Pacific (Papua New Guinea, Solomon Islands, New Caledonia, Vanuatu, and Fiji Islands) have relatively large human populations and are comparatively rich in mineral and forestry resources. Environmental pressures on these islands are linked for the most part to rapid population growth and to complications associated with traditional approaches to land management that have led to land degradation. Other common problems arise from unsustainable deforestation, the depletion of nearshore fisheries, the pollution of rivers and lakes caused by mining and agricultural practices, and the invasion of exotic species.

REGIONALLY SHARED ENVIRONMENTAL PROBLEMS

Coastal and marine problems are among the most common kinds of environmental issues in the Pacific Islands region. The main concerns are coastal erosion; depletion and pollution of mangrove forests, sea grasses, and coral reefs; depletion of shallow-water and coastal marine life; and unsustainable management of offshore fishery resources, including destruction by drift-net fishing, commercial bycatch of seabirds and marine mammals, and whaling.

Certain natural hazards are also common among the islands of the Pacific. The region lies on the western perimeter of the Pacific Rim of Fire, an area of severe seismic activity extending from the Northern Mariana Islands in the north to Vanuatu in the south. Earthquakes, volcanic activity, and tsunamis are persistent major threats, and tropical cyclones, floods, and drought are not infrequent. These extremes have serious environmental consequences when combined with unsustainable land-use practices.

Another problem common to many islands in the region, especially atolls, is the pollution of the limited supplies of fresh water available. With increasing population, accompanied by increasing construction, agriculture, and tourism, islands' water supplies have become contaminated owing to agricultural runoff and inadequate sewage management systems. Tourists place additional stress on local ecosystems and place big demands on water supplies and on waste disposal systems.

Declining biodiversity has been seen on many islands in the region. Many unique species of plants and animals evolved in isolation in the Pacific Islands region, and the specialized habitats to which they adapted are vulnerable to destruction by deforestation, land clearance, fire, agricultural chemicals, and nonnative organisms introduced by visitors to the islands.

The extraction, processing, and transport of mineral resources have caused localized environmental damage on some islands. The major mining centers are found in Papua New Guinea, New Caledonia, and Fiji, where regulations intended to minimize damage from mine tailings, processing fumes, and siltation of streams have had varying degrees of success. Mining for construction material is widespread throughout the Pacific Islands region and is a problem that increases in step with population growth. Small islands suffer the most. The removal of sand (for concrete) from beaches causes coastal erosion, and the dredging of coral reefs and lagoon sands, along with the use of corals for building material, often causes irreversible damage.

C. R. de Freitas

FURTHER READING

Banks, Glenn. "Mining and the Environment in Melanesia: Contemporary Debates Reviewed. " *Contemporary Pacific* 14, no.1 (2002): 39-67.

Lal, Brij V., and Kate Fortune, eds. *The Pacific Islands: An Encyclopedia.* Honolulu: University of Hawaii Press, 2000.

Nunn, Patrick D. Climate, Environment, and Society in the Pacific During the Last Millennium. Oxford, England: Elsevier, 2007.

Rapoport, Moshe. *The Pacific Islands: Environment and Society.* Hong Kong: Bess Press, 1999.

SEE ALSO: Alliance of Small Island States; Biodiversity; Fisheries; Habitat destruction; Introduced and exotic species; Runoff, agricultural; Sea-level changes; Water pollution.

Paclitaxel

CATEGORIES: Human health and the environment; forests and plants

IDENTIFICATION: A potent cancer-fighting drug

SIGNIFICANCE: The discovery of the cancer-fighting properties of paclitaxel illustrated the value for humankind of protecting and nurturing biodiversity, given that many currently unknown species of plants, animals, and insects could be the sources of more such substances. It appeared initially that demand for paclitaxel could pose a threat to the existence of the drug's natural source, the Pacific yew, but alternative sources were found.

Paclitaxel was originally derived from the bark of the Pacific yew tree (*Taxus brevifolia Nutt*), a small-to medium-sized understory tree that occupies Pacific coastal forests from southwestern Alaska to California. The development of paclitaxel, which was originally called taxol, as a drug began in 1962 with the collection in Washington State of a sample of the reddish-purple bark of the Pacific yew tree by a technician working for the National Cancer Institute (NCI). The NCI was employing a "shotgun" approach to cancer research: A wide variety of plant parts of various species were being screened for anticancer properties. Thereafter, several scientists, including Monroe Wall and M. C. Wani at Research Triangle Institute in North Carolina and Susan Horwitz and Peter Schiff of the Albert Einstein College of Medicine in New York, recognized the potential of paclitaxel as an anticancer drug and became intensely interested.

After years of delay, the pharmaceutical company Bristol-Myers Squibb continued testing and production of paclitaxel, but on a larger scale; it eventually marketed the drug under the trade name Taxol. Paclitaxel is able to arrest the growth of cancer cells by attaching to their microtubules, thus preventing cell division. By the late 1980's paclitaxel, despite its high cost, had become the drug of choice for the treatment of a wide range of cancers, but especially ovarian and breast cancer.

In spite of paclitaxel's prominence as a success story in the "herbal renaissance" of the twentieth century, several problems involved in production and use of the drug persisted. For one, the cost of paclitaxel treatment was prohibitive for many who desperately needed it; this remained a problem until generic competitors to Bristol-Myers Squibb's Taxol became available in 2002, when the price of a single dose went down from about $1,000 to $150. Another problem was the large amount of yew bark required to produce the drug—all the bark from a one-century-old tree yields only enough paclitaxel for a 300-milligram dose. This raised fears among environmentalists that continued harvesting could threaten the species. Although the Pacific yew occurs over a wide area, the trees exist only in relatively small numbers. Furthermore, it is a slow-growing species that rarely reaches a height of more than 18 meters (60 feet); stripping the bark kills the tree.

Several means of producing paclitaxel without destroying wild Pacific yews were proposed, and eventually biotechnology techniques enabled the extraction of paxlitaxel from the cell lines of other trees of the *Taxus* species. This method improved on an earlier semisynthetic method that involved the use of hazardous chemicals. Pacific yew tree bark is no longer harvested to produce paclitaxel.

Thomas E. Hemmerly

FURTHER READING

Goodman, Jordan, and Vivien Walsh. *The Story of Taxol: Nature and Politics in the Pursuit of an Anti-cancer Drug.* New York: Cambridge University Press, 2001.

Pierce, Jessica, and Andrew Jameton. *The Ethics of Environmentally Responsible Health Care.* New York: Oxford University Press, 2004.

SEE ALSO: Biodiversity; Coniferous forests; Logging and clear-cutting.

Pandemics

CATEGORY: Human health and the environment

DEFINITION: Outbreaks of infectious diseases that affect large numbers of people over large geographic areas

SIGNIFICANCE: Pandemics have long affected human populations around the world, often killing substantial numbers of people, weakening many others, and causing economic distress. Owing to advances in medicine and disease research, some infectious diseases have become less likely to pose pandemic threats, whereas improvements in transportation and increased global travel have increased the likelihood that other diseases may become pandemics.

The spreads of several infectious diseases—most notably plague, cholera, influenza, typhus, smallpox, malaria, and yellow fever—have reached pandemic levels in the past. During the early twenty-first century acquired immunodeficiency syndrome (AIDS), which is caused by the human immunodeficiency virus (HIV), reached pandemic status, with large numbers of cases appearing in all regions of the world and more than 25 million deaths recorded. In 1918-1919 influenza reached pandemic proportions and is estimated to have killed between 20 million and 40 million people worldwide. In 2009 influenza again reached pandemic proportions, although the death rate was much lower than in the earlier pandemic.

SPREAD OF INFECTIOUS DISEASES

Infectious diseases are either bacterial or viral in nature and spread in different ways. Some, such as smallpox (a virus), are spread through human contact or through material that has come in contact with an infected person. Other diseases, such as influenza (a virus), spread through the air as infected persons cough or sneeze. Bubonic plague (a bacterial disease) is spread to humans through the bites of fleas that have previously bitten infected animals such as rats. (One type of plague—pneumonic plague—is spread through the air from one person to another.) Cholera (a bacterial disease) is spread through drinking water that contains the bacteria *Vibrio cholerae*. The complicated cycle for the spread of malaria (a protozoal disease) involves an infected mosquito biting a person, as is also the case with yellow fever. HIV/AIDS is a viral disease that spreads through contact with bodily fluids of an infected person.

When the global population was small and disaggregated it was difficult for infectious diseases to spread because an infected host might never come in contact with a person susceptible to the disease. The growth of urban areas and improved transportation made it possible to spread infectious diseases readily. If a population has no existing immunity or previous exposure to an infectious disease, a so-called virgin-soil epidemic may result if the population is exposed to the disease. When this occurs, the disease may have a high mortality rate and reach pandemic proportions.

Environmental disturbances such as land clearance, construction of lakes, and climate change may affect the spread of infectious diseases by making it easier for microbes to find hosts. Environmental changes may also expose human populations to new diseases by causing mutations in diseases that enable them to spread from animal hosts to humans, as scientists believe occurred with HIV and some of the hemorrhagic fevers.

As medical knowledge improved during the late nineteenth and early twentieth centuries, it became possible for human beings to prevent the spread of many infectious diseases and to treat and cure those who had been infected. Scientists first identify the agent of a disease, such as mosquitoes for yellow fever and malaria, and then look for ways to control the agent, such as draining standing water to decrease the number of mosquitoes. In some cases it is possible to confer immunity to a disease through vaccination, as occurred with smallpox, which has been eradicated as a disease outside laboratory settings. Other diseases, such as cholera, can be treated with massive rehydration. Malaria can be prevented with the use of chloroquine. A vaccine exists for bubonic plague, and cases can be treated with large doses of antibiotics.

Since the late twentieth century, drug therapy has been successful in preventing many pandemics. Infectious diseases still occur, but in many cases it is possible for medical personnel to cure such diseases readily and prevent their further spread. However, drug-resistant strains of several diseases, such as malaria and plague, have developed, so that potential for new pandemics exists. HIV/AIDS can be controlled with drug cocktails composed of several drugs, but this viral disease mutates rapidly, so eradication is unlikely. Influenza vaccines exist and are administered every year, but the virus mutates rapidly, so that new vaccines are necessary for every flu season. In some parts of the world, such as Africa, drug therapies are too expensive for most people; this makes diseases such as HIV/AIDS and cholera difficult to combat, enabling them to reach pandemic proportions. Wars and natural disasters can also produce conditions that encourage the spread of diseases such as cholera.

HISTORICAL OVERVIEW

Smallpox (caused by the virus *Variola major*) is one of the oldest infectious diseases; it is known to have existed in antiquity. During the second century B.C.E. what is known as the Antonine Plague saw smallpox spread throughout the Italian peninsula from the Near East. The third century Plague of Cyprian, which may have also been smallpox, led to massive numbers of deaths in Roman Italy. During the American Revolution smallpox broke out in Boston in 1776 and gradually spread across the continent, nearly decimating several of the Native American tribes of the Great Plains.

Plague (*Yersinia pestis*) is often considered the classic pandemic disease. The first plague pandemic, the Plague of Justinian, in the sixth century, started in Egypt, and from 541 to 750, bubonic plague spread across the Western world, causing Europe's population to drop by as much as 50 percent. From an origin possibly in Mongolia, plague made another appearance in Europe from 1348 to 1350 and then continued to reappear until the early eighteenth century. The first outbreak, which involved both bubonic and pneumonic plague, killed between 30 and 60 percent of the population in Europe. Combined with other environmental factors, this mortality rate led to a decline in cultivation in some areas and substantially reduced human impacts on the environment.

The third plague pandemic started in western China during the late eighteenth century, spread to coastal cities and India by the late nineteenth century, and reached the United States and South America during the early twentieth century. Although the third plague pandemic did not produce as high a mortality rate as the first two pandemics, it still killed hundreds of thousands of people in China and India. In the early twenty-first century, plague remains endemic in some areas, such as Mongolia, Madagascar, and the American Southwest, raising the possibility of another major outbreak.

Cholera first appeared as a pandemic disease in the early nineteenth century. The disease had apparently existed in India for some time, but it did not reach the level of a pandemic until 1816-1826, when it spread to China, Indonesia, and the Caspian Sea region. Since that time six more cholera pandemics have been re-

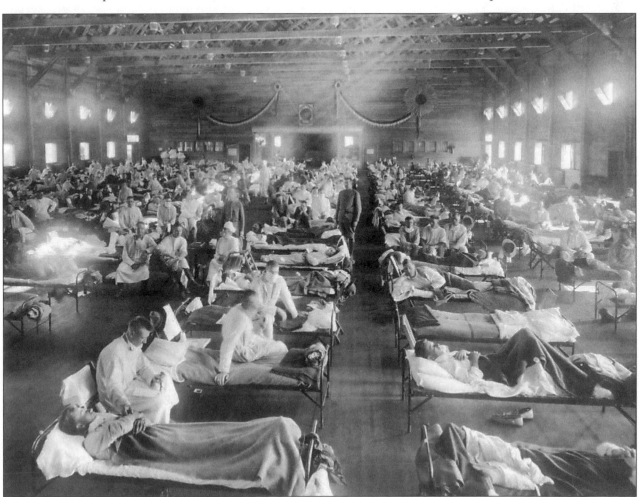

Victims of the influenza epidemic of 1918-1919 fill an emergency hospital facility at Fort Riley, Kansas. (AP/Wide World Photos)

corded. The disease reached England and the United States during the second pandemic (1829-1851). The third pandemic (1852-1860) mainly affected Russia, Japan, and China, and the fourth (1863-1875) affected Europe and Africa. The fifth cholera pandemic (1881-1896) was worldwide, causing at least 250,000 deaths in Europe, 50,000 deaths in the Americas, nearly 300,000 deaths in Russia, and 90,000 deaths in Japan. Advances in public health prevented the sixth pandemic (1899-1923) from having much of an impact on Europe or the United States, but as many as 800,000 people died in India and more than 200,000 in the Philippines. The seventh cholera pandemic, which began in 1962 and continues in the early twenty-first century, introduced a new cholera strain, known as El Tor; this pandemic started in Indonesia and spread to India, parts of the former Soviet Union, Africa, and parts of South America.

It is difficult to identify past influenza pandemics because the symptoms of influenza are similar to those of many other diseases. An influenza pandemic may have affected most of Europe during the late sixteenth century. A major pandemic that may have caused as many as one million deaths spread worldwide in 1889-1890. Some people who contracted the disease and survived had some immunity against influenza during the 1918-1919 pandemic, which appears to have originated in the United States before spreading worldwide, producing particularly high mortality rates in Samoa and Alaska. More than 600,000 people died from the disease in the United States, and it has been estimated that 17 million people may have died from the disease in India.

Influenza reached pandemic proportions again in 1957-1958 and 1968-1969, although it did not produce the large number of deaths that occurred earlier, largely because of improved medical care. Each of these pandemics was caused by a different strain of the influenza virus, which mutates rapidly, making the development of preventive vaccinations difficult. During the early twenty-first century, a new strain of influenza, the H1N1 strain, appeared in China and by 2009 had spread worldwide. Because of improvements in medical care and preventive measures, influenza during the early twenty-first century does not pose the threat of earlier pandemics. Some researchers, however, have voiced concerns that a new "superstrain" of influenza could occur that will be difficult to treat and will cause a high mortality rate.

HIV, the virus that causes AIDS, may have jumped species from monkeys in Africa sometime during the early twentieth century. By the early 1970's AIDS was spreading across the globe, and by 2010 the HIV/AIDS infection rate in parts of sub-Saharan Africa was as high as 25 percent. Some southern and eastern African countries have seen declines in population owing to the high mortality rate produced by AIDS as well as related issues. Although HIV/AIDS can be controlled somewhat through drug therapy, this option is very expensive and thus unavailable to most infected people in Africa and Asia. Researchers have estimated that 90 million to 100 million people will die of AIDS in Africa by 2025; projected AIDS deaths in India and China by 2025 are 31 million and 18 million, respectively.

THE ENVIRONMENT AND PANDEMIC DISEASE

Changes in how human beings relate to the environment have had impacts on the creation and spread of diseases. As humans venture into new environments they encounter new diseases, some of which migrate from other species. In Africa human-made large lakes helped to spread malaria by providing breeding grounds for the mosquitoes that carry the disease, just as controlling standing water all but eliminated the threat of malaria and yellow fever epidemics in the Western world. Climate change, too, may, have an impact on the spread of infectious diseases, as warmer temperatures may make it possible for some tropical diseases to spread to new surroundings.

The high death tolls of some pandemics have had impacts on the environment, although it is difficult to trace these impacts clearly. The high death rate among Native Americans produced by contact with European diseases such as smallpox and even measles seems to have depopulated some areas in the Americas in the sixteenth century. The massive death rates produced by the Black Death (possibly a bubonic plague pandemic) in the fourteenth century decreased the impacts of human activities on the environment by a substantial amount. Other pandemics, such as cholera and at times malaria and yellow fever pandemics, have reduced human populations enough to have produced some environmental impacts. The extremely high number of deaths produced by the influenza pandemic of 1918-1919 had little impact on the environment except in isolated regions because of the large size of the world's population by the early twentieth century.

John M. Theilmann

FURTHER READING

Barnes, Ethne. *Diseases and Human Evolution.* Albuquerque: University of New Mexico Press, 2005.

Crawford, Dorothy H. *Deadly Companions: How Microbes Shaped Our History.* New York: Oxford University Press, 2007.

Crosby, Alfred W. *America's Forgotten Pandemic: The Influenza of 1918.* 2d ed. New York: Cambridge University Press, 2003.

Engel, Jonathan. *The Epidemic: A Global History of AIDS.* Washington, D.C.: Smithsonian Institution, 2006.

Fenn, Elizabeth A. *Pox Americana: The Great Smallpox Epidemic of 1775-82.* New York: Hill & Wang, 2001.

McMichael, Tony. *Human Frontiers, Environments, and Disease: Past Patterns, Uncertain Futures.* New York: Cambridge University Press, 2001.

Orent, Wendy. *Plague: The Mysterious Past and Terrifying Future of the World's Most Dangerous Disease.* New York: Free Press, 2004.

Theilmann, John M., and Frances Cate. "A Plague of Plagues: The Problem of Plague Diagnosis in Medieval England." *Journal of Interdisciplinary History* 38 (2007): 371-393.

SEE ALSO: Antibiotics; Bacterial resistance; Centers for Disease Control and Prevention; Climate change and human health; Globalization; Smallpox eradication; World Health Organization.

Particulate matter. *See* **Airborne particulates**

Passenger pigeon extinction

CATEGORY: Animals and endangered species

IDENTIFICATION: Extinction of a species of migratory pigeon that was native to North America

SIGNIFICANCE: The extinction of the passenger pigeon, a bird species that was once abundant in North America, illustrates how quickly even very large populations of animals can be wiped out by human activity.

During the sixteenth century, giant flocks of passenger pigeons (*Ectopistes migratorius*) flew freely across the North American continent. This pattern

---

Eulogy for the Passenger Pigeon

The famous American preservationist Aldo Leopold offered these thoughts on the extinction of the passenger pigeon in his essay "A Monument to the Pigeon," from A Sand County Almanac *(1949):*

Men still live who, in their youth remember pigeons; trees still live that, in their youth, were shaken by a living wind. But a few decades hence only the oldest oaks will remember, and at long last only the hills will know.

We grieve because no living man will ever see again the onrush of victorious birds, sweeping a path for spring across the March skies, chasing the defeated winter from all of the woods and prairies.

There will always be pigeons in books and in museums but they are dead to all hardships and to all delights. They cannot dive out of a cloud, nor clap their wings in thunderous applause. They know no urge of seasons; they feel no kiss of sun, no lash of wind and weather, they live forever by not living at all.

---

changed rapidly with increased settlement of the continent by Europeans. Even at the start of the nineteenth century, billions of passenger pigeons existed in North America. One hundred years later, however, only one was left. The extinction of the passenger pigeon was caused entirely by human beings. Hunters killed millions of birds each year, but, more important, they disrupted the passenger pigeon's breeding cycle.

Passenger pigeons were particularly vulnerable because they nested en masse; flocks of pigeons would nest so closely together in trees that a hunter could kill more than one hundred birds with a single shotgun blast. In fact, passenger pigeons were so easy to shoot that they were not considered a game bird until the mid-nineteenth century. Hunters achieved especially high yields by using nets baited with "stool pigeons," live passenger pigeons with their eyelids sewn shut. The blinded birds were placed on stools in front of nets in order to lure other pigeons in; a single net could capture more than one thousand birds.

Adult pigeons abandoned the new generation of young pigeons, or squabs, before they could fly, leaving them to wander on the ground. Much fatter than the adult birds, squabs were delicious to eat. Hunters dislodged squabs in nests by setting fire to the bark of trees at the bases, causing the birds to leap to the ground. One observer recalled that the "squabs were so fat and clumsy that they would burst open on striking the ground." The senseless slaughter of squabs sealed the fate of passenger pigeons as a species.

No legislation was created to protect the passenger pigeon. In fact, the majority of laws precluded shooting in the vicinity of nesting areas so as not to interfere with the hunters' nets. It is believed that the last wild passenger pigeon was killed on March 24, 1900. One month later, while arguing for a bill to prevent interstate trade of wild birds, Iowa congressman John Fletcher Lacey noted that the wild pigeon "has entirely disappeared from the face of the earth." Passage of the Lacey Act in May, 1900, was a significant step toward protection of birds and a major precursor to the Endangered Species Act of 1970. Like the buffalo, passenger pigeons were particularly vulnerable because of their great numbers. Similarly, the expansion of railroads and influx of commercial hunters created a short-lived industry that rapidly exterminated millions of birds. Sadly, however, not enough passenger pigeons were preserved in zoos to guarantee survival of the species. The world's last passenger pigeon died on September 1, 1914, at a zoo in Cincinnati.

Peter Neushul

FURTHER READING

Fuller, Errol. *Extinct Birds*. Rev. ed. Ithaca, N.Y.: Cornell University Press, 2001.

Schorger, A. W. *The Passenger Pigeon: Its Natural History and Extinction*. 1955. Reprint. Caldwell, N.J.: Blackburn Press, 2004.

SEE ALSO: Dodo bird extinction; Endangered species and species protection policy; Extinctions and species loss; Hunting; Migratory Bird Act; Whooping cranes.

PCBs. *See* Polychlorinated biphenyls

PEMEX oil well leak

CATEGORIES: Disasters; water and water pollution

THE EVENT: Blowout in an oil drilling well in the Gulf of Mexico that resulted in the spilling of some three million barrels of oil into the sea

DATE: June 3, 1979

SIGNIFICANCE: The oil pollution that resulted from the PEMEX oil well blowout and the nine-month-long spillage of oil that followed destroyed Mexico's shrimping industry and threatened other marine life as well as the tourist industry of coastal areas in the Gulf of Mexico.

On June 3, 1979, an offshore oil operation in the Gulf of Mexico drilled into a high-pressure gas pocket, causing a blowout that ignited the oil well. The accident produced a massive oil slick nearly 65,000 square kilometers (25,000 square miles) in area. According to a July, 1979, issue of the *Christian Science Monitor*, at the time it was considered not only "the world's worst oil spill from an oil well, but also the worst spill ever." The runaway well, Ixtoc I, belonged to Petróleos Mexicanos (PEMEX), the Mexican state-owned oil company. The exploratory well was being drilled in the Gulf of Campeche off the Yucatán Peninsula, a low-lying limestone tableland that separates the Caribbean Sea from the Gulf of Mexico.

Natural gas escaping from the well ignited and destroyed the $22 million drilling rig. The crude oil that issued from the damaged well polluted waters in the region and ruined much of Mexico's shrimping industry. The toxic effects of the crude oil also threatened the population of Atlantic sea turtles that nest along the Mexican coast, as the young of this endangered species normally swim out to seaweed beds after they hatch.

The oil slick further posed a threat to certain regions along the Texas coast, including the Padre Island National Seashore area and resort beaches such as Galveston, Texas. As a preventive measure, floating booms and other types of barriers were placed as shields for bays and estuaries along the south Texas coast by the Department of Texas Water Resources. The south Texas beaches escaped extensive damage; only isolated patches of diluted oil reached the Texas coastline. Some Texas beaches were littered with nontoxic, pancakelike hydrocarbon globs that resembled chocolate mousse.

Immediately after the blowout, PEMEX began

drilling relief wells, named Ixtoc I-A and I-B, in an attempt to intersect the runaway well and divert the escaping oil. Later, the famous oil-field troubleshooter Paul "Red" Adair and his crew were called in to cap the well. They succeeded in temporarily shutting the well down in late June, but it burst out again shortly thereafter. Over the next several months steel balls, lead balls, and gelatin were pumped down the well in repeated attempts to check or stop the oil flow. Although most of the balls that were first pumped in were expelled, a later injection of 108,000 balls reportedly reduced the flow of oil from 20,000 to 10,000 barrels per day on October 12. Meanwhile, several different organizations, including Oil Mop Incorporated and Shell Oil, were using a variety of equipment in attempts to skim oil from the ocean surface at or near the well site.

Ixtoc I-A and I-B ultimately injected drilling mud, water, and cement into the reservoir rock and base of the well, which cut off flow. The well was finally blocked with three cement plugs on March 22, 1980.

The Ixtoc I and other wells drilled in the Gulf of Campeche are evidence of the presence of large quantities of oil in this offshore region. However, the disastrous blowout of Ixtoc I led to questions about PEMEX's credibility as a company, as well as about Mexico's entire oil policy.

Donald F. Reaser

FURTHER READING

Fingas, Merv. *The Basics of Oil Spill Cleanup.* 2d ed. Boca Raton, Fla.: CRC Press, 2001.

Ornitz, Barbara E., and Michael A. Champ. *Oil Spills First Principles: Prevention and Best Response.* New York: Elsevier, 2002.

SEE ALSO: BP *Deepwater Horizon* oil spill; Monongahela River tank collapse; Oil spills; Santa Barbara oil spill.

People for the Ethical Treatment of Animals

CATEGORIES: Organizations and agencies; activism and advocacy; animals and endangered species

IDENTIFICATION: Organization devoted to establishing and defending the rights of animals

DATE: Established in 1980

SIGNIFICANCE: People for the Ethical Treatment of Animals has grown to be one of the most powerful and effective animal rights organizations in the world, despite the fact that the organization's activities are often the subject of controversy.

Alex Pacheco, who founded People for the Ethical Treatment of Animals (PETA) with Ingrid Ward Newkirk in 1980, received initial notoriety for exposing cruelty to monkeys in a Silver Spring, Maryland, research laboratory. Although PETA began its campaign with only two people during the 1980's, by 2010 it had a worldwide membership of some two million people and annual expenditures of more than thirty million dollars. PETA's brand of radicalism and sophisticated use of the media are often the topics of national debate. During the 1980's PETA's major protest activities were centered on vivisection, factory farming, hunting, fishing, zoos, and circuses. Over time, the organization's concerns widened to include the fur industry and the use of animals in product testing, agricultural production, and biomedical research. PETA's campaign against the fur industry convinced designers such as Giorgio Armani, Ralph Lauren, and Calvin Klein, as well as supermodels such as Cindy Crawford and Christy Turlington, not to design or model clothing made of or incorporating fur.

The general principle upon which PETA was founded is simple but potent: Animals are not on earth for humans to eat, wear, or use for entertainment or experimentation. PETA asserts that advances in technology have enabled humans to make incredibly diverse substitutes for animals in all such uses. In advocating the end of all animal abuse, PETA maintains that violating animal rights is similar to violating human rights. In fact, the former is worse because animals cannot speak for themselves.

One tactic that PETA uses in attempting to stop individuals, organizations, or companies that violate animal rights is the undercover investigation. Many of PETA's investigations have revealed patterns of cru-

In a 2009 protest, members of People for the Ethical Treatment of Animals lie on the steps of Canada House in London, England, wearing masks depicting seal faces and white suits smeared with fake blood, to urge the Canadian prime minister to stop the annual seal hunting in Newfoundland and Labrador. (AP/Wide World Photos)

elty that have appalled the public and resulted in widespread public support for change as well as generous contributions of money to the organization. One of PETA's major goals is public education, which it accomplishes through the use of graphic visual images, expert testimony, media events, seminars, workshops, and lectures. PETA also conducts grassroots activities at colleges and universities.

PETA has been criticized for ecoterrorism and monkeywrenching, examples of which include illegal entry, vandalism, and theft of laboratory animals and equipment. Critics contend that PETA's philosophical stance is based on sentimentalism and narrow-minded dogmatism and that many of the organization's practices are akin to terrorism. They charge that PETA manipulates young people's emotions through misinformation disguised as education. Despite active opposition from industry and research communities, however, PETA has proven to be one of the most effec-

tive animal rights groups in the world. It has gained mainstream support from the public and from many powerful people involved in government, the entertainment industry, and humanitarian organizations.

Chogollah Maroufi

FURTHER READING

Best, Steven, and Anthony J. Nocella II, eds. *Terrorists or Freedom Fighters? Reflections on the Liberation of Animals.* New York: Lantern Books, 2004.

People for the Ethical Treatment of Animals. *All Animals Are Equal: The PETA Guide to Animal Liberation.* Washington, D.C.: Author, 2005.

Singer, Peter. *Animal Liberation.* 1975. Reprint. New York: HarperPerennial, 2009.

SEE ALSO: Animal rights; Animal rights movement; Ecoterrorism; Hunting; Monkeywrenching; Singer, Peter; Zoos.

Pesticides and herbicides

CATEGORY: Pollutants and toxins

DEFINITION: Chemicals designed to kill or inhibit the growth of unwanted organisms

SIGNIFICANCE: Although the use of pesticides, including herbicides, has been beneficial to humankind, enabling increased crop yields and helping to prevent disease, many of the chemicals that have been used to kill pests have had detrimental effects on the environment as well as direct negative effects on human health.

The major types of pesticides in common use are insecticides (to kill insects), nematocides (to kill nematodes), fungicides (to kill fungi), herbicides (to kill weeds), and rodenticides (to kill rodents). Although the use of pesticides has mushroomed since the introduction of monoculture (the agricultural practice of growing only one crop on a large amount of land), the application of chemicals to control pests is by no means new. The use of sulfur as an insecticide dates back before 500 B.C.E. Salts from heavy metals such as arsenic, lead, and mercury were used as insecticides from the fifteenth century until the early part of the twentieth century, and residues of these toxic compounds are still being accumulated in plants that are grown in soil where these materials were used. In the seventeenth and eighteenth centuries, natural plant extracts such as nicotine sulfate from tobacco leaves and rotenone from tropical legumes were used as insecticides. Other natural products, such as pyrethrum from the chrysanthemum flower, garlic oil, lemon oil, and red pepper, have long been used to control insects.

In 1939 the discovery of the utility of dichloro-diphenyl-trichloroethane (DDT) as a strong insecticide opened the door for the development of a wide array of synthetic organic compounds to be used as pesticides. Chlorinated hydrocarbons such as DDT were the first group of synthetic pesticides. Other commonly used chlorinated hydrocarbons include aldrin, endrin, lindane, chlordane, and mirex. Because of the low biodegradability and long persistence in the environment of these compounds, their use was eventually banned or severely restricted in the United States. Organophosphates such as malathion, parathion, and methamidophos replaced the chlorinated hydrocarbons. These compounds biodegrade in a fairly short time but are generally much more toxic to humans and other animals than the compounds they replaced. In addition, they are water-soluble and, therefore, more likely to contaminate water supplies. Carbamates such as carbaryl, maneb, and aldicarb have also been used in place of chlorinated hydrocarbons. These compounds biodegrade rapidly and are less toxic to humans than organophosphates, but they are also less effective in killing insects.

Herbicides, which are used specifically to kill or retard the growth of unwanted plant life, are classified according to the ways in which they work rather than their chemical composition. As their name suggests, contact herbicides such as atrazine and paraquat kill when they come into contact with a plant's leaf surface; these herbicides generally work by disrupting the photosynthetic mechanism. Systemic herbicides such as diuron and fenuron circulate throughout the plant after being absorbed. They generally mimic the plant hormones and cause abnormal growth to the extent that the plant can no longer supply sufficient nutrients to support growth. Soil sterilants such as triflurain, diphenamid, and daiapon kill microorganisms necessary for plant growth and also act as systemic herbicides.

PESTICIDE USE

In the United States, approximately 55,000 different pesticide formulations are available, and Americans apply about 500 million kilograms (1.1 billion pounds) of pesticides each year. Fungicides account for 12 percent of all pesticides used by farmers, insecticides account for 19 percent, and herbicides account for 69 percent. These pesticides are used primarily on four crops: soybeans, wheat, cotton, and corn. By the end of the twentieth century, the annual expenditure on pesticides each year in the United States had reached approximately $5 billion, about 20 percent of which was for nonfarm use. On a per-unit-of-land basis, home owners apply approximately five times as much pesticide to their yards as farmers do to their fields. Worldwide, more than 2.5 tons of pesticides are applied each year. Most of these chemicals are applied in developed countries, but the amount of pesticides being used in developing countries is rapidly increasing. Total annual expenditure on pesticides worldwide exceeds $20 billion, a figure expected to continue increasing as the use of pesticides grows in developing countries.

The use of pesticides has had a beneficial impact on the lives of humans by increasing food production and reducing food costs. Even with pesticides, insects and other pests reduce the world's potential food supply by as much as 55 percent. Without pesticides, the losses would be much higher, resulting in increased starvation and higher food costs. Pesticides also increase the profit margin for farmers. It has been estimated that for every dollar spent on pesticides, farmers experience an increase in yield worth three to five dollars. Pesticides appear to work better and faster than alternative methods of controlling pests. These chemicals can rapidly control most pests, are cost-effective, can be easily shipped and applied, and have a long shelf life in comparison with alternative methods. In addition, farmers can quickly switch to different pesticides if the pests they are trying to kill develop genetic resistance to a given pesticide.

Perhaps the most compelling argument for the use of pesticides is the fact that pesticides have saved lives. It has been suggested that since the introduction of DDT, the use of pesticides has prevented approximately seven million premature human deaths from insect-transmitted diseases such as sleeping sickness, bubonic plague, typhus, and malaria. It is likely that even more lives have been saved from starvation because of the increased food production resulting from the use of pesticides. It has been argued that this one benefit far outweighs the potential environmental and health risks of pesticides. In addition, new pesticides are continually being developed, and safer and more effective pest control may be available in the future.

ENVIRONMENTAL CONCERNS

An ideal pesticide would have the following characteristics: It would not kill any organism other than the target pest; it would in no way affect the health of nontarget organisms; it would degrade into nontoxic chemicals in a relatively short time; it would prevent the development of resistance in the organism it is designed to kill; and it would be cost-effective. No pesticide currently available meets all of these criteria, however, and, as a result, a number of environmental

An American farmworker takes health precautions while preparing pesticides for use on crops. (USDA)

problems have developed from the use of pesticides. One of these problems is broad-spectrum poisoning. Most, if not all, chemical pesticides are not selective. In other words, they kill a wide range of organisms rather than just the target pest. The extermination of beneficial insects, such as bees, ladybugs, and wasps, may result in a range of problems, including reduced pollination and explosions in populations of unaffected insects.

When DDT was first used as an insecticide, many people believed that it was the perfect solution for controlling many insect pests. Initially, DDT dramatically reduced the number of problem insects; within a few years, however, a number of insect species had developed genetic resistance to the chemical and could no longer be controlled with it. By the 1990's approximately two hundred insect species had genetic resistance to DDT. Other chemicals were designed to replace DDT, but many insects also developed resistance to these newer insecticides. As a result, although many synthetic chemicals have been introduced into the environment, the pest problem is still as great as it ever was.

Depending on the types of chemicals they contain, pesticides remain in the environment for varying lengths of time. Chlorinated hydrocarbons, for example, can persist in the environment for up to fifteen years. From an economic standpoint, this can be beneficial because the pesticide has to be applied less frequently, but from an environmental standpoint, it can be detrimental. In addition, many pesticides degrade in such a way that their breakdown products, which may also persist in the environment for long periods of time, are often toxic to other organisms.

Pesticides may concentrate in animals as they move up the food chain. All organisms are integral components of at least one food pyramid. While a given pesticide may not be toxic to species at the base, it may have detrimental effects on organisms that feed at the apex because the concentration increases at each higher level of the pyramid, a phenomenon known as biomagnification. With DDT, for example, some birds can be sprayed with the chemical without any apparent effect, but if these same birds eat fish that have eaten insects that contain DDT, they lose the ability to metabolize calcium properly. As a result, they lay soft-shelled eggs, which causes the death of most of their offspring.

Pesticides can also be hazardous to human health. Many pesticides, particularly insecticides, are toxic to humans, and thousands of people have been killed by direct exposure to high concentrations of these chemicals. Many of those who have died have been children who were accidentally exposed to toxic pesticides because of careless packaging and storage. Numerous agricultural laborers, particularly in developing countries where there are no stringent guidelines for the handling of pesticides, have also died as a result of direct exposure to these chemicals. Workers in pesticide factories are also a high-risk group, and many of them have been poisoned through job-related contact with the chemicals. Pesticides have also been suspected of causing long-term health problems such as cancer, and some pesticides have been classified as carcinogens by the U.S. Environmental Protection Agency.

D. R. Gossett

FURTHER READING

Carson, Rachel. *Silent Spring.* 40th anniversary ed. Boston: Houghton Mifflin, 2002.

Connell, Des W. *Basic Concepts of Environmental Chemistry.* 2d ed. Boca Raton, Fla.: CRC Press, 2005.

Miller, G. Tyler, Jr., and Scott Spoolman. "Food, Soil, and Pest Management." In *Living in the Environment: Principles, Connections, and Solutions.* 16th ed. Belmont, Calif.: Brooks/Cole, 2009.

Monaco, Thomas J., Stephen C. Weller, and Floyd M. Ashton. *Weed Science: Principles and Practices.* 4th ed. New York: John Wiley & Sons, 2002.

Ohkawa, H., H. Miyagawa, and P. W. Lee, eds. *Pesticide Chemistry: Crop Protection, Public Health, Environmental Safety.* New York: Wiley-VCH, 2007.

SEE ALSO: Agricultural chemicals; Agricultural revolution; Biopesticides; Dichloro-diphenyl-trichloroethane; Diquat; Green Revolution; Integrated pest management; Malathion; Organic gardening and farming.

PETA. *See* **People for the Ethical Treatment of Animals**

Pets and the pet trade

CATEGORY: Animals and endangered species

DEFINITIONS: Companion animals and the industry in which such animals are bought and sold, whether legally or illegally

SIGNIFICANCE: As the buying and selling of companion animals has grown into a multibillion-dollar industry, a number of abuses have become increasingly widespread. Among these are illegal trafficking in endangered species and the maltreatment of animals as a result of efforts to increase profits.

Pets play an important part in the lives of humans. They provide many benefits, including companionship, love, humor, exercise, a sense of power, and outlets for displacement, projection, and nurturance. Studies have shown that talking to pets reduces stress, promotes feelings of reverie and comfort, and enhances longevity and physical health. Pets also play therapeutic roles, helping sick or elderly persons by involving them in activities of caring, sacrifice, and companionship. Researchers have found that institutionalized elderly persons show great signs of improvement when dogs are introduced to their environment. A 1980-1981 study of patients in the hospital-based care program of a large Veterans Administration medical center showed that pet-owner patients fared better with their health issues than did patients who had no pets.

The multibillion-dollar worldwide pet industry is fed by ever-growing demand for animals, domestic and exotic, legal and illegal. Because of the ongoing high demand for dogs in the United States, and thus the money that can be made from puppy sales, dramatic increases have been seen in the proliferation of dog-breeding operations known as puppy mills. Many of these operations are run by people who know little to nothing about the animals they are breeding. Puppy mills are generally characterized by the presence of unhealthy animals living in filthy conditions; the animals suffer from overcrowding, lack of veterinary care, and unsanitary food and water sources. Another mark of puppy mills is incessant breeding, with many following the unhealthy practice of breeding females twice each year, or every time they come into heat. The puppies that puppy mills sell to retail pet shops are very often malnourished and prone to illness. The operators of puppy mills often ship four-week-old dogs in crates without food or water; many

die, and those that survive to be sold in pet stores commonly have ongoing health and psychological problems.

In the United States, commercial dog kennels are licensed and regulated by the Department of Agriculture, which enforces the Animal Welfare Act of 1966. The regulations specify only limited standards of care, however, and many activists concerned with animal rights and animal welfare have argued that the law is in need of revision.

ILLEGAL TRADE AND REGULATIONS

In addition to the millions of people who cherish their family cats, dogs, birds, and other domesticated animals, many wealthy collectors seek rare pets as status symbols. Illegal trade in such animals is a worldwide phenomenon. The most prominent attempt to regulate such trade is the Convention on International Trade in Endangered Species (CITES), which entered into force on July 1, 1975. This international agreement, which by 2010 had been signed by 175 countries, bans commercial international trade in endangered species and regulates and monitors trade in other threatened species. CITES classifies animal species in three appendixes based on the species' probability of extinction. Appendix I concerns the most vulnerable species; the convention outlaws the commercial trade of live animals listed in this appendix, as well as their parts or derivatives. Animals listed in Appendix II are subject to trade under certain controls, and those in Appendix III may be traded only with export permits or certificates of origin.

Numerous examples exist of violations of the intention of CITES. In one instance, Thailand falsified a quota by issuing inexact statements concerning the number of Appendix II tortoises being shipped out of the country. Investigators found that the number of tortoises in each shipment ranged from seven to seven thousand. The implementation of regulations has increased the value of endangered animals, thus encouraging smugglers to take extreme risks to make money. In February, 1995, a man was arrested for smuggling three young gibbons—a type of ape found in southeastern Asia and the East Indies—in his luggage. Two of the gibbons died of disease, and the third was placed in a zoo. The smuggler, convicted of the crime, was both fined and sentenced to two years in prison. In January, 1996, a man returning to Japan from Thailand was arrested for attempting to bring thirty-nine Appendix II tortoises into Japan. Authorities later discovered that the same man had previously made twenty-six successful smuggling trips between Thailand and Japan.

CITES allows the transportation of animals that have been bred in captivity for commercial purposes, and illegal traffickers take advantage of this loophole, using false documentation to show that transported animals were captive-bred when they were actually taken from the wild. Another problem in upholding the intention of CITES lies in the lack of domestic reinforcement for the terms of the agreement in some nations. In Japan, for instance, pet shops and other sources continue to sell endangered species that have made their way into the country. When the provisions of CITES are not enforced, those who smuggle, buy, and sell endangered animals for profit have no motivation to stop their illegal activities.

The illegal animal trade has tremendous impacts on the animals themselves. For example, during the period 1985-1990, orangutans constituted a large portion of the illegal animal trade in Taiwan; for every one of these animals that survived transportation, three or four died. Illegal transportation is almost never accompanied by proper animal care.

UPHOLDING REGULATIONS

Two cases illustrate how CITES signatory nations have worked together to uphold the convention's regulations. Operation Chameleon and Operation Jungle Trade were two international projects implemented to stop trade in endangered animals; both of them were conducted under the jurisdiction of the U.S. Fish and Wildlife Service.

In September, 1998, Operation Chameleon was instituted to stop the worldwide trade in rare reptiles, most of which were listed in the CITES Appendix II. Among the rare reptile species confiscated during the course of the operation were the Chinese alligator, the Gray's monitor (a lizard native to the Philippines), the false gavial (a crocodile found in parts of Malaysia, Indonesia, Singapore, and southern Thailand), the spider tortoise (Madagascar), and the radiated tortoise (Madagascar). Of these, the Gray's monitor and the spider tortoise are the only reptiles that can be traded legally with permits. At the time of the operation, the Chinese alligator was selling for as much as fifteen thousand dollars; the Gray's monitor could sell for eight thousand dollars. Adult radiated tortoises and false gavials were worth five thousand dollars at the time, and spider tortoises sold for about

two thousand dollars. Investigators were successful in shutting down a major chain of the illegal reptile trade between Asia and North America. Most of the reptiles, which originated in southeast and central Asia, New Zealand, and Madagascar, were destined for the United States.

Operation Jungle Trade sought to bring an end to the illegal trade of exotic birds across the U.S.-Mexico border and into several other countries, including Australia, Belize, Brazil, Costa Rica, Egypt, Ghana, Honduras, New Zealand, Panama, and South Africa. The operation produced forty arrests and the seizure of more than 660 exotic birds and other animals. Most of the confiscated animals were native Mexican species listed under CITES Appendix I. During Operation Jungle Trade, law-enforcement officials from Australia, Canada, New Zealand, and Panama worked together to document criminal activity and gather evidence, carrying out more than forty separate but related investigations all over the world.

U.S. REGULATIONS

In the United States, the Lacey Act of 1900 and the Endangered Species Act of 1973 gave the responsibility of controlling the import and export of wildlife to the Department of the Interior. The United States is a member of CITES through the Endangered Species Act, in which the convention's regulations and guidelines are made law. The Department of the Interior carries out its obligations by designating specific ports of entry for wildlife, providing inspectors for each port to monitor shipments of wildlife, and issuing licenses to commercial wildlife traders. Furthermore, the department cooperates with other U.S. federal organizations and other countries to monitor and investigate possible animal trade violations.

The U.S. Customs Service has primary responsibility for the inspection and clearance of all goods imported to the United States, including animals. Legal support is provided by attorneys from the Wildlife and Marine Resources Section of the Environment and Natural Resources Division of the U.S. Department of Justice. The Animal and Plant Health Inspection Service of the Department of Agriculture is responsible for enforcing regulations regarding quarantine of all birds and specific mammals upon entry to the United States. Quarantine is a critical aspect of the importation of animals. Species from South America, for example, may carry specific strains of diseases or viruses not already present in North America and thus may endanger existing animal populations. Because illegal smuggling operations bypass quarantines and health inspections, they increase the likelihood that diseases may be imported and spread to other animals or humans.

Allison M. Popwell

FURTHER READING

Curnutt, Jordan. *Animals and the Law: A Sourcebook.* Santa Barbara, Calif.: ABC-CLIO, 2001.

Frischmann, Carol. *Pets and the Planet: A Practical Guide to Sustainable Pet Care.* Hoboken, N.J.: John Wiley & Sons, 2009.

Hutton, Jon, and Barnabas Dickson, eds. *Endangered Species, Threatened Convention: The Past, Present, and Future of CITES.* Sterling, Va.: Earthscan, 2000.

Podberscek, Anthony L., Elizabeth S. Paul, and James A. Serpell, eds. *Companion Animals and Us: Exploring the Relationships Between People and Pets.* New York: Cambridge University Press, 2000.

SEE ALSO: Animal rights; Captive breeding; Convention on International Trade in Endangered Species; Endangered Species Act; Endangered species and species protection policy; Poaching.

Photovoltaic cells

CATEGORY: Energy and energy use

DEFINITION: Devices made of materials that create electrical energy directly from absorbed light

SIGNIFICANCE: Multiple photovoltaic cells can be grouped together in an array to form a panel capable of making electricity from sunlight. Most other forms of solar power generation require solar heating of materials, which are then used to do work. Because photovoltaics do not have moving parts, they require very little maintenance, have long service lives, and do not cost anything to operate once installed.

A bank of solar panels made up of photovoltaic cells, which collect solar energy and convert it to electricity. (©Milacroft/Dreamstime.com)

Light can be described as particles of energy called photons. Different wavelengths of light have different amounts of energy. When light of sufficient energy is absorbed by an object, it can excite an electron to sufficient energy to be lost by the atom. This effect, called the photoelectric effect, was described by Albert Einstein. Photovoltaic cells are generally fabricated from semiconductors. Silicon-based photovoltaic cells are most common, but other materials are also used. Electrons at sufficiently high energies in semiconductors can move within the material, creating electricity. The energy of light needed to accomplish this is called the band gap energy. Different materials have different band gaps, and the most efficient photovoltaic cells are constructed of different materials to take best advantage of the various wavelengths of light shining on them.

Not all of the light striking a photovoltaic cell is converted into electrical energy. To compute the overall power efficiency of a photovoltaic cell, one divides the power output by the power of the incoming light. The first photovoltaic cells used commercially in solar panels had efficiencies of about 6 percent, meaning that only 6 percent of the energy of the sunlight striking the cells was converted into electric energy. Some high-efficiency commercially produced solar panels have efficiencies in excess of 20 percent. Research efforts focusing on the creation of very high-efficiency photovoltaics have yielded photovoltaic cell efficiencies in excess of 40 percent under laboratory conditions. Efficiencies are typically lower, however, in less-than-ideal field applications.

Photovoltaic cells can be fairly efficient at converting solar energy into electricity, compared with other methods, and they cost little to operate. The fabrication costs of photovoltaic systems are quite high, however, which reduces their cost-effectiveness. Most large photovoltaic systems cost so much that many years of use are required to recover the cost of installation. Some of the most recently developed photovoltaic cells use thin films, which use fewer raw materials, are more flexible, and are less expensive to build and install than earlier kinds of photovoltaics. Reduction in the amount of materials needed for photovoltaic systems is also beneficial to the environment, because many of the chemicals used in manufacturing photovoltaics are extremely toxic.

The larger a solar panel is, the more sunlight shines on it, increasing the electrical power produced; large solar panels are more expensive than smaller ones, however. One method used to make photovoltaic systems more cost-effective involves lenses or mirrors that focus light onto the photovoltaic cells. These concentrator photovoltaic (CPV) systems increase the intensity of light falling on the solar panels, allowing smaller solar panels to produce as much electrical power as larger panels, thus cutting the cost of photovoltaic solar power generation. Plans for many modern large photovoltaic solar systems include the use of CPV systems. It has been hypothesized that modern CPV systems may be able to produce electricity at prices competitive with other forms of electric generation.

Raymond D. Benge, Jr.

Further Reading

Muñoz, E., et al. "CPV Standardization: An Overview." *Renewable and Sustainable Energy Reviews* 14, no. 1 (2010): 518-523.

Patel, Mukund R. *Wind and Solar Power Systems: Design, Analysis, and Operation.* Boca Raton, Fla.: Taylor & Francis, 2006.

Perlin, John. *From Space to Earth: The Story of Solar Electricity.* Ann Arbor, Mich.: Aatec, 1999.

Wenham, Stuart R. *Applied Photovoltaics.* Sterling, Va.: Earthscan, 2007.

Zweibel, Ken, James Mason, and Vasilis Fthenakis. "Solar Grand Plan." *Scientific American*, January, 2008, 64-73.

See also: Alternative energy sources; Renewable energy; Solar energy; Sun Day.

Pinchot, Gifford

Categories: Activism and advocacy; preservation and wilderness issues

Identification: American conservationist and forester

Born: August 11, 1865; Simsbury, Connecticut

Died: October 4, 1946; New York, New York

Significance: As the first head of the U.S. Forest Service, Pinchot influenced national policy making concerning the conservation of natural resources as well as their management for human use.

After graduating from Yale University in 1889, Gifford Pinchot, whose family's fortune came in part from the lumber business, pursued further studies in forestry. Because no forestry school then existed in the United States, he trained at the French National Forestry School in Nancy in northeastern France for a year. His first forestry projects were for George W. Vanderbilt at the Biltmore estate and W. Seward Webb, Vanderbilt's brother-in-law, in the Adirondack Mountains in New York.

Pinchot believed that all forest resources should be made available for wise use by humans. He was critical of both Yellowstone National Park and the Adirondack State Forest Preserve because they did not permit logging. In his posthumously published autobiography, *Breaking New Ground* (1947), Pinchot states that "conservation is the foresighted utilization, preservation, and/or renewal of forests, waters, lands, and minerals, for the greatest good of the greatest number for the longest time." Conservationists led by Pinchot believed in the wise use of all the nation's resources. They were not preservationists who valued nature for nature's sake. Pinchot's chief ambition was to practice forestry on government land.

In 1891 the U.S. Congress passed a law allowing land to be removed from the public domain and set aside in forest reserves. At that time, the Bureau of Forestry was in the Department of Agriculture, but the forest reserves were administered through the General Land Office in the Department of the Interior. The reserves were run by political hacks holding patronage appointments. In 1897 the secretary of the interior asked Pinchot to travel west and report on the conditions of the reserves. Pinchot's assignment caused the head of the Bureau of Forestry to resign, and Pinchot was appointed to replace him in 1898. To supply their son with trained foresters, Pinchot's parents endowed the Yale Forest School (now the Yale School of Forestry and Environmental Studies), which opened in 1900 as a two-year postgraduate program. Pinchot still had no authority over the forest reserves.

Pinchot had advised Theodore Roosevelt on forestry when Roosevelt was governor of New York. Roosevelt's message to Congress after President William McKinley's assassination in 1901 favored placing the forest reserves under the authority of the Department of Agriculture. The transfer occurred in 1905 when the Forest Service was organized. Pinchot boasted that the forest reserves would be run to benefit poor settlers rather than rich people, a position that ultimately led to Pinchot's downfall.

With Roosevelt's approval, Pinchot brought the concept of conservation to national attention at the

Conservationist and forester Gifford Pinchot. (Library of Congress)

National Governors' Conference in 1907, the meeting of the National Conservation Commission in 1908, and the North American Conservation Conference in 1909. Pinchot's views were not supported by Roosevelt's handpicked successor, President William Howard Taft, or by Taft's secretary of the interior, Richard Ballinger. The resulting feud caused Taft to fire Pinchot. Although Pinchot remained active in public service (he served two terms as governor of Pennsylvania), he never fully regained his former position of national prominence.

Gary E. Dolph

FURTHER READING

Miller, Char. *Gifford Pinchot and the Making of Modern Environmentalism.* Washington, D.C.: Island Press, 2001.

Pinchot, Gifford. *The Conservation Diaries of Gifford Pinchot.* Edited by Harold K. Steen. Durham, N.C.: Forest History Society, 2001.

Wellock, Thomas R. *Preserving the Nation: The Conservation and Environmental Movements, 1870-2000.* Wheeling, Ill.: Harlan Davidson, 2007.

SEE ALSO: Conservation; Conservation policy; Forest and range policy; Forest management; Forest Service, U.S.; National forests; Roosevelt, Theodore.

Planned communities

CATEGORY: Urban environments

DEFINITION: Human settlements for which most design aspects are determined before building begins

SIGNIFICANCE: Among the factors that designers of planned communities take into account is the need to protect and preserve a healthful environment. Through careful placement of roads, retail spaces, and other amenities in relation to living spaces, such communities can minimize detrimental environmental impacts.

Ever since the development of agriculture, enclaves of people have clustered in central locations and dwelled in close proximity to one another. Such ancient cities as Babylon, Nineveh, Jerusalem, Cairo, Athens, and Rome have set the pattern for urban planners through the ages. The typical Greek or Roman city had two main thoroughfares running perpendicular to each other that intersected to form a public area at the city's center. Smaller streets running parallel to the main thoroughfares were lined by dwellings and shops. This configuration created the grid pattern still used by many city planners. The typical medieval city was walled for defense and had narrow streets that curtailed the movement of invaders. These small arteries led to a central square, where inhabitants gathered to socialize and receive information. During the Renaissance, narrow passageways gave way to large boulevards that opened cities to light and space.

The first city in the United States to be built according to a plan was Washington, D.C., designated to replace New York City as the nation's capital. The city planner for Washington, Pierre Charles L'Enfant, borrowed heavily from his mentor Baron Georges-Eugène Haussmann, inventor of Paris's spokes-and-hub plan. L'Enfant superimposed diagonal thoroughfares, all of which led to the Capitol, on a giant rectangle.

A great advance in city planning occurred in 1908 with the creation of Forest Park Gardens outside New York City. The designer of this project, Frederick Law Olmsted, had designed company towns in New England. He was committed to preserving large areas of open space in densely populated residential communities. Olmsted persuaded the commissioners of New York City to purchase large tracts of undeveloped land, which eventually became Central Park. Olmsted's influence is felt in such subsequent projects as the Sunnyside Gardens in the New York City borough of Queens, and Radburn, New Jersey, a bedroom community for Manhattan created in 1929. The designers of many later planned communities, such as Reston in Virginia, Greenbelt and Columbia in Maryland, and Westlake and Laguna Niguel in California, heeded Olmsted's call for the preservation of space for public parks, bicycle lanes, jogging paths, and other such amenities.

SENSITIVITY TO THE ENVIRONMENT

Modern-day city planners consistently pay careful attention to environmental concerns in their planning. Among the matters they consider are the desirability of parks and open spaces, the need for noise abatement, residential areas' proximity to shopping facilities so that automotive pollution is minimized, and zoning regulations that prevent the incursion of factories, airports, and other potential polluters into residential areas.

One planned community that has taken such considerations very seriously is Tapiola, which is located 10 kilometers (6 miles) outside Finland's capital, Helsinki. Tapiola, the brainchild of Heikki von Hertzen, a banker who developed the garden city's plan, has separate roads for automobiles and for bicycles, as well as an elaborate network of footpaths for pedestrians. Major outside roads are routed around Tapiola, minimizing vehicular traffic in the town. Each of Tapiola's dwellings is within 230 meters (755 feet) of a food store, so it is unnecessary for residents to drive within the community to buy groceries. Every building in the town is heated by a central heating plant. Tapiola values people over machines and industry, although some small, nonpolluting industry is permitted inside the town limits to provide work for inhabitants, relieving them of the need to commute to other workplaces.

A portion of Levittown, New York, one of the first planned communities in the United States, as seen from the air shortly after construction was completed in 1948. (AP/Wide World Photos)

U.S. COMMUNITIES

When World War II ended in 1945, hundreds of thousands of American veterans returned to the United States, married, and established families. The nation had an immediate need for affordable housing. This need was met in part by William Levitt, who planned and built communities in the Northeast. By 1950 he had turned a potato patch near New York City into Levittown, a community of 17,500 reasonably priced single-family homes that became a model of city planning for the masses. Levittown's schools, shops, and parks were attractive, but the community was dominated by identical cookie-cutter houses. Levitt overcame this problem somewhat by using different colors of exterior paint and varied landscaping around the homes, lending them some individuality.

As the overall population of the United States began to age, many new communities were established specifically for people over the age of fifty. Among the

notable so-called leisure communities that serve this population are Del Webb's various Sun Cities, Lake Havasu City, and Green Valley in Arizona. A more recent development in planned communities for seniors has been the establishment of retirement communities for gays and lesbians, such as Manasota outside Sarasota, Florida.

Among the most ambitious planned communities in the United States are two near Washington, D.C. Columbia, Maryland, begun in 1963 by John Wilson Rouse, consists of ten villages centered on a business area. The population of Columbia, a self-contained community, is about 97,000, many of whom are employed within the town. At about the time Rouse was establishing Columbia, Robert E. Simon, Jr., planned a community in Virginia, 32 kilometers (20 miles) northwest of Washington. This community, which was eventually named Reston, consists of five villages and a town center that accommodate approximately

56,000 people in a broad variety of dwellings ranging from single-family homes to high-rise apartments and condominiums. Similar planned communities have been established since 1965 in other parts of the United States, including Southern California, Arizona, New York, Pennsylvania, Arkansas, Texas, Hawaii, North Carolina, and Florida.

R. Baird Shuman

FURTHER READING

Forsyth, Ann. *Reforming Suburbia: The Planned Communities of Irvine, Columbia, and the Woodlands.* Berkeley: University of California Press, 2005.

Gause, Jo Allen, ed. *Developing Sustainable Planned Communities.* Washington, D.C.: Urban Land Institute, 2007.

Hardy, Dennis. *From Garden Cities to New Towns: Campaigning for Town and Country Planning, 1899-1946.* New York: Routledge, 1991.

Platt, Rutherford H., ed. *The Humane Metropolis: People and Nature in the Twenty-first-Century City.* Amherst: University of Massachusetts Press, 2006.

Talen, Emily. "Planned Communities." In *New Urbanism and American Planning: The Conflict of Cultures.* New York: Routledge, 2005.

SEE ALSO: Greenbelts; Olmsted, Frederick Law; Open spaces; Urban planning; Urban sprawl.

Planned obsolescence

CATEGORIES: Philosophy and ethics; resources and resource management

DEFINITION: Engineering of a product such that its style or function guarantees the product's replacement after a given period of time

SIGNIFICANCE: Manufacturers' policies of planned obsolescence promote the sales of new products by ensuring that consumers will be prompted to replace the products they own at an optimal rate of consumption. Environmentalists have noted that planned obsolescence harms the environment in several ways, including by increasing waste and contributing to the consumption of resources.

Planned obsolescence was identified as a promising approach for the makers of consumer products as early as the 1920's, when mass production, well under way in the developed world, attracted exacting empirical research that exposed to analysis not only every minute aspect of production processes but also consumer spending practices. In 1932 Bernard London's famed pamphlet *Ending the Depression Through Planned Obsolescence* made a virtue of the process of planned obsolescence by blaming the Great Depression on consumers who used their old cars, radios, and clothing much longer than statisticians had expected they would.

The term "planned obsolescence" entered the popular vocabulary in 1954, when Brooks Stevens, an American industrial designer, used it in the title of an address he gave at an advertising conference in Minneapolis. Stevens defined planned obsolescence as instilling in the buyer the desire to own something a little newer, a little better, a little sooner than is necessary. By 1959, the phrase was in such broad everyday usage that Volkswagen was able to capitalize on it by turning the virtue into a vice, launching what would become a legendary advertising campaign that distinguished Volkswagen as the automobile manufacturer that did not believe in planned obsolescence.

FORMS OF PLANNED OBSOLESCENCE

Planned obsolescence was further sealed in the public consciousness as a vice rather than a virtue in 1960, when the cultural critic Vance Packard published his book *The Waste Makers.* This study of modern society exposed planned obsolescence as the systematic strategy of unscrupulous business leaders who were taking advantage of a mindless public of greedy and wasteful consumers, who in turn were fast becoming debt-ridden in a frantic race to feed their advertising-inflamed desires with products they did not need. Packard distinguished between two forms of planned obsolescence: obsolescence of desirability, or psychological obsolescence, which involves the continual evolution of the aesthetic design of products to manipulate consumers into believing the products are out-of-date, even when they remain functional; and obsolescence of function, which involves the purposeful engineering of products with technical or functional limitations that limit the products' life spans so that customers will need to buy replacements.

A further form of planned obsolescence is systemic obsolescence, which is the practice of altering the systems in which certain products are used so that the products' continued use becomes difficult or impossible within those system settings. Software companies often employ this practice in their design strategies, introducing new software that is incompatible with the previous generation of software and thereby rendering the older software effectively obsolete. The lack of compatibility between generations of software products forces consumers to purchase new software prematurely. Developers of computer hardware products sometimes employ similar strategies, designing new products that are incompatible with older parts and connector plugs. Another way in which manufacturers effect systemic obsolescence is to cease production of replacement parts for older products and to stop offering maintenance services for those products.

INFLUENCE ON CONSUMERS

Planned obsolescence is a tempting policy for manufacturers because it trains the public in consuming practices that have high potential to increase sales of consumer products. Under the influence of planned obsolescence policy, consumers move beyond the limited realm of necessity purchasing and into the infinite realm of desire. From vehicles to deodorants, from real estate to technological products, people are conditioned to replace their belongings, not because their possessions no longer serve them well but simply because those belongings have been around for a while.

As fierce competition in globalizing trade markets puts growing pressure on producers to increase profits, planned obsolescence becomes an increasingly powerful temptation in corporate decision making about product engineering. Manufacturing companies constantly seek to minimize production costs, often using the least expensive components they can find, many of which are produced in countries where regulations on manufacturing processes are weak or nonexistent. Such practices further ensure that products will enjoy life spans that barely exceed their warranty periods.

A worker sorts through electronic equipment at a recycling center in Mumbai, India. Planned obsolescence has contributed to increasingly large amounts of electronic waste throughout the world. (AP/Wide World Photos)

Planned obsolescence has been a powerful force in shaping consumer practices across the globe. Arguments for and against planned obsolescence remain appealing to both producers and consumers. Producers defend planned obsolescence as an essential driving force behind manufacturing innovation and economic growth. They highlight its tendency to promote research and product development, pointing to the many products that have become less expensive, more useful, and better designed as a result of planned obsolescence policies. Although consumers resent the idea that functional everyday products are designed to break easily, they rarely complain about the policy of planned obsolescence as it pertains to the latest fashions and exciting new gadgets.

Although public outcry is muted, there is little doubt that planned obsolescence policies cause significant harms to society and to the environment. In encouraging consumption at the highest levels, planned obsolescence is hardly good for consumers, many of whom go into debt for the sake of owning products they really do not need. Continual replacement of products also results in the depletion of natural resources and increasing rates of environmental pollution, as new products are manufactured to keep up with consumer demand. What to do with old products that are replaced poses a growing challenge as well, as the management of trash has become an immense problem, especially in the cities of industrialized nations. Landfill space has become increasingly hard to find, and some consumer wastes, such as cast-off electronic devices, pose special pollution problems because they contain toxic components.

Environmentalists argue that consumers should seek to reduce their consumption of new products, reuse those already existing products they can, and recycle whatever materials they can from products that are no longer useful in their original forms. In these ways, people can counter the results of manufacturer policies of planned obsolescence.

Wendy C. Hamblet

FURTHER READING

Dobrow, Larry. *When Advertising Tried Harder: The Sixties—The Golden Age of American Advertising.* New York: Friendly Press, 1984.

Frank, Thomas. *The Conquest of Cool: Business Culture, Counterculture, and the Rise of Hip Consumerism.* Chicago: University of Chicago Press, 1997.

Orbach, Barak Y. "The Durapolist Puzzle: Monopoly Power in Durable-Goods Markets." *Yale Journal on Regulation* 21 (2004): 67-118.

Packard, Vance. *The Waste Makers.* 1960. Reprint. New York: Simon & Schuster, 1978.

SEE ALSO: Carbon footprint; Composting; Eco-fashion; Ecological footprint; Electronic waste; Garbology; Globalization; Green marketing; Landfills; Recycling; Sustainable development.

Plastics

CATEGORY: Pollutants and toxins

DEFINITION: Materials that, at some point in their manufacture, can be molded into various shapes

SIGNIFICANCE: Serious environmental issues are raised in both the manufacture and the disposal of plastics. Most plastics are made from oil, the use of which can have negative impacts on the environment, and the available methods of disposing of waste plastic all have detrimental impacts of their own.

Plastics are generally divided into two types: thermoplastics and thermosetting plastics. Thermoplastics can be repeatedly softened by heating and molded into shapes that harden upon cooling. Thermosetting plastics are also soft when first heated and can be molded into shapes, but when reheated, they decompose instead of softening. Both kinds of plastics are used widely around the world to make all kinds of products.

Because the primary raw material for making plastics is oil, the reliance on plastics is part of the world's dependence on the drilling for and transport of oil, both of which are related to environmental problems, among them oil spills, tanker accidents, and air pollution from oil refineries. Another important way in which plastics affect the environment is in the difficulty they pose for disposal. The very properties that make plastics so useful—their stability and resistance to attack by chemicals and bacteria—make them almost indestructible. Unlike metal, wood, and paper, plastics do not corrode or decay. They contribute to unsightly litter along roadsides and can be found floating on the surfaces of the ocean everywhere. Because plastics are light, they account for a large proportion of the volume of the world's solid waste.

For example, while plastics account for only 7 percent of the total weight of solid waste generated in the United States, they make up about 20 percent of its volume.

A number of solutions to the problem of disposing of plastics have been found and tried, but each solution has had its own adverse effects. The plastics industry and governments have therefore been using a combination of solutions to mitigate the effects of the disposal of plastic on the environment. One problem associated with disposal is related to the chemicals that must be added to some plastics when they are manufactured to improve or alter their properties; these chemicals may contaminate the environment when the plastic is discarded. Plastics that would otherwise be hard and brittle are often made soft and pliable through the addition of compounds called plasticizers. For example, polyvinylchloride (PVC) treated with plasticizers can be used to make raincoats and drapes. Plasticizers, as a result of their widespread use, have been detected in virtually all of the world's soil and water ecosystems, including the open ocean.

METHODS OF DISPOSAL

No single way has yet been found to solve the problem of disposal of plastics, but a number of methods have lessened the impacts of plastics on the environment. The first of these is recycling, which helps reduce the amount of new plastic that is made and the energy needed to make it. More than three-fourths of all plastics are thermoplastics that can be recycled through melting and remolding. In order for recycling to be successful, however, used plastic items must first be separated according to type. Different plastics melt at different temperatures, and molding machines work only if they are fed with pure materials. In 1988 the Society of the Plastics Industry developed a uniform coding system to simplify the sorting of waste plastic; most U.S. states require plastics manufacturers to place the codes on the items they make, and the system is now in use internationally.

Uses have been developed for recycled plastic that take advantage of this material's low cost and durability. Plastic products with short service lives, such as packing foam, wrap, and containers, can be made into products with long lives, such as construction materials and plastic pipe. Recycled soft-drink bottles are used to make carpeting and insulation for parkas. Disposable cups and plates are converted to plastic "lumber."

The incineration of plastics to produce energy is another approach to disposal, but this method has the drawback that certain plastics release toxic gases when burned. Proponents of incineration nevertheless assert that it is safe and causes little negative impact to the environment as long as the process is carefully monitored and controlled; they note that this method of disposal can generate much-needed energy while resulting in a large reduction of plastic waste.

The plastics industry is continually seeking ways to make plastics more degradable. Certain elements are introduced into plastics during the manufacturing process to make them susceptible to bacterial attack (biodegradability) or decomposition by light. Modern landfills, however, are constructed to keep out light and bacteria. This prevents any material, even natural wastes, from decomposing while in a landfill. Biodegradable plastics are also not recyclable.

The remaining option for reducing the amount of waste plastic, source reduction, appears simple and direct. The idea is to mitigate the problem by decreasing the quantity of plastics produced and consumed. Even this option is far more complicated than it appears, however. The problem is that in many cases something else must be used to replace the plastic, and this substitution may be more expensive, may pose greater danger to the environment, or may produce more waste than would the plastic. Looking for ways to reduce the amount of packaging used by industry appears to be a solution with fewer problems; this approach has been used successfully in Europe.

David F. MacInnes, Jr.

FURTHER READING

Andrady, Anthony L., ed. *Plastics and the Environment.* Hoboken, N.J.: John Wiley & Sons, 2003.

Hill, Marquita K. "Solid Waste." In *Understanding Environmental Pollution.* 3d ed. New York: Cambridge University Press, 2010.

Snyder, Carl H. *The Extraordinary Chemistry of Ordinary Things.* 4th ed. Hoboken, N.J.: John Wiley & Sons, 2003.

Wolf, Nancy, and Ellen Feldman. *Plastics: America's Packaging Dilemma.* Washington, D.C.: Island Press, 1990.

SEE ALSO: Disposable diapers; Incineration of waste products; Landfills; Ocean dumping; Polychlorinated biphenyls; Recycling; Waste management.

Poaching

CATEGORY: Animals and endangered species

DEFINITION: Illegal or unauthorized hunting, capture, or killing of wild animals, or illegal harvest of protected plant species

SIGNIFICANCE: Poaching has been responsible for dramatic declines in the populations of some animal species, leaving some without genetically viable populations in the wild. Growing awareness of the problem has led to international bans on products derived from endangered species and efforts to educate communities that have relied on poaching for survival.

Poaching occurs for many reasons, most of which are economic. Bears and tigers are slaughtered for their meat, pelts, and internal organs, which are used in some forms of medicine in Asia; birds and monkeys are captured and sold as pets; predators that prey on livestock are killed for obeying their natural instincts; deer and sheep are collected as trophies; and all are, at times, hunted simply for sport.

The concept of poaching dates back to the Middle Ages, when the lands and all they contained were controlled by the ruling class. Killing animals on these lands, even for the sole purpose of putting food on the table, was outlawed and punishable by death. Even well into the twentieth century, many poachers claimed the right to hunt out of season or over prescribed limits in order to subsist.

Nevertheless, people who poach to survive are a small minority; the rest are in it for profit. Experts liken the trade in illegal plants and animals to drug trafficking, with the profits rising dramatically each time the product changes hands. In India, a tiger carcass can bring between one hundred and three hundred dollars, an enormous sum in a region where the average wage is less than one dollar per day. Experienced middlemen separate the carcass into a series of valuable by-products. Almost the entire animal is used: Potions made of tiger bone are thought to promote longevity and cure rheumatism, the whiskers are believed to provide strength, and pills made from the eyes are believed to calm convulsions. An adult tiger can yield up to 23 kilo-grams (50 pounds) of bones, which will in turn sell for hundreds of dollars per pound; its pelt is worth tens of thousands of dollars.

The threat of extinction caused by poaching is real, and the losses are staggering. During the 1970's and 1980's, ivory poachers killed 93 percent of the elephants in Zambia's North Luangwa National Park. In 1988 poachers slaughtered 145 giant pandas for their gallbladders, eliminating one-seventh of the worldwide population of the species in a single year. Studies have indicated that some species, such as orangutans and rhinoceroses, have been killed by poachers at rates two to three times the rates at which they are able to reproduce. As the number of remaining animals of a species drops, their value on the black market rises dramatically.

As a result of poaching, some species, such as the Bengal tiger and the northern white rhinoceros, no longer have genetically viable populations in the wild. Reduced numbers lead to inbreeding, which in turn

Officials view the carcass of a one-horned rhinoceros allegedly killed by poachers at Kaziranga National Park in India. (AP/Wide World Photos)

leads to weaker, more disease-prone animals that are less likely to survive in a diminishing environment. In addition, trophy hunters typically seek the largest, healthiest animals. This not only reduces the quantity of wild animals in the gene pool but also reduces the quality of the herds. More than 38 percent of the elephants in North Luangwa National Park are genetically tuskless, compared to 2 percent in a natural elephant population, because poachers targeted the mature, tusk-bearing animals of breeding age.

The illegal slaughter carried out by poachers also takes its toll on the animals remaining in the wild. Poachers target the largest, most mature animals, which leaves the younger, inexperienced creatures to fend for themselves before they have learned skills essential for survival. In turn, when these inexperienced animals breed, they often lack the parenting skills they need to raise their offspring to maturity.

Marine poaching carries a hidden, and dangerous, side effect. Many types of shellfish, mollusks, and other types of marine life have been declared off-limits not because of any danger of extinction but because their environment has become contaminated by bacteria and other waste products. Shellfish, long a target of enterprising poachers who sell them to restaurants, are especially susceptible to contamination because they feed by filtering up to 38 liters (10 gallons) of water per hour.

An international ban on ivory in 1989, combined with increased law enforcement on Africa's game preserves, led to a slow but steady increase in the wild elephant population. Poaching is also being battled on the economic front, as the booming tourist trade has shown many governments and their citizens that their indigenous wildlife populations may be more valuable alive. Education also plays an important role, as communities that formerly relied on poaching for survival are taught improved farming practices and methods of harvesting that do not endanger native plants and wildlife.

A more controversial approach to conservation of wildlife is "sustainable use," the limited harvest of threatened animals. The benefits are twofold. First, local governments and communities gain economic benefit from the harvest and are therefore less likely to engage in large-scale poaching. Second, the availability of legally obtained animals and their by-products should theoretically drive down the value of these animals on the black market, so that poaching will no longer be worth the risk. Some countries have also begun "farming" endangered animals, not for eventual release into the wild but for slaughter and sale.

P. S. Ramsey

FURTHER READING

Bonner, Raymond. *At the Hand of Man: Peril and Hope for Africa's Wildlife.* New York: Alfred A. Knopf, 1993.

Leakey, Richard, and Virginia Morell. *Wildlife Wars: My Fight to Save Africa's Natural Treasures.* New York: St. Martin's Press, 2001.

Oldfield, Sara, ed. *The Trade in Wildlife: Regulation for Conservation.* Sterling, Va.: Earthscan, 2003.

Woodroffe, Rosie, Simon Thirgood, and Alan Rabinowitz, eds. *People and Wildlife: Conflict or Coexistence?* New York: Cambridge University Press, 2005.

SEE ALSO: Endangered Species Act; Endangered species and species protection policy; Extinctions and species loss; Hunting; Ivory trade; Mountain gorillas; Wildlife refuges.

Polar bears

CATEGORY: Animals and endangered species

DEFINITION: Large white bear species native to the Arctic Circle

SIGNIFICANCE: Polar bears face several threats as a species, including the loss of habitat, some the result of global warming and some caused by human activities related to mineral extraction in the Arctic. Governments, environmental organizations, and individuals have been involved in efforts to preserve the species.

The polar bear (*Ursus maritimus*) is one of the earth's largest land-roving animals. Mature male polar bears may weigh as much as 770 kilograms (1,700 pounds). Polar bears live principally in the Arctic, in about twenty different locations. Their thick fur and blubber beneath their skin enable them to survive in the frigid climate of their habitat. Their principal food source is seals, which they hunt in the icy Arctic waters, but they also eat whale meat, vegetation, and kelp. Polar bears depend on the pack ice of the Arctic as they travel far from land while hunting for food; they also depend on soft snow, in which they prepare their dens for birthing.

Polar bears have been hunted for their white fur in Russia since the fourteenth century. The commercial trade in the fur increased over the centuries as more people moved into Arctic regions and firearms made killing the bears easier. However, in comparison to the hunting of other Arctic species, such as the arctic fox, the commercial hunting of polar bears has never been important for the fur industry. The hunting of polar bears reached a peak during the early twentieth century, when perhaps 1,500 bears were killed annually, and then fell off as the population of bears declined. Later in the century the use of mechanized hunting and hunting from snowmobiles, ice-breakers, and airplanes increased the number of bears killed to a new peak of 1,200 to 1,300 during the late 1960's.

A polar bear emerges from the Arctic sea among a group of ice floes. (©Outdoorsman/Dreamstime.com)

Although hunting and poaching pose problems for the survival of the polar bear as a species, the main threat to the polar bear is global warming. As pack ice has disappeared and snowfall has decreased, the bears have found it increasingly difficult to hunt seals. This problem is compounded by reductions in the seal population, which may also be linked to global warming. The habitat destruction caused by the processes associated with the extraction of oil and coal from Arctic lands has also contributed to the decline of polar bears. In 2008 the International Union for Conservation of Nature estimated the total population of polar bears in the Arctic to be 20,000 to 25,000. The U.S. Geological Survey has projected that if the trends seen in the early twenty-first century continue, two-thirds of the polar bear population will disappear by 2050, and by 2080 only a few will remain in the Arctic Circle.

In 1973 the five nations having the most land in the polar bears' feeding range—the United States, Canada, the Soviet Union, Denmark (in Greenland), and Norway—signed the International Agreement for the Conservation of Polar Bears, which established rules that allowed for the capturing and killing of polar bears only for scientific research purposes or to preserve the ecology of the Arctic. An exception was made for native peoples hunting bears in a traditional manner. The agreement also specified that skins and other objects of value taken from polar bears could not be sold commercially, and it outlawed the hunting of the bears from planes and icebreakers. Norway later banned all hunting of polar bears.

Polar bears are listed in Appendix II of the Convention on International Trade in Endangered Species (CITES), which means that CITES signatory nations are allowed to conduct limited commercial hunting of the animals. Concerned environmental groups have petitioned to have the bears listed instead in Appendix I of the convention, which would prohibit signatory nations from all commercial hunting of polar bears. These groups' efforts to protect polar bears have put them in conflict with some of the indigenous peoples of the Arctic, who want to retain the right to hunt polar bears as part of their traditional cultural practices.

Frederick B. Chary

FURTHER READING

Ellis, Richard. *On Thin Ice: The Changing World of the Polar Bear.* New York: Alfred A. Knopf, 2009.

Kazlowski, Steven. *The Last Polar Bear: Facing the Truth of a Warming World.* Seattle: Mountaineers Books, 2008.

Rosing, Norbert. *The World of the Polar Bear.* Tonawanda, N.Y.: Firefly Books, 2006.

SEE ALSO: Arctic National Wildlife Refuge; Climate change and oceans; Convention on International Trade in Endangered Species; Endangered species and species protection policy; Global warming; Habitat destruction; Hunting; Indigenous peoples and nature preservation; Oil drilling; Wildlife management.

Pollinators. *See* Bees and other pollinators

Polluter pays principle

CATEGORIES: Treaties, laws, and court cases; atmosphere and air pollution; water and water pollution

DEFINITION: Legal rule that the costs of pollution prevention and remediation should be borne by the entity that profits from the process that causes the pollution

SIGNIFICANCE: The polluter pays principle is designed to impose penalties on parties responsible for producing pollution that damages the natural environment and hence society as a whole. Hence the principle is intended to deter activities that pollute the environment.

The polluter pays principle assigns responsibility for the negative impacts arising from polluting activities to the individuals, companies, and other groups that perform those activities. These negative impacts are designated "social negative externalities." The principle is grounded in both the 1992 Rio Declaration, which came out of the United Nations Conference on Environment and Development (also known as the Earth Summit) and states that national governments should "promote the internationalization of environmental costs and the use of economic instruments," and the International Union for Conservation of Nature's Draft International Covenant of Environment and Development, which states that "parties shall apply the principle that the costs of preventing, controlling, and reducing potential or actual harm to the environment are to be borne by the originator."

The Swedish government may have been the first to apply the polluter pays principle, also known as the principle of extended polluter responsibility, in the 1970's. By moving the responsibility for addressing pollution away from taxpayers and governments to the companies and individuals causing the pollution, the principle makes the costs of waste disposal and pollution remediation part of the cost of the process that produces the pollution. Thus companies and individuals have incentive to reduce the amounts of waste and pollution they produce and to increase their efforts to reuse and recycle materials.

The polluter pays principle is applied differently in different situations. Application of the principle tends to be closely related to market-based incentives to regulate consumption or production activity. These policies are broken down between taxes and tradable permit schemes.

SEE ALSO: Accounting for nature; Environmental economics; International Union for Conservation of Nature; Pollution permit trading; Resource recovery; Rio Declaration on Environment and Development; Superfund.

Pollution permit trading

DEFINITION: Incentive-based strategy for pollution control in which government grants companies or other polluters the right to emit specified amounts of pollutants over specified time periods and to trade unused permits

CATEGORIES: Resources and resource management; pollutants and toxins

SIGNIFICANCE: Pollution permit holders may profitably reduce their emissions by selling unused permits to other polluters. Permit trading systems are generally viewed as a more efficient, less costly approach to pollution control than the command-and-control regulation traditionally used by government.

In a pollution permit trading system (also known as a cap-and-trade system), a type of property right is created: the right to produce a certain amount of pollution. A regulatory agency or special commission decides on the number of permits to be issued based on the toxicity, longevity, and other characteristics of the pollutant. If, as is likely, the issuing agency wants to reduce the amount of pollution, the total number of permits circulated account for less than the existing pollution. Each permit gives the holder the right to emit one unit of pollution over a specified period of time, normally one year. The unit—tons, tonnes, pounds, cubic meters, or other measures—depends on the particular pollutant. Each company holds many permits. The firm can choose to emit all the pollution covered by its permits or reduce pollution and sell its unused rights to a different company. The right to pollute is transferable; thus a market is created. Participation in the market is not limited to corporations. An environmental group or governmental unit may buy and hold permits, further reducing the amount of pollution actually created.

The cost of the permits is not set by government but rather by the forces of supply and demand within that particular market. Permit brokers and some form of trading exchange facilitate transactions. The demand in the market comes from new companies beginning operations, the expansion of existing companies, and those companies that face unusually steep pollution-abatement costs. The supply of permits comes from firms going out of business or, more important, from firms that have reduced emissions.

OPTIMAL POLLUTION LEVELS

At first glance, granting ownership to a right to pollute seems harmful to society. It seems as if society would benefit more if pollution were to be completely eliminated rather than traded to someone else. However, one must consider that there are costs involved in reducing pollution. From an economic perspective, zero pollution is not the optimal level of pollution. Because resources are limited, both the abatement cost—the cost of decreasing the pollution—and the monetary damage caused by the pollution must be considered. The optimal level occurs when the abatement cost and the damage cost are balanced. If the cost of reducing the pollution is greater than the damage created by the pollution, society is using too many resources to reduce that pollution. Environmentalists often object to this approach, arguing that not all environmental changes are quantifiable in dollars and that zero pollution is the optimal level of pollution.

Granting firms the ability to buy and sell pollution rights improves the overall efficiency of abatement. Each firm can decide whether it should reduce its pol-

lution and sell more permits to others. Because firms do not use exactly the same materials and production processes, different firms have different abatement costs. Firms with abatement costs above the market price of the requisite number of permits benefit by buying permits rather than spending their resources on pollution control. Firms with abatement costs below the market price benefit by reducing their pollution even more and selling additional permits.

A simple example may help clarify this concept. Suppose XYZ company and ABC company are firms of similar sizes located near each another. Although they produce the same type of good, they use different production techniques. XYZ's production costs increase by $1,000 each time it removes 1 ton of air pollution. ABC spends $200 to remove 1 ton of pollution. Suppose the permitting agency wants to eliminate 20 tons of pollution. If each company is required to reduce pollution by 10 tons, the total cost would be $12,000 (XYZ spends $10,000, and ABC spends $2,000). Using marketable permits, however, ABC could reduce pollution by 20 tons and sell ten permits to XYZ for $500 each. XYZ spends $5,000 on the permits, which is $5,000 less than it would have spent on abatement. ABC makes money: It spent $4,000 on abatement but made $5,000 selling the permits. Society still gets the benefit of a 20-ton reduction in air pollution. Marketable permits thus allow pollution reduction to occur in the most efficient, cost-minimizing way.

A permit trading system also has lower enforcement costs than traditional regulation. Regulation requires a permanent bureaucracy to gather and analyze information, monitor activity, and enforce compliance. Given the complexity accompanying these tasks, regulation often creates litigation. Permit trading relies on the market to achieve efficiency. Corporate decision makers do not have to wait for a government agency to tell them whether they are making proper decisions and do not have to go to court if they think the agency is wrong. The market will reward them for sound decisions and punish them for poor decisions. Permit trading encourages flexibility and rewards innovation by allowing corporations to decide the appropriate technology and techniques as long as the permitted pollution level is not exceeded.

EXAMPLE: TRADING OF SULFUR DIOXIDE PERMITS

Responding to the criticisms economists and corporations aim at traditional command-and-control regulation, the U.S. government began considering permit trading programs in the late 1970's. The Acid Rain Program (ARP) created under Title IV of the Clean Air Act (CAA) amendments of 1990 include the first statutorily mandated national market-based approach to pollution control. Sulfur dioxide (SO_2) emissions contribute to the acid rain problem in Canada and the northeastern section of the United States. The 1990 CAA amendments set a goal of permanently reducing total anthropogenic (human-caused) SO_2 emissions in the United States by 9.1 million tonnes (10 million tons) by the year 2010, down from the 23.5 million tonnes (25.9 million tons) emitted in 1980. Electric power generated by burning coal and other fossil fuels is responsible for about two-thirds of the nation's anthropogenic SO_2 emissions. The ARP set a permanent cap on SO_2 emissions from electric generating units (EGUs) at 8.12 million tonnes (8.95 million tons) by 2010, a level roughly half that of 1980 emissions from the power sector.

During Phase I of the ARP (1995 to 1999), 263 of the nation's largest, highest-emitting coal-fired EGUs participated, along with 135 to 182 additional units each year. The plants involved were major utilities in the East and Midwest. Each permit issued allowed the emission of 0.91 tonne (1 ton) of SO_2. By the time the program was implemented in 1995, pollution control measures had already reduced SO_2 emissions from Phase I sources to 4.8 million tonnes (5.3 million tons), which represented a decrease of 3.7 million tonnes (4.1 million tons) from 1980 levels and 3.1 million tonnes (3.4 million tons) from 1990 levels. When Phase I ended in 1999, emissions had dropped to 4.4 million tonnes (4.9 million tons). Annual emissions during Phase I were well below the caps set for each year, representing a total of 11.62 million allowances.

Phase II, which began in 2000, added the nation's remaining coal-, oil-, and gas-fired EGUs with outputs greater than 25 megawatts. The Phase II cap began at 9.04 million tonnes (9.97 million tons) total for these 2,262 plants. Banked allowances from Phase I were used to offset emissions between 2000 and 2005, which exceeded annual caps. Yearly emissions began to fall in 2006, and by 2009 total annual emissions from the 3,572 EGUs subject to the allowance program were down to 5.2 million tonnes (5.7 million tons), substantially below the year's cap of 8.6 million tonnes (9.5 million tons). The program entered 2010, the year the total number of annual allowances be-

came fixed at 8.95 million, with 12.3 million unused allowances.

The ARP has generally been a successful program. By early 2010 participating EGUs had reduced their annual SO_2 emissions by 67 percent compared with 1980 levels and 64 percent compared with 1990 levels. Emission allowance costs were lower than initially projected, largely because of a reduction in compliance costs resulting from technological innovation and the availability of low-cost, low-sulfur coal. It was estimated that plants needing to buy emission permits would pay between $180 and $981 per ton. However, because so many plants successfully reduced emissions, the price of a permit during Phase I fell to less than $100 and generally traded around $125. During the first decade of the program, prices tended to remain under $200. The likelihood of additional emission reduction requirements under the proposed Clear Air Interstate Rule caused a rise in prices beginning in 2004, with a peak price around $1,550 in late 2005; however, as the market adjusted, prices fell sharply in 2006 and returned to pre-2004 levels by mid-2008.

To participate in the program, all firms are required to install and maintain continuous monitoring devices. Because the government does not have to perform the continuous inspections and litigation typical of traditional regulation, enforcement costs are quite modest. To ensure compliance, the program includes a fine of a few thousand dollars per ton for excess emissions, well above the price per ton determined in the market ($40 as of May, 2010). The fine is nondiscretionary, meaning that any violator must pay the entire fine and offset the violating emission in the following year. The U.S. Environmental Protection Agency (EPA) maintains a central registry of allowances and transfers permits between accounts when a transaction is completed. In late 2001 the EPA introduced an online allowance transfer system, and by 2008 more than 99 percent of all private allowance transfers were conducted through this system.

SPECIAL CONCERNS

The success of permit trading is dependent on several factors. There must be a limited number of allowances in circulation. Existing polluters, understanding that permits will have value in the future, want as many as possible. Thus the first steps in the development of a trading program—deciding what formula to use to distribute permits and deciding whether to auction the permits or give them away—can stir great controversy. The eventual benefits of the tradable permit system are independent of the initial distribution. In other words, the eventual market price will not depend on whether the permits were initially sold or given away. The distribution process is political, not economic, in nature. This creates several problems. If permits are distributed based on current levels of pollution, companies that have already taken steps to reduce their emissions are penalized. If permits are distributed equally, very large and very small companies are affected differently. Some industries may have an easier time reducing pollution than others. Some local economies are completely dependent on single industries, while others are more diversified. What happens if the allocation of permits harms a plant that is the major employer in an economically distressed area? These concerns create considerable political difficulty in determining how to distribute permits.

The European Union (EU) launched the world's largest trading system for carbon dioxide emissions in 2005. It covers more than 11,500 facilities such as coke ovens, coal plants, cement factories, and ironworks within the EU's twenty-seven member nations. Prices were devalued shortly after trading began when it became evident that the EU had distributed too many emissions allowances, which left industries with little motivation to reduce emissions or purchase credits to meet targets. As a result, the period 2005-2006 saw a slight rise in carbon dioxide emissions from major polluters. While emissions trading has been credited with the EU's 3 percent decrease in carbon dioxide emissions in 2007-2008, critics maintain that the system creates more economic benefit for major polluters than it does environmental benefit. To promote economic competitiveness, the EU has continued to give away a substantial number of credits for free.

The major problem facing the development of marketable permits is that many potential pollution markets are hampered by thin markets with only a few buyers and sellers. Thin markets are troublesome because trades are so infrequent that the market price may not be immediately apparent. For example, if only ten cars were sold in Houston each year, would the eleventh buyer know the approximate cost of a car before stepping into the showroom? If a seller loses a potential buyer by setting the price too high, how long

will it be before the next customer arrives? Under these circumstances, firms face considerable transaction costs as buyers and sellers independently try to determine the appropriate price. The thin market problem proved an insurmountable obstacle in several early attempts to develop permit trading programs. The Emissions Trading Program initiated by the EPA and the Fox River, Wisconsin, water-pollution abatement program, for instance, experienced insufficient transactions to establish a market.

Allan Jenkins
Updated by Karen N. Kähler

FURTHER READING

Anderson, Terry L., and Donald R. Leal. *Free Market Environmentalism.* Rev. ed. New York: Palgrave, 2001.

Brohé, Arnaud, Nick Eyre, and Nicholas Howarth. *Carbon Markets: An International Business Guide.* Sterling, Va.: Earthscan, 2009.

Environmental Defense. *From Obstacle to Opportunity: How Acid Rain Emissions Trading Is Delivering Cleaner Air.* New York: Author, 2000.

Field, Barry C., and Martha K. Field. *Environmental Economics: An Introduction.* 5th ed. New York: McGraw-Hill/Irwin, 2009.

Hansjürgens, Bernd, ed. *Emissions Trading for Climate Policy: U.S. and European Perspectives.* New York: Cambridge University Press, 2010.

McLean, Brian J. "Evolution of Marketable Permits: The U.S. Experience with Sulfur Dioxide Allowance Trading." *International Journal of Environment and Pollution* 8 (1997): 19-36.

Napolitano, Sam, et al. "The U.S. Acid Rain Program: Key Insights from the Design, Operation, and Assessment of a Cap-and-Trade Program." *Electricity Journal* 20, no. 7 (2007): 47-58.

Tokar, Brian. "Trading Away the Earth: Pollution Credits and the Perils of 'Free Market Environmentalism.'" In *Taking Sides: Clashing Views on Environmental Issues,* edited by Thomas A. Easton. 13th ed. Boston: McGraw-Hill, 2010.

SEE ALSO: Acid deposition and acid rain; Air-pollution policy; Clean Air Act and amendments; Coal-fired power plants; Environmental economics; Free market environmentalism; U.S. Climate Action Partnership.

Polychlorinated biphenyls

CATEGORY: Pollutants and toxins

DEFINITION: Chemical compounds with biphenyl molecules on which chlorine molecules substitute for two or more of the hydrogens

SIGNIFICANCE: Polychlorinated biphenyls are among the most insidious of environmental pollutants. Although their production has been banned in many countries, widespread use has led to almost universal contamination by these compounds, which are resistant to biodegradation and have a tendency to bioaccumulate.

Polychlorinated biphenyls (PCBs) were first synthesized in 1881 and became readily available during the 1930's. A PCB is a biphenyl molecule on which chlorine molecules substitute for two or more of the hydrogens. A related chemical, polybrominated biphenyl (PBB), contains bromine instead of chlorine. Many different kinds of PCBs form during their synthesis because each ring on the biphenyl molecule can have up to five chlorine atoms. Monsanto Company made PCBs in the United States and sold them under the trade name of Aroclor. Between 1930 and 1975, more than 570 million kilograms (1.26 billion pounds) of PCBs were made in the United States alone.

Some of the properties that made PCBs attractive chemicals were their high boiling points, low water-solubility, and heat resistance. These physical properties meant that PCBs were hard to burn, resisted acids and bases, and were mostly inert. Consequently, PCBs were used in many manufactured products, including adhesives, fluorescent lights, insulation in transformers, high-pressure hydraulic fluids, plasticizers, varnishes and paints, and protective coatings for wood, metal, and concrete. Braided cotton and asbestos in electric wire insulation were also impregnated with PCBs.

The properties that made PCBs useful to industry also made them potentially long-lasting and widespread pollutants. However, PCBs were not suspected of being an environmental problem because they were mostly inert, were not deliberately spread, and were not acutely toxic. One exception was an incident in Japan in 1968 when rice bran oil was accidentally contaminated with PCBs. It was difficult to show chronic PCB toxicity, although symptoms such as chloracne and low birth weights in babies born to exposed mothers were known. The U.S. Environmental Protection Agency (EPA) did not list PCBs as cancer-causing chemicals until they had been in use for a very long time.

In 1966 Sören Jensen, a Swedish chemist, reported observing traces of PCBs in animal tissue that dated as far back as 1944. Scientists in California corroborated his results, and subsequent analyses demonstrated that PCBs were everywhere in the environment: rainwater in England, river water in Japan, seals in Scotland, eagles in Sweden, Baltic Sea cod, mussels in the Netherlands, penguin eggs in Antarctica, pelican eggs in Panama, and, most disturbingly, human hair and fat. All evidence pointed to a global spread of PCBs. The most likely route of spread was determined to be the incineration of PCB-tainted wastes. It is now virtually impossible to find samples of organic material that do not contain some trace amounts of PCBs. Even foods grown in chemical-free fields can receive trace amounts of PCBs from atmospheric deposition of incinerator exhausts.

PCBs are not very soluble in water. They are, however, extremely soluble in lipids and rapidly adsorbed through the walls of the digestive system. PCBs are almost twice as soluble in fats as dichloro-diphenyl-trichloroethane (DDT) and thousands of times more likely than other environmental pollutants to bioconcentrate in animal tissue. Jensen's results indicated that they were starting to bioaccumulate in the food chain. Scientists and environmentalists alike worried that PCBs might ultimately become a greater environmental problem than DDT because more of them had been produced, they lasted longer, and they apparently had wider distribution.

Declines in fish-eating bird populations and evidence of thin-shelled eggs began to increase after World War II. This coincided with a rapid rise in the use of chlorinated chemicals such as PCBs. PCBs cause birds to lay thin-shelled eggs, as does DDT, because these chemicals inhibit enzymes involved in calcium movement. The hormone estrogen controls the calcium level in breeding female birds, and PCBs stimulate enzymes that make estrogen more soluble and readily excreted. When estrogen is low, calcium reserves are low, and little calcium is available for eggshell formation.

Several events ultimately led to the halt of PCB and PBB production in the United States by 1975. In 1974 thousands of cattle and other farm animals in Michigan were quarantined and destroyed because they were contaminated with PBBs. The contamination oc-

A thermal treatment system is used to remove polychlorinated biphenyls from contaminated soil at the Industrial Latex Superfund site in Wallington, New Jersey, in 1999. (AP/Wide World Photos)

curred because a chemical company accidentally mixed bags of a fire retardant containing PBBs with bags of a magnesium oxide mix used in cattle feed. The chemicals were mixed with feed that was distributed around the state. Farmers first began noticing toxicity in their farm animals in late 1973, and contaminated milk and eggs exposed several thousand farm families to PBBs.

It has proved difficult to remove PCBs once they get into the environment. Like all highly chlorinated compounds, PCBs are extremely resistant to biodegradation. In the late 1980's, scientists at the General Electric Research and Development facility in Schenectady, New York, began looking at sediment from sites in the upper Hudson River that were contaminated with up to 268,000 kilograms (591,000 pounds) of PCBs. The PCBs came from a capacitor manufacturing operation at Hudson Falls and Fort Edward, New York, that operated between 1952 and 1971. The scientists reported that highly chlorinated PCBs changed to lower-chlorinated PCBs in anaero-

bic sediments. They suggested that PCBs in the environment biodegrade in a two-step process. The researchers found that chlorine removal in anaerobic aquatic sediments was followed by further decomposition in aerobic environments that eventually caused total PCB destruction. More important, chlorine removal detoxified carcinogenic PCB congeners known to persist in humans.

Mark Coyne

FURTHER READING

Goudie, Andrew. *The Human Impact on the Natural Environment: Past, Present, and Future.* 6th ed. Malden, Mass.: Blackwell, 2006.

Hill, Marquita K. "Persistent, Bioaccumulative, and Toxic." In *Understanding Environmental Pollution.* 3d ed. New York: Cambridge University Press, 2010.

Manahan, Stanley E. "Keeping Water Green." In *Environmental Science and Technology: A Sustainable Approach to Green Science and Technology.* 2d ed. Boca Raton, Fla.: CRC Press, 2007.

Robertson, Larry W., and Larry G. Hansen, eds. *PCBs: Recent Advances in Environmental Toxicology and Health Effects*. Lexington: University Press of Kentucky, 2001.

Walker, C. H., et al. "Major Classes of Pollutants." In *Principles of Ecotoxicology*. 3d ed. Boca Raton, Fla.: CRC Press, 2006.

SEE ALSO: Biomagnification; Bioremediation; Chloracne; Dichloro-diphenyl-trichloroethane.

Population Connection

CATEGORIES: Organizations and agencies; population issues

IDENTIFICATION: U.S.-based nonprofit organization working toward population stabilization in the United States and globally

DATE: Founded in 1968

SIGNIFICANCE: Population Connection emphasizes education, advocacy, voluntary action, and service efforts to lower birthrates in its efforts to promote the achievement of sustainable balances among population levels, resources, and the environment, both in the United States and worldwide.

Over the several decades of its history, Population Connection (originally founded as Zero Population Growth) has evolved an increasingly nuanced and sophisticated approach to issues of human overpopulation. This evolution has moved the organization away from the idea of "population control" (with that term's coercive overtones) to an emphasis on voluntary action for population stabilization based in an environmental and social justice perspective. This perspective emphasizes that overpopulation not only degrades the natural environment (exacerbating problems such as deforestation, wildlife extinction, and climate change) but also degrades the social environment (weakening democratic governments, multiplying urban problems, and increasing competition for agricultural land, fresh water, and a host of other scarce resources).

Key to the organization's change in emphasis has been the realization (arising out of the 1994 International Conference on Population and Development) that slower population growth and reduced family size are strongly dependent on the free choice and empowerment of women. In line with this realization, Population Connection works for the equality of women and men in the United States and worldwide—not only under law and in politics but also in terms of access to education, careers, property, family planning, and health care services, including abortion. The organization is strongly opposed to race-based population-control arguments and also strongly opposed to the use of force or violence as a means toward population stabilization.

Although Population Connection's emphasis on social justice might seem to place it on the political left in the United States, where the organization is based, its "educate, motivate, legislate" approach in fact occupies a middle ground in the spectrum of population responses, between believers in unlimited abundance (and unfettered population growth), who deny that overpopulation is a problem, and those who feel overpopulation is such a tremendous threat to the planet that human population growth must be curbed quickly through the implementation of negative population growth strategies. The pragmatic utopianism of Population Connection's middle way—despite the organization's reliance on the United States to exercise international leadership in population matters—provides perhaps a more realistic appraisal of the global environmental situation and of global political realities than either of the other two ends of the spectrum.

Howard V. Hendrix

FURTHER READING

Ellis, Amanda, Claire Manuel, and C. Mark Blackden. *Gender and Economic Growth in Uganda*. Washington, D.C.: World Bank, 2006.

Engelman, Robert. *More: Population, Nature, and What Women Want*. Washington, D.C.: Island Press, 2008.

Sheehan, Molly. *City Limits: Putting the Brakes on Sprawl*. Washington, D.C.: Worldwatch Institute, 2001.

SEE ALSO: Carrying capacity; Club of Rome; Dubos, René; Ehrlich, Paul R.; Environmental justice and environmental racism; Johannesburg Declaration on Sustainable Development; *Limits to Growth, The*; Population-control and one-child policies; Population-control movement; Population growth; United Nations population conferences; World Fertility Survey.

Population-control and one-child policies

CATEGORY: Population issues

DEFINITIONS: Population-control policies are high-level governmental plans and procedures to limit population growth; one-child policies are population-control policies that limit citizens to one child per couple

SIGNIFICANCE: Many population experts believe that the population explosion is the direct cause of such environmental problems as pollution, ozone and resource depletion, forest destruction, desertification, extinction of plant and animal species, epidemics, and famine. Such experts maintain that for humanity to have a positive future, effective and appropriate population-control systems must be implemented.

By 2010 some 6.8 billion people were living on the earth, more than double the planetary population in 1965. More than 77 million people were being added yearly. Many population experts, such as Paul R. Ehrlich and Anne H. Ehrlich, believe that uncontrolled population growth is a major factor in most important social and environmental problems. In contrast, other researchers, such as Ben Wattenberg, a senior fellow at the conservative American Enterprise Institute, believe that the population problem has been overblown. They note that in Canada, Europe, the former Soviet Union, the People's Republic of China, Australia, and Japan, birthrates have dropped below the replacement level (that is, the level at which current population is maintained). Even in many less developed countries, fertility rates are at or falling toward replacement levels. Loss of population endangers economic prosperity because it means there are fewer producers and consumers. In contrast, those who believe that overpopulation is a real problem agree that birthrates in industrialized nations are below replacement levels but point out that birthrates in many less developed countries are far too high. Also, if the planetary carrying capacity is defined as the population load level at which all people could have their basic needs for food, clothing, housing, health care, and education satisfied, the human population load is already far beyond the earth's maximum carrying capacity.

In discussions of world population, attention is usually focused on the billions of people and the high birthrates in the less developed countries in South America, Africa, and Asia. However, the industrialized world, about 15 percent of the world's population, consumes 80 percent of the planet's resources. The average American consumes roughly twenty times more meat and fish and sixty times more gasoline and diesel than the average Indian.

FAMILY PLANNING

Family-planning programs give contraceptive information and devices to couples so they may freely choose their own family size. Other approaches to the promotion of voluntary population control include encouraging couples to delay marriage, providing sex education programs in schools, making abortion services accessible, and offering welfare benefits that favor small families. Coercive population-control systems might include requiring unwed teenagers to place their children up for adoption, mandatory sterilization after two children, or compulsory abortion of pregnancies after one child.

Government-supported voluntary family-planning programs are popular, but many couples choose to limit family size only after they have more than two children. Family-planning programs thus do not always effectively prevent high population growth rates.

Although compulsory government control of family size is opposed in most countries, projections of increasing environmental destruction and reduced resources caused by population pressures have caused even some democratic countries to consider changing tax laws to favor single adults, childless couples, and small families. In their book *The Population Explosion* (1990), Paul Ehrlich and Anne Ehrlich conclude that population research has found five crucial noncoercive factors that cause significantly reduced levels of pregnancy: adequate nutrition, effective sanitation, basic health care, education of women, and equal rights for women. When the status of women is no longer based on fertility, family size usually declines.

CHINA'S ONE-CHILD POLICY

The People's Republic of China is the only nation that has officially adopted a one-child policy. During the late 1970's Chinese leaders were startled to learn that China's population had surged to more than 1 billion people, which was more than 100 million more than previous estimates. In 1979 the government set a goal of limiting 50 percent of the nation's couples to

A farmer rides past a billboard that promotes China's one-child policy near a village in Shandong Province. The sign urges people to "improve the quality of the population" and "control the population increase." (AP/Wide World Photos)

one child and the other 50 percent to two children. Within five years, the average family size had dropped close to the replacement level of two children per family. Peer pressure was a major motivational force. In certain factories, notices were posted indicating which women workers could become pregnant. If social pressure was not successful, abortions and sterilizations were coerced. However, traditional rural families strongly resisted the one-child limit. Evidence indicating that the first child would be a girl was often followed by abortion, and female infanticide was a frequent occurrence in rural areas. The numbers of infant girls who were abandoned or given up for adoption also increased.

During the 1980's China's one-child policy was relaxed. The need for children to help in farming also pressured the birthrate upward again. Despite its uneven enforcement, the policy resulted in a significant reduction in the Chinese population, so that it fell below the replacement level. It has been estimated that by 2010 China's population would have been greater by at least 300 million people if not for the policy. However, abortion and infanticide practices against girls also left China with a shortage of females and the prospect of 24 million "excess males" by 2020. Additional problems that are anticipated in the future include a shrinking young workforce and a disproportionately large elderly population requiring care.

Vietnam employed a two-child policy during the 1960's and reinstituted it during the 1990's. The government imposes financial sanctions on families that have more than two children. Although the two-child policy is intended to reduce the likelihood of female infanticide and selective termination of pregnancy, enough couples opt for abortion when prenatal testing shows the fetus to be female that a gender imbalance has resulted.

Lynn L. Weldon
Updated by Karen N. Kähler

FURTHER READING

Brown, Lester R., Gary Gardner, and Brian Halweil. *Beyond Malthus: Nineteen Dimensions of the Population Challenge.* Sterling, Va.: Earthscan, 2000.

Eager, Paige Whaley. *Global Population Policy: From Population Control to Reproductive Rights.* Burlington, Vt.: Ashgate, 2004.

Greenhalgh, Susan. *Just One Child: Science and Policy in Deng's China.* Berkeley: University of California Press, 2008.

Hartmann, Betsy. *Reproductive Rights and Wrongs: The Global Politics of Population Control.* Boston: South End Press, 1995.

Riche, Martha Farnsworth. "Low Fertility and Sustainability." *World Watch Magazine,* September/October, 2004, 50-54.

Smail, J. Kenneth. "Global Population Reduction: Confronting the Inevitable." *World Watch Magazine,* September/October, 2004, 58-59.

Tobin, Kathleen A. *Politics and Population Control: A Documentary History.* Westport, Conn.: Greenwood Press, 2004.

SEE ALSO: Ehrlich, Paul R.; Malthus, Thomas Robert; Population Connection; Population-control movement; Population growth; Sustainable development; United Nations Conference on the Human Environment; United Nations population conferences; World Fertility Survey.

Population-control movement

CATEGORIES: Population issues; activism and advocacy

IDENTIFICATION: International development strategy that seeks to impose limits on population growth to improve quality of life

SIGNIFICANCE: The population-control movement is driven by concerns about the negative impacts of overpopulation. The movement has advocated the worldwide spread of family-planning services, particularly the distribution of contraceptives. Cultural and religious resistance can complicate family planning efforts. Some nations have used coercive methods to impose limitations on family size.

Birth control strategies have been used to regulate the timing and spacing of births for millennia, but the idea that humans should limit family size in order to improve quality of life originated with Thomas Malthus, an English cleric and economist who published *An Essay on the Principle of Population, as It Affects the Future Improvement of Society* in 1798. Nineteenth century neo-Malthusians were convinced that overpopulation caused poverty and that contraceptives should be provided to the poor, a position opposed by the medical community of the United States. Physicians triumphed when the Comstock Act was passed in 1873, making the sending of contraceptives and associated information through the U.S. mails illegal.

Support for birth control came from a diverse constituency, including woman suffragists, moral reformers, and advocates of eugenics (the science of improving human hereditary qualities). Some feminists advocated sexual abstinence rather than the use of "unnatural" contraceptives, but most believed that women should have the right to choose when to have a child. Margaret Sanger, a radical feminist and socialist, wrote extensively about birth control, set up clinics where women could seek contraceptive devices and advice, and recruited physicians. Sanger also founded the American Birth Control League in 1921—later renamed the Birth Control Federation of America (BCFA)—and helped organize the first World Population Conference in Geneva, Switzerland, in 1927. At the conference, many eugenicists, including Henry F. Osborn and Frederick Osborn, called for government intervention in birth control and sterilization of the "unfit."

Population-control advocate Margaret Sanger, left, and her sister, Ethel Byrne, in court in Brooklyn, New York, in 1917. Sanger had been charged with maintaining a public nuisance after she opened the first birth control clinic in the United States in Brooklyn. (AP/Wide World Photos)

In 1922 British sociologist and educator Sir Alexander M. Carr-Saunders published a book titled *The Population Problem* that laid the basis for "transition theory," a description of how fertility and mortality rates change during modernization. The evidence from European history indicated that the pattern of high fertility and mortality that preceded nutritional and health improvements was immediately followed by lowered mortality coupled to high fertility and rapid growth. Eventually the increased costs of raising children in an industrial, market economy led to smaller families, but in the poorer regions of Africa, Asia, and South America, large families were still preferred, even after mortality declined. Discovering how to speed the process of fertility decline became an important goal of demographers, eugenicists, and many feminists, such as Sanger and her British counterpart, Marie Stopes.

Eugenicist influence on the population-control movement was evident when the BCFA targeted African American doctors for assistance in spreading the family-planning message in 1939, a policy that was labeled racist by prominent African Americans. This effort followed the founding of birth control clinics in economically depressed Puerto Rico in 1935 by the government's relief agency. However, the Nazi atrocities during World War II tempered the more extreme positions of eugenicists, and the impoverishment of many educated and middle-class families during the Great Depression also confirmed that poverty was not necessarily a function of bad genes. Birth control information was thus added as part of the services provided to the poor by the U.S. government following the New Deal, and making that information more acceptable to a wider married constituency resulted in the renaming of the BCFA, which became the Planned Parenthood Federation of America in 1942.

GENERATING SUPPORT

By the 1950's and 1960's a number of wealthy businesspeople in the United States were waging a battle on behalf of birth control because of what they perceived as the evils of overpopulation. The major contributors to these efforts were the Ford and Rockefeller foundations. In 1952 John D. Rockefeller III founded and became the first president of the Population Council, an institution that aids countries in developing their own population policies by provide scientific research and grants. Hugh Moore, founder of the company that made Dixie Cups, sent his pamphlet *The Population Bomb* to ten thousand prominent citizens to garner support during the late 1950's, a time when most U.S. families practiced family planning and new contraceptive technologies were being developed. The first birth control pill (Enovid) was marketed in 1965.

Bringing the U.S. government into the picture was the big challenge. President Dwight D. Eisenhower formed the Committee to Study the U.S. Military Assistance Program in 1958 because of complaints that

U.S. funds for overseas assistance put too much emphasis on military support at the expense of economic aid. Major General William H. Draper, Jr., the former director of the European Recovery Program, chaired the committee and was encouraged by Moore to study the population problem. Ansley Coale and Edgar Hoover's *Population Growth and Economic Development in Low-Income Countries* (1958) had just been published; this work showed that having too many children overtaxed family resources, reducing private investment. That economic development might be impeded by rapid population growth provided Draper with the rationale for using international assistance for fertility limitation, but his report was later disavowed by President Eisenhower and the Roman Catholic bishops of the United States. Draper later tried to convince President John F. Kennedy to take action, but Kennedy suggested that the private sector, particularly the Ford Foundation, fund population control.

During the 1950's a series of surveys (known as knowledge, attitudes, and practices, or KAP, surveys) measured the "unmet need" for contraception and found that most women would have smaller families if they could. Dramatic publicity about overpopulation during the early 1960's alarmed U.S. voters. Thus, by 1965, during the administration of President Lyndon B. Johnson, the federal government began a tradition of funding population programs through the U.S. Agency for International Development (USAID) and later also through the United Nations.

UNITED NATIONS

In 1962 the United Nations invited member states to formulate population policies, and in 1966 it reemphasized the connection between population and socioeconomic factors. The United Nations also reinforced individual governments' rights in setting policy and the right of families to determine number, timing, and spacing of births. However, Western and Asian countries, particularly India, Sweden, and the United States, pressured the United Nations to take on a leadership role in instituting family-planning programs.

In 1967 United Nations secretary-general U Thant established the U.N. Trust Fund for Population Activities (renamed the U.N. Fund for Population Activities, or UNFPA, in 1987) with the goals of helping developing nations with population-related matters, expanding the UN's role in family planning, and pursuing new programs. He was assisted in his efforts by

Reimert T. Ravenholt, head of USAID, who wanted contraceptives distributed worldwide in order to hasten economic development. By 1971 the trust fund was a recognized hub of the United Nations system, which also included support for population activities by the International Labour Organization, the Food and Agriculture Organization, the U.N. Educational, Scientific, and Cultural Organization (UNESCO), and the World Health Organization.

The first world conferences to study population issues were held in Rome, Italy, in 1954 and Belgrade, Yugoslavia, in 1965. However, the Third World Population Conference held in Bucharest, Romania, in 1974 was the first official government conference with an emphasis on policy rather than research. Delegates agreed that population issues needed to be addressed, but representatives from the Northern Hemisphere and those from the Southern Hemisphere were strongly divided over the relative importance of population planning versus economic development. Delegates from Southern countries believed that an inequitable distribution of resources and the heavy consumption patterns of wealthy Northern nations contributed as much to environmental deterioration as population growth by the poor.

REPRODUCTIVE POLITICS

The Federal Office of Population Affairs was founded in 1970 and continues to provide federal money for family-planning services (except abortions) for poor Americans. In 1972 the Commission on Population Growth and the American Future advocated population planning as important for world stability. The commission believed that efforts should include access to abortion and limits to illegal immigration. President Richard Nixon would not approve the document, and debate intensified when the U.S. Supreme Court prohibited interference with a woman's right to an abortion during the first three months of pregnancy in *Roe v. Wade* (1973). At the 1984 Population Conference in Mexico City, the U.S. delegation reversed its support for family-planning measures because of concerns about coercive population-control measures, such as forced abortion and involuntary sterilization in some countries, notably the People's Republic of China. However, the World Plan of Action written at Bucharest was revised by delegates representing nations in general agreement that family-planning programs are useful whether economic development takes place or not.

Out of the 1984 conference came the Mexico City Policy, or Global Gag Rule, a U.S. policy implemented under President Ronald Reagan that denies federal funding to nongovernmental organizations that provide or promote abortions overseas. Under this policy, the United States no longer funded the International Planned Parenthood Federation or the UNFPA, a position that was not reversed until President Bill Clinton took office in 1993. One of President George W. Bush's first actions after his inauguration in 2001, however, was to reinstate the policy. Yet another reversal came almost exactly eight years later under President Barack Obama, who rescinded the policy a few days after being sworn in.

Abortion is a particularly contentious topic in the United States. While forced abortion and other coercive, abusive population-control measures overseas have been cited as the reason for the United States to withhold funding from international family-planning organizations, U.S. antiabortion forces tend to oppose the voluntary termination of pregnancies as well. The antiabortion movement that arose in the United States in the wake of the 1973 *Roe v. Wade* decision was initially led by the U.S. Catholic Church, but from the 1980's onward the movement increasingly became associated with Christian fundamentalism. The belief that life begins at conception and that abortion is therefore synonymous with murder motivates the most passionate opposition to the termination of pregnancies. Some even regard contraception as an immoral choice that defies the will of God. For these reasons, U.S. funding of international family-planning interests is a politically charged subject. Abortion and contraception tend to be similarly controversial in countries that are predominantly Roman Catholic or Muslim.

ONGOING DEBATES

No consensus exists on the interrelationships among population growth, environmental degradation, and economic development. Some feel population growth is a cause of poverty, while others would reverse that relationship. Women with the lowest education and wages and fewest opportunities outside the household tend to have the largest families. The perspective that gender inequality lies at the heart of the problem in the regions of the world that continue to exhibit high fertility has gained momentum, particularly evident at the International Conference on Population and Development in Cairo, Egypt, in 1994.

Many Southern Hemisphere nations have subsistence economies requiring labor-intensive strategies, and larger families are a rational choice where savings are difficult to put aside and children are productive assets who also provide security during old age. Children in rural India as young as six years old can look after domestic animals and younger siblings, and help with other tasks. In extended-family households, the costs of raising children are also shared because access to common lands expands as the household becomes larger. However, increasing urbanization and market pressures are forcing changes in traditional ownership of land so that these shared lands are decreasing in spite of ever-larger demands that put pressure on local environments.

Few doubt the importance of family-planning measures, and most Southern nations support and feel responsible for providing services to individual couples. However, reformers have tried to shift attention to women's health and empowerment because the population-control approach is believed to have resulted in ethical violations and coercive abuses. Furthermore, it is not always successful where children are an important source of labor. In the twenty-first century, population control efforts have placed emphasis on women's reproductive health, with increased sensitivity to local context. In particular, effecting societal change within communities in the developing world that practice child marriage has become a priority. Young women who receive an education, enjoy some financial independence, and marry later in life tend to be healthier, better mothers, give birth to healthier children, live less isolated lives, and be less likely to become trapped in physically abusive marriages.

Joan C. Stevenson
Updated by Karen N. Kähler

FURTHER READING

Brown, Lester R., Gary Gardner, and Brian Halweil. *Beyond Malthus: Nineteen Dimensions of the Population Challenge.* Sterling, Va.: Earthscan, 2000.

Eager, Paige Whaley. *Global Population Policy: From Population Control to Reproductive Rights.* Burlington, Vt.: Ashgate, 2004.

Harkavy, Oscar. *Curbing Population Growth: An Insider's Perspective on the Population Movement.* New York: Plenum Press, 1995.

Hartmann, Betsy. *Reproductive Rights and Wrongs: The Global Politics of Population Control.* Boston: South End Press, 1995.

Huggins, Laura E., and Hanna Skandera, eds. *Population Puzzle: Boom or Bust?* Stanford, Calif.: Hoover Institution Press, 2004.

Tobin, Kathleen A. *Politics and Population Control: A Documentary History.* Westport, Conn.: Greenwood Press, 2004.

See also: Ehrlich, Paul R.; Malthus, Thomas Robert; Population Connection; Population-control and one-child policies; Population growth; Sustainable development; United Nations Conference on the Human Environment; United Nations population conferences; World Fertility Survey.

Population growth

Category: Population issues

Definition: Increase in the numbers of human beings inhabiting the earth

Significance: Population growth challenges the world's economy, social structures, and environment. The precise value of the theoretical maximum population capacity of the planet depends on the availability of natural resources, the acceptable quality of life, the role of technology, and the values underlying the human relationship with nature.

The world population passed the 6 billion mark in 1999, according to the U.S. Bureau of the Census, triple the 1940 population figure. By 2010 the United Nations estimated world population to be 6.8 billion, and experts predicted that some 70 to 80 million people would be added to the count every year for at least another generation. This population explosion is unprecedented in human history. The estimated population of the world in 8000 B.C.E. was about 5 million. It did not reach 500 million until around 1650, but then the rate of increase accelerated. By the mid-nineteenth century, the planet held 1 billion people; by 1975, the number was 4 billion.

Even though the fertility rate—the average number of children born per woman—declined after the late 1960's, vigorous growth is expected to continue through the twenty-first century, and perhaps beyond. The U.S. Bureau of the Census has predicted that in 2050 the world population will be 9.34 billion. The United Nations has projected 7.8 to 21.2 billion, de-

pending on whether the fertility rate stabilizes or increases again. Whatever the actual total, the distribution of the population will change as people migrate to urban areas, with the world's population shifting so that, for the first time in history, more people live in cities than in rural areas. Most of the total population growth and urban growth will occur in developing countries, particularly those in Africa.

How much population increase can the earth's environment sustain? Answers to that question are often biased by ideology, religious tenets, and interpretive methods, so that estimates vary by nearly four orders of magnitude: from 500 million to 1 trillion. Most, however, fall between 6 and 10 billion, which would mean that the planet's carrying capacity has been exceeded or will be exceeded within one century. The degradation of the global environment—or at least the locales of the greatest population concentrations, the sea coasts—could endanger billions of people and far greater numbers of wildlife.

Resources

Humans affect natural resources by displacing or consuming them. Many resources are renewable, such as forests, which can be regrown. Nevertheless, scientists fear that increased demands for such raw materials may surpass the rate of replacement. For example, increasing use of wood could cause loggers to cut down forests faster than new trees can grow; eventually, the amount of wood available would diminish. Moreover, spreading urban areas and agriculture could take over land that once supported forests.

Some resources are gone forever once used, leaving none for future consumers and necessitating replacement by other materials if the enterprises dependent on the originals are to continue. For example, if petroleum reserves are depleted, propane or natural gas might serve as replacements for heating and vehicle fuel. Not all such nonrenewable resources are replaceable, however. The biodiversity of nature is the critical example, although some commentators insist that biodiversity is not a resource for exploitation but a heritage that should be safeguarded. In any case, proponents of the preservation of biodiversity believe that the destruction of natural habitats and the attendant extinction of species obliterate much-needed new sources of food and medicines.

Species die out naturally. All the species now living on earth amount to only 2 to 4 percent of those that

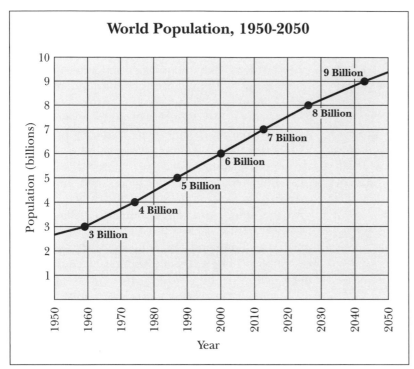

World Population, 1950-2050

World population increased from 3 billion in 1959 to 6 billion by 1999, a doubling that occurred over forty years. The U.S. Census Bureau's 2006 projections suggest that population growth will continue into the twenty-first century, although more slowly. The world population is projected to grow from 6 billion in 1999 to 9 billion by 2042, an increase of 50 percent in forty-three years.

ever existed. Nonetheless, biologists believe that because of human expansion, pollution, and increasing energy consumption, a mass extinction is under way that is as serious as the one that killed the dinosaurs during the Jurassic period. The current extinction rate is calculated to be ten thousand times higher than the average rate before the rise of humans. Of the five to ten million species thought to exist, ten thousand to twenty-five thousand disappear every year. Most are unrecorded insect species, but since 1650 about 1 percent of recorded bird and mammal species have become extinct, and at least another 25 percent are threatened with extinction.

People also change the character of the land and its fresh water. Land cleared for wood or farming is subject to erosion, and soil fertility declines with heavy agricultural use, especially because of monoculture crops, overgrazing, and some irrigation methods. According to the United Nations, every year about 27.5 million hectares (68 million acres) of arable land turn into desert because of human enterprise; 35 percent of all cropland is threatened. Innovations in agricul-

tural methods, known as the Green Revolution, multiplied crop yields and more than made up for the loss in arable land during the late twentieth century, but accelerating growth and desertification are expected to strain farmers' ability to feed the population.

Little can be done about the diminishing supply of fresh water except conservation and antipollution measures. In 1998 some 500 million people lived without access to potable water; by 2025 the number is projected to rise to 2.8 billion, about one in every three people. Polluted water threatens health and agriculture, especially in developing countries, where populations are increasing the fastest. Political scientists worry that competition among nations for declining water resources could escalate into armed conflict in Africa, Central Asia, the Middle East, and South America.

Air pollution also increases with expanding industry and transportation. Changes in atmospheric chemistry from particulates, hydrocarbons, carbon monoxide, nitrogen oxides, and sulfur oxides cause acid rain and smog in urban regions; some pollutants also trap solar energy, creating a greenhouse effect, or destroy ozone high in the atmosphere. The increase in average temperatures and ultraviolet radiation because of these effects will further threaten crops, human health, and wild flora and fauna.

POLITICS AND ECONOMICS

Proposals for dealing with population growth tend to emphasize either restricted fertility or economic development, and sometimes both. Nationalism and international politics, however, complicate the establishment of policies and make implementation difficult.

Attempts to restrict fertility take various forms. A government may specify by law the number of children each couple may have. The People's Republic of China is the best-known example of this approach: Its one-child policy, introduced in 1978, is intended to control the world's largest national population— more than 1.3 billion in 2010. Although recognizing that China desperately needs to check growth, humanitarian groups have denounced the Chinese government for violating human rights in carrying out the policy. Most observers doubt that such a restrictive population-control policy could be effective in a country without a dominating central government like China's.

Other nations and international organizations encourage limiting family size through two other methods. One is to teach women family-planning techniques and offer them contraceptive devices or drugs. The other is to raise the educational level of women in general to encourage women to enter the workforce. Studies in all countries show that educated, working women have fewer children on average than do uneducated women who stay at home. However, great obstacles confront such efforts. Some religions prohibit or discourage the use of contraceptives. Also, the governments of developing nations are frequently suspicious of family-planning programs financed by rich countries or the United Nations, afraid that the programs are really covert political sabotage. Most often, however, cultural traditions forbid women to compete with men intellectually or economically.

Proponents of economic development argue that resources can be further developed and more effi-ciently used to accommodate population growth, but criticism also plagues their efforts. Conservationists throughout the world, especially in industrialized regions such as North America and Europe, are the leading opponents, especially if the development entails exploitation of nonrenewable resources, anticipation of new technologies, or increased pollution. The governments of developing nations join in denouncing the economies of developed countries, accusing them of overconsumption and unwillingness to control their own pollution. Wealthy nations account for only one-sixth of the world's population but control three-fourths of the world's gross product and trade. Moreover, the industrialized nations pollute far more than do developing nations; the United States alone, with only 5 percent of the world's population, produces 25 percent of the carbon emissions that contribute to global warming.

Some political theorists assert that the effects of population growth may be mitigated by more equitable distribution of wealth and political power. They argue that less wasteful consumption will decrease pollution and husband resources, while freer trade and better distribution of land and the capital to develop it wisely will end such harmful practices as slash-and-burn agriculture. Stronger international institutions,

People line up in New Delhi, India, for a 2008 convention of village councils in preparation for a national census. India has the second-largest national population in the world. (B. Mathur/Reuters/LANDOV)

many propose, could administer the needed economic and political changes. The twentieth century saw the globalization of markets, encouraged by such treaties as the General Agreement on Tariffs and Trade (GATT), which may, over time, help equalize use of labor and resources. However, resistance from national governments to new worldwide political institutions and economic systems is daunting.

PHILOSOPHICAL ISSUES

Deep philosophical differences also trouble the debate on population growth. The Roman Catholic Church, for instance, treats the environment and population matters as separate issues and teaches that contraception is immoral because it is contrary to the family's purpose to procreate. The Church vigorously objects to government interference in family life and permits birth control by rhythm method (sexual abstinence during ovulation) alone. Some radical environmentalists, by contrast, want to halt human reproduction altogether until the population returns to pre-Industrial Revolution levels or lower.

These antithetical positions reflect two fundamental questions. First, is humanity to be considered part of nature? Much Western philosophy and theology holds humanity to be superior to nature according to divine law or distinct from it because humans alone possess intelligence. On the other hand, many non-Western thinkers and environmentalists consider humanity to be an integral part of nature, or the "web of life"—in this view, species are mutually dependent on each other, and none is superior. Second, should humans take from the environment whatever they want when they want it, or should they consume only in such a way that biodiversity is not threatened?

Even if all of humankind were to agree on answers to these questions, the environment itself might settle the problem of population growth. Some scientists fear that the global environment is so intricate, complex, and fragile that if it is stressed too much, a "jump effect" could occur in which the earth's ability to support humans and other large organisms would decline precipitously—faster than human technology or conservation could compensate.

Roger Smith

FURTHER READING

Carter, Neil. *The Politics of the Environment: Ideas, Activism, Policy.* 2d ed. New York: Cambridge University Press, 2007.

Cartledge, Bryan, ed. *Population and the Environment.* New York: Oxford University Press, 1995.

Cohen, Joel E. *How Many People Can the Earth Support?* New York: W. W. Norton, 1995.

De Steiguer, J. E. "Paul Ehrlich and *The Population Bomb.*" In *The Origins of Modern Environmental Thought.* Tucson: University of Arizona Press, 2006.

Hardin, Garrett. *Living Within Limits: Ecology, Economics, and Population Taboos.* New York: Oxford University Press, 1993.

Mazur, Laurie, ed. *A Pivotal Moment: Population, Justice, and the Environmental Challenge.* Washington, D.C.: Island Press, 2010.

Moffett, George D. *Critical Masses: The Global Population Challenge.* New York: Viking Press, 1994.

Tietenberg, Tom, and Lynne Lewis. "The Population Problem." In *Environmental and Natural Resource Economics.* 8th ed. Boston: Pearson Addison-Wesley, 2009.

SEE ALSO: *Limits to Growth, The*; Population Connection; Population-control and one-child policies; Population-control movement; Resource depletion; United Nations population conferences.

Positive feedback and tipping points

CATEGORY: Weather and climate

DEFINITIONS: Positive feedback is a process in which some ongoing change in a system amplifies itself; tipping points are rapid transitions from one stable state in a system to another stable state

SIGNIFICANCE: Positive feedback can exacerbate the effects of small changes in climate and force a major climate system to a tipping point beyond which it alters quickly and unpredictably.

The term "tipping point" is often used loosely for several types of rapid change in climate. Scientists, however, restrict the usage to two related effects in a rapidly evolving system. "Positive feedback," by contrast, has a generally accepted, specific usage: It refers to a self-amplifying effect. When in the1990's scientists found evidence that regional and global climates were altering because of pollution, some worried that change would come not gradually but unpredictably, perhaps suddenly. The concepts of positive feedback and tipping points are used to explain some aspects of sudden change. They also exemplify the complexity of cause and effect in climate.

The term "feedback" in general refers to a direct relation between the input of a process and its output. The idea comes from electronics. Amplifiers in particular can be made to reduce (negative feedback) or increase (positive feedback) amplification by redirecting some part of the output signal back to the input. In climate science the input and output are heat energy. A salient example comes from the Arctic Ocean. The packed ice that has covered it for millennia reflects much more of the sun's light (and heat energy) than does open water. As Arctic atmospheric temperature rises (the input of the system) from heat retained in the atmosphere because of the greenhouse effect, more and more ice melts (an output of the system), exposing open water. The water absorbs more heat energy than is typical of an icebound ocean, and that heat melts more ice, exposing yet more open water. The process continues until some equilibrium point has been reached among air temperature, water temperature, and ice cover (if any).

A tipping point entails a loss of equilibrium or stability in a system so that it inevitably evolves from one state to another. In a mundane example, a man may trip and begin to fall, and the point at which his balance is irrecoverably lost is the tipping point that takes him from one state (standing) to another (fallen). In some cases a tipping point, once past, removes some essential capacity from the system—a second meaning for the term. If the man is so obese that he no longer can stand up again on his own after he falls, he loses an essential capacity. Climatic tipping points are difficult to calculate, yet climate-determined systems, if sufficiently pressured by external forces, may pass a tipping point and lose stability. By the late twentieth century increasing average air temperatures became a powerful forcing agent. Among many systems affected, the seasonal melting on the ice sheet covering Greenland accelerated. If air temperatures continue to climb, scientists believe, at some point Greenland's ice will melt faster during summers than it can form during winters. At that tipping point the loss of the ice sheet into the ocean becomes unstoppable, even though it requires centuries to complete. This tipping point, however, would not be a "point of no return." A new Greenland ice sheet could form if average summer temperatures were to drop low enough.

Roger Smith

FURTHER READING

Flannery, Tim. *The Weather Makers: How Man Is Changing the Climate and What It Means for Life on Earth.* New York: Atlantic Monthly Press, 2005.

Gladwell, Malcolm. *The Tipping Point: How Little Things Can Make a Big Difference.* Boston: Back Bay Books, 2002.

Pearce, Fred. *With Speed and Violence: Why Scientists Fear Tipping Points in Climate Change.* Boston: Beacon Press, 2007.

SEE ALSO: Climate change and oceans; Climatology; Global warming; Greenhouse effect; Sea-level changes.

Powell, John Wesley

CATEGORY: Land and land use
IDENTIFICATION: American geologist and explorer
BORN: March 24, 1834; Mount Morris, New York
DIED: September 23, 1902; Haven, Maine
SIGNIFICANCE: Powell contributed significantly to scientific knowledge of the American West in the mid-nineteenth century, and his ideas regarding environmental policy are recognized as being ahead of their time.

Best known for his explorations of the western United States, John Wesley Powell contributed to both the romantic and the scientific views of those lands. Many stories have been told of his exploits in leading surveying expeditions through uncharted territory, especially his trip down the Grand Canyon in 1869. Newspapers across the United States reported on the progress of Powell's travels as he led that first exploration of the Grand Canyon by European Americans. In the twenty-first century, people preparing for recreational white-water rafting trips still read his accounts of the challenges he faced on the Colorado and Green rivers. The personal drama of the stories was heightened by the fact that Powell had lost his right arm in the Battle of Shiloh during the Civil War.

Powell's first contributions to scientific knowledge of the American West were the materials he returned to the Smithsonian Institution in Washington, D.C., and the reports he made to the U.S. Congress. Among the most enduring of his environmental science con-

Geologist and explorer John Wesley Powell. (NPS)

tributions was his initiative in generating topographic maps of the United States. He envisioned a set of maps that would cover the entire country, but the project was too big for his lifetime. Still, that initiative continues to have direct impacts on people who enter public lands, whether for recreational or commercial purposes.

Wallace Stegner's 1954 biography *Beyond the Hundredth Meridian: John Wesley Powell and the Second Opening of the West* brought Powell's views back to the attention of environmental management officials. Stegner presented Powell not only as a leading figure in the organization of science in government but also as an environmental policy visionary who was ahead of his time. Powell's *Report on the Lands of the Arid Region of the United States* (1878), written a few years before he became director of the U.S. Geological Survey in 1881, advised against the use of the checkerboard grid—the standard approach to land surveying then being used in the eastern United States—in the surveying of the West. Arguing that the lack of water in the West needed to be the dominant factor in dividing land for agricultural purposes, Powell promoted an approach in which 65 hectares (160 acres) would not be stan-

dard. He saw 32.4 hectares (80 acres) as a sufficient amount of irrigable land for a homestead and thought that 1,036 hectares (2,560 acres) might be needed for pasturage. Though the National Academy of Sciences supported his recommendations, they were not accepted by Congress.

The value of Powell's influence on federal land policy remains in dispute, but in the late twentieth century the Bureau of Reclamation called him "the father of irrigation development," and at least two secretaries of the U.S. Department of the Interior (Stewart Udall and Bruce Babbitt) have claimed that they were inspired by Powell.

Larry S. Luton

FURTHER READING

Aton, James M. *John Wesley Powell: His Life and Legacy.* Salt Lake City: Bonneville Books, 2010.

Worster, Donald. *A River Running West: The Life of John Wesley Powell.* New York: Oxford University Press, 2001.

SEE ALSO: Conservation; Forest and range policy; Grand Canyon; Irrigation; Land-use policy; Range management; Watershed management.

Power plants

CATEGORY: Energy and energy use

DEFINITION: Facilities that generate electrical power through any of a variety of methods

SIGNIFICANCE: Many of the practices used in the generation of electrical power are environmentally destructive. In particular, power plants that rely on fossil fuels contribute to greenhouse gas emissions, acid precipitation, urban smog, and toxic mercury emissions. Power plants in general often also affect water quality, disrupt land use, and generate a variety of hazardous and toxic wastes.

Electric power is produced from a number of sources. Of the 4,119 million megawatt-hours of electricity generated in the United States in 2008, coal contributed 48.2 percent; natural gas, 21.4 percent; nuclear, 19.6 percent; hydroelectric, 6.0 percent; other renewable resources such as solar, geothermal, wind, and biomass, 3.1 percent; petroleum, 1.1 percent; and other sources, 0.6 percent. In other countries, the relative utilization of fuels varies depending on what local resources are available. France has little coal and oil, so for most of its electricity (more than 75 percent) it depends on nuclear power. China's increasing industrialization is largely fueled by its vast coal resources, from which it generates roughly 80 percent of its electricity.

Large-scale electric power facilities and related infrastructure tend to present certain environmental challenges regardless of the fuels they consume. Massive centralized power generation facilities require substantial amounts of land, and environmentally irresponsible operations can leave the sites badly polluted and the neighboring land significantly devalued. Power transmission lines have a negative aesthetic impact, particularly in natural areas. These large structures can also disrupt wildlife habitats, and where they cross agricultural land they have the potential to impede farming operations. If a power plant diverts water from a river to serve as a coolant and then returns the heated water to the river, the resulting thermal pollution can cause a localized reduction in the river's dissolved oxygen levels and otherwise negatively affect aquatic organisms and ecosystems.

Coal-fired power plants such as London's Eggborough Power Station emit high levels of greenhouse gases into the atmosphere. (AP/Wide World Photos)

POWER GENERATION USING FOSSIL FUELS

Coal is the most abundant of the world's fossil fuels. Mining, transportation, and combustion of coal create a host of environmental problems. Surface mining of coal defaces the landscape and causes erosion. Deep mining is a hazardous occupation for miners because of long-term lung effects; the coal-mining industry also has a history of deadly underground explosions. Whenever coal is transported or stored, care must be taken to minimize the risk of spontaneous combustion.

Oil drilling, be it on land or offshore, can result in massive accidental releases to the environment, a well-known example being the BP *Deepwater Horizon* oil spill in the Gulf of Mexico that continued for months during the spring and summer of 2010. Oil is commonly transported by pipelines and ocean tankers, both of which have a history of ruptures and spills. Environmental releases of oil can harm aquatic organisms, waterfowl and seabirds, other wildlife, and their ecosystems. Natural gas pipelines are also problematic, as leaks can lead to damaging and deadly explosions such as the one that occurred in San Bruno, California, in 2010.

When coal, oil, or natural gas is burned at an electric power plant, high-pressure steam is produced to turn a turbine that is connected to a generator. Some coal contains as much as a 10 percent sulfur impurity, which is converted to gaseous sulfur dioxide during burning. Oxygen and nitrogen from the air also combine during combustion to form gaseous oxides. When these gases are released into the atmosphere, they combine with water to form sulfuric acid and nitric acid, which eventually return to earth as acid rain. The harmful effects of acid rain include loss of fish in acidified lakes, deterioration of forests, and surface erosion of marble buildings and monuments. Acid rain can be reduced through the use of cleaner fuel, such as natural gas or low-sulfur coal. Also, sulfur dioxide can be removed before it leaves the smokestack by means of chemical "scrubbers," but such technology adds expense. For some older coal plants, it makes more sense from an economic standpoint to phase out the plant than to retrofit it with scrubber technology.

Pollution-control measures can also minimize other harmful air pollutants generated during combustion, including ozone, mercury, and particulate matter. Ozone (the main component of smog) and particulates can have detrimental impacts on the lungs of humans and animals, in addition to reducing visibility. Trace amounts of mercury released from fossil fuels during burning can bioaccumulate in the environment to harm humans and ecosystems.

When fossil fuels are burned, carbon combines with oxygen from the air to form gaseous carbon dioxide. In the last decades of the twentieth century, concerns arose that increased carbon dioxide levels would cause the earth's atmosphere to act like the glass panels of a greenhouse, allowing solar radiation to enter the atmosphere but preventing infrared heat from escaping. Global warming from this greenhouse effect could melt polar ice caps and snow cover, which in turn could raise the levels of the oceans and cause disastrous flooding of coastal communities. Also, a hotter and drier climate could create worldwide problems for agriculture. The only way to reduce the production of carbon dioxide is to burn less coal, oil, and natural gas.

POWER GENERATION USING OTHER SOURCES

Nuclear power plants offer one alternative to fossil-fuel-burning plants. The major benefit of nuclear power is that no chemical burning takes place, so no fly ash, acid rain, or greenhouse gases are produced. A common objection to this power generation technology stems from the possibility that an accident could release radioactivity into the environment. Such a release occurred during the 1986 Chernobyl disaster in Ukraine, which remains the worst nuclear reactor accident in the history of the industry. Opponents of nuclear power also point to the problem of long-term

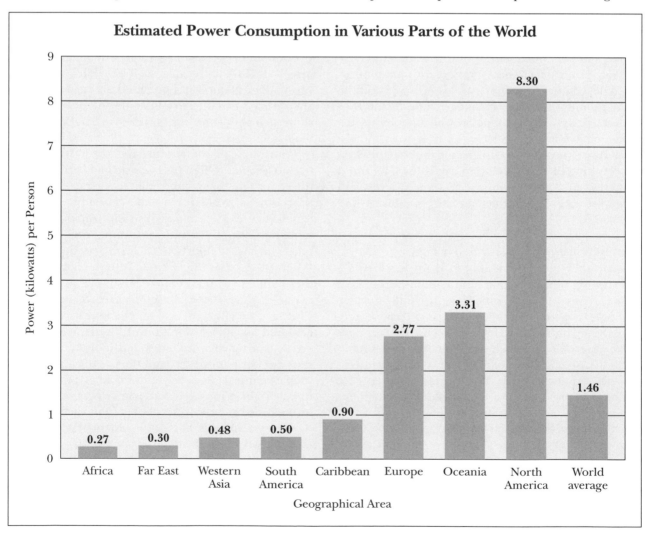

Estimated Power Consumption in Various Parts of the World

nuclear waste storage, which presents both technical and sociopolitical challenges.

Hydroelectric power plants use water, a renewable, nonpolluting resource, to generate electricity, but they also have environmental disadvantages. When a new hydroelectric dam is built, the stored water often fills a scenic canyon or floods farmland near the river. During extended droughts, as happened in the Pacific Northwest during the 1980's, consumers of hydroelectric power are faced with electric power shortages. The impact of dam construction on downstream water supplies can cause friction among communities, states, or nations. The greatest hazard of a dam is the large quantity of water stored behind it. If a dam experiences catastrophic failure, as China's Banqiao Reservoir did in 1975, the resulting loss of life and property downstream can be staggering.

Geothermal energy from underground can be harnessed anywhere, but major electric power generation projects have generally focused on locations coinciding with geothermal hot spots near the earth's surface. Tidal power technologies have been limited geographically to places on coastlines where large differences exist between high and low tides; new developments may make power production at other locations possible. Wind power, generated by the motion of wind turbines, requires an average wind speed of at least 16 kilometers (10 miles) per hour to be practical. Problems inherent to wind power include the difficulty of maintaining a constant turbine rotation speed regardless of wind speed variation in order to produce electricity at a constant frequency, the difficulty of preventing damage to turbine blades from high-velocity wind gusts and ice in winter, and the unavoidable noise pollution created by the rotating blades. Biomass power plants, which use materials such as wood, agricultural wastes, and gaseous fuels derived from them, generate greenhouse gases, nitrogen oxides, and particulates as part of the combustion process.

Solar energy generation, which uses sunlight, produces no air pollution or waste products and so seems to be an ideal method. However, since solar radiation is not available at night, an energy storage system is needed. One solar technology employs reflectors that concentrate sunlight to heat a liquid or gaseous medium, which in turn creates steam or pressurized gas that drives a turbine to generate electricity. Another solar technology uses photovoltaic cells that convert sunlight directly into electricity. Some environmen-

talists have proposed that the future of solar energy may lie not in the building of traditional centralized power plants but rather in the development of small collector systems for individual communities.

Hans G. Graetzer
Updated by Karen N. Kähler

FURTHER READING

Ayres, Robert U., and Ed Ayres. *Crossing the Energy Divide: Moving from Fossil Fuel Dependence to a Clean-Energy Future.* Upper Saddle River, N.J.: Wharton School Publishing, 2010.

McCully, Patrick. *Silenced Rivers: The Ecology and Politics of Large Dams.* Enlarged ed. London: Zed Books, 2001.

Murray, Raymond L. *Nuclear Energy: An Introduction to the Concepts, Systems, and Applications of Nuclear Processes.* 6th ed. Burlington, Vt.: Butterworth-Heinemann/Elsevier, 2009.

National Research Council. *Electricity from Renewable Resources: Status, Prospects, and Impediments.* Washington, D.C.: National Academies Press, 2010.

Visgilio, Gerald Robert, and Diana M. Whitelaw, eds. *Acid in the Environment: Lessons Learned and Future Prospects.* New York: Springer, 2007.

SEE ALSO: Coal-fired power plants; Fossil fuels; Geothermal energy; Hydroelectricity; Nuclear power; Renewable resources; Solar energy; Thermal pollution; Tidal energy; Wind energy.

Precautionary principle

CATEGORY: Philosophy and ethics

DEFINITION: Heuristic device for judging an action or policy, usually in the public realm, that stresses the need to minimize harm in the context of uncertainty about the effects of that action or policy

SIGNIFICANCE: The precautionary principle is usually applied in the decision-making processes of institutions with a responsibility to prevent harm, including international organizations such as the United Nations and the European Union, national governments, and other regulatory bodies. The principle is often used in decisions regarding actions or policies that may have impacts on the environment.

The precautionary principle might be summarized simply as "Prevention is better than cure" or "Better safe than sorry." The principle is invoked when some scientific evidence arises that a particular action or policy results in harm, but the evidence is inconclusive and there is still time for preventive action. The harm might be to people, the environment, animal life, or plant life, and it might be irreversible, as in the case of species loss. The practical significance of the precautionary principle is its denial that a lack of certainty is adequate reason to refrain from action.

When the principle is applied by decision makers considering whether to prohibit, cease, or outlaw an action or policy, several factors are important: the possible severity of the alleged harm, the likelihood of such harm occurring, the costs of preventing such harm, and economic costs in general—including the costs of adopting alternatives. The principle is applied in many areas of public policy, including those concerned with global warming and air pollution, environmental protection and the maintenance of biodiversity, genetically modified foodstuffs and food safety issues, consumer protection law, and technological development. The precautionary principle is, for example, incorporated into the 1982 United Nations World Charter for Nature, and it is one of the twenty-seven principles of the Rio Declaration made at the 1992 Earth Summit.

Stronger and weaker formulations of the principle have been expressed. The strongest versions of the principle hold that the mere possibility of harm to human well-being or the environment is sufficient grounds for intervention, even if the evidence is speculative and the potential costs of intervention are high. Weaker versions hold that action in the face of scientific uncertainty is permissible but not required. According to weaker versions, the burden of proof necessary for intervention is greater and tends to fall on those demanding precautionary action.

The precautionary principle has both supporters and critics. Supporters of the principle cite its flexibility and range of applications, but many worry that a lack of political will can result in a failure to uphold the principle. Major policy decisions involve many different interest groups, such as corporations, environmentalists, and consumers, and lack of consensus about what counts as harm can lead to the principle's being interpreted broadly so as to excuse inaction. Critics of the principle worry that it is applied too readily, without agreement on what counts as reason-

able scientific evidence of harm. As a result, they argue, policies or technologies might be abandoned prematurely. All new innovations carry some degree of risk, critics assert, and this must be considered alongside and offset against the costs of not adopting new innovations and technology, as well as the costs associated with alternatives.

Andrew Lambert

FURTHER READING

Goklany, Indur M. *The Precautionary Principle: A Critical Appraisal.* Washington, D.C.: Cato Institute, 2001.

Sunstein, Cass R. *Laws of Fear: Beyond the Precautionary Principle.* New York: Cambridge University Press, 2005.

Whiteside, Kerry H. *Precautionary Politics: Principle and Practice in Confronting Environmental Risk.* Cambridge, Mass.: MIT Press, 2006.

SEE ALSO: Benefit-cost analysis; Best available technologies; Climate change and human health; Earth Summit; Ecological economics; Genetically modified foods; Global warming; Nanotechnology; Polluter pays principle; Rio Declaration on Environment and Development; Risk assessment.

Predator management

CATEGORY: Animals and endangered species

DEFINITION: Efforts to control, maintain, and sometimes reintroduce wild populations of predatory species without presenting undue risk to people, domestic livestock, other wildlife, or ecosystems

SIGNIFICANCE: Predator management involves weighing the needs of humans and their domesticated animals against the intrinsic value of predators that are assuming their natural place in the ecosystem. This aspect of wildlife management is a great source of controversy among conservationists, hunters, ranchers, and animal rights activists.

Humans have always competed with their fellow members of the animal kingdom for food and survival. This competition intensified with the advent of ranching, where large herds of domesticated animals pose a ready food source for wild predators, particularly wolves and coyotes. For much of human history, the concept of predator management has been limited to killing on sight. Well into the twentieth century, predators were seen as "nuisance animals" with no intrinsic value to the environment and, as such, were trapped, snared, shot, and poisoned. In some cases, this overzealous action against natural predators led to a population explosion among prey species, which in turn resulted in habitat degradation.

Some people in remote areas of the United States, including Native Americans, rely on big-game animals for food and economic survival. However, the primary reason for protecting nonendangered big-game species from predators is to ensure that hunters have something to shoot during hunting season. Many hunters liken themselves to large predators with regard to their place in the food chain. The difference is that, while natural predators bring down the old and the injured, hunters seek only prime specimens, which weakens the gene pool of the game animals instead of supporting natural selection.

PRACTICES IN THE UNITED STATES

The U.S. Congress began providing predator-control assistance to ranchers in 1915, though it did not become a federal obligation until the Animal Damage Control Act of 1931. Still in effect, this act authorizes government agents to kill thousands of predators every year. In fiscal year 2008, nonbird predators that were killed included coyotes (89,300), gray wolves (396), red foxes (2,447), gray foxes (2,350), black bears (395), mountain lions (373), bobcats (1,883), and badgers (581). Most of the slaughter of these so-called nuisance animals occurs in the western states in the name of livestock protection. According to figures from the U.S. Humane Society, in 1994 the government spent more than $38 million, nearly as much as it spent to protect endangered species, to kill predators in the United States. In an attempt to combat negative publicity, in 1997 Animal Damage Control, the federal program charged with managing these predators, was renamed Wildlife Services. This program is run by the Animal and Plant Health Inspection Service (APHIS) of the U.S. Department of Agriculture (USDA).

According to figures from the U.S. National Agricultural Statistics Service, predators throughout the

A hunter in Monkton, Vermont, tags a coyote he shot as part of an organized bounty hunt. Coyotes are often targeted by predator management programs because they are considered nuisance animals that threaten livestock. (AP/Wide World Photos)

United States killed 224,200 sheep and lambs in 2004. Coyotes, the primary culprit, were responsible for 60.5 percent of the predation deaths. These adaptive creatures, once described by writer Mark Twain as "a long, slim, sick and sorry-looking skeleton, with a gray wolf-skin stretched over it," now inhabit Canada, Mexico, and every state except Hawaii. Wolves are also a problem for ranchers, and their 1995 reintroduction into Yellowstone National Park in Wyoming remains a controversial issue.

Large-scale predator kills continued in the United States into the 1990's. Wolves and other large predators are still killed in Alaska and the Pacific Northwest to protect big-game animals such as moose, elk, and caribou. In 1994 Alaskan governor Tony Knowles halted the state's wolf-kill program after a nationwide television broadcast showed trapped wolves suffering in snares. While wildlife groups enthusiastically endorsed the move, hunters claimed that the state had bowed to outside pressure. Subsequent political shifts have led to policies authorizing lethal predator control against the state's wolves and bears. Such practices are an ongoing source of controversy, particularly the state's exploitation of a loophole in the Federal Airborne Hunting Act of 1971 that results in private citizens being allowed to conduct aerial hunting in the name of wildlife management. Another practice in Alaska that draws outrage from animal rights activists is the gassing of wolf pups in their dens. Organizations such as the American Society of Mammologists have repeatedly expressed concern that Alaska's predator-control programs fail to meet scientific standards for sound wildlife management.

NONLETHAL METHODS AND IMPLICATIONS

Not all solutions to predator problems are lethal. Live capture and relocation is another option—provided that there is a suitable habitat available and the animal has not become a nuisance killer. The problem with relocation is that some of these animals have become accustomed to humans and the easy meals that come from living in their company. This is especially a problem with black bears, which are true omnivores that are just as willing to forage through backyards and garbage cans as through their natural habitats.

Sterilization, by either surgical or chemical means, is another viable option. Surgical sterilization is especially effective in the case of wolves. As a result of the hierarchical nature of their society, only the dominant (or alpha) pair of wolves in each pack mate and bear pups, so only two animals from each pack need to be sterilized to prevent the group from breeding. Another effective, if unusual, method of protecting livestock is the use of guard animals such as donkeys or llamas. This method is particularly effective against coyotes, which instinctively fear anything new in their environment.

While challenging, managing predators within a framework of environmental cooperation is the ideal approach. During the late 1970's, after a century of wide-scale hunting had left the Mexican wolf on the brink of extinction, the United States and Mexico established a binational captive-breeding program. In 1998, release of captive-bred Mexican wolves began within the Blue Range Wolf Recovery Area in the American Southwest. A total of ninety-two individual wolves were released over the next decade. Reintroduction, as well as monitoring and managing the reintroduced predators, is a multiagency effort involving the U.S. Forest Service, the U.S. Fish and Wildlife Service, USDA-APHIS Wildlife Services, the Arizona Game and Fish Department, the New Mexico Department of Game and Fish, and the White Mountain Apache Tribe. Although the Mexican wolf is an endangered subspecies of gray wolf, the reintroduced population has been designated "nonessential experimental" to provide more flexibility in controlling the animals in case of livestock depredations or nuisance behavior. Ranchers are permitted to kill Mexican wolves on private or tribal land only if the predators are discovered in the act of attacking livestock, and the kill must be reported immediately thereafter. A recovery plan approved in 1982 includes a compensation fund to reimburse ranchers for losses and incentives that reward landowners when pups are born on private property.

P. S. Ramsey
Updated by Karen N. Kähler

FURTHER READING

Clark, Tim W., Murray B. Rutherford, and Denise Casey, eds. *Coexisting with Large Carnivores: Lessons from Greater Yellowstone.* Washington, D.C.: Island Press, 2005.

Fascione, Nina, Aimee Delach, and Martin E. Smith, eds. *People and Predators: From Conflict to Coexistence.* Washington, D.C.: Island Press, 2004.

Gittleman, John L., et al., eds. *Carnivore Conservation.* New York: Cambridge University Press, 2001.

Lopez, Barry Holstun. *Of Wolves and Men.* 25th anniversary ed. New York: Scribner Classics, 2004.

Rawson, Timothy. *Changing Tracks: Predators and Politics in Mount McKinley National Park.* Fairbanks: University of Alaska Press, 2001.

Ray, Justina C., et al., eds. *Large Carnivores and the Conservation of Biodiversity.* Washington, D.C.: Island Press, 2005.

Sellars, Richard. *Preserving Nature in the National Parks: A History.* New ed. New Haven, Conn.: Yale University Press, 2009.

Stolzenburg, William. *Where the Wild Things Were: Life, Death, and Ecological Wreckage in a Land of Vanishing Predators.* New York: Bloomsbury, 2009.

Young, Christian C. *In the Absence of Predators: Conservation and Controversy on the Kaibab Plateau.* Lincoln: University of Nebraska Press, 2002.

See also: Balance of nature; Captive breeding; Endangered species and species protection policy; Extinctions and species loss; Forest and range policy; Grazing and grasslands; Hunting; Kaibab Plateau deer disaster; Wildlife management; Wolves; Yellowstone National Park.

Preservation

Categories: Preservation and wilderness issues; philosophy and ethics

Definition: Maintenance of wilderness areas in an undisturbed state

Significance: While debates continue regarding the best uses of public lands, it is clear that proponents of preservation have played a large role in preventing the conversion of many wilderness areas to development, logging, mining, or other uses.

The concept of preservation of wilderness emerged in the United States in the nineteenth century as a response to the large-scale disposal of public lands then taking place and to such economic activities as mining and logging, which had altered much of the western landscape. Preservation is typically contrasted with conservation, which allows managed exploitation of resources for economic purposes. John Muir, who is usually cited as the first American preservationist, condemned the common perception of wilderness as an economic resource. Muir and other preservationists argued that the American wilderness possesses an inherent value that must be protected from commercial exploitation. Most historians point to the battle over the damming of Yosemite's Hetch Hetchy Valley at the beginning of the twentieth century as the first major conflict between conservationists and preservationists.

Although preservationists successfully excluded most commercial development from the national park system, conservation remained the dominant ethic in land-use management in the United States for the first half of the twentieth century. However, preservation gained a powerful new constituency in the years after World War II. A rising standard of living allowed growing numbers of Americans the opportunity to pursue leisure activities. Many such Americans, who typically lived in urban or suburban areas, supported preservation because it afforded them unspoiled settings in which to enjoy outdoor activities such as camping.

The growing demand for wilderness areas influenced government officials charged with setting land-use policy, culminating in 1964 with the passage of the Wilderness Act. Throughout the 1960's and 1970's, politicians responded to the American public's continued support for preservation. However, during the 1980's and into the 1990's, government officials increasingly favored multiple-use management of public lands over preservation.

The history of preservation shows that it served as a response to the success of capitalism as an economic system. Ironically, resource exploitation allowed Americans the income and free time to enjoy the outdoors while at the same time significantly altering the environment. For many Americans, the negative consequences of the exploitation of natural resources began to outweigh the benefits that development brought to society. Environmentalists articulated this concern, which even appeared in the wilderness legislation of the 1960's. Politicians, including President Lyndon B. Johnson, began discussing the importance of beauty and the value of nature.

Critics have contended that the preservation ethic is an expression of middle- and upper-class Americans who already enjoy the benefits of economic exploitation of resources; the needs of lower-class citizens who would profit from the continued development of natural resources are disregarded in favor of leisure activities for the affluent. However, preservation constitutes more than an approach to land-use management that calls for the maintenance of natural playgrounds for middle-class Americans. It also serves as more than a muddled critique of capitalism. It expresses a value system, and as such it could be counted as a philosophy that has important political and social implications.

Preservation Versus Conservation

In discussions of preservation's philosophical import, preservation is usually contrasted with conservation. Whereas conservation understands nature in terms of resources that have value to human economies, preservation regards nature as possessing additional value to humans, bringing aesthetic and even spiritual qualities to human life. Conservationists perceive forests, minerals, and wildlife as separate categories, whereas preservationists argue that such categorization is artificial. Preservationists assert that nature must be understood as a whole in which the apparently individual parts are intimately interconnected in ways that humans, who are a part of the natural system, cannot fully understand. The failure to preserve wilderness thus has consequences that ripple through the entire ecosystem, including human society.

Some environmentalists argue that the distinction between conservation and preservation, while useful for defining advocacy groups and management plans, is simplistic and misleading. Rather than perceiving the two as mutually exclusive, adversarial positions, they maintain that preservation and conservation are two tools for understanding the human relationship to the environment. Some radical environmentalists dismiss preservation on the grounds that, like conservation, it assesses nature according to human values, albeit a wider range of such values than are applied by conservation. They maintain that preservation ultimately fails as a philosophy because it does not understand that nature exists wholly apart from any value system. It is thus faulty because it does not lead to the complete reform of human perceptions necessary to end the destruction of the environment.

Despite such criticisms, preservation remains an important environmental ethic at the beginning of the twenty-first century. In the United States and elsewhere, preservationists can claim that their efforts have saved wilderness regions that otherwise would have been devastated by mining, logging, or other activities. These victories, however, also require preservationists to address difficult issues regarding the management of wilderness areas; lively debates continue around such issues as scientific study within wilderness preserves, fire suppression, and the reintroduction of wildlife species, such as wolves. Moreover, preservationists often must work to ensure the establishment of standards concerning air and water quality outside preserved areas in order to protect the integrity of the wilderness.

Thomas Clarkin

FURTHER READING

Allin, Craig W. *The Politics of Wilderness Preservation.* 1982. Reprint. Fairbanks: University of Alaska Press, 2008.

Cawley, R. McGreggor. *Federal Land, Western Anger: The Sagebrush Rebellion and Environmental Politics.* Lawrence: University Press of Kansas, 1993.

Davis, Charles, ed. *Western Public Lands and Environmental Politics.* 2d ed. Boulder, Colo.: Westview Press, 2001.

Lewis, Michael. *American Wilderness: A New History.* New York: Oxford University Press, 2007.

Nash, Roderick. *Wilderness and the American Mind.* 4th ed. New Haven, Conn.: Yale University Press, 2001.

Oelschlaeger, Max. *The Idea of Wilderness: From Prehistory to the Age of Ecology.* New Haven, Conn.: Yale University Press, 1991.

SEE ALSO: Conservation; Conservation movement; Conservation policy; Deep ecology; Muir, John; National parks; Nature preservation policy; Wilderness Act; Wilderness areas.

Price-Anderson Act

CATEGORIES: Treaties, laws, and court cases; nuclear power and radiation

THE LAW: U.S. federal law that limits the liability of nonmilitary nuclear facilities in the United States

DATE: Enacted on September 2, 1957

SIGNIFICANCE: Passage of the Price-Anderson Act enabled the commercial development of nuclear energy in the United States by making it possible for nuclear power plants to purchase liability insurance.

The Price-Anderson Act (also known as the Price-Anderson Nuclear Industries Indemnity Act) limits the extent of liability of a nuclear power plant for a nuclear accident in the United States. Before passage of the act, no insurance company would insure a nuclear power plant, and thus no such plants could function, as the possibility existed that claims resulting from any accidents would bankrupt the facilities. Many believed that nuclear power would never be developed in the United States unless some insurance plan was made available, and the Price-Anderson Act was designed to remedy this problem. The act required each nuclear facility to purchase the maximum amount of available insurance, only $60 million in 1957. The government would then be liable for another $500 million. Plants would not be allowed to deny claims even if they could show that they were not at fault in any accidents; claims were to be paid whether a plant was negligent or not.

The act was originally limited to ten years because it was thought that after nuclear plants had shown they could operate safely, commercial insurance would become available. In 1966, insurance was still not available, however, and the act was extended until 1976. In addition to extending the act, Congress added a provision that made it easier for damaged parties to make claims against a plant; the provision defined a nuclear accident in such a way that a claimant only had to show a monetary value for injury or property damages owing to radiation exposure. The act was extended another ten years in 1975, when the insurance plan was changed again, to the plan that has since remained in place. The act was renewed with new monetary values both in 1988 and in 1998.

In 2005, the act was renewed for twenty years with the following values, which are to be adjusted for inflation. If an accident occurs, the insurance pays the first $300 million in damages. The remaining damages are paid from a pool of funds supplied by the nuclear power plants; each plant pledges $111.9 million to the pool to be paid at a rate of $17.5 million per year if an accident occurs. In 2008 the pool was about $11.6 billion. Damages that exceed the insurance and pool values are guaranteed by the government, but Congress must decide who is to make the payments. As of 2008, only $151 million in claims had been paid out, and all of that had been paid by the insurance.

Critics of the Price-Anderson Act argue that it is a government subsidy to the nuclear power industry, which could not exist without the subsidy. The act thus favors nuclear power at the expense of other energy sources. Another complaint is that the plants are protected even in cases of negligence or misconduct. Many have objected also that the amount of the insurance is not enough—a large accident would result in much higher costs than could be covered by the funds available.

C. Alton Hassell

FURTHER READING

Collier, Charles S. *Are the "No Recourse" Provisions of the Price-Anderson Act Valid or Unconstitutional?* Washington, D.C.: Jarboe, 1965.

Greenberg, Michael R. "Time and Flexibility Criterion: Nuclear Power Revisited." In *Environmental Policy Analysis and Practice.* New Brunswick, N.J.: Rutgers University Press, 2007.

Murray, Raymond L. "Reactor Safety and Security." In *Nuclear Energy: An Introduction to the Concepts, Systems, and Applications of Nuclear Processes.* 6th ed. Burlington, Vt.: Butterworth-Heinemann/Elsevier, 2009.

U.S. Senate Committee on Environment and Public Works. Subcommittee on Transportation, Infrastructure, and Nuclear Safety. *Price-Anderson Act Reauthorization: Hearing Before the Subcommittee on Transportation Infrastructure and Nuclear Safety of the Committee on Environment and Public Works, United States Senate, One Hundred Seventh Congress, Second Session on January 23, 2002.* Washington, D.C.: Government Printing Office, 2003.

SEE ALSO: Atomic Energy Commission; Breeder reactors; Chalk River nuclear reactor explosion; Chernobyl nuclear accident; Hanford Nuclear Reservation; Nuclear accidents; Nuclear and radioactive waste; Nuclear power; Nuclear Regulatory Commission; Nuclear regulatory policy; Radioactive pollution and fallout; Rocky Flats, Colorado, nuclear plant releases; Three Mile Island nuclear accident.

Privatization movements

CATEGORIES: Philosophy and ethics; land and land use

DEFINITION: Efforts to protect individual property by weakening governmental policies and programs intended to protect the environment

SIGNIFICANCE: Privatization movements, which have both intellectual and advocacy components, are concerned with changing the relationship between government and society. They have had some successes and have raised some important questions. In spite of the efforts of some privatization advocates, these movements have not always exerted positive influences on the environment.

Proponents of environmental privatization emphasize the primacy of individual property rights in the American political tradition. Privatization movements generally take one of two related approaches. One group argues that privatization can provide better protections for the environment than can existing governmental regulatory programs. The second group simply contends that Americans have the right to do what they want with their own property; those in this latter group often display scant concern for the environment.

All efforts to privatize the environment are infused with the views of the seventeenth century English philosopher John Locke concerning the primacy of individual property rights. One group, however, maintains that protecting the environment is also an important goal. Proponents of this view argue that common-pool resources such as water or wilderness are subject to the "tragedy of the commons" because individuals have no incentive not to exhaust (or pollute) resources that belong to everyone. One response to the tragedy of the commons is extensive governmental regulation.

Advocates of privatization decry this approach on two grounds. First, they argue that it flies in the face of American political traditions regarding the importance of private property that go back to the founding of the nation. Second, they contend that governmental regulation will only exacerbate tendencies to use up or pollute a resource because if a resource belongs to everyone, it belongs to no one, and no one will take responsibility for protection. Privatizing a resource will guarantee protection by the property owner. For example, an owner of a forest will be disinclined to see it harmed by pollution and will take appropriate action to protect it. Such privatization advocates often consider themselves ardent environmentalists and contend that privatization can protect the environment more efficiently than can governmental regulation. Although they contend that humans have primacy over nature, they maintain that protecting the environment is good for humankind and should be done.

A second group of advocates of privatizing the environment also starts from the Lockean perspective of the primacy of individual property rights. More so than those in the first group, they are often suspicious of government, especially the national government. This group, however, views the environment as something to be exploited for the benefit of individual property owners. They do note that a wise owner might wish to leave something for future generations, but this is an individual decision. Some advocates of this form of privatization regard natural resources as inexhaustible or easily substitutable through the operation of market forces. In their view, governmental regulation of the environment is wrong because it interferes with individual property rights, not because it

may be an ineffective means of protecting the environment.

Both groups of advocates of privatization are opposed to the command-and-control policies of environmental protection used in the United States since the 1970's. They argue that governmental regulations such as the Clean Air Act and efforts at preserving species interfere with Americans' rights to do whatever they wish with their property. During the late 1970's and early 1980's, a movement in the western United States known as the Sagebrush Rebellion often resorted to court actions (and sometimes violence) to try to weaken various environmental laws and regulations, such as the Endangered Species Act, that limited the exercise of property rights. Others described a philosophy known as "wise use," the view that natural resources should be used wisely, which usually means providing a profit for their owners. This movement achieved some success during the presidency of Ronald Reagan as agencies such as the Environmental Protection Agency scaled back enforcement of environmental regulations. Some advocates of privatization called for massive sales of federal lands in the western states, saying that the national government should not own natural resources.

These efforts at privatization generally had negative impacts on the environment. What land sales occurred benefited large businesses, not small property owners. Meanwhile, the attack on environmental regulation slowed efforts to achieve a cleaner environment. One aspect of the privatization efforts of the 1980's that has had a more positive impact has been the attempt to improve environmental quality through market-based solutions that take property rights into account. These programs are based on the concept that individual property owners will be more effective and efficient in protecting the environment, as they enhance their own property, than a program based on governmental regulation of industry. While programs such as emissions trading have not always been completely successful, they represent an innovative means of dealing with environmental issues such as air pollution.

Movements to privatize the environment have often been part of larger movements characterized by suspicion of government and the desire to decrease its impact. These movements thus are directly opposed to the philosophy of the early twentieth century Progressives, such as Theodore Roosevelt, who maintained that the protection of the environment is the responsibility of everyone, including the government. Advocates of privatization rely on the workings of the market to protect the environment but do not readily accept that the market does not always work in favor of the environment. Their narrow interpretation of the views of America's Founding Fathers on property rights sometimes verges on distortion. The Founders were concerned with protecting individual property, but they were also concerned with achieving the common good, even if this required governmental action.

John M. Theilmann

FURTHER READING

Anderson, Terry L., and Donald R. Leal. *Free Market Environmentalism*. Rev. ed. New York: Palgrave, 2001.

Cawley, R. McGreggor. *Federal Land, Western Anger: The Sagebrush Rebellion and Environmental Politics*. Lawrence: University Press of Kansas, 1993.

Davis, Charles, ed. *Western Public Lands and Environmental Politics*. 2d ed. Boulder, Colo.: Westview Press, 2001.

Echeverria, John D., and Raymond Booth Eby, eds. *Let the People Judge: Wise Use and the Private Property Rights Movement*. Washington, D.C.: Island Press, 1995.

Ostrom, Elinor, ed. *The Drama of the Commons*. Washington, D.C.: National Academy Press, 2002.

Vig, Norman J., and Michael E. Kraft, eds. *Environmental Policy: New Directions for the Twenty-first Century*. 7th ed. Washington, D.C.: CQ Press, 2010.

SEE ALSO: Antienvironmentalism; Land-use policy; Property rights movement; Sagebrush Rebellion; Wise-use movement.

Property rights movement

CATEGORIES: Land and land use; philosophy and ethics

IDENTIFICATION: Grassroots movement that seeks to reaffirm property rights of landowners against governmental regulations or intrusions in the name of environmental protection

SIGNIFICANCE: The property rights movement has achieved substantial affirmation in the courts, academia, and legislatures and has had a significant impact on the U.S. government's environmental policies.

The apparent paradox that the property rights movement brings to light is worth pondering: The strongest property rights movement in the world has emerged and developed in the United States, the country that has arguably the world's most vigorous protection and defense of private rights and private property. The groups most closely identified with the movement oppose most federal regulations and intrusions on land that is privately owned and held, especially in cases where involvement of the federal government comes in the form of environmental laws or regulations that limit the owners' full or partial use of land for development. Despite the fact that these groups unequivocally uphold the constitutionally enshrined idea of private property, they generally recognize and accept that there are and will always be limits set by government to the use of privately owned land in the name of the public good. The basis of their discontent is arguably about the ways in which similar plots of land are treated differently, because land-use rules and regulations are usually formulated and enforced after substantial and at times excessive development has already taken place.

BASIS OF THE MOVEMENT

While the movement's position has some initial plausibility, and may even seem to be underwritten by the equal protection clause of the U.S. Constitution, it is based on a particular conception of property rights viewed *in space* rather than *in time*. Thus one may raise the following question: Is the issue at stake whether similar plots (in space) are treated similarly, or is it rather about whether circumstances affecting land use and development have changed *over time*, and whether these new circumstances subsequently warrant that some limitations be placed on the use of privately owned land? If it is the latter, then it is easy to understand how the courts may be compelled to consider the validity and legality of a time-sensitive argument in support of limits, laws, and regulations of the use of private land for development purposes for the sake of a greater public good, namely, environmental protection or conservation of biodiversity. Some supporters of this movement have even sought to cast the issue in broader political terms by making it about the role and obligations of government in the balancing act between private and public good.

Although those involved in the property rights movement can be said to be among numerous individuals, groups, and organizations that fall under the classification of "environmental opposition," the movement must be distinguished from the so-called wise-use movement, which grew out of the Sagebrush Rebellion of the mid-1970's and during which state legislators in western U.S. states sought to transfer federal public lands to state control. Advocates of wise use support an antigovernmental regulatory agenda related to the use of public land and resources. In contrast, the property rights movement is about the use of privately owned and held lands.

IMPACTS

The property rights movement achieved significant impacts on the nation's environmental policy agenda in the 1980's during the presidency of Ronald Reagan and then gained even more ground in the 1990's when it resurfaced with the rebellion of local grassroots organizations comprising individuals seeking to develop their own property, usually by building homes, clearing out trees or brush, or draining wetlands. Many property owners did not know about federal regulations and laws that could thwart their projects, such as provisions of the Clean Water Act and the Endangered Species Act. After being prohibited from developing their property by the federal government, they often joined other frustrated property owners in their areas who had similarly been prohibited from doing what they wanted with their land. Typically, they shared their respective grievances against the government based on their individual disputes.

These individuals were and continue to be most active in eastern and southern states, where titles to land owned have often been in the names of the owners' families for many generations. The assumption has been historically that the right to control land belongs to the title holder, regardless of changes in the law or public policy. Many of the individuals who have been active in the property rights movement have been farmers, ranchers, and owners of rural or beachfront property who did not know much about the ecological value of their land until they decided to develop it. This gave rise to a nationwide debate about competing land-related interests—that is, the compelling rights of property owners to use their land as they see fit versus the government's compelling interest in controlling pollution, protecting biodiversity, preserving endangered species and their habitats, and managing ecosystems or even other landowners' property.

INTERPRETATIONS OF PROPERTY RIGHTS

From a historical point of view, property rights are based on English common law and the Magna Carta. However, there has been an evolution in legal interpretation of these rights since the 1920's. Most of the legal cases since the late twentieth century have dealt with the concept of federalism and, more specifically, with the Fifth Amendment to the Constitution. One of its clauses refers to "takings," a requirement that the government cannot take privately owned land for public use without compensating the owner for the adjusted value of the land. By properly placing the takings clause in the context of wilderness designations, endangered species, and wetlands protections, one could provide an intellectual and legal basis for the claimants of the property rights movement against the government.

Private property owners have often invoked the takings clause in seeking compensation from the government by filing lawsuits before the U.S. Court of Federal Claims or the U.S. Supreme Court. Since 1987 the courts have generally ruled that the government must compensate property owners for the loss of the use of their land when that loss is the result of federal regulations, such as the Clean Water Act, that deny the owner the economically viable use of the land. President Reagan further expanded the rights of property owners by executive order and supported the enactment of regulations requiring government agencies to evaluate the risk of unanticipated takings. Since 1988 the government has even been required to budget funds for takings impact analyses that protect the constitutional rights of property owners.

Although the law on these issues is bound to evolve and change over time as activists on both sides continue to press for their respective agendas, it is unlikely that the property rights movement will wither and disappear altogether from the scene. Those in favor of stronger and more stringent environmental protections will have to contend with this fact for the foreseeable future.

Nader N. Chokr

FURTHER READING

Dana, David A., and Thomas W. Merrill. *Property: Takings*. New York: Foundation Press, 2002.
Davis, Charles, ed. *Western Public Lands and Environmental Politics*. 2d ed. Boulder, Colo.: Westview Press, 2001.
Eagle, Steven J. "The Birth of the Property Rights

Movement." *Policy Analysis* 404 (June 26, 2001): 1-40.

Epstein, Richard. *Takings: Private Property and the Power of Eminent Domain.* 1985. Reprint. Cambridge, Mass.: Harvard University Press, 1998.

Freyfogle, Eric. *The Land We Share: Private Property and the Common Good.* Washington, D.C.: Island Press, 2003.

Yandle, Bruce, ed. *Land Rights: The 1990's Property Rights Rebellion.* Lanham, Md.: Rowman & Littlefield. 1995.

See also: Antienvironmentalism; Clean Water Act and amendments; Eminent domain; Endangered Species Act; Environmental Protection Agency; Land-use planning; Land-use policy; *Massachusetts v. Environmental Protection Agency*; National Environmental Policy Act; Natural Resources Conservation Service; Privatization movements; Resource Conservation and Recovery Act; Sagebrush Rebellion; Wise-use movement.

Public opinion and the environment

Categories: Philosophy and ethics; activism and advocacy

Significance: Public opinion is one of the primary factors determining the degree of environmental protection in a society.

A clean environment is a highly desirable amenity, and on occasion very important in terms of long-term health. It can also be expensive in many ways. As a result, except in extreme cases, environmental protection is a luxury rather than an immediate necessity. Everyone wants a clean local environment, as long as the price they pay personally is reasonable. This makes spectacular events, such as the Cuyahoga River fires of 1952 and 1969 and large oil spills, especially significant.

Historical Examples

It has long been noticed that more affluent populations have cleaner environments. A 1992 World Bank study reported that concentrations of airborne particulates started to decline with a per-capita gross national product (based on purchasing power parity) of $3,280, sulfur dioxide at $3,670, and fecal coliform bacteria in river water at $1,375. Access to safe water and adequate sanitation is thus more immediately important than access to clean air (which is less visible). Similar data reveal that U.S. sulfur dioxide emissions per capita peaked in 1920, particulates and carbon monoxide have declined steadily since 1945, and nitrogen oxides peaked in 1980. A 1992 study by the World Health Organization found that air pollutants were generally lower in most megacities (except for Los Angeles) in the noncommunist developed world than in developing nations.

This situation is caused in part by the fact that more affluent societies can afford expensive water treatment, plumbing, and air-pollution controls. For example, from 1972 to 1994 annual pollution-control expenditures in the United States (in 1986 dollars) increased steadily from $26 to $127 billion. This grew especially rapidly starting in 1970 (when the Environmental Protection Agency was created), the year after the Santa Barbara, California, oil spill and the last Cuyahoga River fire. Part of the difference between affluent and developing societies (especially noticeable in pollutant levels), however, results from the fact that less affluent societies have more urgent concerns.

After the fall of the Communist dictatorship of the Soviet Union, Green parties did well in many places in the former Soviet countries, especially in local elections. Environmental quality there was often atrocious, demonstrated by such notorious problems as the massive shrinking of the Aral Sea and catastrophic levels of air and water pollution from heavy industry that led to high levels of infant mortality and birth defects, food contamination, and environmentally induced illnesses. In addition, attempts to fix old Soviet industrial facilities often led to their closure, at least temporarily, which resulted in serious loss of production even of necessary supplies (such as pharmaceuticals) and led to a serious backlash against environmental protection (much of it by the authorities, but ultimately also by voters).

Twenty-first Century Issues

The struggle to save endangered species reveals the important of public opinion. An estimated 95 percent of endangered species are tiny, obscure plants (including algae and fungi) or invertebrates (insects, worms, and mollusks), mostly tropical. Groups that seek to protect endangered species, however, generally use more popular animals, especially birds (such as the spotted owl and the whooping crane) and large mammals (such as polar bears and various whale species), as emblems.

Another example of the importance of public opinion can be seen in the effects of the severe economic downturn that started in 2008. During this crisis, environmental activism lost much of its popularity with the American public. A spill as spectacular as the one created by the blowout of the BP *Deepwater Horizon* oil rig in 2010 may in previous decades have led to strong popular demand to end offshore oil drilling (as happened after the 1969 spill off the shores of Santa Barbara). Instead, polls showed strong popular support for continued drilling, not only from those involved in the drilling but also often from those whose occupations were harmed by the spill.

The most significant environmental dispute in the United States involves global warming, with debates concerning whether the cause is primarily anthropogenic (human-created) or natural and whether the results will be catastrophic. Both sides rely heavily on emotional appeals rather than on rational discussion of the science. Skeptics primarily discuss the expense of measures to reduce the greenhouse gas emissions linked with global warming, a tactic that is especially effective among those (such as coal miners) most affected by proposed preventive measures. Some supporters of taking action to reduce greenhouse gases greatly exaggerate the likely effects of global warming and ignore (and even smear as corrupt) any scientific dissent; some have been accused of falsifying data.

Cautionary Notes

Acid rain provides a very instructive lesson regarding the influence of public opinion on environmental policy. Starting around 1980, acid rain was a major theoretical concern. The fear was that acid rain was causing significant damage to forests, eroding buildings, and causing increasingly acidic lakes (often leading to the disapperance of fish). In response to public concerns, the United States set up in the National Acidic Precipitation Assessment Project, which after ten years of study concluded that acid rain is primarily an aesthetic concern. Acidic lakes result primarily from acidic deposits from land; many such lakes were found to have been only temporarily neutral or alkaline as a result of heavy lumbering, which had since ceased. Acid rain causes only minor damage to high-altitude forests, while actually fertilizing the soil with additional nitrogen and sulfur. (Later, many global warming scientists concluded that sulfate haze had also helped cool the atmosphere, explaining why global temperatures had increased less than expected.)

During the 1980's, however, the news media operated in standard crisis mode, making the unproven theory of serious damage caused by acid rain appear to be a major proven crisis. As a result, significant pollution controls were put into place to manage a relatively minor problem. It was certainly desirable for these controls to be instituted, but the resources available for pollution control are not unlimited, and it is possible that those resources could have been better used to deal with other, more serious, environmental problems.

Timothy Lane

Further Reading

Bailey, Ronald, ed. *The True State of the Planet.* New York: Free Press, 1995.

Feshbach, Murray, and Alfred Friendly, Jr. *Ecocide in the USSR: Health and Nature Under Siege.* New York: Basic Books, 1992.

Lomborg, Bjørn. *The Skeptical Environmentalist: Measuring the Real State of the World.* New York: Cambridge University Press, 2001.

Mooney, Chris. *Storm World: Hurricanes, Politics, and the Battle over Global Warming.* Orlando, Fla.: Harvest Books, 2008.

Murray, Iain. *The Really Inconvenient Truths.* Washington, D.C.: Regnery, 2008.

See also: Acid deposition and acid rain; Airborne particulates; Aral Sea destruction; Cuyahoga River fires; Global warming; Oil spills; Polar bears; Smog; Sulfur oxides; Whaling; Whooping cranes.

Public trust doctrine

Categories: Treaties, laws, and court cases; land and land use

Definition: Legal principle that government is the custodian of common properties for the use of the public

Significance: The public trust doctrine, an important concept in environmental law, has been the basis of legal arguments regarding access of the public to beaches and streams on private land. The doctrine also limits the uses of privately held wetlands.

The origins of the public trust doctrine can be traced to Roman law, and the development of the doctrine can be seen in English common-law decisions. The public trust, in common with the better-known private trust, involves an asset managed by a trustee in accordance with the wishes of the trustor for named beneficiaries. Unlike the private trust, however, the public trust is based not on a written document but on the legal fiction that the creator gave trust assets to government to hold for the people.

In *Shively v. Bowlby* (1894), the U.S. Supreme Court concluded that the original states held the tidewaters and their lands in trust and that states subsequently admitted to the union held the same title. In the hands of state and federal judges, the public trust doctrine has become a continually growing and sometimes conflicting body of state common law that has defined the meaning of trust assets and terms, the extent of state governments' trustee duties, and the identity of trust beneficiaries.

Judicial decisions have expanded the scope of state trust assets beyond tidal waters and their use for navigation. Now the public trust assets include all tidal waters, not just navigable ones, and their use extends to recreation, fishing, and wildlife habitat. The state trust assets also include rivers, wetlands, lakes, parks, trees, and wildlife, and may even include entire ecosystems. The public trust also lies latent within private property rights and provides the public with access to beaches and streams on private land. The doctrine also limits the uses of privately held wetlands. In *Marks v. Whitney* (1971), the California Supreme Court held that the state's public trust easement for navigation, commerce, and fisheries forbids an owner to fill and develop such wetlands.

State governments, as trustees, have an obligation to regulate, manage, develop, and preserve trust property for the benefit of the public. State legislatures and agencies may not make decisions that merely reflect current political, economic, and social needs because they have a fiduciary duty to a present and future public constituency. This duty does not, however, restrict trust assets to their present use but requires state governments to identify the impacts on trust assets in their planning processes and balance competing trust interests in their decision making. In *National Audubon Society v. Superior Court* (1983), the California Supreme Court held that the Los Angeles Department of Water and Power did not have the unrestricted right to appropriate the waters of streams that flowed into Mono Lake and to divert them to the Los Angeles Aqueduct. The state Water Resources Board had a duty to take public trust interests into account in the planning and allocation of water resources in a manner that would protect the lake's trust uses.

State governments may violate their trust responsibilities in three ways. First, they may alienate trust assets by selling them to a private party. In *Illinois Central Railroad v. Illinois* (1892), the U.S. Supreme Court held that the state legislature had violated its public trust duty when it made a grant to the railroad of 405 hectares (1,000 acres) of submerged Chicago waterfront property. When the legislature subsequently withdrew the grant, it did not take private property without just compensation because the railroad had held only a revocable title.

State governments may also divert ownership of public trust assets from one public use to another by transferring them from one government agency holding public trust lands to another agency with a mission to develop public property. In *Robbins v. Department of Public Works* (1968), the Massachusetts Supreme Judicial Court held that the state park agency's transfer of the Fowl Meadows wetlands to the state department of public works for highway use violated the agency's responsibility to protect the parklands for the public.

Finally, state agencies may derogate trust assets by destroying them. The Los Angeles Department of Water and Power could not destroy Mono Lake by its continued appropriation of lake waters, nor may private people or corporations pollute trust assets. When a privately owned oil barge dumped thousands of gallons of crude oil into coastal marshes, killing wildlife and destroying ecosystems, the court in *In re Steuart Transportation Company* (1980) held that Virginia had the sovereign right, derived from the public trust doctrine, to protect the public interest in the preservation of wildlife resources.

Public trust doctrine and its principles have been incorporated into state constitutional provisions and environmental statutes. Pennsylvania's state constitution, like the constitutions of ten other states, recognizes that the state government is the trustee of the state's public resources, with the power to maintain and preserve the natural environment. The Michigan Environmental Protection Act, which has served as a model for similar statutes in other states, provides that the government hold in trust the air, water, and public resources of the state and bestows upon the courts the authority to decide whether those resources have been or are likely to be polluted, impaired, or destroyed. The public trust doctrine may even apply to the federal government because it holds in trust the public lands and waters of the nation, and, as *In re Steuart Transportation Company* has suggested, it has a duty to protect and preserve the public's interest in those natural resources.

William Crawford Green

FURTHER READING

Archer, Jack, et al. *The Public Trust Doctrine and the Management of America's Coasts.* Amherst: University of Massachusetts Press, 1994.

Goldstein, Robert Jay. "The Social Evolution of American Real Property Law." In *Ecology and Environmental Ethics: Green Wood in the Bundle of Sticks.* Burlington, Vt.: Ashgate, 2004.

Hart, John. *Storm over Mono: The Mono Lake Battle and the California Water Future.* Berkeley: University of California Press, 1996.

Light, Andrew, and Avner De-Shalit, eds. *Moral and Political Reasoning in Environmental Practice.* Cambridge, Mass.: MIT Press, 2003.

Organ, John F., and Gordon R. Batcheller. "Reviving the Public Trust Doctrine as a Foundation for Wildlife Management in North America." In *Wildlife and Society: The Science of Human Dimensions*, edited by Michael J. Manfredo et al. Washington, D.C.: Island Press, 2009.

SEE ALSO: Intergenerational justice; Land-use policy; Nature preservation policy; Privatization movements; Sustainable forestry; Water use.

Pulp and paper mills

CATEGORY: Pollutants and toxins

DEFINITION: Operations where raw materials are processed for the manufacture of paper

SIGNIFICANCE: The chemicals used in the primary industrial processes in pulp and paper mills present multiple environmental problems, including noxious odors, potentially toxic solid wastes, and polluted effluents.

The pulp and paper industries constitute an important segment of the world's economy. In the manufacture of paper, vegetable matter is reduced to a liquid slurry, excess liquid is removed from this pulp,

and the remaining fibers are formed into a mat. Materials used to produce paper include wood, cotton, hemp, and flax, depending on the desired product, but all require similar processing involving large amounts of water. Most kinds of paper contain at least a portion of wood, and if wood is used, the cellulose fibers in the wood that can be formed into paper must be separated from the lignin that surrounds them. Once the fibers are successfully separated, the resulting pulp mat may require bleaching to produce a paper that is sufficiently white for commercial use.

All the processes involved in the manufacturing of pulp and paper present potential environmental hazards, but the kraft process—the process used to reduce wood fibers to pulp—may present the greatest dangers. Prior to the invention of the kraft process, pulp mills used mechanical methods to crush plant matter and used rags to separate the unwanted material from the fibers necessary to make paper. The runoff from such mills was often dirty and thus posed some hazards to the environment, but because the processes used were mechanical rather than chemical, no new chemical compounds were formed.

AOX COMPOUNDS

In contrast with mechanical methods, the kraft process uses chemicals to break down the lignins in wood. Industry experts report that more than 70 percent of the world's pulp is produced using the kraft process. Well into the late twentieth century, many pulp and paper mills discharged untreated wastewaters from the pulping and bleaching processes directly into rivers and lakes. While the kraft process has allowed production of paper from woods once considerable unsuitable for paper, the resulting pulp is generally more difficult to bleach than pulps produced using other methods. The preferred bleaching agent has been chlorine dioxide, an acid that reacts easily with the residual lignin compounds in the pulp to form chlorinated organic compounds. Scientists refer to such compounds as adsorbable organic halogens (AOX), while the general public more commonly knows them as organochlorines and dioxins.

Researchers have shown that many of these compounds have mutagenic and carcinogenic properties; that is, excessive exposure may lead to birth defects, reproductive difficulties, and some forms of cancer. AOX compounds pose special threats because many accumulate in the fatty tissues of animals rather than passing through the body, and therefore they tend to

U.S. Paper and Paperboard Production
(millions of short tons)

	2003	2004	2005	2006
Total paper	40.37	41.82	41.40	41.81
Total paperboard	48.02	50.08	49.71	50.41
Unbleached kraft	21.73	22.67	22.58	23.41
Semichemical	6.10	6.53	6.41	6.22
Bleached kraft	5.36	5.65	5.66	5.70
Recycled	14.83	15.24	15.05	15.07

Source: American Forest and Paper Association, *Monthly Statistical Summary of Paper, Paperboard, and Woodpulp.*

become concentrated as they move up the food chain (a process known as biomagnification). Residents of the Great Lakes region of the United States, particularly children and women of childbearing age, have been advised to limit their consumption of lake trout, a fish that is known to have high concentrations of AOX in its fat.

Because of the environmental dangers associated with organochlorines, many countries, including the United States, have implemented AOX emissions standards. Some experts believe that the only acceptable standard would be zero discharge of AOX emissions. While some pulp and paper manufacturers have argued that achieving zero emissions would be prohibitively expensive, particularly if they are forced to modify existing mills, others have explored alternative manufacturing techniques, such as the use of closed-loop mills. In a closed-loop facility, no liquid waste leaves the factory. Effluents are treated on-site and cycled back through the system. Another option is conversion to oxygen bleaching, or totally chlorine-free (TCF) bleaching. Mills in European countries that converted to TCF bleaching found that their operating costs were actually lower than with previously used chlorine processes. In addition to posing less of a threat to the environment, oxygen-based bleaching compounds are less corrosive. This means that the equipment within a mill lasts significantly longer before requiring replacement.

OTHER ENVIRONMENTAL HARMS

In addition to the well-publicized hazards of AOX, pulp and paper mills present a number of other envi-ronmental problems. People living near pulp mills often complain about noxious odors. Strong odors are often a by-product of the chemical digesting process used in pulping wood, but to date researchers have been unable to find any harmful effects of these odors other than psychological discomfort and, thus, lowered quality of life for those exposed. The solid wastes produced by mills may also have negative impacts on the environment. The exact composition of these wastes may be difficult to determine, particularly when mills use recycled materials. Most pulp and paper mills routinely use the waste sludge and liquids from the digester as boiler fuel for steam turbines to generate electricity. This reduces the volume of potential solid waste but cannot completely eliminate it. The industry attempts to recycle as much of the solid waste from mills as possible, but optimum efficiency has yet to be achieved in this area.

The world's ever-growing appetite for paper means that hundreds of thousands of hectares of forest are harvested annually to be processed into pulp. In addition to the obvious environmental harms of deforestation, possible erosion, and destruction of unique ecosystems and wildlife habitat, harvesting exposes timber industry workers and the general public to dust, debris, and exhaust fumes from logging equipment.

Nancy Farm Männikkö

FURTHER READING

Clay, Jason. "Wood Pulp." In *World Agriculture and the Environment: A Commodity-by-Commodity Guide to Impacts and Practices.* Washington, D.C.: Island Press, 2004.

Miller, G. Tyler, Jr., and Scott Spoolman. *Living in the Environment: Principles, Connections, and Solutions.* 16th ed. Belmont, Calif.: Brooks/Cole, 2009.

Rodd, Tony, and Jennifer Stackhouse. *Trees: A Visual Guide.* Berkeley: University of California Press, 2008.

Wright, David A., and Pamela Welbourn. *Environmental Toxicology.* New York: Cambridge University Press, 2002.

SEE ALSO: Biomagnification; Birth defects, environmental; Dioxin; Odor pollution; Water pollution.

R

Race-to-the-bottom hypothesis

CATEGORY: Resources and resource management

DEFINITION: Theory that the mobility of international capital pits workers and governments against one another in a competition to underbid and underregulate that results in increasingly primitive working conditions and increasingly lax environmental laws

SIGNIFICANCE: If the race-to-the-bottom hypothesis is correct, the expected outcomes include widespread environmental deregulation and consequent degradation, especially in the developing world.

According to the race-to-the-bottom hypothesis, globalization dilutes not only the bargaining power of laborers but also the ability of governments to regulate industrial pollution. Desperate to build and maintain their tax bases, states in the developing world compete to appease capital sources with increasingly business-friendly and environment-hostile standards.

Critics of the hypothesis argue that while it may be the case that environmental regulations, working conditions, and pay may have weakened in some regions in the short run, some investment is almost always better than no investment, and the trend will inevitably reverse when market equilibrium has been neared, unemployment has sufficiently diminished, and a requisite degree of regulations have been enacted in an adequate number of states. Some also assert that the process is a necessary growing pain of industrialization, and the developing world's quality of life and environmental integrity will continue to improve and eventually match that of the global North.

The empirical evidence is unclear. Anecdotal cases of capital flight can be cited, as well as Ireland's notable attempt to attract capital by lowering environmental standards, and some studies have shown a negative correlation between regulation and investment. Other studies, however, have found environmental regulatory improvement in so-called export processing zones, where one might expect any actual race to the bottom to be most evident. Proponents of the theory, however, maintain that given the profit motive of investors and unquestionably looser restrictions in poorer regions, it is at least reasonable to expect market forces to encourage a suboptimal hovering near the middle, if not an all-out race to the bottom.

Matt Deaton

SEE ALSO: Ecological economics; Environmental economics; Free market environmentalism; Globalization.

Radioactive pollution and fallout

CATEGORIES: Nuclear power and radiation; pollutants and toxins

DEFINITIONS: Radioactive pollution is environmental contamination resulting from the release of radioactive materials; fallout is the radioactive particles that fall to earth after detonation of nuclear weapons

SIGNIFICANCE: The use of radioactive materials for civilian and military applications has helped to disseminate radioactivity throughout the environment. Because of the long half-lives of some radioactive pollutants, their detrimental effects on ecosystems and human health can persist for generations and pose major challenges for safe, long-term disposal efforts.

Many people are well aware of the use of radioactive elements in nuclear weapons and reactors. Likewise, the general public understands that nuclear war, nuclear weapons testing, and nuclear reactor accidents such as the disastrous Chernobyl plant meltdown of 1986 can have profound, far-reaching, and long-lasting environmental impacts. It may be less well known, however, that radioactive elements are sometimes used for their physical or chemical properties rather than their radioactivity, and that these less familiar applications can cause small amounts of radioactive pollution. For example, uranium and thorium compounds have been used for centuries to give ceramic glazes brilliant orange and yellow hues. (Although pieces produced in the United States after about 1950 are generally safe to use, it is recommended that other ceramics with uranium or thorium glazes be left for occasional show unless they are tested and known to be safe for use.)

In addition, until the 1980's trace amounts of uranium were used to color porcelain teeth, and tiny amounts of uranium or thorium are present in some tinted contact lenses, eyeglass lenses, and other optical glass. Gas lantern mantles have long used thorium to produce a bright, white glow. Although the hazard to the public is believed to be negligible, nonthorium mantles are now available as an alternative. Tungsten electrodes for arc welding may contain 2 percent thorium for easier starting and greater weld stability. The resulting radiation dose to both the welder and the public is very small. Ionization smoke detectors use tiny amounts of radioactive americium to ionize air in a small chamber. This allows a current to flow. Smoke particles reduce this current and trigger the detector. Americium has a 458-year half-life, so it will remain radioactive long after the smoke detector's service life is over. While a smoke detector poses little threat when installed and used properly, its radioactive content can present a health hazard once the smoke detector is discarded. In order to keep americium out of the waste stream, many U.S. state and local governments conduct roundups of ionization smoke detectors or encourage consumers to return used smoke detectors to their manufacturers.

Three days after the accident at the Chernobyl nuclear power plant in April, 1986, a specialist takes air samples in Warsaw, Poland, to test for radiation left behind by a radioactive cloud that passed over Poland from the Chernobyl site in Ukraine. (AP/Wide World Photos)

INDUSTRIAL, MEDICAL, AND SCIENTIFIC USES

Because small amounts of radioactivity are easily detected, both industries and the medical field use radioactive elements as tracers. For example, radioactive technetium may be injected into a patient's vein so that its progress can be followed with radiation detectors. Imaging systems can then reveal constrictions in the patient's heart or arteries. The technetium used has a six-hour half-life. The nondecayed portion of technetium is eventually eliminated from the body and passes into the sewage system. No special precautions are taken because the radioactivity quickly decays and is greatly diluted with normal waste. Worldwide, more than 30 million diagnostic procedures using various radioactive elements are performed each year.

Radioisotope thermoelectric generators (RTGs) were developed to supplement the power from solar cells or replace them on space missions where sunlight is too weak. The Jupiter mission's *Galileo* spacecraft carried two RTGs, while the Saturn mission's *Cassini* spacecraft carried three. A standard RTG uses the heat from the radioactive decay of 11 kilograms (24 pounds) of plutonium dioxide to produce electricity. Because the process involves no moving mechanical parts, it is very reliable. RTGs are built to withstand the explosion of the spacecraft during launch as well as the heat of reentry.

Although the United States has used RTGs in many space missions, it has had only three accidents involving these power sources. In 1964 a navigational satellite failed to reach orbit and burned up over the Indian Ocean. Its RTG was of an earlier design that also burned up, as was then intended. The resulting plutonium oxide dust settled out of the stratosphere over the next several years. Greatly diluted across the globe, it barely increased background radiation. In 1968 a spacecraft was destroyed after launch by the range safety officer, and its RTGs were recovered intact. In 1970 the lunar module of the damaged Apollo 13 spacecraft reentered the atmosphere over the

South Pacific. Its RTG plunged into the Tonga Trench, which is 6 kilometers (3.7 miles) deep. Although it was never recovered, surveys have shown no release of radioactivity.

The Soviet Union used RTGs not only in spacecraft but also as power sources for lighthouses and other navigational beacons in remote locations. Since the breakup of the Soviet Union in late 1991, many of these RTGs have been unattended and have fallen into disrepair. Looters hoping to strip metal parts and sell them to recyclers have stolen unsecured RTGs and received high doses of radiation. Stolen RTGs have been abandoned in forests, dumped in the sea, and in one case left at a bus stop. The United States has taken in interest in helping to decommission Russian RTGs, as they are a source of radioactive material that could be exploited by terrorists.

DEPLETED URANIUM

Uranium is a widely distributed trace element. Its estimated concentration in the earth's crust is 2.7 parts per million. Pure uranium is a lustrous, silver-white metal, and although it is radioactive, the activity consists of alpha particles and low-energy gamma rays that can be readily shielded to safe levels. The ease with which uranium can be safely handled under favorable conditions has led to its increasing use. Many view this as dangerous.

Natural uranium consists of three different forms, or isotopes: about 0.7 percent is uranium 235, 99.3 percent is uranium 238, and only a trace amount is uranium 234. Weapons-grade uranium must be enriched to at least 90 percent uranium 235, while reactor fuel is generally enriched to 3 to 5 percent uranium 235. Extraction processes leave uranium that has only 0.2 to 0.3 percent uranium 235. This is called depleted uranium. Depleted uranium is only 60 percent as radioactive as natural uranium, but there is a lot of it. The United States entered the twenty-first century with an estimated 480,000 metric tons of depleted uranium; the total world inventory at the turn of the century was more than 1.1 million metric tons.

Rather than provide storage for it as low-level radioactive waste, the U.S. Department of Energy actively seeks uses for depleted uranium. Almost twice as dense as lead, it makes a good radiation shield, and it is used to shield radioactive isotopes shipped to hospitals. Ducrete, a special concrete containing depleted uranium, is used as radiation shielding in shipping casks for spent reactor fuel.

Depleted uranium packs a great deal of weight into a small volume; therefore, it is used to make counterweights for commercial aircraft and the tips of Tomahawk cruise missiles. Its density and hardness led to its use in the armor of the M1A1 Abrams tank and armor-piercing ammunition. Powdered uranium is pyrophoric (that is, it burns spon-

Soldiers with a peacekeeping force in Kosovo measure radiation levels in 2001 near a Yugoslav army tank that was destroyed during the 1999 bombing campaign conducted by the North Atlantic Treaty Organization (NATO) in the Balkans. Concerns had arisen that cases of cancer among peacekeeping soldiers stationed in Kosovo were linked to depleted uranium contained in the bombs used by NATO during the campaign. (AP/Wide World Photos)

taneously in air). Some of the depleted uranium in an armor-piercing round burns upon impact with a hard surface. This makes the round more effective against tanks because it helps the round to penetrate armor and often ignites a secondary explosion. While most of the uranium is expected to remain within several meters of the target, a significant amount may not. Some uranium oxide particles formed from burning uranium are smaller than 5 microns and can be carried by the wind for long distances. Such small radioactive particles can also lodge in the lungs of organisms and are potentially hazardous.

Hundreds of metric tons of depleted uranium ammunition have been used in Middle Eastern conflicts such as the 1991 Persian Gulf War and the 2003 invasion and subsequent occupation of Iraq. Depleted uranium ammunition was also employed during the 1992-1995 Bosnia-Herzegovina conflict and the 1999 Balkan war. While depleted uranium rounds are highly effective, their potential impact on the environment and the health of soldiers and civilians makes them a controversial weapon. During the mid-1980's, concerns about possible radiation exposure of ship crews led the U.S. Navy to switch from depleted uranium to tungsten rounds for the Phalanx guns used to defend military vessels against planes and missiles. The long-term health effects of depleted uranium use during wartime have yet to be established. The World Health Organization recommends that monitoring and cleanup operations be carried out in impact zones following military conflict if there is a reasonable possibility that depleted uranium could enter the food chain or groundwater in sufficiently high quantities to pose a health and environmental hazard.

FALLOUT FROM NUCLEAR WEAPONS

Fallout is the name given to radioactive particles that rain down from the debris cloud of a nuclear explosion. Whether the fallout is local or global depends chiefly on the yield of the weapon. Nuclear weapon yields are measured in terms of how many tons of the high-explosive trinitrotoluene (TNT) would be required to release the same energy. The bomb used in the American attack on the Japanese city of Hiroshima during World War II, for example, had a yield of approximately 13 kilotons. The largest nuclear weapons built by the United States had an estimated yield of 25 megatons (25,000 kilotons).

Yields of less than 100 kilotons produce local fallout, which consists of particles that fall to the ground within twenty-four hours of the explosion. Local fallout is generally most intense near ground zero; however, winds may carry local fallout hundreds of kilometers or more. Larger weapons loft debris higher into the air, and small particles may drift for several days before falling to the ground. Particles lifted into the stratosphere may remain there for months or longer and be carried around the world. The radioactivity of fallout decreases with time, so that the longer it remains aloft, the less dangerous it is. Therefore, local fallout is far more hazardous than global fallout.

When a nuclear weapon explodes, it instantly becomes an expanding fireball of radioactive vapor. The radioactivity chiefly comes from the debris of the nuclear fission of uranium. The explosion also produces a torrent of neutrons that can transform some normal elements into radioactive elements. Wherever the fireball touches the ground, dirt and debris are sucked into the air. As the fireball expands and cools, radioactive vapor condenses into radioactive particles and debris that are pulled into the fireball. Because hot air rises, the fireball rises and forms the hallmark mushroom-shaped cloud. Fallout begins within minutes as the heaviest radioactive pebbles rain down near the stem of the mushroom cloud. Fine particles are carried downwind from ground zero and continue to fall to the earth over the next several hours.

If the wind is steady, the radioactivity of the fallout accumulated on the ground could be described with a series of concentric, elongated ovals. The near ends of all of the ovals would touch at ground zero. The innermost oval would have the highest radioactivity. The next oval outward would mark lower activity, and its far end would extend farther from ground zero. For a 1-megaton bomb, the oval that contains a lethal dose of fallout during the first forty-eight hours would be 1,000 square kilometers (386 square miles) in area. However, fallout radioactivity decays relatively quickly, so that after one year the lethal oval would cover only 1 square kilometer (0.38 square mile). At first, 20,000 square kilometers (7,722 square miles) would be contaminated badly enough that an unprotected person might show signs of radiation sickness within two weeks. After one year, that area would decrease to about 20 square kilometers (7.72 square miles).

Depending on such factors as targeting strategies, timing, and weather, fallout from a large-scale nuclear attack could kill tens of millions of people and endanger hundreds of millions more. Because fallout is radioactive dust, it accumulates most readily on horizontal surfaces such as the ground and the roofs of buildings. If caught in a fallout zone, the best strategy would be to get as far away from the fallout as possible and place as much mass between the person and the fallout as possible. For example, the shelter of a simple basement can reduce the radiation dose to 5 or 10 percent of that of a person in the open.

Charles W. Rogers
Updated by Karen N. Kähler

FURTHER READING

Bréchignac, François, and Brenda J. Howard, eds. *Radioactive Pollutants: Impact on the Environment.* Les Ulis, France: EDP Sciences, 2001.

Edelstein, Michael R., Maria Tysiachniouk, and Lyudmila V. Smirnova, eds. *Cultures of Contamination: Legacies of Pollution in Russia and the U.S.* Amsterdam: Elsevier JAI, 2007.

Leopold, Ellen. *Under the Radar: Cancer and the Cold War.* Piscataway, N.J.: Rutgers University Press, 2009.

Miller, Alexandra C., ed. *Depleted Uranium: Properties, Uses, and Health Consequences.* Boca Raton, Fla.: CRC Press, 2007.

Till, John E., and Helen A. Grogan. *Radiological Risk Assessment and Environmental Analysis.* New York: Oxford University Press, 2008.

SEE ALSO: Bikini Atoll bombing; Hiroshima and Nagasaki bombings; Limited Test Ban Treaty; Nuclear accidents; Nuclear and radioactive waste; Nuclear regulatory policy; Nuclear testing; Nuclear winter; Yucca Mountain nuclear waste repository.

Radon

CATEGORIES: Atmosphere and air pollution; nuclear power and radiation

DEFINITION: Radioactive gas that occurs naturally in rocks as the decay product of radium

SIGNIFICANCE: Although it accounts for approximately 50 percent of the normal background radioactivity in the environment, radon can pose a health hazard if it accumulates in houses and other buildings.

Radon and Its Link to Lung Cancer

Breathing low levels of radon, a known carcinogen, can lead to lung cancer. Below are some more facts, provided by the Environmental Protection Agency, about the health risks of radon:

Lung cancer kills thousands of Americans every year. The untimely deaths of Peter Jennings and Dana Reeve have raised public awareness about lung cancer, especially among people who have never smoked. Smoking, radon, and secondhand smoke are the leading causes of lung cancer. Although lung cancer can be treated, the survival rate is one of the lowest for those with cancer. From the time of diagnosis, between 11 and 15 percent of those afflicted will live beyond five years, depending upon demographic factors. In many cases lung cancer can be prevented; this is especially true for radon.

Smoking is the leading cause of lung cancer. Smoking causes an estimated 160,000 deaths in the U.S. every year (American Cancer Society, 2004). And the rate among women is rising. On January 11, 1964, Dr. Luther L. Terry, then U.S. Surgeon General, issued the first warning on the link between smoking and lung cancer. Lung cancer now surpasses breast cancer as the number one cause of death among women. A smoker who is also exposed to radon has a much higher risk of lung cancer.

Radon is the number one cause of lung cancer among non-smokers, according to EPA estimates. Overall, radon is the second leading cause of lung cancer. Radon is responsible for about 21,000 lung cancer deaths every year. About 2,900 of these deaths occur among people who have never smoked.

Unsafe levels of radon have been detected in structures built over soils and rock formations containing uranium. One of the radioactive products of uranium is radium, which decays directly to radon. Every 3 square kilometers (1.2 square miles) of soil to a depth of 15 centimeters (6 inches) contains about 1 gram (0.035 ounces) of radon-emitting radium. Certain regions across the United States and around the world contain comparatively high concentrations of radium in their rocks and soils. One such area is the Reading Prong, which stretches from southeastern Pennsylvania to northern New Jersey and portions of New York.

Three forms of radon are generated in the decay of uranium in rocks and soils. The potential health risks are posed by the radon isotope with an atomic mass of 222 (radon 222), which has a 3.8-day half-life. Radon 220 and radon 219 also form in rocks and soils, but these isotopes have half-lives of 56 seconds and 4 seconds, respectively. The shorter half-lives of these isotopes compared to radon 222 give them a much greater chance to decay within rocks and soils before they can become airborne; thus they are of lesser radiological significance.

Radon is chemically inert, and within its 3.8-day half-life, the gas can become airborne and enter buildings through small fissures in the foundations. Indoor radon levels are typically four or five times more concentrated than outdoor levels, because air dilution occurs in outdoor settings. Contributions to indoor radon levels also come from building materials, well water, and natural gas.

Airborne radon itself poses little hazard to health. As an inert gas, inhaled radon is not retained in significant quantities by the body. The potential health risk arises when radon in the air decays, producing nongaseous radioactive products. These products can attach themselves to dust particles or aerosols. When inhaled, these particles can be trapped in the respiratory system, causing irradiation of sensitive lung tissue. Sustained exposure may result in lung cancer. The U.S. Environmental Protection Agency (EPA) has estimated that more than twenty thousand deaths from lung cancer each year are attributable to radon products. The EPA recommends remediation measures if radon levels in a building exceed 4 picocuries per liter of air. Remediation techniques to relieve indoor radon pollution usually involve ventilating basements and foundation spaces to outside air.

Anthony J. Nicastro

FURTHER READING

Hill, Marquita K. "Pollution at Home." In *Understanding Environmental Pollution*. 3d ed. New York: Cambridge University Press, 2010.

McKinney, Michael L., Robert M. Schoch, and Logan Yonavjak. "Air Pollution: Local and Regional." In *Environmental Science: Systems and Solutions*. 4th ed. Sudbury, Mass.: Jones and Bartlett, 2007.

Pipkin, Bernard W., et al. *Geology and the Environment*. 5th ed. Belmont, Calif.: Thomson Brooks/Cole, 2008.

SEE ALSO: Environmental illnesses; Indoor air pollution; Radioactive pollution and fallout.

Rain forests

CATEGORY: Forests and plants

DEFINITION: Ecosystems of diverse plant and animal species that develop in regions receiving average annual rainfall of 50 to 800 centimeters (20 to 315 inches)

SIGNIFICANCE: Rain forests are the world's most biologically diverse terrestrial ecosystems. They also play an indispensable role in the preservation and maintenance of global climate. Logging, mining, agriculture, urbanization, and a host of other pressures related to human activity have drastically reduced the extent of rain forests, and their area continues to decline dangerously.

The world's largest rain forest, located mostly in Brazil, covers most of the Amazon River basin. Other countries with large rain forests include Indonesia, Cameroon, and Malaysia. Countries with smaller rain forests include Australia, China, Côte d'Ivoire, and Nigeria and countries of the Caribbean Islands, northern South America, and the Pacific Islands. Biodiversity in these forests is very high for both plants and animals. In the tropical rain forests, a few hundred species of plants may be found within a hectare of land. Many scientists believe that the tropical rain forests, which cover only 6 to 7 percent of the earth's surface, may house more than half of the earth's plant and animal species. Because of this, tropical rain forests constitute important laboratories for scientists worldwide. Many chemicals with important pharmaceutical properties have been found in rain-forest plants. These include quinine, long used in the treatment of malaria, and vincristine and vinblastine, used against pediatric leukemia and Hodgkin's disease.

Trees in tropical rain forests may grow to more than 90 meters (300 feet) in height. These tall trees, along with their middle-layered or middle-storied plants, have leaves that spread like a sea to form what is generally called the forest canopy. Large buttress roots such as those of the kapok trees of Brazil and the fig trees in Sabah, Malaysia, anchor the large trees while at the same time functioning to absorb essential nutrients and moisture from the soil. Nutrient recycling is critical to the luxuriant growth of plants in tropical rain forests. It is also essential for the numerous epiphytes (plants that grow nonparasitically on trees) present in the lower to middle stories of the rain forests.

Although temperate rain forests have not received the same level of media attention as tropical rain forests, they are just as fascinating and are similarly threatened with regard to loss of biodiversity. These forests, which occur in places such as the Pacific Northwest and southeastern United States, are characterized by deciduous tree species (species that shed their leaves during the winter months) or by evergreen pine tree species. Among the many important species of trees are the large redwoods (sequoias) of Northern California. These giant trees can grow to heights of more than 90 meters and are among the oldest trees in the world. Other temperate rain forests may be found in central and northern Europe, temperate South America, and New Zealand.

IMPORTANCE AND DESTRUCTION

The luxuriant growth of the plants in tropical rain forests gives the impression that rich and fertile soils underlie them. However, as slash-and-burn agriculture and other production techniques have demon-

strated, the soils are generally infertile. The luxuriant growth is the result of the tremendous amount of nutrient recycling that takes place among species in a humid environment. Thus the removal of native plants to clear space for agriculture depletes necessary biomass for recycling and eventually exposes the soils to rapid erosional forces. After a few years of high crop productivity, yields decline dramatically, leading to further clearing of more forests.

Despite this productivity problem, the tropical rain forests play significant roles in the lives of the people who inhabit them, and they are a source of important materials for economic development. Indigenous peoples have developed the means to cohabit with other living organisms in these unique environments. However, civilization, industrialization, and population growth are gradually threatening the homelands and the very existence of these native peoples.

Rain forests exert a major influence on rainfall patterns, so that land clearance in such forests causes these areas to become drier. Perhaps most important,

Rain forests such as this one, in El Yunque Caribbean Recreation Area in Puerto Rico, are some of the most biodiverse places on the earth. (AP/Wide World Photos)

the abundant plant life in rain forests enhances the global environment through photosynthesis, an ongoing biological process that removes carbon dioxide from the atmosphere and produces oxygen. This process is an important factor in the prevention of global warming.

Rain-forest destruction has been occurring for many centuries. The process, which first started in the temperate rain forests of Europe and North America, greatly intensified in tropical areas during the latter half of the twentieth century. Rain forests are believed to have covered more than 14 percent of the earth's surface at one time, but they are now vanishing at such an alarming rate that countries such as Thailand, Nigeria, Indonesia, Côte d'Ivoire, and Malaysia have already lost more than half of their original forestland. Between August, 2006, and July, 2007—during a period of declining deforestation rates—Brazil lost 1.1 million hectares (2.7 million acres) of rain forest in the Amazon. In the Congo basin, rain-forest losses have threatened the very existence of several indigenous peoples. In Indonesia, which is second only to Brazil in the amount of tropical rain forest land it contains, the deforestation rate in the period 2000-2005 was 1.8 million hectares (4.4 million acres) each year.

Factors Contributing to Destruction

Many factors are responsible for the loss of tropical rain forests. These include population growth and associated development, fuelwood collection, industrialization, and mineral exploitation. Agriculture has also taken its toll, particularly large-scale cattle ranching and the production of cash crops such as soybeans and palm oil.

Human population growth—and the accompanying demand for habitable and agricultural land and other resources—has had a major impact on rain forests. Migration of large human populations has also taken its toll. For example, following the Sahelian drought of the early 1980's, Côte d'Ivoire received an estimated 1.5 million refugees, thereby putting significant pressures on its resources. The situation in Nigeria, with one of the highest population densities in Africa, is similar. By 2010 only 10 percent of Nigeria's forest cover remained, the rest having fallen to logging, agriculture, and petroleum exploration. It has been estimated that at the rate of destruction seen in the first decade of the twenty-first century, Nigeria may be without rain forests by 2020. Meanwhile, as cities expand and new megalopolises are created in in-

dustrialized countries, temperate rain forests also continue to lose ground.

Related to population growth is the growing need for energy. Many countries that still have abundant rain forests use wood as their main source of energy. Therefore, along with intensive logging for export, cutting wood as biomass for fuel has accelerated the destruction of the rain forests and led to severe erosional problems.

Another important factor contributing to rain forest destruction is slash-and-burn agriculture, in which land is cleared, burned, and subjected to crop and livestock production for a few years, followed by new slashing and burning of previously undisturbed forestlands. While crop yields are initially high, they rapidly decline after the second and third years. Subsistence agriculture, upon which slash-and-burn agriculture is based, was originally associated with small populations. Land could be allowed to go fallow for many years, leading to full regeneration of secondary forests under natural conditions. Increased population, along with other factors, has greatly reduced the length of time land can be allowed to go fallow, and more forests have been destroyed to keep up with demand.

Halting Destruction

Reforestation efforts have been under way for many years in Europe and North America to restore lost temperate rain forests. However, such reforestation activities have often been approached from an industrial perspective. Plantations of only one or two species are grown, giving rise to monocultures. This is different from a natural regeneration approach that aims eventually to restore an area's original plant and animal biodiversity and watershed functions. One 2010 study in Australia found that rain-forest restoration projects involving a diverse mix of trees were more efficient at capturing carbon than were plantations with monocultures or mixed tree species. Restoration projects that strive to mimic natural ecosystems are also more likely to withstand diseases and insect infestations than are monoculture plantations.

One way in which the destruction of tropical rain forests is being halted or slowed is through the use of agroforestry systems—sustainable integrations of agriculture and forestry—to complement and reduce slash-and-burn agriculture. The collection of fuelwood can be decreased during sunny months through the use of inexpensive solar cookers, which are nonpolluting and cost nothing to operate. Another ap-

Aerial view of deforested areas of the Brazilian rain forest in 2005. (AP/Wide World Photos)

proach is the establishment of nature preserves, national parks, and large botanical gardens, as has been done successfully in Costa Rica. Promoting rain forests as destinations for sustainable tourism creates an economic incentive against deforestation. Nature preserves have also been established for a number of temperate rain forests in the United States and elsewhere. Yet another strategy is the debt-for-nature swap, in which a national debt is forgiven in exchange for environmental preservation efforts. The United States has forgiven billions of dollars in loans to a number of Central and South American countries, with the condition that these countries set aside large tracts of forestland as nature preserves.

The research organization Bioversity International, which began as an agricultural agency of the United Nations Food and Agriculture Organization, works to preserve the genetic resources of forest plants in addition to those of traditional crop resources. Facilities such as the Kew Royal Botanic Gardens in Great Britain participate in the Millennium Seed Bank, a partnership that conserves seeds for plants outside their native habitats. This program has

successfully banked some 10 percent of the planet's wild plant species, focusing its efforts on plants and regions, such as rain forests, that are most at risk of being obliterated by human activity and climate change.

Participants in the 1992 Earth Summit in Rio de Janeiro, Brazil, adopted a set of principles regarding sustainable forestry, and in 2007 the United Nations General Assembly adopted the Non-legally Binding Instrument on All Types of Forests to bring all stakeholders together in sustainable forest management efforts. The first international instrument devoted to sustainable forest management, its purpose is to strengthen and provide a framework for political commitment, action, and cooperation. In 2008 the United Nations launched the United Nations Collaborative Initiative on Reducing Emissions from Deforestation and Forest Degradation, or UN-REDD. This program provides developing forested countries with financial incentives to protect and better manage their forests, thereby contributing to the fight against global climate change.

Oghenekome U. Onokpise
Updated by Karen N. Kähler

FURTHER READING

Bush, Mark B., and John Flenley. *Tropical Rainforest Responses to Climatic Change*. Berlin: Springer, 2007.

Forsyth, Adrian, Michael Fogden, and Patricia Fogden. *Nature of the Rainforest: Costa Rica and Beyond*. Ithaca, N.Y.: Comstock, 2008.

Kirk, Ruth, Jerry F. Franklin, and Louis Kirk. *The Olympic Rain Forest: An Ecological Web*. Updated ed. Seattle: University of Washington Press, 2001.

London, Mark, and Brian Kelly. *The Last Forest: The Amazon in the Age of Globalization*. New York: Random House, 2007.

Marent, Thomas, with Trent Morgan. *Rainforest*. London: Dorling Kindersley, 2006.

Primack, Richard B., and Richard Corlett. *Tropical Rain Forests: An Ecological and Biogeographical Comparison*. Malden, Mass.: Blackwell, 2005.

Weber, William, et al., eds. *African Rain Forest Ecology and Conservation: An Interdisciplinary Perspective*. New Haven, Conn.: Yale University Press, 2001.

SEE ALSO: Biodiversity; Debt-for-nature swaps; Deforestation; Mendes, Chico; Old-growth forests; Rainforest Action Network; Slash-and-burn agriculture; South America.

Rain gardens

CATEGORIES: Water and water pollution; urban environments

DEFINITION: Garden areas designed to capture, slow, and filter rainwater runoff

SIGNIFICANCE: Rain gardens provide a number of environmental benefits, including the filtering and neutralization of water pollutants, the reduction of stormwater flooding, and the creation of small islands of natural habitat in the midst of urban areas.

The basic idea behind rain gardens is the replication, on a small scale, of the natural conditions that existed before urbanization. In forests and on farmland, rainwater soaks into the soil and percolates slowly through it. Water that is not recycled by immediate evaporation or used by plants eventually works its way through the soil to end up replenishing underground aquifers. The average suburban lawn is relatively impervious to water infiltration by this process, because the grass has shallow roots and many soils, es-

pecially clay, are not porous enough to drain well. A well-sited rain garden planted with native shrubs, perennials, and other hardy plants will allow at least 30 percent more water from rainstorms or snowmelt to seep into the ground than a lawn does.

Although the basic idea is surprisingly simple, rain gardens as a way to counter environmental damage were not widely recognized as a workable concept until the early 1990's. Larry Coffman, the environmental official in charge of creating a plan for handling stormwater in Prince George's County, Maryland, coined the term. The use of rain gardens in some large-scale projects, such as one in Maplewood, Minnesota, in 1996, helped to popularize the practice. Maplewood officials, led by landscape architect Joan I. Nassauer, sponsored the building of rainwater-gathering garden strips along the edges of suburban streets.

Most rain-garden projects in the United States have been small in scale, created by individual home owners or by communities around schools and parks with drainage problems, rather than large initiatives by developers or municipalities. The rain garden's acceptance as a tool to combat water pollution has been rapid, however, and relatively conflict-free. European countries, with their traditions of centralized planning, have been more innovative in using rain gardens than have North American nations.

ENVIRONMENTAL BENEFITS

As cities grow, their land area becomes almost entirely covered with the hard surfaces of buildings, streets, and sidewalk pavements. In order to move rainwater quickly out of the way, most modern cities have built gutters and storm sewers that ultimately drain rainfall into rivers and lakes. In the journey from the city, the water picks up many toxic substances from the urban landscape through which it flows. Oil products, organic wastes, pesticides, and other residues of industrial processes are all carried along in storm runoff. The U.S. Environmental Protection Agency estimates that 70 percent of all water pollution is the result of rainwater runoff. Such water is seldom even treated before it is discharged into lakes and rivers, and any attempts at treatment are likely to be prohibitively expensive.

When stormwater runoff is received by rain gardens, most of the pollutants are filtered out. The soil is enriched by the movement of nitrogen, phosphorus, and other compounds through the plants' root systems and soil processes. As an added benefit, rain gardens' comparatively slow absorption rate helps reduce downstream erosion from the rapid runoff that can come with heavy storms. The existence of garden spaces in urban areas also helps in a small way to combat the heat island effect, which results when heat is absorbed by and then radiates off of the unrelieved hard surfaces of buildings and pavements.

So long as those who create rain gardens follow a few basic rules—such as not putting a rain garden so close to a house that its foundations are undermined by excess water—there is virtually no downside to this conservation concept. Rain gardens bring a touch of nature to urban and suburban residents while helping to purify water resources.

Practical Considerations

Rain gardens do not necessarily look different from purely ornamental gardens, but the creation of rain gardens requires special attention to location and soil preparation. Because water should flow into a rain garden from impermeable surfaces such as roofs and driveways as well as from lawn areas subject to flooding, the rain garden needs to be somewhat lower than the surrounding ground. Occasionally a natural depression can be used, but normally some digging—to a depth of about 20 to 25 centimeters (8 or 10 inches)—is necessary. The digging also enables the replacement of the original soil with "rain-garden soil," an optimal mix for root establishment and permeability. A mixture of half sand and 20 to 30 percent each of topsoil and compost is usually suggested.

A downspout with a drainpipe or shallow troughs usually need to be installed so that water is directed from roof or driveway to the rain garden. At the upper or higher end of the garden, a border of grass can keep water from entering the garden too fast on stormy days. A berm or even a low wall can keep it from overflowing at the downslope border. With all this, the object is not to create a pond or small swamp. An effective rain garden will drain rain within forty-eight hours after a heavy rainfall.

Effective rain gardens contain a variety of plants that can thrive under both wet and dry conditions.

The edge of this rain garden in Fairville Park in Harrisburg, Pennsylvania, is slightly bermed to contain water just long enough to allow the soil in the saucer-shaped garden to absorb rainwater within a few hours. (©George Weigel/ The Patriot News/LANDOV)

Enough perennials and ornamental foliage fit this description that rain gardens have the potential to be pleasing landscape design elements. Once they are well established, rain gardens often become places where birds and other wildlife shelter, providing additional touches of nature in the city.

Emily Alward

Further Reading

Dunnett, Nigel, and Andy Clayden. *Rain Gardens: Managing Water Sustainably in the Garden and Designed Landscape.* Portland, Oreg.: Timber Press, 2007.

Kinkade-Levario, Heather. *Design for Water: Rainwater Harvesting, Stormwater Catchment, and Alternate Water Re-Use.* Gabriola Island, B.C.: New Society, 2007.

Woelfle-Erskine, Cleo, Laura Allen, and July Oskar Cole, eds. *Dam Nation: Dispatches from the Water Underground.* New York: Soft Skull Press, 2007.

SEE ALSO: Aquifers; Erosion and erosion control; Greenbelts; Open spaces; Renewable resources; Runoff, urban; Stormwater management; Urban ecology; Urban sprawl; Water pollution.

Rainforest Action Network

CATEGORIES: Organizations and agencies; activism and advocacy; forests and plants

IDENTIFICATION: Nonprofit organization that works to protect tropical rain forests and other endangered forests and the rights of the peoples native to those forests

DATE: Founded in 1985

SIGNIFICANCE: Through its programs and projects, in particular consumer boycotts and letter-writing campaigns, the Rainforest Action Network raises awareness of the environmental impacts of deforestation and puts pressure on governments and corporations to end practices that endanger the world's forests.

The Rainforest Action Network (RAN) publicizes the environmental dangers associated the destruction of rain forests and old-growth forests, focusing public attention on the actions of companies involved in the degradation of the world's forests. RAN achieves its conservation mission through education and direct grassroots activities, with support from activists working in countries with rain forests. RAN organizes product boycotts to influence corporate executives and uses letter-writing campaigns, petition drives, and nonviolent demonstrations to influence public policy makers. RAN also develops coalitions among scientific, environmental, and grassroots organizations worldwide; holds conferences and seminars; and provides technical and financial assistance to native communities and nongovernmental organizations (NGOs) in rain-forest countries.

At the international level, RAN members participate in letter-writing campaigns targeting the leaders of countries that permit the destruction of rain forests. Information sharing and coordination of activities is facilitated through RAN's cooperative alliances with other environmental and human rights groups in more than sixty countries. At the local level, RAN members are organized within grassroots organizations known as Rainforest Action Groups. Members of these local organizations are encouraged to write letters to policy makers, coordinate nonviolent demonstrations, organize product boycotts, and participate in educational and direct-action campaigns. Through

Protesters from the Rainforest Action Network and Greenpeace, dressed as animals, block the entrance to Pacific Bell's headquarters in San Francisco with phone books during a 1995 demonstration against Pacific Bell's practice of purchasing paper for its phone books from companies that destroy rain forests. (AP/Wide World Photos)

its Protect-an-Acre program, RAN supports local organizations within rain-forest countries that initiate projects to protect the ecological or cultural integrity of forest communities.

In its first notable success, RAN led a boycott of Burger King fast-food restaurants across the United States to raise public awareness concerning Burger King's purchase of beef from companies involved in expanding pastureland for cattle at the expense of rain forests. This campaign, which ran from 1985 to 1987, led to a 12 percent drop in Burger King's sales and prompted company officials to cancel $35 million in contracts for beef raised in Central America and discontinue the company's purchase of beef fed on former rain-forest lands. In 1998, RAN's success in leading a boycott of products produced by the Mitsubishi Corporation encouraged corporate executives to discontinue company practices that were harmful to rain forests and their native cultures.

The public pressure resulting from other RAN campaigns has also influenced many large companies to change their policies concerning the purchase and resale of old-growth wood products from U.S. forests. For example, in 1999 Home Depot stores made a commitment to stop selling old-growth redwood. Other companies from which RAN has won concessions include Scott Paper, Boise Cascade (a manufacturer of paper and other wood products), Occidental Petroleum, and Goldman Sachs investment bank. RAN has also encouraged organizations such as the World Bank to deny funding for companies involved in ecologically destructive activities within rain forests. Because timber harvesting is the leading agent in rain-forest destruction, RAN has become a strong advocate for the use of sustainable alternatives to pulp, paper, and tropical wood in furniture and building construction.

Thomas A. Wikle

FURTHER READING

Gunther, Marc. "The Mosquito in the Tent: A Pesky Environmental Group Called the Rainforest Action Network Is Getting Under the Skin of Corporate America." *Fortune*, May 31, 2004, 158-162.

Holzer, Boris. "Transnational Protest and the Corporate Planet: The Case of Mitsubishi Corporation Versus the Rainforest Action Network." In *Environmental Sociology: From Analysis to Action*, edited by Leslie King and Deborah McCarthy. Lanham, Md.: Rowman & Littlefield, 2005.

Place, Susan E., ed. *Tropical Rainforests: Latin American Nature and Society in Transition*. Rev. ed. Wilmington, Del.: Scholarly Resources, 2001.

SEE ALSO: Deforestation; Logging and clear-cutting; Old-growth forests; Rain forests; Slash-and-burn agriculture; Sustainable forestry.

Rainwater harvesting

CATEGORIES: Resources and resource management; water and water pollution

DEFINITION: The collection and storage of rainwater to meet freshwater needs

SIGNIFICANCE: The increasing urbanization throughout the planet requires an ever-increasing supply of water. Rainwater is a free resource that can be collected through a variety of relatively inexpensive methods, and rainwater harvesting can provide water for drinking, irrigation, and the refilling of aquifers.

Because many natural water sources around the world have been depleted or even completely exhausted, rainwater harvesting is becoming increasingly important to the well-being of some populations. Rainwater can be collected from roofs or from areas at ground level. Rainwater collected from roofs is often fit for drinking without any processing, but collected rainwater usually must undergo some processing before it is suitable for human or animal consumption. Harvested rainwater is often suitable for irrigation, flushing toilets, and laundering clothes without processing, however. Harvested rainwater can serve to supplement available water sources; in some areas, collected rainwater is the only readily available source of water. For example, small islands with high annual rainfall may be sustained totally by rainwater harvesting. In urban areas, rainwater harvesting can supplement city water supplies and reduce the likelihood of flooding.

METHODS

Rainwater harvested from roofs can be channeled into storage tanks through systems of gutters and pipes. The gutters must have an incline so that water does not stand in them and must be strong enough to handle peak flow. Pollutants that may be found in rooftop water include pesticides, dust particles, animal and bird feces, minerals, and dissolved gases such as carbon dioxide, sulfur dioxide, and nitrous oxide. The level of pollutants is commonly highest in water from the first rainfall after a dry spell; such water must therefore be discarded or put through some form of decontaminating treatment, such as boiling or chemical treatment.

Ground collection of rainwater involves a number of techniques that can improve the runoff capacity of a land area: clearing or altering vegetation, increasing the land slope, and decreasing soil permeability (penetration) through compaction or the application of chemicals (or both). Ground collection methodologies provide opportunities to collect water from larger surface areas than rooftop collection allows. The placement of a subsurface dike can obstruct the natural flow of groundwater through an aquifer (underground water channel), raising the groundwater level as well as increasing the amount of water stored in the aquifer (this process is known as groundwater recharge).

Harvested water is commonly collected in tanks; these tanks must be covered to prevent evaporative loss, algae growth, and mosquito breeding. The damming of small creeks and streams, including those that contain water only during heavy rains, is an inexpensive method of storing water during dry spells. Sometimes, however, large portions of water stored in this way are lost through infiltration into the ground. Usually this water is unfit for human consumption; thus it is often solely used for irrigation.

WORLDWIDE PRACTICES AND REGULATIONS

Rooftop rainfall harvesting is widespread in China and Brazil, where the water is consumed by humans and animals; it is also used for crop irrigation and replenishment of groundwater stores. In India, where urbanization is growing at a rapid rate, rooftop harvesting is also widespread; in this nation it represents an essential component of the water supply. In island nations such as Bermuda and the U.S. Virgin Islands, rainwater harvesting is commonly practiced. In rural portions of New Zealand, the only source of water for all household activities is the rooftop collection of rainwater. In the United Kingdom, it is a common practice to collect rainwater for irrigating gardens.

In some parts of the world, no statutes are in place to regulate rainwater harvesting or the allowable uses of the water harvested. In the United States, regulations vary among state and local governments. California, for example, has no state laws related to rainwater harvesting, but some local governments in the state have created regulations. San Francisco, for instance, allows the use of rainwater only for toilet flushing and irrigation. California's neighbor to the north, Oregon, which receives high levels of rainfall annually, has established statutes regulating rainwater harvesting, and the city of Portland specifically permits the use of rainwater for human consumption. For a long time Colorado had laws prohibiting rainwater harvesting; the argument for these statutes was that harvesting could infringe on the water rights of those living downstream. The laws were changed after a 2007 study found that in an average year, almost all of the precipitation that fell in the southern suburbs of Colorado never reached a stream; it either evaporated or was absorbed by plants.

Robin L. Wulffson

FURTHER READING

Banks, Suzy, with Richard Heinichen. *Rainwater Collection for the Mechanically Challenged.* 2d ed. Dripping Springs, Tex.: Tank Town, 2006.

Davis, Allen P. *Stormwater Management for Smart Growth.* New York: Springer, 2005.

Dunnett, Nigel, and Andy Clayden. *Rain Gardens: Managing Water Sustainably in the Garden and Designed Landscape.* Portland, Oreg.: Timber Press, 2007.

Gould, John, and Erik Nissen-Petersen. *Rainwater Catchment Systems for Domestic Supply: Design, Construction, and Implementation.* London: Intermediate Technology Publications, 1999.

Kinkade-Levario, Heather. *Design for Water: Rainwater Harvesting, Stormwater Catchment, and Alternate Water Re-Use.* Gabriola Island, B.C.: New Society, 2007.

Lancaster, Brad. *Rainwater Harvesting for Drylands and Beyond.* 2 vols. Tucson, Ariz.: Rainsource Press, 2006-2007.

SEE ALSO: Clean Water Act and amendments; Drinking water; Groundwater pollution; Rain gardens; Runoff, agricultural; Runoff, urban; Stormwater management; Water conservation; Water pollution; Water quality.

Ramsar Convention on Wetlands of International Importance

CATEGORIES: Treaties, laws, and court cases; preservation and wilderness issues

THE CONVENTION: International agreement that provides a framework for the conservation and wise use of wetlands

DATE: Opened for signature on February 2, 1971

SIGNIFICANCE: As the first international agreement to address issues surrounding the conservation of natural resources, the Ramsar Convention broke new ground in global environmental efforts.

During the 1960's concerns began to grow regarding the decline in populations of waterfowl in many parts of the world. One of the major factors causing this trend was the reduction in the number and size of the world's wetlands, which are habitats heavily used by waterfowl. In order to address these interrelated environmental issues, representatives of various nations came together to develop the Convention on Wetlands of International Importance. This agreement, reached by delegates from eighteen nations on February 2, 1971, in Ramsar, Iran, broke new ground in global environmental efforts; it was the first international treaty to address the conservation and wise use of a natural resource.

The Ramsar Convention came about because the importance of wetlands to the global environment and the needs of humans, as well as the need for an international approach to deal with wetlands, had become apparent. Wetlands are important because of their high biodiversity and their role in water purification, water storage, flood abatement, and groundwater recharge. Many wetlands extend across national boundaries; for example, fish may hatch in the wetlands of one country but be caught as adults in those of another country. Also, many birds migrate hundreds or thousands of

A great blue heron on a jetty in Chesapeake Bay. The Chesapeake Bay estuarine complex is one of the Ramsar Convention's named wetlands of international importance. (©Dreamstime.com)

kilometers twice each year and need to rest, feed, and breed in the wetlands of many countries.

The Ramsar Convention, which entered into force on December 21, 1975, has been ratified by more than 150 nations, and the list of areas designated as "wetlands of international importance" has grown to more than 1,800. Key issues of the convention include urging member nations to develop management plans for wetlands that include the involvement of local communities and indigenous peoples, to promote wetlands education, and to establish monitoring programs that have the ability to detect changes in the ecological character of wetlands. Under the treaty, each country is obligated to implement the convention in four basic ways: It must designate at least one wetland for inclusion in the list of wetlands of international importance, it must include wetlands conservation and wise use as a major focus within its national land-use planning, it must promote wetlands conservation by establishing nature reserves on wetlands and promoting wetlands education, and it must consult with other countries concerning the implementation of the convention.

The policy-making body of the convention is the Conference of the Contracting Parties. Each member nation sends representatives to a conference every three years for the purpose of receiving and reviewing reports on the work of the convention and for approving the work and budget of the convention for the next three years. The Ramsar Convention is administered by the Ramsar Bureau, located in Gland, Switzerland. The bureau is advised on a regular basis by the Standing Committee and the Scientific and Technical Review Panel, each of which is composed of representatives from various member nations. The work of the convention is funded by contributions from the member nations.

Roy Darville

FURTHER READING

Hunt, Constance Elizabeth. *Thirsty Planet: Strategies for Sustainable Water Management.* New York: Zed Books, 2004.

Smardon, Richard C. *Sustaining the World's Wetlands: Setting Policy and Resolving Conflicts.* New York: Springer, 2009.

SEE ALSO: Biodiversity; Convention on Biological Diversity; Wetlands; Wildlife refuges; World Heritage Convention.

Range management

CATEGORY: Land and land use

DEFINITION: Policies governing the uses of regions that are not suitable for cultivation but serve as sources of forage by free-ranging domesticated and wild animals

SIGNIFICANCE: Rangelands provide tangible products such as wood, water, and minerals and intangibles such as natural beauty, open space, and wilderness; accordingly, range management requires consideration of these multiple uses.

Rangelands, which make up about one-half of the earth's surface, include a variety of types, such as temperate grasslands, tropical savannas, Arctic and alpine tundras, desert shrublands, shrub woodlands, and forests. While most are semiarid, rangelands often feature riparian zones and can include wetlands. Rangeland plants coevolved with the herbivores that depended on them for food, which were in turn eaten by carnivorous predators. A balance of nature resulted, subject to annual variation and unpredictable natural catastrophes, that led to relative stability of rangeland plant and animal species and populations. Prior to the development of agriculture ten thousand years ago, this balance of nature permeated the entire earth. Since then, one-half of the land has been used for the agriculture, industry, and habitation of humans.

Rangelands remain unsuitable for crop cultivation because of physical limitations, such as inadequate precipitation, rough terrain, poor drainage, or cold temperatures. They have been homes to numerous nomadic peoples and their animals around the world. While not subject to intensive use, they are affected by human activities. Because rangelands support an estimated 80 percent of livestock production worldwide, provide habitat for many wild animals, and yield numerous tangible and intangible products, their viability and sustainability are important.

OVERGRAZING

Mismanagement of rangelands is commonly caused by overgrazing by domesticated or wild herbivores. Continued heavy grazing leads to denuding of the land, erosion by the elements, and starvation of the animal species. Because decreased plant cover changes the reflectance of the land, climatic changes can follow that prevent the regeneration of plant life, lead-

American bison graze on rangeland in Crook County, Wyoming. (USDA/Ron Nichols)

ing to desertification. Semiarid regions are particularly prone to overgrazing because of low and often unpredictable rainfall; however, these are the areas of the world where the greatest numbers of livestock have been relegated.

Overgrazing has contributed to environmental devastation worldwide. Largely uncontrolled grazing by livestock is partly responsible for the formation of the desert of the Middle East, the degradation of rangeland in the American West in the late nineteenth and early twentieth centuries, and the devastation of parts of Africa and Asia. Feral horses have damaged environments in the western United States and the Australian Outback. Overgrazing by wildlife can be deleterious as well. Removal of predators can lead to overpopulation of wildlife, excessive grazing, starvation, and large die-offs, such as that seen among mule deer on Arizona's Kaibab Plateau during the 1920's and 1930's.

PREVENTING DETERIORATION

Proper management can prevent the deterioration of rangelands. Managing rangelands involves control-

ling the numbers of animals and enhancing their habitats. The land's carrying capacity—that is, the number of animals that can be supported indefinitely on a given unit of land—must not be exceeded. Optimizing, instead of maximizing, the number of animals in a rangeland area will sustain a healthy plant community, referred to as good range condition. For private land, optimizing livestock numbers is in the long-term self-interest of the landowner, although it is not always seen as such. For land that is publicly owned or owned in common, or that has unclear or disputed ownership, restricting animals to the optimum level is particularly difficult to achieve. As Garrett Hardin describes in his 1968 essay "The Tragedy of the Commons," the pursuit of personal, short-term benefits often leads to long-term disaster.

Restricting livestock through herding and fencing is physically easy but can be politically difficult and expensive. Controlling charismatic feral animals, such as horses, or wildlife when natural predators have been eliminated and hunting is severely restricted is much more problematic. As for habitat improvement, various approaches can increase carrying capacity for either domesticated or wild herbivores. Removal of woody vegetation through controlled burning or mechanical means will increase grass cover, fertilization can stimulate forage growth, and reseeding with desirable species, often native plants, can enhance the habitat. It may be necessary also to control plant pests (noxious weeds) and animal pests (such as grasshoppers and rabbits). Effective rangeland management requires matching animals with the forage on which they feed.

Riparian areas—linear strips of land along the sides of rivers and streams—are particularly susceptible to overgrazing. Animals naturally congregate in areas with water, lush vegetation, and shade, and they can seriously damage these areas by preventing grasses from regrowing and young trees from taking root, as well as by compacting the soil and fouling the watercourse. The ecosystem can be devastated, leaving the land subject to erosion and the survival of plant and animal species threatened. Herding and fencing can be used to control animals in these areas, but a less expensive way to encourage their movement away from rivers or streams is to distribute sources of water and salt, which range animals crave and will seek.

Effective range management also accounts for the multiple demands placed on rangelands. The balance between livestock and wildlife is an acute source of controversy between ranchers and environmentalists in the American West, especially because much of the land used for grazing livestock is publicly held. Issues include the low cost of grazing permits on public versus private land; the Emergency Feed Program, which reimburses ranchers for one-half the cost of feed during droughts and other disasters; the lack of control for wild animal populations; the livelihood of ranchers; and the contribution of range livestock to the country's food supply. Other uses of rangeland include extraction of lumber, minerals, and energy, as well as recreational uses.

James L. Robinson

Further Reading

Cheeke, Peter R. *Contemporary Issues in Animal Agriculture.* 2d ed. Upper Saddle River, N.J.: Prentice Hall, 2004.

Chiras, Daniel D., and John P. Reganold. *Natural Resource Conservation: Management for a Sustainable Future.* 10th ed. Upper Saddle River, N.J.: Benjamin Cummings/Pearson, 2010.

Heady, Harold F., and R. Dennis Child. *Rangeland Ecology and Management.* 2d ed. Boulder, Colo.: Westview Press, 2002.

Holechek, Jerry L., Rex D. Pieper, and Carlton H. Herbel. *Range Management: Principles and Practices.* 6th ed. Upper Saddle River, N.J.: Prentice Hall/ Pearson, 2011.

See also: Balance of nature; Erosion and erosion control; Grazing and grasslands; Hardin, Garrett; Kaibab Plateau deer disaster; Soil conservation; Tragedy of the commons.

rBST. *See* Bovine growth hormone

Reclamation Act

Categories: Treaties, laws, and court cases; land and land use; agriculture and food

The Law: U.S. federal law concerned with the building of irrigation projects to convert arid lands in the American West into productive agricultural regions

Date: Enacted on June 17, 1902

Significance: The Reclamation Act created the federal agency that would become the Bureau of Reclamation and established a fund to be used to construct irrigation works so that marginal desert lands in the American West could be used for agriculture. The projects carried out under the act had major impacts on the environment, both by changing landscapes and by making further development of the West possible.

The Reclamation Act of 1902 (also known as the Newlands Act for its author, Congressman Francis Griffith Newlands of Nevada) laid the foundation for the Reclamation Service, which was created in July, 1902. This agency, initially administered by the U.S. Geological Survey, became the Bureau of Reclamation in 1907, and its administration moved to the Department of the Interior. The Reclamation Act provided that all moneys not previously earmarked for education that were received from the sale or disposal of public lands in the sixteen western states and territories (Arizona, California, Colorado, Idaho, Kansas, Montana, Nebraska, Nevada, New Mexico, North Dakota, Oklahoma, Oregon, South Dakota, Utah, Washington, and Wyoming) would be set aside in a "reclamation fund" in the U.S. Treasury. This fund would be used to finance the construction and maintenance of irrigation works for the storage, diversion, and development of waters "for the reclamation of arid or semiarid lands."

The act enabled the settlement of extensive new land for farms in the West. The secretary of the interior withdrew from public entry the federal lands required for each irrigation project, assigned construction contracts, and reported to Congress annually on each water project. Owners of irrigated reclaimed lands were assessed water use fees, which were paid into the reclamation fund to repay costs and to build and maintain additional projects. Bona fide landowners held water rights that could not be sold.

As a result of the Reclamation Act, large-scale reservoirs, dams, canals, and diversion channels were constructed that developed the West. The major river systems affected were the Colorado, Columbia, and Missouri. Population boomed in the region when water was applied to desert scrublands and hydroelectricity was generated. Vast new areas of land were brought under cultivation as new farmers and ranchers complied with homestead laws and used at least half of their newly reclaimed land for agriculture.

Ownership of the federally funded irrigation projects largely remained with the federal government, which protected and operated them. The Reclamation Act did not interfere with the laws of any state or territory relating to the control, appropriation, use, or distribution of water to be used to irrigate arid lands, to make them productive and profitable, thereby increasing state or territorial revenues by increasing population and tax bases. As far as was practicable, the major portion of funds arising from the sale of federal lands within each state or territory was reinvested in that same state or territory. Texas was added under the Reclamation Act's purview in 1906 and projects were extended to the Rio Grande system.

The Reclamation Act marked the entry of the U.S. government into a herculean effort to assist farmers and ranchers by constructing major irrigation projects, including Boulder Dam (later renamed Hoover Dam), Shasta Dam, and Grand Coulee Dam. The act's irrigation successes greatly enlarged the capacity of America to feed itself, led to increased development of U.S. transportation and communication networks, increased electricity production, provided flood control, created recreational facilities, and enlarged and redistributed U.S. population. After the enactment of the National Environmental Policy Act of 1969, awareness of the negative environmental impacts associated with dam construction led to a reassessment of the goals of the Reclamation Act with regard to safeguarding the environment.

Barbara Bennett Peterson

FURTHER READING

Pisani, Donald J. *Water and American Government: The Reclamation Bureau, National Water Policies, and the West, 1902-1935.* Berkeley: University of California Press, 2002.

Rowley, William D. *Reclaiming the Arid West: The Career of Francis G. Newlands.* Bloomington: Indiana University Press, 1996.

Surhone, Lambert M., Miriam T. Timpleton, and Susan F. Marseken. *Salt River Project: Arizona, Phoenix Metropolitan Area, Phoenix Metropolitan Salt River, National Reclamation Act of 1902, Theodore Roosevelt Dam.* Beau Bassin, Mauritius: Betascript, 2010.

SEE ALSO: Bureau of Land Management, U.S.; Department of the Interior, U.S.; Environmental economics; Environmental law, U.S.; Glen Canyon Dam; Grand Coulee Dam; Homestead Act; Hoover Dam; Irrigation; National Environmental Policy Act.

Recycling

CATEGORY: Waste and waste management

DEFINITION: Processing of used materials into fresh supplies of the same materials for new products or salvaging of certain materials from more complex products

SIGNIFICANCE: Recycling constitutes the primary component of modern waste management. It benefits the environment by reducing waste, conserving resources, saving energy, and reducing air and water pollution that can result from waste disposal methods, thus protecting human health. In addition, recycling can provide fiscal benefits, such as by expanding manufacturing jobs.

In 2009 the United States produced 251 million tons of trash, with 82 million tons (32.5 percent) of those materials recycled. In the first decade of the twenty-first century, total recycling in the United States in-

A mountain of plastic beverage bottles, aluminum cans, and glass bottles sits outside a recycling center in Yokohama, Japan. (AP/Wide World Photos)

creased approximately 100 percent. At the national level, the U.S. Environmental Protection Agency (EPA) oversees waste management, including hazardous wastes and landfills, and sets recycling goals. However, no national laws for recycling exist; instead, individual state and local governments have created their own laws concerning recycling.

Several U.S. states have laws establishing deposits and refunds for beverage containers. Other states ban the deposition of recyclable materials into landfills. Some cities, including New York and Seattle, have passed laws that include fines for failing to recycle certain materials. Several organized voluntary and educational programs have been established to increase recycling where it has not been mandated by law. Recycling education is usually integrated into science or social studies classes at the elementary, middle school, and high school levels. November 15 is celebrated as America Recycles Day, which is dedicated to raising awareness about the importance of recycling and to encouraging Americans to recycle and buy recycled products.

For recycling to be economically feasible, efficiently managed, and environmentally effective, adequate recyclable materials must be available, a system must be put in place to extract those materials from the waste stream, and a facility must be available locally where the materials can be reprocessed. In addition, there must be demand for the recycled products.

Recyclable materials are generally collected in three ways: through curbside pickup at consumer homes and businesses, through consumer delivery to drop-off centers, and through consumer return through deposit or refund programs. After collection, recyclables are sent to materials recovery facilities, where they are sorted and prepared for processing into marketable items made entirely or partially of recycled content.

In 2008 the United States generated some 250 million tons of municipal solid waste (MSW), commonly called trash or garbage. Organic materials such as yard trimmings, paper products, and food wastes make up more than two-thirds of human trash. This organic waste could become useful compost instead of being deposited in landfills. In composting, natural aerobic bacteria break down organic materials into fertile topsoil for plant cultivation. Humus and its nutrients in compost help regenerate and enrich soils, help remediate contaminated soils, prevent pollution, and provide economic benefits by, for example, reducing the need for fertilizers and pesticides.

COMMONLY RECYCLED MATERIALS

Paper is the most common material in municipal solid waste, approximately 35 percent of the total. Americans recycled more than 50 percent of paper

used in 2008, but that percentage could be increased. Approximately 80 percent of the paper mills in the United States are designed to depend on paper recycling; they reduce recycled paper to pulp, which is then combined with pulp from newly harvested wood. Wood fiber can be recycled only up to five times, however, since damage to the fibers with each recycling process decreases the quality of the resulting product; hence, more new fibers must be added with each cycle.

Recycled fabric items are sorted into grades; some are then processed to make industrial wiping cloths, whereas others are made into filling products or used

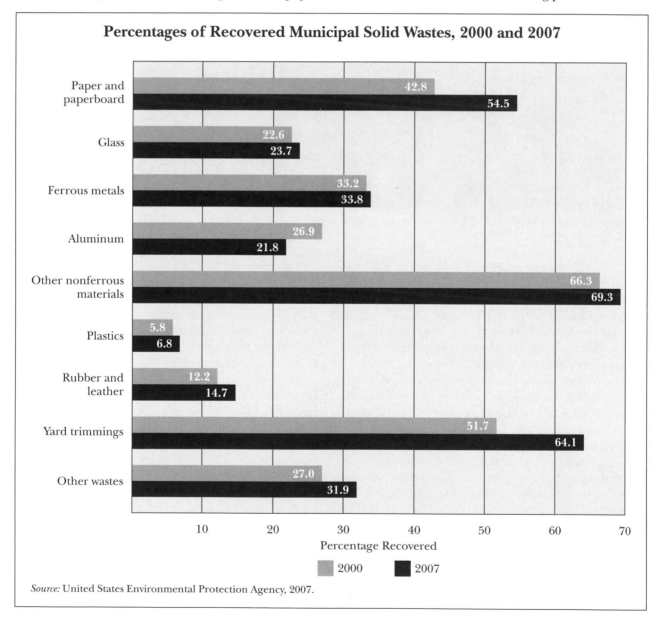

Percentages of Recovered Municipal Solid Wastes, 2000 and 2007

Paper and paperboard — 2000: 42.8, 2007: 54.5
Glass — 2000: 22.6, 2007: 23.7
Ferrous metals — 2000: 33.2, 2007: 33.8
Aluminum — 2000: 26.9, 2007: 21.8
Other nonferrous materials — 2000: 66.3, 2007: 69.3
Plastics — 2000: 5.8, 2007: 6.8
Rubber and leather — 2000: 12.2, 2007: 14.7
Yard trimmings — 2000: 51.7, 2007: 64.1
Other wastes — 2000: 27.0, 2007: 31.9

Percentage Recovered

2000   2007

Source: United States Environmental Protection Agency, 2007.

in the manufacture of paper. Discarded clothing items often end up in landfills in the United States, but many unwanted but still wearable clothes can be recycled through charitable giving. This practice helps to reduce unwanted waste and conserves the resources needed to make new clothing, while providing clothing to persons in need.

Plastic recyclable items are sorted and sent to facilities where they are washed and ground into small flakes that are dried, melted, filtered, and formed into pellets used to manufacture new plastic products. In 2008 the U.S. MSW stream included some 13 million tons of plastics—about 12 percent, an increase from less than 1 percent in 1960. The recycling of plastics has been found to reduce energy use in the United States by 26 percent. Given that 70 percent of the plastics made in the United States are made by factories that use domestic natural gas, the energy saved is freed up for other uses, such as heating and cooling homes.

When glass bottles and jars are recycled, the containers are first smashed and the broken glass, or cullet, is examined for purity and cleaned of contaminants at a glass recycling plant. The cullet is then further crushed and placed in a melting furnace with a mix of raw materials. The resulting material is then mechanically blown or molded into new bottles, jars, and other glass items. Of all the glass recycled in the United States, 90 percent goes into the making of new containers. Glass cullet is also used for aggregate and "glassphalt" for road construction (glassphalt contains 30 percent recycled glass). Unlike paper, glass can be recycled over and over again, as reprocessing does not affect glass's structure. Because cullet costs less than the raw materials needed to make new glass, recycling glass saves manufacturers money at the same time it conserves natural resources; it also helps glass-making furnaces last longer because it melts at a relatively low temperature. Of the 12.2 million tons of glass that were part of the U.S. MSW stream in 2008, 2.8 million tons, or 23 percent, was recovered for recycling. That figure represents a significant increase from 750,000 tons in 1980.

Iron and steel, the world's most commonly recycled ferrous metals, are generally separated from the waste stream through the use of magnets. Recycled steel is processed at steelworks, where scrap steel is remelted and forged into new products. Like glass, steel can be recycled repeatedly with no reduction in quality.

Aluminum is one of the most widely recycled non-ferrous metals and one of the most efficiently processed. Approximately 40 percent of aluminum in the average aluminum can is recycled material. Aluminum beverage containers are the largest source of aluminum in the U.S. MSW stream; in 2008 they accounted for 3.4 million tons (1.3 percent) of MSW, up from only 0.4 percent in 1960. The aluminum recycling process uses 92 percent less energy than is needed to produce aluminum from bauxite ore, in part because the temperature necessary to melt recycled aluminum is 600 degrees Celsius (1,112 degrees Fahrenheit), in contrast to the 900 degrees Celsius (1,652 degrees Fahrenheit) needed to extract mined aluminum from bauxite. The processes of recycling and reuse produce no changes in aluminum, so it can be recycled repeatedly.

Electronic wastes pose problems of energy cost and release of toxic substances that can be reduced by recycling. In many parts of the world the disposal of computers, televisions, mobile phones, and other electronic devices into the general waste stream is forbidden because of the toxic contents of certain components in these products. In the United States, the EPA advises consumers and local governments regarding the safe recovery and recycling of components of electronic waste. Electronic waste recycling plants process discarded electronic items by taking them apart and separating the various metal, plastic, and other components, collecting those that can be reused.

Batteries are sometimes recycled, but because of wide variations in sizes and types, recycling is not always practical. Those that cannot be recycled must be disposed of carefully, as batteries contain heavy metals that can pollute soil and water. The lead-acid batteries used in motor vehicles are easily recycled, and in the United States many local and state governments require businesses that sell such batteries to receive and recycle used batteries from consumers. The recycling rate for automotive batteries in the United States is 90 percent; new batteries generally contain up to 80 percent recycled material.

Tires are frequently recycled at an industrial level. Used tires are ground up and the resulting material is used for a variety of purposes, including insulation and road surfacing. Rubber mulch made from old tires is used to soften the surfaces of playgrounds. Tires are also sometimes recycled into consumer products. Recycled rubber from tires is used to make items such as doormats, trash cans, and even messenger bags; also, a number of companies make shoes that have soles made of recycled tire treads.

Technologies are advancing to produce improved materials for faster recycling, such as completely biodegradable packing materials. An example is a plant-based, all-natural material that breaks down into inert proteins when it comes into contact with water and is consumed by soil bacteria to produce a product used as lawn fertilizer. Scientists at Sony discovered that expanded polystyrene foam, such as that used as packing "peanuts" for cushioning fragile items inside boxes, completely dissolves at room temperature when sprayed with limonene (a natural oil extracted from the skins of citrus fruits) and can be processed for reuse.

DEBATES

Some critics of the promotion of recycling as a solution to environmental problems have suggested that the modern waste management system has fundamental flaws that call for reexamination—they have therefore added a "fourth R," rethink, to the "three R's": reduce, reuse, and recycle. They argue that the costs of collecting and transporting recyclable materials for processing far outweigh the costs of production processes using raw materials. They also assert that more jobs are lost through the reduced collection of raw materials than are created by the recycling industry, which, additionally, pays low wages and offers poor working conditions. Critics argue further that when all processes are considered, the production of recycled products consumes more energy than would the traditional landfill disposal of the recycled materials used to make the products.

Recycling proponents counter that the benefits of recycling compensate for any higher monetary costs it may create. Landfilled wastes pollute groundwater and waterways and contribute significantly to global warming through the release of methane into the atmosphere, producing long-term financial costs of pollution remediation. Proponents also argue that the workers that would be needed to gather amounts of virgin materials equal to the amounts provided through recycling would be working in jobs, such as timber harvesting and ore mining, that have much more dangerous workplace conditions than are found in the recycling industry. Recycling proponents note also the reductions in energy needs represented by the recycling of such materials as paper and aluminum compared with the processing of raw materials. The EPA strongly supports recycling, emphasizing that by saving energy, recycling reduces emissions of carbon dioxide, a greenhouse gas linked with global warming.

Samuel V. A. Kisseadoo

FURTHER READING

Ackerman, Frank. *Why Do We Recycle? Markets, Values, and Public Policy.* Washington, D.C.: Island Press, 1997.

Loeffe, Christian V., ed. *Trends in Conservation and Recycling of Resources.* Hauppauge, N.Y.: Nova Science, 2006.

McKinney, Michael L., Robert M. Schoch, and Logan Yonavjak. "Resource Use and Management." In *Environmental Science: Systems and Solutions.* 4th ed. Sudbury, Mass.: Jones and Bartlett, 2007.

Tammemagi, Hans. *The Waste Crisis: Landfills, Incinerators, and the Search for a Sustainable Future.* New York: Oxford University Press, 1999.

Weeks, Jennifer. "Future of Recycling: Is a Zero-Waste Society Achievable?" *CQ Researcher* 17, no. 44 (December 2007): 1033-1060.

Williams, Paul T. "Waste Recycling." In *Waste Treatment and Disposal.* 2d ed. Hoboken, N.J.: John Wiley & Sons, 2005.

SEE ALSO: Composting; Electronic waste; Hazardous waste; Incineration of waste products; Landfills; Ocean dumping; Resource recovery; Solid waste management policy; Waste management.

Refuse-derived fuel

CATEGORIES: Energy and energy use; waste and waste management

DEFINITION: Solid fuel material created from the processing of municipal solid waste

SIGNIFICANCE: The burning of refuse-derived fuel has some advantages over the burning of coal in terms of reducing damage to the environment, but this alternative energy source is not completely clean.

Raw municipal solid waste (MSW) is a notoriously poor fuel because of its high moisture and low heat content. In addition, mass-burn incineration of MSW produces a broad range of atmospheric pollutants, and the ash produced may become concentrated in potentially toxic elements, such as cadmium or arsenic. One goal of the production of refuse-derived fuel (RDF) is to improve the combustion of MSW by producing a fuel of lower moisture content, more uniform size, greater density, and lower ash content than raw MSW. Another goal is to reduce the amount of material in landfills.

A typical process for the production of RDF involves passage of raw wastes through a screen to remove small, inert materials (such as stones, soil, and glass), pulverization of larger particles in a shredding device, separation of ferrous metals by magnetic extraction, and segregation of the lightweight, mostly organic fraction in an upward airstream (this step is known as air classification). The shredded organic waste fraction can be used directly as a fuel (fluff RDF), or it can be compressed into high-density pellets or cubettes (densified RDF). The latter material is popular because it is easy to transport and store and because of its adaptability in handling and combustion.

RDF can be utilized as a cofuel with coal or fired separately. A major advantage of RDF over coal is that its sulfur content is markedly lower (0.1 to 0.2 percent compared with 5 percent or more for some coal samples), as is its nitrogen content. Both sulfur and nitrogen are among the more notorious atmospheric precursors to acid rain. Also, as a result of processing, RDF contains smaller amounts of potentially toxic metals than does MSW.

RDF possesses only 50-60 percent of the calorific value and 65-75 percent of the density of typical bituminous coals. As a consequence, considerably larger weights of RDF must be burned to obtain performance similar to that of coal. The use of RDF in a boiler may therefore have an adverse impact on the performance of the boiler's systems for air-pollution control and ash removal. In addition, the chlorine content of RDF is higher than typical coal. Other problems sometimes associated with the use of RDF concern odors and dust production in storage and particulate, carbon monoxide, and hydrogen chloride discharges during combustion.

John Pichtel

Further Reading

Hickman, H. Lanier, Jr. "Refuse-Derived Fuel and Energy Recovery: Fulfilling the Resource Recovery Promise." In *American Alchemy: The History of Solid Waste Management in the United States.* Santa Barbara, Calif.: Forester Communications, 2003.

Niessen, Walter R. "Refuse-Derived Fuel Systems." In *Combustion and Incineration Processes.* 3d ed. New York: Marcel Dekker, 2002.

See also: Alternative energy sources; Alternative fuels; Coal; Coal-fired power plants; Landfills; Resource recovery; Waste management.

Renewable energy

Category: Energy and energy use
Definition: Energy derived from natural, unlimited, and replenishable sources
Significance: The burning of fossil fuels such as coal, natural gas, and petroleum releases emissions that contain greenhouse gases and cause air pollution and acid rain, whereas most forms of renewable energy are nonpolluting. In addition, because the earth has a finite supply of fossil fuels, the development of renewable energy sources is important to the long-term future of humankind.

The environmental movement and the oil crises of the 1970's led to interest in the development of energy sources that would offer alternatives to the use of fossil fuels. Fossil fuels are limited resources, and the burning of fossil fuels to generate energy creates emissions of carbon dioxide, toxic chemicals, and air pollutants that harm the environment and human health. Because renewable, or clean, energy systems use natural, local sources that are inexhaustible and such systems have fewer negative impacts on human life and the environment, governments have provided increasing support for the development of renewable energy technologies.

Biomass

The oldest renewable energy source is biomass, which is organic animal and plant material and waste. Biomass resources include grass crops, trees, and agricultural, municipal, and forestry wastes. Since the discovery of fire, humans have burned biomass to release its chemical energy as heat. For example, wood has been burned to cook food and to provide heat. Biomass energy has also been used to make steam and electricity. Biomass oils can be chemically converted into liquid fuels or biodiesel, a transportation fuel. Ethanol, another transportation fuel, comes from fermented corn or sugarcane. Crops such as willow trees and switchgrass are also cultivated for biomass energy generation.

Biomass energy has many environmental benefits when compared with fossil-fuel energy. It contributes little to air pollution, as it releases 90 percent less carbon dioxide than do fossil fuels. Energy crops, such as prairie grasses, require fewer pesticides and fertilizers than do high-yield food crops such as wheat, soy beans, and corn, so they cause less water pollution. Energy crops also add nutrients to the soil. About 4 percent of the energy used in the United States is biomass energy.

SOLAR ENERGY

One of the most promising and popular kinds of renewable energy is solar energy, which uses radiant energy produced by the sun. Solar energy was used as early as the seventh century B.C.E., when a magnifying glass was used to concentrate sunlight to light fires. In 1767, Swiss scientist Horace-Bénédict de Saussure invented the first solar collector, a device for storing the sun's radiation and converting it into a usable form, such as by heating water to create steam. In 1891, American inventor Clarence Kemp patented the first commercial solar water heater.

Sunlight can be converted directly into electricity at the atomic level by photovoltaic (PV) cells, also called solar cells. The photovoltaic phenomenon was first noted in the eighteenth century and became more practical with the use of silicon for the cells in the twentieth century. The cells are joined together in panels, often connected together in an array. They can be placed on rooftops and connected to a grid. In the twenty-first century, solar cells are used worldwide in home and commercial electrical systems, satellites, and various consumer products.

Solar energy has numerous environmental benefits. Photovoltaics produce electricity without gaseous or liquid fuel combustion or hazardous waste by-products. Decentralized PV systems can be used to provide electricity for rural populations, saving more expensive conventional energy for industrial, commercial, and urban needs. By providing electricity for remote and rural areas, solar energy also reduces the use of disposable lead-acid cell batteries, which can contaminate water and soil if they are not disposed of properly. The use of solar energy in rural areas also reduces

air pollution by decreasing the use of diesel generators and kerosene lamps.

Since solar energy depends on sunlight, the efficiency and performance of solar energy systems are affected by weather conditions and location. Nevertheless, the amount of inexhaustible solar energy that could be generated around the globe exceeds the amount needed to meet the world's energy requirements. It has been estimated that if PV systems were installed in only 4 percent of the world's deserts, they could supply enough electricity for the entire world. As solar technologies improve and costs decrease, solar energy has the potential to be the leading alternative energy source of the future.

WIND ENERGY

One of the fastest-growing types of renewable energy during the 1990's, wind energy has been used by humans for centuries. Windmills appear in Persian drawings from 500 C.E., and they are known to have been used throughout the Middle East and China. The English and the French built windmills during the twelfth century, and windmills were indispensable for pumping underground water in the western and Great Plains regions of the United States during the nineteenth and early twentieth centuries. These windmills converted wind into mechanical power.

The modern windmills used to convert wind energy into electricity are called wind turbines or wind generators. In 1890 Poul la Cour, a Danish inventor, built the first wind turbine to generate electricity. Another Dane, Johannes Juul, built the world's first alternating current (AC) wind turbine in 1957. In the twenty-first century, large wind plants are connected to local electric utility transmission networks to relieve congestion in existing systems and to increase reliability for consumers. Wind energy is also used on a smaller scale by home owners in what is known as distributed energy; home-based wind turbines, with batteries as backup, can lower electricity bills by up to 90 percent.

The use of wind energy has long-term environmental benefits. Unlike nuclear and fossil-fuel electricity generation plants, wind generation of electricity does not consume fuel, cause acid rain and greenhouses gases, or require waste cleanups. For example, it has been estimated that when Cape Wind, America's first offshore wind farm under development in Nantucket Sound, becomes fully operational, its 130 wind turbines will reduce greenhouse gas emissions by 734,000 tons annually.

In 2010 the annual wind energy generating capacity of the United States was more than 35,000 megawatts, enough electricity to power 9.7 million homes. This amount of electricity generated by fossil-fuel-burning plants would have released some 62 million tons of carbon dioxide; avoiding the release of so much carbon dioxide is the equivalent of keeping 10.5 million cars off the roads.

Wind energy technology has advanced to the point that wind power is affordable and can compete successfully with fossil fuels and other conventional energy generation. According to a report by the U.S. Department of Energy, wind energy could provide 20 percent of the U.S. electricity supply by 2030. Wind power as a commercial enterprise has been established in more than eighty countries, and in 2009 wind energy capacity increased by 31 percent, with the main markets in Asia, North America, and Europe.

The main disadvantage of wind energy is that it is intermittent, because wind velocities are inconsistent even in areas of strong winds. In addition, some environmentalists have objected to the establishment of wind farms because of their potential to harm wildlife and the aesthetic damage they do to natural landscapes.

HYDROPOWER

Long before electricity was harnessed, in about 4000 B.C.E., ancient civilizations used hydropower, or energy from moving or flowing water, in the waterwheel, the first device employed by humans to produce mechanical energy as a substitute for animal and human labor. Running water in a stream or river moves the wooden paddles mounted around a waterwheel, and the resulting rotation in the shaft drives machinery. The earliest waterwheels were used to grind grain, and the technology went on to be used worldwide for that purpose, as well as to supply drinking water, irrigate crops, drive pumps, and power sawmills and textile mills.

In the nineteenth century, the water turbine replaced the waterwheel in mills, but then the steam engine replaced the turbine in mills. The hydraulic turbine reemerged, however, to power electric generators in the world's first hydroelectric power stations during the 1880's. By the early twentieth century, 40 percent of the U.S. electricity supply was hydroelectric power. Modern large hydropower plants are attached to dams or reservoirs that store the water for

turning the turbines and are connected to electrical grids or substations that transmit the electricity to consumers.

Hydropower is the leading renewable energy source for generating electricity. It has both negative and beneficial effects on the environment. Building dams and reservoirs changes the environment and can harm native habitats and their fish, animal, and plant life. In addition, reservoirs sometimes emit methane, a greenhouse gas. Water is a natural and inexpensive energy source, however. No fuel combustion takes place in the generation of hydropower, so the process does not pollute the air, and energy storage is clean. Hydroelectric power accounts for 24 percent of electricity used worldwide, the electricity consumption of more than 1 billion people. It provides electricity for more than 35 million American households, equivalent to about 500 million barrels of oil a year.

GEOTHERMAL ENERGY

Geothermal energy comes from heat produced deep inside the earth. Deep wells and pumps bring underground hot water and steam to the earth's surface to heat buildings and generate electricity. Some geothermal energy sources come to the surface naturally, including hot springs, geysers, and volcanoes. The ancient Chinese, Native Americans, and Romans used hot mineral-rich springs for bathing, heating,

Top Consumers of Geothermal Energy, 2005

	MEGAWATT CAPACITY	GIGAWATT-HOURS PER YEAR
United States	7,817	8,678
Sweden	4,200	12,000
China	3,687	12,605
Iceland	1,844	6,806
Turkey	1,495	6,900
Japan	822	2,862
Hungary	694	2,206
Italy	607	2,098
New Zealand	308	1,969

Note: Worldwide installed capacity for direct use increased from 8,604 megawatts in 1995 to 28,268 megawatts in 2005. Yearly direct use increased from 31,236 gigawatt-hours per year in 1995 to 75,943 gigawatt-hours per year in 2005.

and cooking. Food dehydration became the major industrial use of this form of energy. In 1904, the first electricity from geothermal energy was generated in Larderello, Italy.

Although not as popular a renewable energy source as wind or solar energy, geothermal energy has significant advantages and benefits for the environment. Because the earth's heat and temperatures are basically constant, geothermal energy is reliable and inexhaustible; it is also not affected by changes in climate or weather. It is very cost-efficient as well; heat pumps can be operated at relatively low cost. The steam and water used in geothermal systems are recycled back into the earth.

Geothermal plants are environmentally friendly. Because they do not burn fuel to generate electricity, they release little or no carbon dioxide and other harmful compounds. Geothermal plants produce no noise pollution and have minimal visual impacts on the surrounding environment, because they do not occupy large surface areas.

The U.S. Environmental Protection Agency and the Department of Energy support the use of geothermal heat pumps. The American Recovery and Reinvestment Act of 2009 (ARRA, also known generally as the Stimulus) provided for grants and tax incentives worth $400 million to the industry, which added 144 geothermal energy plants in fourteen states at the beginning of 2010. ARRA measures benefited the development of all renewable energy sources; it included a Treasury Department grant program for renewable energy developers, increased funding for research and development, and a three-year extension of the production tax credit for many renewable energy facilities.

It has been predicted that by 2070, 60 percent of all global energy will come from renewable energy sources. The World Bank, the World Solar Decade, and the World Solar Summit have designated $2 billion for projects focused on renewable energy resources and the environment.

Alice Myers

FURTHER READING

Craddock, David. *Renewable Energy Made Easy: Free Energy from Solar, Wind, Hydropower, and Other Alternative Energy Sources*. Ocala, Fla.: Atlantic, 2008.

Da Rosa, Aldo Vieira. *Fundamentals of Renewable Energy Processes*. 2d ed. Boston: Elsevier Academic Press, 2009.

Langwith, Jacqueline, ed. *Renewable Energy*. Detroit: Greenhaven Press, 2009.

MacKay, David J. C. *Sustainable Energy—Without the Hot Air*. Cambridge, England: UIT Cambridge, 2009.

Nelson, Vaughn. *Wind Energy: Renewable Energy and the Environment*. Boca Raton, Fla.: CRC Press, 2009.

Pimentel, David, ed. *Biofuels, Solar, and Wind as Renewable Energy Systems: Benefits and Risks*. New York: Springer, 2008.

SEE ALSO: Alternative energy sources; Alternative fuels; Biomass conversion; Fossil fuels; Geothermal energy; Hydroelectricity; Photovoltaic cells; Solar energy; Tidal energy; Wind energy.

Renewable resources

CATEGORIES: Resources and resource management; energy and energy use

DEFINITION: Natural resources that are capable of replenishing themselves for future use

SIGNIFICANCE: Nonrenewable resources such as coal, oil, gas, and mineral deposits regenerate themselves too slowly to keep up with human demand; once consumed, they are gone. By contrast, sustainably managed renewable resources can meet current needs and still provide for generations to come. Misuse or overuse, however, can tax renewable resources beyond their ability to recover.

The term "renewable resources" is often used interchangeably with "renewable energy." However, water, soil, wildlife, forests, plants, and wetlands are also types of renewable resources. Renewable energy sources, such as wind energy and hydroelectricity, are mainly derived from solar energy in one form or another. Direct solar radiation is usually converted into heat, which can be used for such purposes as heating homes or water. Solar water heating has been used in the southern United States since at least the early twentieth century. In mild climates such as southern Florida, it can easily furnish all the hot water requirements of a typical home. It has also been widely used in tropical countries throughout the world. As of 2008 solar photovoltaic systems and wind had become the fastest-growing renewable energy sectors in the United States.

TYPES OF SOLAR ENERGY

Passive solar heating, at its most basic, is the heating of a building by solar radiation that enters the building through south-facing windows (north-facing in the Southern Hemisphere). A properly designed passive solar home must have enough interior heat capacity, usually in the form of concrete floors or walls, to be able to keep the house from overheating on a sunny day and to store excess heat for release at night. In many parts of North America and Europe, passive solar homes have proved to be economical, because the passive system is part of the house itself (its windows, walls, and floors) and thus adds little or no extra cost. "Passive solar" can also refer to solar water-heating systems that involve no moving parts and consume no electricity. Using local water pressure, cold water flows into a collector, where solar energy heats it; the heated water, which rises to the top of the collector, then flows to a storage tank.

Active solar heating, by contrast, uses air or liquid solar collectors that convert solar radiation into thermal energy, which is stored and distributed using a mechanical system (fans or pumps). Active systems are more complex, cost more, and are more resource-intensive; however, they offer greater efficiency and can be used to retrofit an existing building. Two common types of active solar systems are used to heat water for household use: closed loop, in which a solar collector heats an antifreeze solution which in turn heats a water tank containing potable water; and open loop, a simpler, less expensive scheme in which potable water is routed through the solar collector to be heated before flowing into the tank.

Direct solar radiation can also be used to produce electricity. Concentrated solar power systems employ an array of mirrors that reflect sunlight onto a collector, where the solar heat is stored or converted into mechanical energy. In 2007 the capacity of concentrated solar power plants in the United States reached 419 megawatts. Ocean thermal conversion, a concept that has been partially tested, generates electricity using the difference in temperature between the warm, solar-heated upper portions of the ocean and the colder water farther down. No complete ocean thermal conversion plant has yet been built.

Photovoltaic or solar cells are semiconductor devices that generate electricity directly from solar (or other electromagnetic) radiation. Originally developed for artificial satellites after World War II, photovoltaic cells work well and are used to power everything from solar calculators, radios, battery rechargers, and patio lights to electric fences, traffic signal controls, field-deployed scientific monitoring equipment, and corrosion prevention systems on metal bridges. However, their price has kept them from gaining widespread use for generating electricity in homes.

OTHER RENEWABLE RESOURCES

A less direct type of renewable solar energy is hydroelectricity—electricity generated by water turbines that are turned by water flowing down a river or dropping from a dammed water reservoir. The kinetic energy of the moving water is derived from the gravitational potential energy of water at greater heights, and that energy is ultimately derived from solar radiation that evaporated the water from the oceans, allowing it to rain down in the mountains. An older form of water power was the waterwheel used by millers until well into the nineteenth century. Ocean energy in the form of tidal movement, wave action, and marine currents is another source of hydropower that is gaining increasing attention. Because of the great expense of harnessing tidal energy, the tidal power plant that has operated at the mouth of the Rance River in France since 1966 has remained the only major facility of its kind. Smaller commercial tidal power plants operate in Canada, Russia, and China, however, and by 2010 several experimental facilities for producing tide- or wave-generated power were online or under construction around the globe.

Wind energy is also a renewable solar energy resource, because it is the uneven heating of the earth's land and water areas by the sun that causes winds. Wind has long been used as an energy source: It powered the sailing ships that explored the globe, and it powered the windmills used in Asia and Europe since the Middle Ages to grind grain and pump water. Since the 1920's wind turbines have been used to generate electricity in rural areas of the United States. In 2008 the generation capacity of the world's wind power facilities reached 120.8 gigawatts.

Geothermal energy is the heat energy produced beneath the earth's surface by the decay of naturally occurring radioactive elements. According to the U.S. Geological Survey, if only 1 percent of the thermal energy contained within the uppermost 10 kilometers (6.2 miles) of the planet's crust could be harnessed, it would provide five hundred times the energy represented by the world's known oil and gas reserves. While present everywhere beneath the earth's surface, geothermal energy is commercially exploitable primarily in areas of active or geologically young volcanoes. The western United States accounts for most of the world's installed geothermal electricity capacity and generation.

Biomass is often defined as the total mass of living organisms in an ecosystem, including both plants and animals, but the term is also used for nonliving biological materials, such as wood from dead trees. The energy stored in biomass is solar energy that has been stored by photosynthesis. Biomass is an important resource not only for human life, because all human food is biomass in one form or another, but also for society in general because biomass materials can be used as a source of energy and organic chemical compounds, including therapeutic drugs. In 2008 biomass accounted for roughly 45 percent of the total renewable electricity generation (excluding hydropower) in the United States.

Biomass energy resources include solid, liquid, and gaseous fuels. The solid fuels include wood (the major energy resource used in the United States until about 1880) and agricultural wastes such as corn stover (the leaves and stalks left behind after a corn harvest) and sugarcane bagasse (the pulp remaining after juice extraction). These are increasingly being used for industrial electric power generation and home heating. Liquid biomass fuels include methanol and ethanol, both of which can be used in motor vehicle engines. The major gaseous biomass fuel is methane, the main constituent of the fossil fuel called natural gas; methane is generated by the anaerobic (oxygen-starved) decomposition of manure and other organic materials.

Laurent Hodges
Updated by Karen N. Kähler

FURTHER READING

Graziani, Mauro, and Paolo Fornasiero, eds. *Renewable Resources and Renewable Energy: A Global Challenge.* Boca Raton, Fla.: CRC Press, 2007.

Kelly, Regina Anne. *Energy Supply and Renewable Resources.* New York: Checkmark Books, 2008.

National Renewable Energy Laboratory. *2008 Renewable Energy Data Book.* Golden, Colo.: U.S. Dept. of Energy, Office of Energy Efficiency and Renewable Energy, 2009.

National Research Council. *Electricity from Renewable Resources: Status, Prospects, and Impediments.* Washington, D.C.: National Academies Press, 2010.

Pimentel, David, ed. *Biofuels, Solar, and Wind as Renewable Energy Systems: Benefits and Risks.* New York: Springer, 2008.

Sharpe, Grant William, John C. Hendee, and Wenonah F. Sharpe. *Introduction to Forests and Renewable Resources.* 7th ed. Long Grove, Ill.: Waveland Press, 2009.

Wengenmayr, Roland, and Thomas Bührke, eds. *Renewable Energy: Sustainable Energy Concepts for the Future.* Weinheim, Germany: Wiley-VCH, 2008.

Young, Anthony. *Land Resources: Now and for the Future.* New York: Cambridge University Press, 2000.

SEE ALSO: Alternative energy sources; Alternative fuels; Biomass conversion; Ethanol; Geothermal energy; Hydroelectricity; Photovoltaic cells; Solar energy; Tidal energy; Wind energy.

Resource Conservation and Recovery Act

CATEGORIES: Treaties, laws, and court cases; waste and waste management; resources and resource management

THE LAW: U.S. federal legislation concerning the protection of human health and the environment from hazardous wastes

DATE: Enacted on October 21, 1976

SIGNIFICANCE: The Resource Conservation and Recovery Act aims to protect the environment by prohibiting the open dumping of wastes on land and by requiring federal agencies to establish programs to reduce the amounts of waste materials generated. The act also addresses the conservation of resources by requiring the recovery or recycling of certain kinds of waste materials.

As human industries and technologies have advanced, increasing numbers of materials have been deposited into the air and land that are dangers to human health. Many consumer products—including electronic devices and batteries—contain harmful components that can leach into the soil and contaminate it if these items are not disposed of safely. Gases and soot emitted from factories can pollute the air and cause difficult breathing as well as permanent damage to lungs. Radioactive materials used in nuclear plants or in hospitals can have deleterious effects if not handled properly.

During the 1970's, the U.S. Congress became aware that the disposal of solid and hazardous wastes in and on land without careful planning and management can present a danger to human health and the environment. The Resource Conservation and Recovery Act of 1976 (RCRA) was one result of this awareness. RCRA authorizes environmental agencies to order the cleanup of contaminated sites and promotes improvement in management techniques for solid and hazardous wastes. The law mandates that such wastes be handled by competent authorities from the time the wastes are generated until the time of their disposal, or "from cradle to grave."

Characteristics of Hazardous Waste

The Environmental Protection Agency provides the following explanations of the characteristics that define wastes as hazardous under the Resource Conservation and Recovery Act.

- **Ignitability:** Ignitable wastes create fires under certain conditions or are spontaneously combustible, or have a flash point less than 60 degrees Centigrade (140 degrees Fahrenheit).

- **Corrosivity:** Corrosive wastes are acids or bases (pH less than or equal to 2 or greater than or equal to 12.5) that are capable of corroding metal containers, such as storage tanks, drums, and barrels.

- **Reactivity:** Reactive wastes are unstable under "normal" conditions. They can cause explosions, toxic fumes, gases, or vapors when mixed with water.

- **Toxicity:** Toxic wastes are harmful or fatal when ingested or absorbed. When toxic wastes are disposed of on land, contaminated liquid may drain (leach) from the waste and pollute groundwater. Toxicity is defined through a laboratory procedure called the toxicity characteristic leaching procedure.

DISPOSAL OF SOLID AND HAZARDOUS WASTES

The Solid Waste Disposal Act was promulgated in 1965 by President Lyndon B. Johnson, who created the Office of Solid Waste within the U.S. Public Health Service. In 1970, the U.S. Environmental Protection Agency (EPA) took over responsibility for solid waste management. RCRA updated this law and made it clear that taking care of trash is a matter of public health. RCRA defines solid waste as any garbage, refuse, or sludge from a wastewater treatment plant, water-supply plant, or air-pollution-control facility. The law does not include household waste, but it does include waste generated by dry cleaners, auto repair shops, hospitals, exterminators, photo-processing centers, chemical manufacturers, electroplating companies, and petroleum refineries, as well as any other businesses that generate solid waste.

Those who handle solid waste must have an RCRA site identification number if they handle 100 kilograms (220 pounds) per month or accumulate more than 1,000 kilograms (2,200 pounds) of dangerous waste at any one time. Owners and operators of these sites must report each year on their waste management activities. Public participation in reporting is encouraged, and anyone can bring a suit in a federal district court to report anyone who is violating the act.

Hazardous waste is a specific type of solid waste, defined in RCRA as anything that is ignitable (burns readily), corrosive, reactive (explosive), or toxic. Hazardous wastes take many physical forms and may be solid, semisolid, or liquid. These wastes can be damaging to the eyes, the skin, and the tissue under the skin, and they may be poisonous when ingested or inhaled. When hazardous wastes are improperly treated, stored, transported, or disposed of, or otherwise mismanaged, they can pose a substantial hazard to the environment. Before the passage of RCRA, some thought that U.S. law allowed the use of hazardous wastes as fertilizer. Hazardous wastes were sometimes mixed with other fertilizing materials to lessen the wastes' potency, and the resulting mixes were thought to be safe for growing crops. RCRA expressly ended this method of disposal.

REGULATORY ASPECTS OF THE LAW

RCRA regulates commercial businesses as well as federal, state, and local government facilities that generate, transport, transfer, treat, store, recycle, or dispose of hazardous wastes. Owners and operators of municipal solid waste landfills must keep track of these wastes from the moment they are generated until their ultimate disposal or destruction. Generators of toxic substances must maintain thorough records and must clearly label and use appropriate containers for the wastes they generate. RCRA also requires them to have waste minimization programs in place.

RCRA requires all waste treatment, storage, and disposal facilities to meet certain standards for the location, design, and construction of the facilities and for their operating methods and practices. Groundwater monitoring is required on an ongoing basis, and the act also calls for federal or state inspection of all treatment, storage, and disposal facilities at least every two years. Failure to comply with these regulations carries a penalty of a fine of $25,000 per day of noncompliance, one year in prison, or both.

Winifred O. Whelan

FURTHER READING

Blewett, Stephen, with Mary Embree. *What's in the Air: Natural and Man-Made Air Pollution.* Ventura, Calif.: Seaview, 1998.

Gano, Lila. *Hazardous Waste.* San Diego, Calif.: Lucent Books, 1991.

Harris, Glenn, and Leah Nelson. "Revisiting a Hazardous Waste Site Twenty-five Years Later." *Journal of Evvironmental Health* 69, no. 9 (2007): 36-43.

Leone, Bruno, ed. *Garbage and Waste.* San Diego, Calif.: Greenhaven Press, 1997.

Rizzo, Christopher. "RCRA's 'Imminent and Substantial Endangerment' Citizen Suit Turns Twenty-five." *Natural Resources and Environment* 23, no. 2 (2008): 50-51.

Wiseman, Joseph F. "Solid Waste Program Benefits Community in More Ways than One." *Public Works* 131 (2000): 46-52.

SEE ALSO: Environmental law, U.S.; Hazardous and toxic substance regulation; Hazardous waste; Land pollution; Landfills; Lead; Mercury; Nuclear and radioactive waste; Soil contamination; Solid waste management policy.

Resource depletion

CATEGORY: Resources and resource management

DEFINITION: Consumption of a resource faster than new supplies of that resource can be found or faster than it can naturally be replenished

SIGNIFICANCE: With the global population approaching 7 billion people during the early twenty-first century, the demand for higher living standards worldwide has accelerated resource depletion. In addition, modern technology has become dependent on a large number of scarce metal resources with rapidly dwindling known reserves.

Throughout history, humans have exploited resources until the resources have been used up. The discovery of the Americas by Europeans at the end of the fifteenth century began an unprecedented and unsustainable harvesting of natural resources that persisted for five centuries. During this period, vast stands of virgin timber were cut and large numbers of animals were slaughtered for pelts, trophies, and food. For centuries, human beings found new sources of various commodities when they had depleted known existing sources. European countries depended on obtaining supplies of scarce resources from African and Asian colonies during the Industrial Revolution, for example; this access to cheap raw materials waned rapidly after World War II, however.

Worldwide population growth in the nineteenth and twentieth centuries spurred waves of immigrants to the Western Hemisphere and accelerated resource consumption around the world. Demand for higher standards of living worldwide, along with technological advances and the availability of cheap energy, accelerated resource exploitation.

By the early twenty-first century, the developed countries, with their increasing reliance on high technology, had become vulnerable to exploitative limits on specific metals needed to maintain and expand their current standards of living. Certain commodities, such as clean drinking water, had come to be viewed as basic rights of citizens, with needs to be met by governments. (In many emerging countries, clean drinking water had yet to be achieved.) Access to readily available sources of cheap natural gas, fuel oil, gasoline, and electricity were also viewed as necessities, and costs of energy were subsidized by governments in both developed and developing countries.

SUPPLY AND DEMAND

When the cost of a commodity drops below its production price, exploitation of that resource ceases, unless a government or organization subsidizes continued production costs. Resource exploitation has always been labor-intensive. The advent of mechanization in the nineteenth century began reducing labor costs, and the availability of cheap energy from fossil fuels allowed greater exploitation of ever-dwindling resources to be profitable.

At times in history, resource exploitation has responded to paradigm shifts. When the cost of production, in nonmonetary value, exceeds what a society deems acceptable, a resource may no longer be exploited, or limits may be set on how much of the resource may be taken over a specified time period. The nonmonetary values of certain resources became increasingly important during the twentieth century, after developed countries had achieved relatively high standards of living. World exploitation of carbon-based fossil fuels may be significantly reduced over the course of the twenty-first century, as global consensus builds that greenhouse gas emissions pose a significant risk to future generations.

After World War II speculation began regarding how long diminishing supplies of essential minerals, metals, petroleum, and other nonrenewable sources of energy would last. By the beginning of the twenty-first century, many poor nations had begun to use their reserves of useful minerals to improve their standards of living by encouraging infrastructure investment rather than exporting raw materials to developed countries. Some of these useful minerals are limited in supply and are also needed by developed nations to maintain certain technologies equated with high standards of living. Future gains in living standards may be limited by real scarcity, or by nations deliberately denying access to raw materials.

PLANT AND ANIMAL RESOURCES

Although plants and animals are usually considered renewable resources, individual species of plants and animals can be consumed or otherwise destroyed faster than they can be naturally replaced. Lumbering practices of the past, such as clear-cutting, have permanently destroyed natural environments worldwide. When all the trees in an area of forest are cut down, the wildlife habitat is also destroyed, and soil erosion follows rapidly. When rivers become laden with sediment from eroded soil, riverine ecosystems are destroyed, and fishing industries are eliminated. The depletion of forest resources is accompanied by the depletion of the resources of adjacent ecosystems.

The growth in the numbers of bird and animal species considered to be endangered, along with the increasing numbers of species extinctions, shows that animal populations are often depleted. Attempts to place limits on fish catches have generally proved ineffective worldwide. It has been estimated that 75 percent of the earth's fish stocks are either fully exploited or overexploited. Overfishing for cod off the North American coast near Newfoundland depleted this resource; in 1992, no cod appeared at the start of the season.

Public concerns about endangered animal species often focus on "iconic" animals, such as elephants. Slaughtered for their ivory tusks prior to the international ban on the ivory trade signed by more than one hundred nations in 1990, only an estimated ten thousand African bush elephants remained in 2010, as poaching continued in spite of the ban.

FOSSIL FUELS

The high standard of living in developed countries has been achieved in large part because of the cheap cost of energy in the twentieth century. During the late twentieth century, periods of artificial scarcity of petroleum caused global political concerns. Coal, natural gas, and oil were readily available worldwide, however, and production costs remained small for these fossil fuels. The International Energy Agency (IEA) has estimated that petroleum production will peak in 2020. Some estimates suggest that proven oil reserves may last until 2060; the exploitation of as-yet unproven and prospective reserves and unconventional sources could extend supplies considerably. New techniques for drilling and oil recovery continue to increase oil production from older oil fields. (Large oil fields may produce more than 50 billion barrels of oil over their exploitable lifetimes. Estimated reserves increase after a field has been in production for a while.)

Known coal reserves are expected to last into the twenty-second century, but supplies of petroleum and natural gas will be exhausted sooner if rates of consumption do not decrease. Because the burning of cheap fossil fuels has led to increased concentrations of greenhouse gases in the atmosphere, limits to carbon consumption may be set by governments and by international agreements at some future time. (Developing countries argued at the United Nations Climate Change Conference in Copenhagen, Denmark, in late 2009 that they should not be required to limit their "carbon footprint.")

Eventually, biofuels may replace petroleum. Ethanol, which is most often derived from corn, has been determined to be a costly alternative to gasoline in the United States, but biodiesel continues to be investigated. "Sidestream products" from biofuels production have considerable value.

Some developed countries have opted to construct nuclear reactors to supply the energy they need. The member nations of the Organization for Economic Cooperation and Development (OECD) produce about 300 gigawatts of energy from nuclear reactors, enough to meet about 25 percent of these nations' demand for electricity. France and Sweden are among the nations that obtain more than half their electricity from nuclear power plants.

ELEMENTS ESSENTIAL TO HIGH TECHNOLOGY

Most modern electronic devices, including computers and cell phones, use an array of scarce metals, including lithium, tantalum, indium, platinum, and rare earth elements. Demand for these scarce metals has accelerated along with increasing demand for green technologies, and shortages may limit the production of clean energy.

Indium and gallium are essential components in light-emitting diodes (LEDs) and flat-screen displays, and in the construction of solar panels. Commercially obtained as by-products of zinc mining, both metals are thought to have limited reserves; it is estimated that about 6,000 tons of indium exist on the earth, and about 1 million tons of gallium. Unless more deposits are found or alternatives to these metals are developed, the ability to produce solar panels will be limited.

The developed countries need lithium for lightweight car batteries if electric vehicles are to become viable in the long term. The largest known supply of lithium (more than 73 million tons) is in Bolivia, which has resisted exporting raw materials to Japan, the United States and Europe, preferring to attract infrastructure investment so that Bolivia can exploit the lithium itself. Lithium mines are in operation in Chile and Argentina, and much smaller deposits have been found in Tibet and Canada; exploration for the metal is ongoing.

Tantalum is an element necessary for the high-resistance capacitors used in cell phones, personal computers, and automobile electronics. It is mined in Australia, Brazil, Ethiopia, Mozambique, Rwanda, and the Democratic Republic of the Congo. It has been alleged that the mining of tantalum ore (columbite-tantalite, or coltan) in the Congo endangers elephants, lowland gorillas, and other wildlife, and provides funds for the ongoing civil war there. Tantalum ore deposits are also known to exist in Saudi Arabia and Egypt. Estimates of known tantalum reserves are

in the region of 40 million kilograms (44,000 tons); in 2009 about 25 percent of the tantalum being used was obtained through recycling.

Rare earth elements are used for catalysts, ceramics, magnets, electronics, and in the chemical industry. Exploitable rare earth deposits (largely from the mineral monazite) are known to exist in the United States, Canada, Brazil, Australia, India, South Africa, Russia, Vietnam, and China. By the early twenty-first century, about 95 percent of commercially available rare earth elements were derived from China. On September 1, 2009, China set a limit on the export of rare earths at 35,000 tons per year for the next five years. This limit was set to encourage foreign companies to produce high-technology items in China.

Anita Baker-Blocker

FURTHER READING

Chiras, Daniel D., and John P. Reganold. *Natural Resource Conservation: Management for a Sustainable Future*. 10th ed. Upper Saddle River, N.J.: Benjamin Cummings/Pearson, 2010.

Cohen, David. "Earth Audit." *New Scientist*, May 26, 2007, 34-41.

Edwards, Andres R. *The Sustainability Revolution: Portrait of a Paradigm Shift*. Gabriola Island, B.C.: New Society, 2005.

Larmer, Brook, and Randy Olson. "The Real Price of Gold." *National Geographic*, January, 2009, 34-61.

McKinney, Michael L., Robert M. Schoch, and Logan Yonavjak. "People and Natural Resources." In *Environmental Science: Systems and Solutions*. 4th ed. Sudbury, Mass.: Jones and Bartlett, 2007.

Managi, Shunsuke. *Technological Change and Environmental Policy: A Study of Depletion in the Oil and Gas Industry*. Northampton, Mass.: Edward Elgar, 2007.

Runge, Ian C. *Mining Economics and Strategy*. Littleton, Colo.: Society for Mining, Metallurgy, and Exploration, 1998.

SEE ALSO: Coal; Commercial fishing; Deforestation; Electronic waste; Energy conservation; Ethanol; Fisheries; Fossil fuels; Oil crises and oil embargoes; Oil shale and tar sands; Overconsumption; Recycling; Resource recovery; Strip and surface mining.

Resource recovery

CATEGORY: Waste and waste management

DEFINITION: Reclamation of useful materials through methods such as recycling and management of waste products

SIGNIFICANCE: Resource recovery offers substantial environmental rewards in that it slows the use of both nonrenewable and renewable resources and also lessens the amount of solid waste that goes to landfills. The economic rewards for resource recovery can be significant for industries, but many have nevertheless been slow to adopt resource recovery methods.

Spurred by wartime necessity, the United States practiced various forms of recycling during World War II. In the postwar years a "throwaway" culture developed in which the convenience of simply discarding used or unwanted materials was reinforced by the relative inexpensiveness of increasing numbers of goods. From the 1970's onward, however, this approach came under fire from environmental groups and government. In particular, the problems associated with municipal solid waste, such as pollution and finding landfill space, led many local governments to take steps to encourage recycling. Some passed bottle- or container-deposit laws (by 2010 eleven U.S. states had such laws), and many initiated requirements for the sorting of recyclables (such as paper, glass, and certain types of plastics) by residents for curbside pickup. Other examples of resource recovery include landfills capturing and selling the methane gas generated as a result of the decomposition of organic wastes and the use of waste incinerators to generate electric power.

RECYCLING

Recycling operates at two levels. Primary recycling occurs when the original waste material is remade into the same material; examples include newspapers recycled to make newsprint and glass that is melted down to make new glass products. In secondary recycling, the original waste material is made into something else. Tires, for example, can be shredded and incorporated into asphalt, and some plastics can be reused in some types of fabrics used for outdoor clothing or as carpet fibers.

Recycling programs have met with varying degrees of success. In some cases the market prices for particular materials have declined and made recycling programs costly to operate. By the late 1990's more than 50 percent of paper waste was being recycled in the United States, either by industry itself or by consumers. In other areas the record was not as good, with a 20 percent recovery rate for glass, a 12 percent recovery rate for old tires, and nearly 40 percent recovery for metals. The environmental and economic savings generated by recycling can be substantial; making new aluminum by using recycled scrap aluminum, for example, takes only 10 percent of the energy required to make aluminum from virgin ore. Internationally, some automobile makers, notably Saab and Volvo, encourage recycling of the components of used cars.

Many consumers are resistant to recycling, preferring the convenience of simply throwing materials away without having to sort them. In the United States, some cities and counties provide economic incentives to recycle; for instance, some charge inhabitants for trash pickup according to the volume of trash generated. Industries, too, often oppose recycling. Although glass bottles are easier to recycle, the soft-drink industry uses plastic bottles, which are cheaper to transport, and generally opposes bottle-deposit laws.

In 1976 the U.S. Congress passed the Resource Conservation and Recovery Act (RCRA) to address the problems associated with municipal solid waste. The bulk of this legislation deals with requirements for the operation of landfills, but it also provides incentive to local governments and industries to use alternatives to landfilling waste, such as resource recovery. RCRA encourages local governments and industries to use fewer materials and to reuse materials where possible.

INDUSTRIAL ECOLOGY

Some firms have found that there are substantial economic rewards to resource recovery. According to the concept of industrial ecology, waste should not be seen as a disposable product but rather as potential raw material for future use. Such an approach diminishes the strain put on renewable and nonrenewable resources. The primary incentive is increased profitability, but this approach also has a positive environmental impact by promoting sustainability. Advocates even contend that some wastes that are not currently usable should be stored with an eye to future use. The chemical industry is one example of an industry that makes extensive use of manufacturing by-products and routinely recovers and recycles materials. For industrial ecology to be successful, however, industries need to find or create markets for the products they make from former waste materials. For instance, a hard-goods manufacturer has no incentive to become a maker of polymer feedstock unless a ready market exists for that product.

The positive potential for resource recovery to reduce pollution and slow the use of natural resources is great. It is difficult to achieve resource recovery because many governments, firms, and individuals operate in the present, in which it is more economical or convenient to throw away materials than to reuse them. Solutions are slowly coming, and they often involve a variety of approaches. In some cases, command-and-control governmental regulations, such as RCRA, provide incentive for action. At the other end of the scale, voluntarism, as seen in community-based recycling programs that emphasize the achievement of improved environmental quality, can be helpful in achieving resource recovery as well as in raising public consciousness. Market-based incentives, whether in the form of profits or the avoidance of the costs of acquiring new resources or disposing of waste, also play an important role in encouraging resource recovery.

The concept of industrial ecology emphasizes the economic incentives of resource recovery for firms and governments in both industrialized and developing nations. Individual recycling of postconsumer waste is a good starting point for resource recovery, but it will take industrial action directed toward waste minimization to achieve a full-scale program of resource recovery. Ultimately, there are limits to how many times a resource can be recycled, but without a program of resource recovery, resources will be used up more quickly, and more pollution will be generated.

John M. Theilmann

FURTHER READING

Graedel, T. E., and B. R. Allenby. *Industrial Ecology and Sustainable Engineering.* Upper Saddle River, N.J.: Prentice Hall, 2010.

Loeffe, Christian V., ed. *Trends in Conservation and Recycling of Resources.* Hauppauge, N.Y.: Nova Science, 2006.

McKinney, Michael L., Robert M. Schoch, and Logan Yonavjak. "Municipal Solid Waste and Hazardous Waste." In *Environmental Science: Systems and Solutions.* 4th ed. Sudbury, Mass.: Jones and Bartlett, 2007.

Socolow, R., et al., eds. *Industrial Ecology and Global Change.* New York: Cambridge University Press, 1994.

Weeks, Jennifer. "Future of Recycling: Is a Zero-Waste Society Achievable?" *CQ Researcher* 17, no. 44 (December 2007): 1033-1060.

SEE ALSO: Landfills; Recycling; Refuse-derived fuel; Solid waste management policy; Sustainable development; Waste management.

Restoration ecology

CATEGORY: Ecology and ecosystems

DEFINITION: Science of returning ecosystems that have been modified or degraded by human activity to a state approximating their original condition

SIGNIFICANCE: It has become increasingly common for governments to require that the parties responsible for disturbances to the natural environment—such as mining or logging companies—work to restore the environment when the disturbance-causing activities are concluded. Such projects are guided by the science of restoration ecology.

In the United States various federal laws require some form of ecological restoration following strip mining, construction, and other activities that alter the natural landscape. In the management of natural areas that have been disturbed to some degree, several options are available. One is to do nothing but protect the property, allowing nature to take its course. In the absence of further disturbances, one would expect the area to undergo the process of ecological succession. Theoretically, an ecosystem similar to that typical of the region, and including an array of organisms, would be expected to return.

One might ask, therefore, why ecological restoration is mandated. For one thing, succession is often a process requiring long periods of time. As an example, the return of a forest following the destruction of the trees and removal of the soil would require more than one century. Also, ecosystems resulting from succession may be lacking in species typical of the region. This is true when succession is initiated in an area

A black-necked stilt walks on the shore of the Kissimmee River near Okeechobee, Florida. In 2004 federal, state, and local partners celebrated two decades of environmental restoration along the river. (AP/Wide World Photos)

where many exotic (nonnative) species are present or where certain native species have been eliminated. Succession can produce a new ecosystem with a biodiversity comparable to the original one only if there is a local source of colonizing animals and seeds of native plants. Also, satisfactory recovery by succession is unlikely if the soil has been heavily polluted by heavy metals or other substances as a result of industrial land use.

Once it has been decided that a given ecosystem is to be restored, success requires that a plan be designed and followed. Although the specifics may vary greatly, the managers of all ecological restoration projects should follow five basic steps: Envision the end result, consult relevant literature and solicit the advice of specialists, remove or mitigate any current disturbances to the site, rehabilitate the physical habitat, and restore indigenous plants and animals.

Much can be learned from restoration projects that have been conducted in various parts of the world involving a wide variety of ecosystems. A classic ecological restoration of a prairie was conducted in Wis-

consin beginning in the 1930's. Because most North American prairies have been converted to agricultural uses, many opportunities exist for prairie restoration. In such projects it is often necessary to eliminate exotic plants by mechanical means or by application of herbicides. Native prairie grasses and forbs can be established by transplantation or from seed. It may also be necessary to introduce native fauna from nearby areas. Periodic prescribed burning is often necessary to simulate the natural fires common in prairies.

After decades of loss of wetlands in the United States, the federal government established a policy of "no net loss." Thus, when a wetland is destroyed by development, a new wetland must be created as compensation. In the creation of a new wetland, the hydrology of the new site must be altered before native flora and fauna are introduced.

Thomas E. Hemmerly

FURTHER READING

Chiras, Daniel D. "Principles of Ecology: Self-Sustaining Mechanisms in Ecosystems." In *Environmental Science.* 8th ed. Sudbury, Mass.: Jones and Bartlett, 2010.

Falk, Donald A., Margaret A. Palmer, and Joy B. Zedler, eds. *Foundations of Restoration Ecology.* Washington, D.C.: Island Press, 2006.

Van Andel, Jelte, and James Aronson, eds. *Restoration Ecology: The New Frontier.* Malden, Mass.: Blackwell Science, 2006.

SEE ALSO: Bioremediation; Ecology as a concept; Ecosystems; Reclamation Act; Wetlands.

Revised Statute 2477

CATEGORIES: Treaties, laws, and court cases; land and land use

THE LAW: U.S. federal law that granted rights-of-way across public lands for the purpose of developing a system of highways

DATES: Enacted on July 26, 1866; repealed on October 21, 1976

SIGNIFICANCE: The controversy over Revised Statute 2477 demonstrates the changing and competing demands in the United States for development and conservation, for private and public property rights, and for attracting and limiting access to wilderness.

Revised Statute 2477, which reads, "The right of way for the construction of highways across public lands, not otherwise reserved for public purposes, is hereby granted," was approved by the U.S. Congress as part of the 1866 Homestead and Mining Act to encourage settlement of the Western states and territories. The law gave states, counties, and individuals the right to create roads across federal lands, making it easier for settlers to travel across the West and for miners to reach new claims. Under the statute, which became known simply as RS 2477, any person or government could create a road across public land to a town, homestead, or mine.

Controversy eventually arose regarding RS 2477 in the 1970's, partly over the definition of the word "highways" in the statute. Clearly, the word did not carry the same meaning in the nineteenth century, before pavement and automobiles, as it did by the late twentieth century. Did the holder of a valid right-of-way for a dirt track through what was now a national park have the right in 1970 to convert it to a paved road? The statute had been enacted to make it possible for more people to get to wild sections of the West, but by the 1970's it seemed more desirable to limit the numbers of people entering the wilderness, and to direct their traffic once there.

In 1976 Congress passed the Federal Land Policy and Management Act, which repealed RS 2477. The new law stopped the granting of rights-of-way for new highways but specifically stated that existing rights-of-way would continue. This did not end the controversy, in large part because few records had been made of the rights-of-way already granted. Congress attempted in 1984 and again in 1996 to resolve questions about where roads could go on federal lands, who could build them, and what they could be used for, but debate continued about the proper balance between preserving wilderness and granting access to recreation and commercial areas, and no resolution was reached.

In the West, where public land was homesteaded in the nineteenth century and became private property, it remains unclear whether the public still has a right to use the land. Roads created under RS 2477 still exist, and riders of off-road vehicles have claimed that they retain the right to use the roads on these lands. Environmental groups such as the Southern Utah Wilderness Alliance have argued that allowing old claims for right-of-way to develop into paved roads for recreational use will threaten wilderness areas in na-

tional parks, weaken private property rights, and weaken Native American tribal property rights. In Alaska debates continue between those who want to increase the building of roads across federal lands to make access to oil and gas fields easier and those who want to protect the Tongass National Forest and the Arctic National Wildlife Refuge from being fragmented by roads.

Cynthia A. Bily

FURTHER READING

Havlick, David G. *No Place Distant: Roads and Motorized Recreation on America's Public Lands.* Washington, D.C.: Island Press, 2002.

Keiter, Robert B. "Collaborative Conservation." In *Keeping Faith with Nature: Ecosystems, Democracy, and America's Public Lands.* New Haven, Conn.: Yale University Press, 2003.

SEE ALSO: Arctic National Wildlife Refuge; Environmental law, U.S.; Federal Land Policy and Management Act; National parks; Property rights movement; Road systems and freeways; Roadless Area Conservation Rule.

Rhine River

CATEGORIES: Places; water and water pollution

IDENTIFICATION: European river arising in Switzerland and flowing generally north to the North Sea

SIGNIFICANCE: Long one of the most polluted rivers in Europe, the Rhine received increased attention in 1986 when a major chemical spill resulted in the destruction of millions of fish and other wildlife. Through the Rhine Action Programme, overseen by the International Commission for the Protection of the Rhine, major environmental improvements have taken place in the Rhine basin.

The Rhine River originates in Switzerland and flows northward through Liechtenstein, Austria, France, Germany, and the Netherlands. Italy, Luxembourg, and Belgium have territory in the Rhine basin. The Rhine is the major river of Western Europe, and with the Danube it forms the most important waterway in the region, cutting across the continent from the North Sea to the Black Sea. The Rhine is one of the most important geographic features of the European continent, both historically and economically. It has dozens of tributaries, and scores of important cities are located along its route.

In the nineteenth and twentieth centuries pollution of the Rhine not only killed plant and animal species dependent on the river but also presented a health hazard to the people who lived along the shore. The river's water was not completely potable, and it caused damage to the infrastructure of the countries and cities through which it flowed. Cooperative efforts to clean the Rhine go back to the nineteenth century. In 1946 the governments of the countries along the river formed the International Commission for the Protection of the Rhine (ICPR).

During the 1980's the Rhine remained one of the most polluted rivers in Europe. In the fall of 1986, as the result of a fire at a chemical storage facility in Basel, Switzerland, the river turned red when agricultural pesticides containing mercury poured into it. The deadly poisons moved northward along the course of the river and killed millions of fish and other wildlife. The public outcry over this incident forced the governments of the riparian countries to form the Rhine Action Programme (RAP) in 1987.

One lofty stated goal of the RAP was to bring salmon back to the Rhine by the year 2000; salmon had disappeared from the river in the 1930's. The RAP's more immediate goals were to cut in half the discharge of dangerous pollutants into the river, to raise safety standards, to take measures to allow fish to swim upstream and spawn in the Rhine's tributaries, and to restore shoreline ecosystems to bring back natural fauna and flora. The RAP program was not incorporated into European Council law, but the ICPR executed and supervised the RAP mission. The member states agreed to enact legislation concerning the discharge of wastes into the river, to require permits for factory emissions, to build purification and measurement stations, and to tax both individuals and factories for environmental protection. They also agreed to cooperate on cleaning the river and conserving and restoring its natural surroundings.

In a 2009 report the ICPR stated that almost all native species had returned to the Rhine. Salmon, the key species, had made a partial return to the river, but more needed to be done. A new target date of 2015

Protesters wearing gas masks and carrying a puppet symbolizing possible death take part in a demonstration in Basel, Switzerland, following the deadly pollution of the Rhine with agricultural chemicals. The sign behind them translates as "Today dead fish, tomorrow it's us." (AP/Wide World Photos)

was set to improve the habitat for salmon by improving access for the fish into the river tributaries. The largest improvement was the elimination of direct dumping of chemicals and pollutants into the river by factories in the member states. The governments have also cooperated in protecting and treating the river water and the basin's groundwater to make it safe for consumption.

Frederick B. Chary

FURTHER READING
Hernan, Robert Emmet. "Rhine River, Switzerland, 1986." In *This Borrowed Earth: Lessons from the Fifteen*

Worst Environmental Disasters Around the World. New York: Palgrave Macmillan, 2010.

Mellor, Roy E. H. *The Rhine: A Study in the Geography of Water Transport.* Aberdeen, Scotland: University of Aberdeen, 1983.

Pomeranz, Kenneth. "The Rhine as a World River." In *The Environment and World History,* edited by Edmund Burke III and Kenneth Pomeranz. Berkeley: University of California Press, 2009.

SEE ALSO: Agricultural chemicals; Danube River; Europe; Fish kills; Habitat destruction; Hazardous and toxic substance regulation; Mercury; Waste management; Water pollution; Water-pollution policy.

Rhinoceroses

CATEGORY: Animals and endangered species

DEFINITION: Large herbivorous mammals native to Africa and southern Asia

SIGNIFICANCE: Efforts are ongoing to conserve and protect the world's remaining populations of rhinoceros, several species of which have been hunted to near extinction.

The term "rhinoceros" is used to refer to a group of five species in the family Rhinocerotidae, one of the largest remaining megafauna. The rhinoceros has a thick, protective, armorlike skin and can grow to weigh more than one ton. The most distinctive physical characteristic of the rhinoceros is the large horn on its nose. Unlike the horns of other mammalian species, the horn of the rhinoceros lacks a bony center and is composed only of keratin, the same protein that makes up human hair and fingernails. Some rhinoceros species, the Indian and Javan rhinoceros, have a single horn, while the African and Sumatran species have two horns.

Of the five existing species of rhinoceros, two are native to Africa and three are native to southern Asia. The two African species are the black rhinoceros (*Diceros bicornis*) and the white rhinoceros (*Ceratotherium simum*). The three Asian species are the Indian rhinoceros, also known as the greater one-horned rhinoceros (*Rhinoceros unicornis*); the Javan rhinoceros (*Rhinoceros sondaicus*); and the Sumatra, or Asian two-horned, rhinoceros (*Dicerorhinus sumatrensis*). Three of the five species (Javan, Sumatran, and black) are listed as "critically endangered" on the International Union for Conservation of Nature's Red List of Threatened Species, with the Javan rhinoceros one of the most rare and endangered large mammals worldwide. The white rhino is Red Listed as "near threatened" and the Indian rhinoceros is listed as "vulnerable."

Despite their large size and fierce reputation, rhinoceroses are vulnerable to poaching, especially during their daily drink from a water hole. The rhinoceros horn is highly valued by practitioners of traditional Asian medicine, who use derivatives of the

White rhinoceroses at Lake Nakuru in Kenya. (©Jumoku/Dreamstime.com)

horn to treat fevers and convulsions. Rhinoceros horns are also used in making dagger handles. Poaching levels in some African countries in the early twenty-first century threatened to jeopardize gains that had been made in rhinoceros populations, with poaching estimated to equal annual population growth. Poaching has caused a greater than 96 percent decrease in Africa's black rhino population, and it appears to be increasing in some countries, as the illegal horn trade has progressively worsened since 2006, while efforts to protect the rhinoceros have become increasingly ineffective.

Many plans for the conservation and recovery of rhinoceros species have been put in place and have been effective in increasing rhinoceros populations in certain areas. Although the Indian rhino was nearly extinct by the early twentieth century, by 2010 the population of this species exceeded twenty-four hundred as a result of concentrated protection efforts. Conservation efforts directed toward the southern subspecies of white rhino, which had been nearly decimated at the end of the nineteenth century, have elevated this population so that it is the most abundant rhinoceros subspecies. The entire population of the northern subspecies of white rhino, which numbered only about thirty in 2010, is located within Garamba National Park in the Democratic Republic of the Congo, where the animals are protected by guards. In addition to efforts to protect rhinoceros species in the wild, zoos have created species survival plans for rhinoceroses that include breeding programs, education programs, and plans for the introduction of captive-bred animals into secure habitats.

The installation of high-voltage fencing around some preserves has curbed the poaching of rhinoceroses somewhat, but the practice persists, in part because of poor enforcement of the laws against hunting these animals and in part because of the increasing availability of automatic weapons to potential poachers. In areas that have experienced the most poaching, efforts have been made to relocate rhinoceros populations to safer regions.

C. J. Walsh

FURTHER READING

Cunningham, Carol, and Joel Berger. *Horn of Darkness: Rhinos on the Edge.* New York: Oxford University Press, 1997.

Enright, Kelly. *Rhinoceros.* London: Reaktion Books, 2008.

Hamilton, Garry. *Rhino Rescue: Changing the Future for Endangered Wildlife.* Buffalo, N.Y.: Firefly Books, 2006.

SEE ALSO: Convention on International Trade in Endangered Species; Endangered species and species protection policy; Poaching; Wildlife management; Wildlife refuges; World Wide Fund for Nature.

Right-to-know legislation

CATEGORY: Human health and the environment

DEFINITION: Laws requiring dissemination to the public of information about hazards posed by pollution and toxic substances

SIGNIFICANCE: Right-to-know laws provide members of the public with information they may need to avoid exposure to possible environmental hazards and also serve to protect employees in particular industries.

During the 1970's and 1980's the U.S. government enacted legislation to address such environmental problems as air pollution, water pollution, and health threats from toxic substances. Much of this legislation includes provisions for informing the general public about hazards posed by pollution and toxic substances.

The Comprehensive Environmental Response, Compensation, and Liability Act (CERCLA), commonly known as Superfund, was signed into law in December, 1980. Its primary purpose was to address the problem of inactive hazardous waste sites. The mandates of CERCLA were controversial, largely because they required the companies responsible for hazardous wastes to perform extensive and costly cleanups no matter how long ago the wastes were discarded or how well the companies had exercised responsibility in their original handling of the wastes.

Sections 102 and 103 of CERCLA provide for reporting procedures. Section 102 outlines reportable quantities set by the Environmental Protection Agency (EPA): Any hazardous substances that have not been assigned a reportable quantity under the Clean Water Act must be reported if they are found in quantities of 1 pound or more. Section 103 stipulates that the National Response Center must be notified as soon as a person in charge of a vessel or facility learns of any release of a hazardous substance that is equal to or greater than the reportable quantity for that substance. Failure to report a release carries a maximum criminal penalty of three years in prison for a first offense and a five-year sentence for repeat offenses; a maximum civil penalty equal to $25,000 per day may be assessed.

The original Superfund act expired in September, 1985. On October 17, 1986, the U.S. Congress reauthorized $9 billion and made major revisions in CERCLA with the Superfund Amendments and Reauthorization Act (SARA). Title III of SARA is the Emergency Planning and Community Right-to-Know Act (EPCRA). EPCRA has four major components, three of which address reporting requirements. Section 302 requires any facility that produces, uses, or stores any substances on the EPA's List of Extremely Hazardous Substances in quantities that equal or exceed the specified limits to notify the State Emergency Response Commission (SERC). Section 304 provides for emergency release notification to the community emergency coordinators for the Local Emergency Planning Committee (LEPC) of areas likely to be affected and to the SERC of any state likely to be affected.

Sections 311 and 312 form the major right-to-know components of EPCRA and are closely intertwined with the Hazard Communication Standard of the Occupational Safety and Health Administration (OSHA). Using the basic framework of the OSHA standard, these sections require that the general public and employees in facilities that produce or use hazardous chemicals be given information regarding the chemicals' presence. Certain items are exempt, such as food, food additives, drugs, cosmetics, gasoline used in a family car, petroleum products used for household purposes, and substances used for medical and research purposes.

Further provisions are made in clauses designated Tier I and Tier II. As of March 1, 1988, and annually thereafter, industries are required to submit chemical inventory forms for the material safety data sheet (MSDS) substances that are regulated by OSHA. Tier I information must be submitted to the LEPC, the SERC, and the fire department with jurisdiction over a given facility. Tier II information is more detailed: It must include the location of each hazardous chemical and a description of how the chemical is stored, unless a facility is excused on the basis of a trade-secret plea. Any facility submitting Tier I or Tier II statements must, on request, allow on-site inspections by the local fire department. While the regulations are silent as to availability of Tier I information, any person may obtain Tier II information by submitting a written request. Any citizen may also request an MSDS on any hazardous chemical, even if the facility does not use it in reportable amounts.

The purpose of section 313 is to ensure that communities are provided with information about chemical releases from local facilities. Owners and operators of certain manufacturing facilities must submit annual reports to the EPA and designated state officials on environmental releases of listed toxic chemicals. This requirement applies to routine and permitted releases as well as accidental releases. Furthermore, certain suppliers must notify their customers if products they distribute contain chemicals that are subject to section 313 mandates. Failure to submit MSDS's or lists of MSDS chemicals are punishable by civil penalties of up to $10,000 per day for each violation. Noncompliance with the annual inventory requirements may result in a penalty of up to $25,000 for each violation.

In addition to EPCRA's right-to-know title, some other legislation includes reporting provisions. Examples are the Occupational Safety and Health Act of 1970, which mandates that employees be apprised of hazards to which they are exposed and that they receive information regarding symptoms of contamination and appropriate emergency treatment, and the Toxic Substances Control Act of 1976, which includes notice provisions for products for domestic use.

Victoria Price

FURTHER READING

Andrews, Richard N. L. *Managing the Environment, Managing Ourselves: A History of American Environmental Policy.* 2d ed. New Haven, Conn.: Yale University Press, 2006.

Ashford, Nicholas A., and Charles C. Caldart. "The Right to Know: Mandatory Disclosure of Information Regarding Chemical Risks." In *Environmental Law, Policy, and Economics: Reclaiming the Environmental Agenda.* Cambridge, Mass.: MIT Press, 2008.

Cox, Robert. "Public Participation in Environmental Decisions." In *Environmental Communication and the Public Sphere.* 2d ed. Thousand Oaks, Calif.: Sage, 2010.

Moore, Gary S. "Environmental Laws and Compliance." In *Living with the Earth: Concepts in Environ-*

mental Health Science. 2d ed. Boca Raton, Fla.: CRC Press, 2002.

SEE ALSO: Clean Air Act and amendments; Environmental law, U.S.; Environmental Protection Agency; Hazardous and toxic substance regulation; Hazardous waste; Superfund.

Rio Declaration on Environment and Development

CATEGORIES: Treaties, laws, and court cases; resources and resource management

THE DECLARATION: Statement of principles designed to guide sustainable development around the world

DATE: June 13, 1992

SIGNIFICANCE: The Rio Declaration offers a bold plan to mobilize local, national, and global action and so represents a unique step forward on the road toward sustainable development.

The Rio Declaration on Environment and Development is a relatively short statement of twenty-seven principles produced at the 1992 United Nations Conference on Environment and Development, also known as the Earth Summit, a milestone event that brought together more heads of state and chiefs of government than any other meeting in the history of international relations. Other participants in the conference included government officials and senior diplomats from around the world, delegates from United Nations agencies and other international organizations, and thousands of representatives of nongovernmental organizations.

The Rio Declaration sought to reaffirm the principles that had been laid out in the 1972 Stockholm Declaration and to build on them. The Earth Summit also issued a document known as Agenda 21, a vast program for the twenty-first century that was approved by consensus among world leaders representing more than 98 percent of the world's population. This document upholds the idea that environmental, economic, and social development are not isolated from one another and builds on the 1987 report by the Brundtland Commission, *Our Common Future*, which placed the concept of "sustainable development" as an urgent imperative on the global agenda. In its preamble, the Rio Declaration recognizes "the integral and interdependent nature of the Earth, our home."

Some of the declaration's principles have come to be regarded by legal scholars as the formulation of the "third generation of human rights." These rights are generally regarded as aspirational soft laws, which are hard to enact in a legally binding way because of the principle of state sovereignty and the preponderance of would-be offenders. They include a broad and disparate range of rights: group and collective rights, the right of self-determination, the right to economic and social development, the right to a healthy environment, the right to natural resources, the right to information and communication, the right to participation in cultural heritage, and the right to intergenerational equity and sustainability. These are typically contrasted with civil and political rights (known as first-generation human rights) and social, economic, and cultural rights (second-generation human rights). First-generation human rights deal essentially with liberty and participation in political life; they include freedom of speech, the right to a fair trial, freedom of religion, and voting rights. Second-generation human rights are those concerned with equality and ensuring that members of the citizenry enjoy equal conditions and treatment. They include the right to be employed, the right to housing and health, the right to social security and unemployment benefits, and so on.

Among the Rio Declaration's principles, the following are particularly noteworthy. Principle 1 states that "human beings are at the centre of concerns for sustainable development. They are entitled to a healthy and productive life in harmony with nature." Principle 2, concerning the right to development, says that this right "must be fulfilled so as to equitably meet the developmental and environmental needs of present and future generations." Principles 5 and 6 address, respectively, the eradication of poverty and the priority to be given to the least developed countries. Principle 15 focuses on the application of the precautionary principle as a way of ensuring protection of the environment: "Where there are threats of serious and irreversible damage, lack of full scientific certainty shall not be used as a reason for postponing cost-effective measures to prevent environmental degradation." Principles 20, 21, 22, and 23 deal, respectively, with the vital roles of women, of youth, of indigenous peoples, and of people under oppression.

Principle 25 of the declaration addresses the interdependent nature of peace, development, and environmental protection.

Nader N. Chokr

FURTHER READING

Adede, A. *International Environmental Law from Stockholm 1972 to Rio de Janeiro 1992: An Overview of Past Lessons and Future Challenges.* Maryland Heights, Mo.: Elsevier Science, 1992.

Blewitt, John. *Understanding Sustainable Development.* Sterling, Va.: Earthscan, 2008.

Hens, Luc, and Bhaskar Nath. *The World Summit on Sustainable Development: The Johannesburg Conference.* New York: Springer Science & Business, 2005.

Porter, Gareth, Janet Welsh Brown, and Pamela Chasek. "The Emergence of Global Environmental Politics." In *Global Environmental Politics.* Boulder, Colo.: Westview Press, 2000.

Speth, James Gustave, and Peter M. Haas. "From Stockholm to Johannesburg: First Attempt at Global Environmental Governance." In *Global Environmental Governance.* Washington, D.C.: Island Press, 2006.

SEE ALSO: Agenda 21; Convention on Biological Diversity; Earth Summit; Indigenous peoples and nature preservation; Intergenerational justice; Johannesburg Declaration on Sustainable Development; Kyoto Protocol; *Our Common Future*; Precautionary principle; Sustainable development; United Nations Commission on Sustainable Development; United Nations Conference on the Human Environment.

Riparian rights

CATEGORY: Water and water pollution

DEFINITION: Benefits associated with the use of water for owners of land bordering on or adjacent to bodies of water

SIGNIFICANCE: Concerns regarding environmental issues and water scarcity have led to some modifications of the traditional doctrine of riparian rights as it is applied in the eastern United States.

The roots of the U.S. system of riparian rights for water usage can be traced back to English common law. Riparian rights and other water laws attempt to reconcile the various demands of users and potential users with the size of the supply of clean water. Increasing demands for water use, coupled with the relative scarcity of adequate water resources and the growing recognition of the importance of habitat protection, have resulted in changing and evolving systems of water law and regulation in the United States.

Two basic water law systems exist for dealing with water-use conflicts in the United States. The eastern states, historically viewed as having adequate water resources, predominantly follow the doctrine of riparian rights. Under this doctrine, water-use rights are based on landownership and require equal sharing among users in times of shortage. Riparian rights typically include the right to hunt and fish and the right to use water for irrigation. The western states, historically seen as lacking abundant water resources, have adopted a system based on the doctrine of prior appropriation, which establishes priorities among competitive users. Under this doctrine, earlier settlers have greater rights to water use. The western water law system developed as a means of encouraging economic development by allowing water rights to be separated from landownership. The right to consume water accessible from one's land is the primary difference between riparian rights and appropriative rights.

Riparian rights treat water resources as a common property, which permits individual users to use the resource freely while spreading the cost of additional resource use to all owners of the resource. As Garrett Hardin warned, overuse of common property resources can result in a "tragedy of the commons." Increasing demands for water, as well as demands that water be used more efficiently and responsibly, have created tensions within the existing schemes of water rights and management. Erratic precipitation and rising per capita water use have led to a number of water emergencies. Likewise, many states now recognize the importance of letting river waters remain untapped to protect fish and wildlife, as well as to preserve aesthetic and recreational values.

Environmental concerns and issues of water scarcity have led to significant modifications of water law systems. The only restriction on water use under the traditional riparian system is a prohibition against "unreasonable harm" to another riparian user. The absence of an efficient, systemwide mechanism to determine what constitute unreasonable uses and to manage water resources successfully has led eastern states to reconsider how water should be allocated among competing uses, and most eastern states have instituted administrative permit systems to regulate riparian rights. Under such a system, a single agency manages issues of water quality and allocation. The agency is also charged with defining and maintaining some minimum water quality and flow. Such modifications to riparian right systems attempt to protect private values and further public interests. As the evolution of riparian rights suggests, successful methods of allocating water must balance a watershed's consumptive water resources needs with its ecological needs.

Michael D. Kaplowitz

FURTHER READING

Andrews, Richard N. L. *Managing the Environment, Managing Ourselves: A History of American Environmental Policy.* 2d ed. New Haven, Conn.: Yale University Press, 2006.

Ferrey, Steven. "Rights to Use Water." In *Environmental Law: Examples and Explanations.* 5th ed. New York: Aspen, 2010.

Vandas, Stephen J., Thomas C. Winter, and William A. Battaglin. *Water and the Environment.* Alexandria, Va.: American Geological Institute, 2002.

SEE ALSO: Drinking water; Hardin, Garrett; Irrigation; Tragedy of the commons; Water conservation; Water rights; Water use; Watershed management.

Risk assessment

CATEGORY: Philosophy and ethics

DEFINITION: Estimation of the potential undesirable and harmful consequences of various activities and substances along with likelihood of those negative things happening

SIGNIFICANCE: Risk assessment is carried out in relation to a growing list of items and activities, including those that may have impacts on the environment. The methods of risk assessment are based on the important goal of reducing unwanted consequences and harmful outcomes. At the same time, the complexity and difficulty of risk assessment is increasingly acknowledged.

Risk assessment is the process of examining potential sources of harm and the likelihood of those harms occurring. Risk assessment is part of a broader risk management process that also examines the strategies that are, or should be, in place to reduce risks. The degree to which risk assessment and risk management can be separated from one another is the subject of much debate, however. Many issues are involved in this debate, but an important element centers on the degree to which risk assessment relies on objective facts or subjective values.

QUANTITATIVE ASPECTS

Quantitative risk assessment uses a variety of formulas to calculate a numerical value for a risk. The methods of quantitative risk assessment began with gamblers calculating odds and insurance companies estimating their risk of having to make payments on insured items. The practice of risk assessment has since expanded into areas such as environmental protection, public health, and workplace health and safety.

The results of quantitative risk assessment are often expressed in such form as someone's risk of getting cancer being one in so-many million or someone has a certain percentage risk of losing an investment. This method of risk assessment may be applied to any hazard—that is, an item or situation that could cause harm. Risk assessment can focus on the risks from hazards to employees of a particular company, to various groups, to citizens of a country, or to the whole planet. Assessment can focus on the various risks from one technology, such as all the risks of permitting genetically modified foods into the market, or the risks of a particular type arising from many different sources, such as the risk of developing skin cancer.

Quantitative risk assessment involves calculating the magnitude of a potential loss or harm and the probability that the loss or harm will occur. Both of these factors can be very difficult to measure. Part of the challenge is that risk assessment involves predictions about future events that are highly complex and involve much uncertainty. Many different factors contribute to whether or not something will occur. A chemical factory may contain hazardous materials, but the risk that a chemical leak will occur depends on many different elements, including human factors (ranging from errors to terrorism), structural issues with buildings, machinery, geological issues (such as earthquakes), and other factors known and unknown.

For quantitative risk assessment to produce valid results, the calculations must be kept as objective as possible. Some experts maintain that an actual risk number can be calculated with increasing precision for any substance, technology, event, or activity as more data are gathered and more scientific information about the hazard itself becomes available. Risk also involves the probability of a particular harm occurring from a hazard. The assessment of such probability relies on data on the impacts of the hazard on people, living things, and the environment, and also on economic, political, and social systems. Such data allow the analyst to have a better understanding of what the hazard could lead to.

Then the probabilities of various events must be calculated. Risk assessment analysts may use information from toxicology and epidemiology studies to get estimates of the likely incidences of various kinds of health effects. Historical records may be consulted for average failure rates for machinery or the environmental consequences of the release of various hazards. Fault tree analysis is widely used in risk assessment to understand past failures. At the top of the fault tree is the event that caused harm, such as the release of a toxin into the atmosphere. Various pathways are traced back to identify the roots of the damaging event and to clarify the contributions of human error, equipment failure, or other causes.

SUBJECTIVE ASPECTS

Although scientific data are very important in risk assessment, questions have been raised about the degree of objectivity in the process. A number of studies have found that when different teams conduct risk assessment on the same process, their risk estimates can vary considerably. Social scientists who have examined the process of risk assessment have pointed out the important roles that assumptions and values play in that process. A consensus has emerged that risk assessment involves both technical calculations based on scientific data and subjective value judgments.

The role of values in the risk assessment process is particularly obvious when questions are asked about whether or not certain risks are acceptable or what strategies should be adopted to reduce risks. However, assumptions and values are involved from the beginning. For example, risk assessments were carried out on coal mining between 1950 and 1970. Calculations based on the number of accidental deaths per ton of coal showed that coal mines were less risky places to work than did calculations based on accidental deaths per employee, which showed the same mines to be slightly more risky. While both assess-ments were mathematically accurate, the more appropriate one involved value judgments. From the perspective of the miners, the latter assessment would be much more significant.

While risk assessment aims to estimate risk, it also seeks to evaluate the significance of the risk. This is where the human component of risk assessment is most apparent and where subjective elements are important. This can be seen in the informal risk assessments that people carry out personally when they decide whether some action they might take is safe enough or too risky. Some people are willing to accept the risks of activities that others would view as unacceptably risky.

Many factors influence how people perceive risk, and these have been studied extensively for a number of decades. Several factors have been shown to drive perceptions of risk, and these must be taken into account in risk assessment. People's perceptions of the risk related to new technologies or products are high when they see the new technologies or products as beyond their control, unfamiliar, or invisible, or if they otherwise evoke negative feelings. Whereas earlier work on risk assessment focused on the rational and cognitive dimensions, more recent work has emphasized the role of emotion and affect. Affect is a subtle form of emotion, leading to feelings of like or dislike for an object or action. For example, in assessing the risks or benefits of various activities, people often include affective terms and whether they like or dislike the activities. Factors other than cognitive information enter into risk assessment and related decision making.

When people conduct personal risk assessments, they accept activities or products if they perceive the benefits as outweighing the risks. However, a number of studies have shown that people's assessments of risks and benefits are linked to their affective reactions to the items being assessed. This is called the affective heuristic of risk assessment. When people like something, they tend to view its benefits as outweighing its risks; when they dislike something, the reverse occurs. The same kinds of responses have been found in studies in which the subjects were scientists and in studies with the general public. Risk assessment in real life is thus not just emotional; rather, it involves a complex interaction between emotional reactions and reason-based analysis.

Social values are also involved in risk assessment. Objects associated with voluntariness, controllability, and equity are generally perceived as lower in risk, while those associated with catastrophic potential, fatality, and threat to future generations are perceived as more risky. Thus flying in an airplane is often viewed as riskier than riding in an automobile, because an airplane crash is catastrophic—in spite of the fact that more people die in road traffic accidents than in airplane crashes. These findings point to the importance of communication about risk. The words and images that become associated with an object or technology can have a major influence in how the risk of the object or technology is perceived and assessed.

Other factors affecting risk assessment are internal to the assessors. One consistent finding in studies is that white males perceive the risks associated with items to be lower than do women and men of other races. Why this is the case is not clear. One of the leading hypotheses is that risk perception is tied to a person's worldview. Those who accept a hierarchical social order have a lower perception of risk compared to those with a more egalitarian view of social order. Hence those who are more involved in managing and controlling hazardous items may benefit from hierarchical structures and perceive less risk in those items. Those who have less control and power over the items may perceive them as more risky. While much remains unclear in this area, the research that has been conducted has shown the importance of taking people's different perspectives and backgrounds into consideration in the risk assessment process.

Risk assessment is recognized as an important part of the evaluation of how objects and processes may negatively affect the environment and many other dimensions of life. While complex mathematical procedures allow the estimation of numerical risk, the role of human values and assumptions is increasingly recognized. As these are made more transparent, their impacts on decisions can be taken into account. This will not remove all the challenges of accurately assessing risk, but it can lead to greater accountability and open dialogue about possible risks.

Dónal P. O'Mathúna

FURTHER READING

Beck, Ulrich. *Risk Society: Towards a New Modernity.* 1992. Reprint. Translated by Mark Ritter. London: Sage, 2009.

Bostrom, Nick, and Milan M. Ćirković, eds. *Global Catastrophic Risks.* New York: Oxford University Press, 2008.

Hansson, Sven Ove. "Fallacies of Risk." *Journal of Risk Research* 7, no. 3 (2004): 353-360.

Hestor, R. E., and R. M. Harrison, eds. *Risk Assessment and Risk Management.* Cambridge, England: Royal Society of Chemistry, 1998.

Hurst, Nick W. *Risk Assessment: The Human Dimension.* Cambridge, England: Royal Society of Chemistry, 1998.

Slovic, Paul. *The Perception of Risk.* Sterling, Va.: Earthscan, 2000.

SEE ALSO: Benefit-cost analysis; Genetically modified foods; Nanotechnology; Precautionary principle.

Road systems and freeways

CATEGORY: Land and land use

DEFINITION: Paved thoroughfares created for motor vehicle traffic

SIGNIFICANCE: The construction and heavy use of highways can have severe impacts on the environment, including air and water pollution, land degradation, and loss of open space. Since motorized transportation is a common means of travel, compromise is often necessary to balance the need for additional roads and the need to protect the environment.

The era of the freeway began during the 1950's, when lobbying groups in the United States encouraged a political vision of a nationwide high-standard, high-speed road network. Retired U.S. Army general Lucius D. Clay led a committee that studied transportation needs across the United States and advised President Dwight Eisenhower that the nation needed what came to be called the National System of Interstate and Defense Highways. The costs of creating and maintaining the more than 69,000 kilometers (43,000 miles) of interstate highways built after 1956 have been shared by federal and state governments on a 90/10 (federal/state) matching basis. The federal share comes from the Highway Trust Fund, which receives revenues from federal taxes on fuels, lubricants, vehicles, and vehicle parts. Although the interstate system accounts for only 1 percent of the total road miles in the United States, it carries 20 percent of the traffic.

Studies of the U.S. road and highway systems as a

whole—including local roads and services—have found that motor vehicle user fees cover only two-thirds of public expenditures, not including the substantial nonmonetary external costs of environmental impacts. Because many of the costs of using private vehicles are hidden, many Americans perceive driving their own cars to be less expensive than using public-transit alternatives. U.S. highway statistics for 2008 indicated the existence of more than 248 million registered motor vehicles (including 137 million automobiles) and more than 4 million miles of public roads.

Traffic congestion is most severe in areas experiencing rapid growth in both total population and number of vehicles in use. In fact, rapid population growth tends to offset the beneficial impacts of remedies adopted to reduce traffic congestion. Low-density settlement generates more total automotive vehicle trip miles per day, which consume more energy and cause greater emission of pollutants. One important development that has helped to reduce congestion, and thus motor vehicle emissions, on some urban and suburban freeways since the 1970's is the so-called carpool lane—a lane reserved for the use of buses and other vehicles carrying more than one or two passengers.

ENVIRONMENTAL IMPACTS

Air pollution constitutes the most serious environmental impact caused by highway transportation. Developments that may generate traffic, such as parking lots for shopping centers, may be classified as indirect sources of pollution. Internal combustion and diesel engines are the principal sources of carbon monox-

Among the environmental impacts of the construction of road systems and freeways are the disturbance of plant and animal habitats and increased air pollution from motor vehicle exhaust. (©Olivier Le Queinec/Dreamstime.com)

ides and hydrocarbons, account for nearly one-half of the nitrogen oxides, and are the chief source of particulate lead in the atmosphere. Highway emissions are directly related to traffic volume and density, vehicle type, speed, and mode (idle, acceleration, cruise, or deceleration). Increased speed produces a demand on an engine for increased power, which leads to more fuel consumed and greater emissions, but the vehicle also passes through an area more quickly. Long trips by motor vehicle and traffic congestion both increase the emissions discharged into the atmosphere.

The building of roads and highways also consumes open space, affecting plant and animal life, as well as climate and water runoff. Highways facilitate the spread of urban areas and often lead to low-density developments, which are difficult to provide with services. Highways that connect developed areas usually follow valleys and other areas with flat terrain, and consequently highways are often built in close proximity to streams, lakes, and wetlands. Until very late in the twentieth century, hydrologic features that blocked proposed roads were seen primarily as obstacles to be bridged, filled, or moved at lowest cost. Laws that were put in place to protect endangered animal and plant species changed this approach, requiring highway developers sometimes to employ routes or construction procedures that are more expensive than those they would have used in the past.

Among the more subtle and probably more serious impacts of road construction are changes in local hydrologic patterns, such as changes in the water table that affect vegetation. Erosion and sedimentation are also associated with road construction activities. Another environmental concern related to roads and highways involves the runoff from street surfaces into waterways; such runoff can include salt (spread on roads to melt ice in winter) and petroleum products that contaminate the soil.

Highway noise has also been identified as an environmental problem. Noise—excessive or unwanted sound—has been found to be more than simply an annoyance; in the extreme, the sleep disturbances and other negative impacts of noise can be considered a danger to public health. Highway noise is troublesome to control, but some attenuation can be achieved through proper planning. Buffer zones and acoustical barriers, modifications of highway alignment, and traffic management measures can reduce the exposure to noise for those who live or work near highways. The U.S. Federal Highway Administration specifies a methodology that should be used for the evaluation of traffic noise and abutting land use and provides guidance on when noise abatement should be considered.

In balancing transportation needs and environmental goals, engineers must distinguish between mobility (simple movement) and access (the ability to reach destinations and desired services). Increasing mobility through more roads, more vehicles, and more traffic may actually reduce access over the long term. For example, older neighborhoods tend to have stores, schools, and transit services within walking distance of residential areas, whereas newer auto-dependent neighborhoods tend to be lower in density with few local services and often no pedestrian facilities, so more trips require driving. Hence mobility increases but access declines.

Stephen B. Dobrow

FURTHER READING

Goddard, Stephen B. *Getting There: The Epic Struggle Between Road and Rail in the American Century.* New York: Basic Books, 1994.

Gonzalez, George A. *The Politics of Air Pollution: Urban Growth, Ecological Modernization, and Symbolic Inclusion.* Albany: State University of New York Press, 2005.

Hester, R. E., and R. M. Harrison, eds. *Transport and the Environment.* Cambridge, England: Royal Society of Chemistry, 2004.

Lay, M. G. *Ways of the World: A History of the World's Roads and of the Vehicles That Used Them.* New Brunswick, N.J.: Rutgers University Press, 1992.

Lewis, Tom. *Divided Highways: Building the Interstate Highways, Transforming American Life.* New York: Viking Press, 1997.

SEE ALSO: Air pollution; Automobile emissions; Black Wednesday; Environmental impact assessments and statements; Land-use policy; Noise pollution; Smog; Urban sprawl.

Roadless Area Conservation Rule

CATEGORIES: Treaties, laws, and court cases; preservation and wilderness issues

THE LAW: Federal policy initiative designed to protect national forests from commercial development

DATE: Issued on January 12, 2001

SIGNIFICANCE: To ensure the natural condition of the wilderness forests under federal jurisdiction and to maintain their integrity as complex ecosystems, Bill Clinton's presidential administration issued the Roadless Area Conservation Rule, a landmark federal environmental initiative, declaring part of the national forest preserves as "roadless," thus protected from any new mining, lumbering, drilling, and housing development interests.

Following World War II the U.S. Forest Service routinely approved private corporation projects aimed at extracting the reserves of natural resources—principally mining and logging interests—in otherwise protected national forests, a little more than 57 million hectares (140 million acres). During the mid-1990's, amid growing concerns over global warming and emerging scientific evidence that indicated the critical importance of protecting forests, Bill Clinton's presidential administration commissioned extensive public hearings into shaping a responsible plan to protect the national forests. Literally in its closing days, the administration issued the Roadless Area Conservation Rule, which essentially prohibited all development—logging, mining, drilling—on more than 23.5 million hectares (58 million acres) of forests and grasslands, roughly one-third of the national preserve, mostly in the West. Administration policy experts argued that the blanket protection would create secure habitat for wildlife, birds, and plant life and help protect the watershed supply for more than one-fifth of the country. In addition, it would provide for public recreation, hiking, kayaking, biking, and camping. The initiative permitted only emergency actions, such as clearing trees against a wildfire or cutting small-diameter trees only as necessary for habitat preservation.

Although environmentalists hailed the bold initiative, development interests and a phalanx of conservative western activists and politicians objected to the initiative as an unconstitutional overreach of federal power that would put at risk state economies. Quickly, the incoming presidential administration of George W. Bush delayed the initiative, calling for additional investigation into its scope. The Bush administration sought to give states authority to designate areas as roadless. In short order, a number of western states, most prominently Alaska, filed suit against the Clinton rule. After four years of contentious litigation, the Bush administration repealed the rule in May, 2005, and announced it would base decisions for setting aside public lands on state-by-state petitions to the Forest Service. That rule was quickly challenged in federal district court by several states and environmental activists (spearheaded by the environment-focused law firm Earthjustice). In September, 2006, the courts ruled the Bush protocol illegal, saying that a state-by-state plan violated the mandates of the National Environmental Policy Act of 1969 and the Endangered Species Act of 1973. The original Clinton rule was reinstated. Shortly after, contracts for oil and gas projects (more than three hundred) approved by the Forest Service during the years when the rule had been suspended were summarily voided. In August, 2008, however, a Wyoming federal district court once again invalidated the original Roadless Rule.

Seeking a plan that would mediate the controversy, and acting under pressure from both environmental activists and congressional representatives concerned about the open expansion by development interests, President Barack Obama's administration, while filing an appeal of the Wyoming decision, moved in May, 2009, to place decisions on a case-by-case basis solely with the secretary of agriculture for one year. That directive was renewed for an additional year in May, 2010, as the administration, confronting the catastrophic BP *Deepwater Horizon* oil spill in the Gulf of Mexico, began shaping a comprehensive environmental policy initiative.

Joseph Dewey

FURTHER READING

Bevington, Douglas. *The Rebirth of Environmentalism: Grassroots Activism from the Spotted Owl to the Polar Bear.* Washington, D.C.: Island Press, 2009.

Turner, Tom. *Roadless Rules: The Struggle for the Last Wild Forests.* Washington, D.C.: Island Press, 2009.

Vaughan, Jacqueline, and Cortner, Hannah J. *George W. Bush's Healthy Forests: Reframing the Environmental Debate.* Boulder: University Press of Colorado, 2005.

See also: Ecosystems; Endangered Species Act; Forest and range policy; Forest management; Forest Service, U.S.; Logging and clear-cutting; National Environmental Policy Act; National forests; Watershed management; Wildlife refuges.

Rocky Flats, Colorado, nuclear plant releases

Categories: Nuclear power and radiation; water and water pollution

The Events: Short- and long-term releases from the Rocky Flats nuclear plant in central Colorado that caused low-level radioactive contamination of off-site reservoirs and land areas

Dates: Notable incidents in 1957, 1969, and 1973; other long-term releases during operational years 1952-1989

Significance: One of thirteen nuclear weapons production facilities operating in the United States during the Cold War, Rocky Flats was the site of a number of radioactive and toxic releases during its four-decade operational history. Releases from Rocky Flats had the potential to be a major public health hazard because of the facility's proximity to a large population center.

The Rocky Flats nuclear plant site stands 26 kilometers (16 miles) northwest of downtown Denver. From 1952 to 1989 the plant's primary mission was to produce nuclear weapon trigger assemblies. This required the machining of plutonium, uranium, beryllium, and other metals. Plutonium was also recovered from obsolete weapons and recycled at the plant. Radioactive americium 241, a decay product of plutonium, was separated and recovered in this process.

Over the four decades of the plant's operation, emergencies threatened the surrounding population and environment. On September 11, 1957, and May 11, 1969, for example, plutonium stored inside glove boxes at the plant spontaneously combusted, allowing plutonium dust and smoke to escape to the outside environment through the ventilation systems.

During the late 1950's and the 1960's more than five thousand drums of waste oil accumulated on-site that had become contaminated with plutonium and the solvent carbon tetrachloride during machining operations. In 1967 and 1968 the barrels—many of them corroded and leaking—were removed, leaving the soil on which they had stood exposed to the elements. Subsequent windstorms blew the contaminated soil east and southeast of the plant. In 1969 the former storage area was asphalted over to contain the remaining contamination.

In 1973 the radioactive isotope tritium entered the plant's waste stream during scrap plutonium processing. It was discharged in wastewater to an on-site holding pond. From there, it entered Walnut Creek, which carried the radioactive contaminant to Great Western Reservoir, a source of drinking water for the Colorado community of Broomfield. Fortunately, radionuclide concentra-

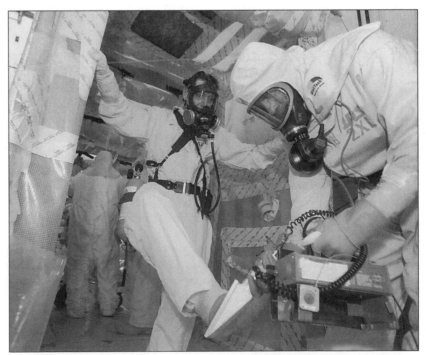

Workers check for alpha and beta radiation on their protective clothing after they helped dismantle glove boxes as part of the cleanup of the Rocky Flats plant site in 2002. (AP/Wide World Photos)

tions were not detected above levels believed to present a health concern.

Some plant workers went public with their concerns that proper safety guidelines were being disregarded. One infamous example involved a substance known as pondcrete. Until the mid-1980's the plant discharged hazardous chemical wastes mixed with low-level radioactive wastes to solar ponds to reduce the volume of the wastes through evaporation. Beginning in 1985, as the evaporation ponds were being phased out, sludge was dredged from them, mixed with Portland cement, and placed in large plastic-lined cardboard boxes. The pondcrete was supposed to form solid blocks that could be shipped elsewhere for burial, but much of the radioactive, toxic mixture remained mushy. The blocks were stacked in a parking lot and left exposed to the weather for three years, and the decomposing pondcrete was washed into the soil by rain and snowmelt.

Requests from the public for specific information on hazards at the plant were frequently denied by officials, who claimed that such records were classified. A series of investigations finally resulted in a 1989 raid by Federal Bureau of Investigation (FBI) agents and Environmental Protection Agency (EPA) staff, who seized plant records. Evidence of the mishandling of hazardous materials at the plant resulted in an $18.5 million fine for Rockwell International, the company that managed the plant for the U.S. Department of Energy (DOE), and termination of its contract. That same year, Rocky Flats was placed on the National Priorities List (NPL), making it a Superfund cleanup site.

The next management company, EG&G, was also accused of safety violations, despite its expenditure of $50 million on repairs. In 1993, Secretary of Energy James Watkins announced the end of nuclear production at Rocky Flats.

Studies of tissue and urine samples from people living near the plant found that they had received radiation doses much lower than those normally caused by natural background radiation. As part of the cleanup and remediation processes, a new water supply for Broomfield was constructed, along with catch basins and diversion ditches to control contaminated runoff. Radioactive and hazardous chemical wastes, as well as decontaminated construction materials from demolished industrial facilities, were transported offsite to licensed repositories. Still-useful nuclear materials were shipped to other DOE facilities. On-site landfills were covered to meet final closure criteria, and groundwater treatment systems were installed as part of the cleanup.

In late 2005 cleanup of chemical and radiological contamination within the 1,619-hectare (4,000-acre) buffer zone surrounding the former plant site was declared complete. The EPA deleted this portion of the Rocky Flats site from the NPL in 2007 and handed the land over to the U.S. Fish and Wildlife Service for use as a national wildlife refuge. The 529-hectare (1,308-acre) central area, formerly the location of the industrial facilities, remained on the NPL as an active Superfund site under DOE control for monitoring and maintenance purposes.

Charles W. Rogers
Updated by Karen N. Kähler

FURTHER READING

Ackland, Len. *Making a Real Killing: Rocky Flats and the Nuclear West.* Albuquerque: University of New Mexico Press, 2002.

Layzer, Judith A. "Government Secrets at Rocky Flats." In *The Environmental Case: Translating Values into Policy.* 2d ed. Washington, D.C.: CQ Press, 2006.

Rood, Arthur S., and John E. Till. *Estimated Exposure and Lifetime Cancer Incidence Risk from Routine Plutonium Releases at the Rocky Flats Plant: Independent Analysis of Exposure, Dose, and Health Risk to Off-Site Individuals.* Neeses, S.C.: Radiological Assessments Corporation, 1999.

U.S. Department of Energy. *Second Five-Year Review Report for the Rocky Flats Site, Jefferson and Boulder Counties, Colorado.* Grand Junction, Colo.: Author, 2007.

SEE ALSO: Atomic Energy Commission; Nuclear accidents; Nuclear and radioactive waste; Nuclear weapons production; Radioactive pollution and fallout; Superfund.

Rocky Mountain Arsenal

Categories: Waste and waste management; pollutants and toxins

Identification: Former U.S. military production facility located near Denver, Colorado

Date: Established in 1942

Significance: Rocky Mountain Arsenal produced, stored, and decommissioned a host of highly toxic and hazardous substances, including chemical weapons, incendiary munitions, and rocket fuels. Shell Chemical Corporation leased part of the facility and manufactured agricultural chemicals there. Both Shell and the U.S. government have participated in a decades-long cleanup program to transform the site into a wildlife refuge.

After the United States officially entered World War II in December, 1941, the military's urgent need for munitions led to the construction of the Rocky Mountain Arsenal (RMA). In 1942, the U.S. Department of Defense (DOD) selected approximately 43 square kilometers (27 square miles) of prairie and farmland roughly 14.5 kilometers (9 miles) northeast of downtown Denver, Colorado, as the site for the facility, which began production about a year after the country's entry into the war.

Chemical weapons manufactured at the facility included chlorine, a lung irritant, and mustard, a blister agent. White phosphorus and other incendiary munitions were also produced. During the war, RMA made 87,000 tons of chemical, intermediate, and toxic products and 155,000 tons of incendiary munitions.

After the war ended, the DOD placed the relatively young arsenal on standby. Rather than allow its chemical production facilities to stand idle during peacetime, it leased portions of the site to commercial interests. Among the lessees was Julius Hyman and Company, which began pesticide production at RMA in 1946. Shell Chemical Corporation acquired Hyman in 1952. For the next three decades, Shell used RMA facilities to manufacture pesticides, herbicides, and agricultural chemicals. Shell products made at RMA included the organochlorine pesticides aldrin, dieldrin, and endrin, neurotoxins that proved so persistent in the environment and so bioaccumulative in the food web that, decades later, their use was banned by international treaty.

After the United States became involved in the Korean War in 1950, the military reactivated RMA. It began constructing a plant to manufacture a more modern and deadlier class of chemical weapon: the nerve agent. In 1953, the same year hostilities in Korea ended, the new facility began production. It manufactured and filled weapons with the organophosphate nerve agent designated GB (sarin), one of the most toxic substances ever synthesized. GB production ended in 1957, but munitions-filling operations continued until 1969.

During the late 1950's and early 1960's RMA worked on the development of a biological agent called TX (a weaponized form of wheat rust, a fungal disease that attacks crops). In 1959 the arsenal constructed a facility for blending hydrazine and unsymmetrical dimethylhydrazine to produce Aerozine-50, a rocket fuel that was used in lunar missions. During the latter half of the 1960's RMA activities supported the war effort in Vietnam by manufacturing land mines and white phosphorus incendiary devices. Other wartime operations included emptying existing munitions filled with the chemical warfare agents phosgene and cyanogen chloride so they could be refilled off-site with high explosives.

Environmental Contamination

RMA's various military and civilian operations generated a vast array of highly hazardous and toxic wastes. More than 750 chemicals were handled or generated at the site, including volatile organic compounds and heavy metals. Environmental contamination occurred not only from deliberate activities such as burying toxic waste and using basins for disposal of hazardous liquid wastes but also from unintended releases such as leaks in tanks, pipes, and sewer lines, wind dispersion, and accidental spills that sometimes involved tens of thousands of gallons of chemicals.

In the arsenal's early days, standard practice was to discharge liquid industrial wastes into centrally located, unlined basins dug into the earth. When farms north of RMA began to complain of crop damage and sickened livestock during the early 1950's, it became clear that chemical wastes from the disposal basins had percolated through the underlying soil and been carried off-site in groundwater. In 1956, RMA installed a lined evaporative basin covering 37.6 hectares (93 acres). Its lining, a layer of asphalt covered with soil, was intended to keep additional contaminants from reaching the groundwater. However, the wastes discharged to this basin soon degraded the asphalt liner, and by the early 1960's it had begun to leak. All of RMA's waste disposal basins were deadly to waterfowl that mistook them for lakes. The toxic basins were particularly attractive to ducks and geese flying overhead during the colder months, as the chemical wastes did not freeze.

The ongoing problem of effective and safe waste disposal drove RMA to explore alternative technologies. In 1962 a waste disposal well was drilled to a depth of more than 3,600 meters (12,000 feet) at the arsenal. More than 662 million liters (175 million gallons) of treated waste material were eventually dis-

A 1954 photo shows two Rocky Mountain Arsenal workers wearing protective gear and standing in front of remote-control devices they used in the manufacture of the deadly nerve gas GB. (©Bettmann/CORBIS)

posed of in this deep-injection well. The month after fluid injection began, however, the Denver area began to experience a series of earthquakes that would continue for years. The pumping facilities were shut down in 1966 amid concerns that the pressurized fluid injection was triggering the seismic activity. In 1968, the military decided to dispose of RMA's stockpile of excess and obsolete chemical munitions by dumping the containerized wastes at sea. Public and congressional concern over this operation's safety led to its cancellation before it began.

During the early 1970's RMA commenced on-site disposal operations for mustard and related blister agents, the nerve agent GB, the anticrop agent TX, and explosives. This major undertaking—at the time, the largest of its kind in U.S. Army history—included chemically neutralizing or incinerating extremely dangerous substances. The military had to take extreme precautions to avoid releasing any chemical agents into the atmosphere, where they could be carried offsite into surrounding communities. These demilitarization operations continued into the early 1980's.

During the mid-1970's, after crop damage had again been observed north and northwest of RMA, a Colorado Department of Health (CDH) investigation found that off-site groundwater was contaminated with diisopropylmethyl-phosphonate, a by-product of the GB manufacturing process, and dicyclopentadiene, a chemical used by Shell in insecticide manufacture. The CDH ordered the Army and Shell to clean up the sources of the contaminants, prevent future releases, and monitor the groundwater on- and off-site. Systems to intercept and treat contaminated groundwater were subsequently installed. Follow-up investigations revealed industrial solvents and the pesticides dibromochloropropane and dieldrin in a shallow aquifer. Surface soils across much of the site were found to contain aldrin and dieldrin. Pesticide residues were also found in the tissues of local wildlife.

SITE CLEANUP

The Army and Shell ended all manufacturing activities at RMA in 1982. In July, 1987, the site became one of the first military installations to be placed on the National Priorities List (NPL) of Superfund sites. The arsenal's only mission became cleanup of contaminated soil, surface water, groundwater, and structures, a complex, decades-long undertaking involving both the Army and Shell, as well as federal, state, and local regulatory agencies and a series of legal battles. Shell and the Army ultimately agreed to split the cleanup costs. Citizens' concerns that potentially hazardous levels of toxic chemicals might be released into the air during remediation operations led to the implementation of a medical monitoring program for communities surrounding RMA to provide periodic health checkups for detection and treatment of any illnesses related to chemical releases.

Cleanup of the lined waste basin proved particularly challenging. Nearly 41,640 kiloliters (11 million gallons) of liquid chemical wastes were pumped from it into temporary storage tanks. The remaining sludge, liner, and underlying contaminated soil were to be excavated and air-dried. Shortly after cleanup operations began in 1988, however, off-site residents and workers north and northwest of RMA complained of noxious odors, headaches, nausea, eye irritation, and rashes. After half a year of trying unsuccessfully to control air emissions, the Army halted excavation activities and capped the remaining materials in place. The liquid wastes were later destroyed by incineration.

In 1986, after a roost of bald eagles was discovered at RMA, the U.S. Fish and Wildlife Service stepped in to manage the site's wildlife. In 1992 the U.S. Congress designated the site as a future national wildlife refuge with the passage of the Rocky Mountain Arsenal National Wildlife Refuge Act. The act specifies that the U.S. government must retain title to RMA. Wildlife habitat is to be preserved and maintained to protect endangered species, while residential or industrial development, agricultural use, consumption of fish or game, and use of groundwater or surface water for drinking purposes are all prohibited. As cleanup has progressed, allowing portions of the site (excluding the underlying groundwater) to be deleted from the NPL, the Army has transferred that land to the Fish and Wildlife Service to establish and expand the wildlife refuge. The refuge will ultimately cover 6,070 hectares (15,000 acres), making it one of the country's largest urban refuges. The former military facility serves as a sanctuary for more than 330 species of animals, including deer, coyotes, foxes, bald eagles, burrowing owls, geese, ducks, badgers, prairie dogs, and a herd of wild bison introduced in 2007. The Army will retain possession of waste disposal trenches, landfills, and areas with groundwater treatment systems.

Alvin K. Benson
Updated by Karen N. Kähler

FURTHER READING

Gascoyne, Stephen. "Slipcovering a Superfund Site." *Bulletin of the Atomic Scientists* 49, no. 7 (1993): 33-37.

Hoffecker, John F. *Twenty-seven Square Miles: Landscape and History at Rocky Mountain Arsenal National Wildlife Refuge.* Denver, Colo.: U.S. Fish and Wildlife Service, 2001.

Konikow, Leonard F., and Douglas W. Thompson. "Groundwater Contamination and Aquifer Reclamation at the Rocky Mountain Arsenal, Colorado." In *Groundwater Contamination,* edited by the Commission on Physical Sciences, Mathematics, and Applications of the National Academies. Washington, D.C.: National Academy Press, 1984.

Natural Resource Trustees for the State of Colorado. *Natural Resource Damage Assessment Plan for the Rocky Mountain Arsenal, Commerce City, Colorado.* Denver: Colorado Department of Public Health and Environment, 2007.

Rocky Mountain Arsenal. *History of Rocky Mountain Arsenal, Commerce City, Colorado.* Ft. Belvoir, Va.: Defense Technical Information Center, 1980.

U.S. Fish and Wildlife Service. *Rocky Mountain Arsenal National Wildlife Refuge.* Washington, D.C.: Author, 2003.

SEE ALSO: Agricultural chemicals; Aquifers; Best available technologies; Groundwater pollution; Hazardous waste; Incineration of waste products; Pesticides and herbicides; Soil contamination; Superfund; Water pollution; Wildlife refuges.

Roosevelt, Theodore

CATEGORIES: Activism and advocacy; preservation and wilderness issues

IDENTIFICATION: American politician and conservationist who served as governor of New York and president of the United States

BORN: October 27, 1858; New York, New York

DIED: January 6, 1919; Oyster Bay, New York

SIGNIFICANCE: During his years as U.S. president, 1901-1909, Roosevelt did more to boost conservation efforts in the United States than any other president before him.

Born into an aristocratic family in New York City, Theodore Roosevelt was a sickly child. In his youth he collected animals both dead and alive, and

Theodore Roosevelt, left, with naturalist and preservationist John Muir at Glacier Point above Yosemite Valley. (Library of Congress)

he considered becoming a biologist while a student at Harvard. During the 1880's, after his first wife's death, Roosevelt sought consolation in the Dakota Badlands as a hunter and a rancher. Hunting was a passion with Roosevelt, but it was combined with a love of nature and considerable scientific knowledge about birds, animals, and plants. In 1887 he was a founding member and the first president of the Boone and Crockett Club, which, while devoted to hunting, also became one of America's earliest conservation organizations.

If the outdoors was Roosevelt's avocation, politics was his vocation. He held a number of political positions and became a national figure while leading the famous Rough Riders cavalry regiment during the Spanish-American War of 1898. After serving as governor of New York, he was elected vice president of the United States in 1900; when President William McKinley was assassinated in September, 1901, Roosevelt became the youngest president in U.S. history up to that time.

As president, Roosevelt made momentous strides in the field of conservation. In fact, the word "conservation" in the sense it is understood today came into use only during his presidency. It was a crucial time, for many of the nation's natural resources that had seemed inexhaustible to Thomas Jefferson one century earlier had been greatly depleted. Roosevelt had many friends who were equally devoted to nature, and

while he was president he camped with John Burroughs in Yellowstone and John Muir in Yosemite. Unlike Muir, however, Roosevelt was not an environmental preservationist. Roosevelt agreed that much of the wild should be preserved in its natural state, but he and Gifford Pinchot, his chief forester and major conservation adviser, also believed that conservation means wisely using nature's resources to benefit later generations.

Promoting water reclamation in the West through the building of dams and irrigation projects was among Roosevelt's successful presidential acts, as was the establishment of millions of hectares of forest reserves (over the opposition of many in Congress) and the creation of fifty-one wildlife refuges. Five new national parks were added during Roosevelt's time in office, including Colorado's Mesa Verde, Oregon's Crater Lake, and South Dakota's Wind Cave Park. Through the Antiquities (or National Monuments) Act of 1906, Roosevelt created eighteen national monuments, including Devils Tower in Wyoming, the Petrified Forest and Grand Canyon in Arizona, and Muir Woods in California.

In 1908, toward the end of his second term as president, Roosevelt organized a three-day White House conservation conference for the nation's governors. This unprecedented event not only gave considerable publicity to the conservation cause but also resulted in the establishment of conservation commissions in thirty-six U.S. states. In February, 1909, Roosevelt hosted the North American Conservation Conference and proposed an international conservation conference, but that dream died when he left the presidency.

Eugene Larson

FURTHER READING

Brinkley, Douglas. *The Wilderness Warrior: Theodore Roosevelt and the Crusade for America.* New York: HarperCollins, 2009.

Morris, Edmund. *Theodore Rex.* New York: Random House, 2001.

SEE ALSO: Burroughs, John; Conservation; Grand Canyon; Hunting; Muir, John; National parks; Pinchot, Gifford; Wildlife refuges.

Rowell, Galen

CATEGORY: Preservation and wilderness issues
IDENTIFICATION: American nature photographer
BORN: August 23, 1940; Oakland, California
DIED: August 11, 2002; Bishop, California
SIGNIFICANCE: One of the best-known American nature photographers, Rowell was also a serious mountaineer and helped to develop the field of participatory photography.

Born in Oakland, California, Galen Rowell grew up in nearby Berkeley, where his father was a professor at the University of California and his mother was a concert cellist. Rowell was a good student but underperformed for his level of ability. He was curious and very bright. His family's devotion to the outdoors, and particularly California's Sierra Nevada mountain range had a profound influence on his life. From infancy he spent a few weeks of every summer camping in the Sierras, where he developed a love for the beauty of the mountains and the wilderness experience. He was also influenced by the rock-climbing section of the Sierra Club and was himself soon climbing mountains. While undertaking increasingly challenging mountaineering expeditions, he became an accomplished photographer.

Like his hero, photographer Ansel Adams, Rowell believed light was everything in a picture, and, like Adams, he was very patient and would go way out of his way to take a photograph at the moment when the lighting was perfect. In contrast to Adams, who worked only in black and white, using huge cameras with big negatives, Rowell used color and shot in 35mm most of his career. Whereas Adams's pictures were usually staged and shot from a tripod, many of Rowell's were close-ups, and many were taken as the photographer stood on a cliff or dangled from a rope. Rowell developed into one of the world's foremost mountain photographers. Although he is particularly known for his work in Yosemite and the High Sierra, he also produced highly respected photographs of the Himalayas and many other mountain ranges all over the world.

Having dropped out of college to work in his own small automobile shop to support his climbing and photographic interests, Rowell finally began to pursue his chosen career full time in 1972. His fame exploded in 1973 and 1974, when *National Geographic* magazine commissioned him to do a feature photo-

graphic article on climbing Yosemite's Half Dome. He recruited two other excellent rock climbers who had photography skills, and in 1973 they made the first "clean climb" of Half Dome—that is, they ascended the sheer monolithic cliff using no pitons. They trusted their lives to their hands, feet, and ability, using small metal wedges that they placed in existing cracks in the granite to support their ropes.

With the success of the *National Geographic* article, Rowell began to be sought out by other magazines, various societies, and individuals to lead groups and produce books. An articulate writer as well as a talented photographer, he eventually published more than twenty books and numerous articles and book chapters. His books were beautifully illustrated with his own photographs of mountain scenery. Perhaps his most highly regarded book is *Mountain Light: In Search of the Dynamic Landscape* (1986); other notable books include *In the Throne Room of the Gods* (1977)

Nature photographer Galen Rowell takes part in a climbing competition in 1990. (Getty Images)

and an edition of John Muir's *The Yosemite* illustrated by Rowell's photographs (2001).

In 1984 Rowell received the Ansel Adams Award for his contributions to the art of wilderness photography. By that time he was devoting much of his time to teaching outdoor photography to small groups of both professional and amateur photographers on guided field trips. He was noted for his enthusiasm and patience with the amateurs. Rowell's life was cut short when he and his wife died in a small-plane crash in 2002.

C. Mervyn Rasmussen

FURTHER READING

Rowell, Galen. *High and Wild: Essays and Photographs on Wilderness Adventure.* Bishop, Calif.: Spotted Dog Press, 2002.

_____. *Mountain Light: In Search of the Dynamic Landscape.* 2d ed. San Francisco: Sierra Club Books, 1995.

SEE ALSO: Adams, Ansel; Muir, John; Yosemite Valley.

Rowland, Frank Sherwood

CATEGORY: Weather and climate
IDENTIFICATION: American chemist
BORN: June 28, 1927; Delaware, Ohio
SIGNIFICANCE: Rowland was the first person to discover that chlorofluorocarbons released into the atmosphere were destroying the protective ozone layer, and he was influential in the eventual move to ban production of these compounds.

Frank Sherwood Rowland received his Ph.D. in 1952 from the University of Chicago, where he studied under Willard F. Libby, recipient of the 1960 Nobel Prize in Chemistry for developing carbon-14 dating. Rowland became the first chairperson of the Chemistry Department at the University of California, Irvine, in 1964. During the 1970's, he began to think about expanding his research into new areas. The use of the stable compounds chlorofluorocarbons (CFCs) as refrigerants and propellants in aerosol cans had become widespread during the 1960's. Rowland reasoned that because of their stability, CFCs should persist in the atmosphere for a long time. That led to the question of what happens to

Chemist Frank Sherwood Rowland. (AP/Wide World Photos)

them as they accumulate in the atmosphere and enter the protective ozone layer at an altitude of about 12,000 meters (40,000 feet).

About this time, Mario Molina joined Rowland's research group as a postdoctoral fellow and undertook the CFC study. Rowland and Molina quickly discovered that the ultraviolet (UV) light streaming through the atmosphere has enough energy to break a bond in the CFCs, releasing the reactive chlorine atom. Chlorine is known to destroy ozone. Rowland and Molina published a paper in 1974 in the journal *Nature* warning that the increasing use of CFCs could result in the destruction of the ozone layer, allowing damaging UV radiation to reach the earth's surface.

Rowland and Molina had no direct proof that the ozone layer was being destroyed, but they called for a ban on CFC production. The production of CFCs had grown to become a huge industry, making some $8 billion per year, so Rowland and Molina knew that their proposal would be controversial. The Du Pont Corporation was the first and largest producer of CFCs, and it led the defense of the industry. Du Pont argued that such a large industry should not be dismantled based on an unproven hypothesis. As warnings about the possible environmental harm increased, however, public concern led many aerosol packagers to cease using CFCs.

The first evidence of ozone layer destruction came from measurements made by Joseph Farman, an English scientist. Farman's data led to the discovery of the seasonal ozone hole over Antarctica. Satellite data gathered by the National Aeronautic and Space Administration (NASA) confirmed what Farman had reported. The CFC industry continued to resist a ban until 1988, when measurements showed a 6 percent overall decrease in the ozone layer. By this time Du Pont had developed alternatives to CFCs and agreed to cease production. An international agreement reached in Montreal led to a worldwide ban on CFC production effective January 1, 1996. It had been fourteen years since Rowland's initial warnings about ozone destruction. Rowland had not been content to report his scientific results but had actively advocated for the CFC ban before numerous commissions and government committees. For his work, Rowland received the Nobel Prize in Chemistry in 1995 jointly with Molina and Paul Crutzen, who had shown that nitrogen oxides also destroy ozone.

Francis P. Mac Kay

FURTHER READING

Hill, Marquita K. "Stratospheric Ozone Depletion." In *Understanding Environmental Pollution*. 3d ed. New York: Cambridge University Press, 2010.

Joesten, Melvin D., John L. Hogg, and Mary E. Castellion. "Chlorofluorocarbons and the Ozone Layer." In *The World of Chemistry: Essentials*. 4th ed. Belmont, Calif.: Thomson Brooks/Cole, 2007.

Miller, G. Tyler, Jr., and Scott Spoolman. "Climate Change and Ozone Depletion." In *Environmental Science: Problems, Concepts, and Solutions*. 13th ed. Belmont, Calif.: Brooks/Cole, 2010.

SEE ALSO: Chlorofluorocarbons; Greenhouse effect; Molina, Mario; Montreal Protocol; Ozone layer.

RS 2477. *See* Revised Statute 2477

Runoff, agricultural

CATEGORY: Water and water pollution

DEFINITION: Water that flows into rivers, lakes, and other bodies of water from agricultural land and operations

SIGNIFICANCE: Runoff from land used for agriculture, which often contains the residues of chemical and organic fertilizers and pesticides, is one of the major sources of water pollution around the world.

The water that flows into streams from farmland after rain or snowmelt carries the residue of the herbicides, fungicides, insecticides, and fertilizers that farmers have used on the land. Many of the wastes produced by cattle, hogs, sheep, and poultry that are raised on feedlots flow into nearby streams. The water that returns to nearby rivers and lakes after it is used to irrigate farmland may be polluted by salt, agricultural pesticides, and toxic chemicals. These organic materials and chemicals that are carried with soil eroding from agricultural land and transported by water runoff degrade the quality of streams, rivers, lakes, and oceans.

Agricultural runoff enters rivers and lakes from farmland spread over large areas. Since such runoff is a nonpoint source of water pollution, it is more difficult to control than discharges from factories and sewage treatment plants. The chemicals in agricultural runoff have contaminated surface water in many areas of the United States. Rivers pick up sediments and dissolved salts from agricultural runoff as they flow to the oceans. Salt concentration in the Colorado River, for example, increases from about 40 parts per million to 800 parts per million as the river flows down from its headwaters to Mexico.

Nutrients such as potassium, phosphates, and nitrogen compounds from organic wastes and fertilizers are carried by agricultural runoff into rivers and lakes. In the process known as eutrophication, excessive nutrients in bodies of water stimulate the growth of plants such as pond weeds and duck weeds, plantlike organisms called algae, fish and other animals, and bacteria. As more of these organisms grow, more also die and decay. The decay process uses up the oxygen in the water, depriving the fish and other aquatic organisms of their natural supply of oxygen. Some types of game fish, such as salmon, trout, and whitefish, cannot live in water with reduced oxygen. Fish that need less oxygen, such as carp and catfish, will replace them. If all the oxygen in a body of water were to be used up, most forms of life in the water would die. Eutrophication that results from human activities, such as agriculture, is known as cultural eutrophication. In the late 1950's Lake Erie, 26,000 square kilometers (10,000 square miles) in area, was reported to be eutrophic. Thanks to stringent pollution-control measures, however, the lake has improved steadily since that time.

MITIGATING ENVIRONMENTAL IMPACTS

The 1987 amendments to the Clean Water Act represent the first comprehensive attempt by the U.S. government to control pollution caused by agricultural activities. In 1991 the U.S. Geological Survey began to implement the full-scale National Water-Quality Assessment (NAWQA) program. The long-term purposes of NAWQA are to describe the status and trends in the quality of the nation's water resources and to provide a scientific understanding of the factors affecting the quality of these resources. In Octo-

ber, 1997, an initiative intended to build on the environmental successes of the Clean Water Act was announced. It focused on runoff from farms and ranches, city streets, and other diffuse sources. The plan called for state and federal environmental agencies to conduct watershed assessments every two years.

Irrigation is necessary for survival in many developing countries. Governments struggle to build advanced agricultural systems, and developments in agriculture have improved food production around the world. However, the growing reliance on fertilizers and other agricultural chemicals has contributed to the pollution of rivers and lakes. Therefore, interest has shifted to farming with reduced use of chemicals. Scientists have turned their attention to developing organic ways to grow food that require less fertilizer and fewer pesticides, and many farmers rotate their crops from year to year to reduce the need for chemical fertilizers. Instead of spraying their crops with harmful pesticides, some farmers combat damaging

Bulldozers remove massive amounts of decaying algae from the beach of Hillion in Brittany, France, in August, 2009. The growth of the algae was a result of agricultural runoff from intensive farming operations. (AP/Wide World Photos)

insects by releasing other insects or bacteria that prey upon the pests. Scientists have also developed genetically engineered plants that are resistant to certain pests. Other strategies for minimizing pollution caused by agricultural runoff include maintaining buffer zones between irrigated cropland and sites where wastes are disposed, restricting application of manure to areas away from waterways, avoiding application of manure on land subject to erosion, reusing water used to flush manure from paved surfaces for irrigation, constructing ditches and waterways above and around open feedlots to divert runoff, constructing lined water-retention facilities to contain rainfall and runoff, applying solid manure at a rate that optimizes the use of the nitrogen it contains for a given crop, and allowing excess wastewater to evaporate by applying it evenly to land.

Watersheds are areas of land that drain to streams or other bodies of water. Most nonpoint pollution-control projects focus their activities around watersheds because watersheds integrate the effects that land use, climate, hydrology, drainage, and vegetation have on water quality. In the United States, the National Monitoring Program was initiated in 1991 to evaluate the effects of improved land management in reducing water pollution in selected watersheds. Federal agencies involved in this program include the Environmental Protection Agency (EPA), the Department of Agriculture, the U.S. Geological Survey, and the U.S. Army Corps of Engineers. A composite index was constructed and published to show which watersheds had the greatest potential for possible degradation of water quality from combinations of pesticides, nitrogen, and sediment runoff. The EPA sets criteria on water quality to help states set their own site-specific standards to control nutrient pollution and thus reduce nutrient loading to rivers and lakes.

G. Padmanabhan

Further Reading

Copeland, Claudia. *The Clean Water Initiative*. Washington, D.C.: Congressional Research Service, Library of Congress, 1998.

Hill, Marquita K. "Water Pollution." In *Understanding Environmental Pollution*. 3d ed. New York: Cambridge University Press, 2010.

Hunt, Constance Elizabeth. *Thirsty Planet: Strategies for Sustainable Water Management*. New York: Zed Books, 2004.

Sullivan, Patrick J., Franklin J. Agardy, and James J. J. Clark. "Water Pollution." In *The Environmental Science of Drinking Water*. Burlington, Mass.: Elsevier Butterworth-Heinemann, 2005.

See also: Agricultural chemicals; Clean Water Act and amendments; Cultural eutrophication; Erosion and erosion control; Eutrophication; Irrigation; Organic gardening and farming; Pesticides and herbicides; Runoff, urban; Water pollution; Watershed management.

Runoff, urban

Category: Water and water pollution

Definition: Water that flows into streams, rivers, and other bodies of water from lawn irrigation, rainfall, and snowmelt in cities and other developed areas

Significance: In urban areas, runoff can contribute to environmental problems such as flooding and water pollution.

When rain falls on natural landscapes, much of it is caught by vegetation or soaks into the ground. A coniferous forest, for example, can intercept as much as 50 percent of the rain that falls on it annually. The rain that reaches the ground percolates into the soil and makes its way into the groundwater or travels slowly through the soil to reach the nearest stream hours, days, or even months later.

In developed areas, rainwater falls on impervious surfaces—roofs, roads, and other nonporous materials—where it is prevented from soaking into the ground. The runoff, often referred to as stormwater, then flows across those surfaces in large quantities, collecting and transporting pollutants, until it reaches a storm sewer, stream, or natural area. This large amount of water flowing into streams almost immediately after a storm can cause local flooding. Research has shown that the number of small floods increases up to ten times when a watershed reaches 20 percent urbanization. Conversely, since less water is allowed to enter the soil and groundwater, stream flows are greatly reduced in the dry season. This rapid fluctuation of water levels causes stream erosion and siltation and destroys habitat for fish and other aquatic life.

The pollution carried by urban runoff is a form of nonpoint source pollution—pollution that is not eas-

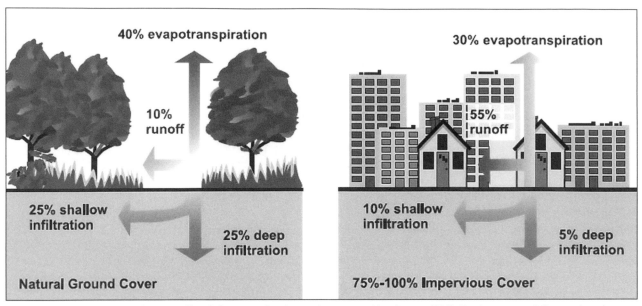

The relationship between surface runoff and the natural or impervious state of land cover. (U.S. Environmental Protection Agency [EPA])

ily traced to a particular point, such as the end of a pipe, but instead comes from all over the environment. Automobiles, combined with impervious surfaces, are responsible for a large part of the pollution in urban runoff. In addition to motor oil that is improperly dumped down storm drains, oil and other automotive fluids leak onto parking lots and roads and are picked up by runoff. Heavy metals, such as zinc and copper, accumulate from the dust of tire and brake wear and can be major pollutants in urban streams. Lawns can be relatively impervious as well, and heavy storms or excessive watering can wash pesticides, herbicides, and fertilizers into streams, disrupting the already damaged aquatic environment. Airborne pollutants that fall on impervious surfaces, substances dumped on roads or in storm drains, bird and pet wastes, and street litter also contribute to the toxic mix of stormwater runoff.

An additional runoff problem is caused by the erosion of disturbed soil in areas of active building and development. Sediment that is washed into streams, either through storm drains or directly, can cover gravel that is critical to aquatic insects and fish and can fill stream channels, contributing to downstream flooding.

STORMWATER MANAGEMENT

The first attempts at urban stormwater management consisted of the construction of gutters and the digging of ditches to convey excess water quickly to the nearest natural waterway. In 1869 famed landscape architect Frederick Law Olmsted was the first to design a system of underground pipes to convey muddy, horse-manure-laden runoff from the streets of the new community of Riverside, Illinois, to the nearby Des Plains River. Many cities soon thereafter developed systems that combined storm sewers with sanitary sewers, but as communities grew and the amount of land covered by impervious surfaces increased, these systems often became overloaded and then overflowed, sending stormwater and raw sewage into rivers and streams. These cities later had to work to separate the storm and sanitary sewer systems, with engineering emphasis placed on storing or detaining runoff during storms and then releasing it to the streams later to reduce flooding.

Large-scale stormwater management relies on swales, ponds, and wetlands. Runoff from a town or development can be routed through a swale, a depression in the landscape that directs the runoff to another place or holds it long enough for it to evaporate or soak into the soil. Vegetation in the swale slows the water and also filters out some of the pollutants. A pond or wetland also detains the water and allows some of the pollutants to settle or be filtered by the plants. Some wetland organisms can actually break down oils and other pollutants into harmless elements.

In 1988 the reauthorization of the Clean Water Act required municipalities in the United States to do more to control nonpoint source pollution, most of which reaches waterways through stormwater. The listing of various species of fish as endangered or threatened under the Endangered Species Act also forced some cities to improve fish habitats by limiting pollution and sedimentation in local streams.

Because nonpoint source pollution, and urban runoff itself, accumulates drop by drop from many small sources, the small actions of many people throughout a watershed can help to solve the problem. In many cities, residents are encouraged to disconnect their roof drains, where appropriate, and divert stormwater to their own landscapes. In addition, rain barrels and cisterns can detain stormwater to be used for landscape watering and other purposes. Naturescaping, or gardening with native plants that are naturally adapted to the local climate and resistant to native pests, requires fewer chemical applications and less supplementary watering. The careful use and disposal of landscape and household chemicals and automotive fluids can also reduce pollutants in stormwater. In these ways, residents of a watershed can become part of the solution to the problem of urban flooding and local stream pollution.

Joseph W. Hinton

FURTHER READING

American Society of Civil Engineers. *Urban Runoff Quality Management.* Reston, Va.: Author, 1998.

Debo, Thomas N., and Andrew J. Reese. *Municipal Stormwater Management.* 2d ed. Boca Raton, Fla.: Lewis, 2003.

Hill, Marquita K. "Water Pollution." In *Understanding Environmental Pollution.* 3d ed. New York: Cambridge University Press, 2010.

Hunt, Constance Elizabeth. *Thirsty Planet: Strategies for Sustainable Water Management.* New York: Zed Books, 2004.

Sullivan, Patrick J., Franklin J. Agardy, and James J. J. Clark. "Water Pollution." In *The Environmental Science of Drinking Water.* Burlington, Mass.: Elsevier Butterworth-Heinemann, 2005.

SEE ALSO: Automobile emissions; Clean Water Act and amendments; Floods; Runoff, agricultural; Stormwater management; Water pollution; Watershed management.

Rural gentrification

CATEGORY: Land and land use

DEFINITION: Development and settlement by nonagriculturists of land typically associated with farming, wilderness areas, or other nonurban uses

SIGNIFICANCE: The transformation of rural lands by urban settlers causes economic and social changes that affect populations living in those areas. Rural gentrification also affects environmental conditions, often damaging or obliterating plant and animal habitats; polluting soil, water, and air; and disrupting vital natural pollination cycles. In some cases, however, rural gentrification can serve to protect environmental resources.

Since the late twentieth century, rural gentrification has occurred when affluent people who have lived in urban areas seek residences in the country. These urban migrants represent various groups, including retirees, investors, and commuters. In the United States, agricultural economic crises in the 1980's resulted in foreclosed farmland being sold at low prices, and some city residents renovated farmhouses as second homes. During the 1990's, the Internet enabled increasing numbers of people to telecommute to their jobs from homes in rural areas. Many baby boomers purchased rural land, anticipating relocating after retirement. By the early twenty-first century, demographic profiles revealed that some rural U.S. counties had experienced increased populations after decades of losing residents to cities.

Rural gentrification can quickly dominate a location. Commercial residential developers of rural land who hope to attract urban migrants often select wildlife and plant habitats, such as beach dunes or timber, that are especially desirable to property buyers. Many rural lands suddenly shift from production of resources for practical uses to appropriation of places for leisure. Rural land targeted for urban newcomers sometimes sells for as much as $100,000 per acre, compared to $2,500 or less per acre for land intended for farming. Developers and newcomers to rural areas often do not recognize the agricultural production possibilities and the natural resources offered by the lands they acquire.

Many conservationists are critical of rural gentrification. They stress that such developments are frequently detrimental, threatening native species and altering ecological resources and ecosystems. Rural golf courses and ski resorts displace vulnerable animal and plant populations when bulldozers clear lands, stripping them of topsoil, grasses, trees, and microorganisms crucial for maintaining ecological balance. Developers often build groups of condominiums or houses in fields as part of planned communities. Acreage that previously sustained crops is landscaped with ornamental grasses, trees, and other decorative plants, many of which are not indigenous. Removal of native flora affects existing biodiversity, and the introduction of alien species can harm the environment.

Commercial development fractures ecosystems by dividing fields and woods into separate properties, impeding wildlife movement with fences or other artificial boundaries and access to essential resources. Rural gentrification can interfere with native animal and plant species' natural cycles of reproduction, hibernation, and other fundamental activities, such as pollination by bees and butterflies, and can have negative impacts on the health of diverse organisms, including lichen, insects, amphibians, and reptiles.

Rural gentrification increases the amounts of litter, noise, and light pollution in the areas in which it takes place. Increased vehicle traffic produces more fuel emissions and often crosses wildlife paths, resulting in collisions causing human and animal deaths. Paved surfaces impede the soil's absorption of rainwater. Critics note that the expanded demand for energy and water caused by rural gentrification often overwhelms existing supplies. In addition, the landscaping techniques used by urban newcomers to rural areas sometimes include fertilizers and pesticides that are not environmentally compatible with the local ecosystem. Excess fertilizer can cause some plants to become brittle, chemical residues from landscaped lawns occasionally contaminate water resources, and herbicides can poison birds and wildlife.

Some aspects of rural gentrification can be considered environmentally beneficial, however. For example, the plants and trees used for landscaping capture carbon. The U.S. Forest Service has noted that, in many cases, rural gentrification does not damage the environment as much as some other uses of the same land would; development of the land for residential use can protect rural areas from the problems that arise from clear-cutting, mining, or soil exhaustion caused by repeated plantings of the same crops.

Elizabeth D. Schafer

FURTHER READING

Bryson, Jeremy, and William Wyckoff. "Rural Gentrification and Nature in the Old and New Wests." *Journal of Cultural Geography* 27, no. 1 (February, 2010): 53-75.

Johnson, Elizabeth A., and Michael W. Klemens, eds. *Nature in Fragments: The Legacy of Sprawl.* New York: Columbia University Press, 2005.

Whitson, Dave. "Nature as Playground: Recreation and Gentrification in the Mountain West." In *Writing Off the Rural West: Globalization, Governments, and the Transformation of Rural Communities*, edited by Roger Epp and Dave Whitson. Edmonton: University of Alberta Press and Parkland Institute, 2001.

SEE ALSO: Air pollution; Biodiversity; Ecosystems; Habitat destruction; Landscape architecture; Light pollution; Noise pollution; Urban sprawl; Water pollution.

S

Sacramento River pesticide spill

CATEGORIES: Disasters; water and water pollution

THE EVENT: Train derailment near Dunsmuir, California, that spilled thousands of gallons of the pesticide metam sodium into the Sacramento River

DATE: July 14, 1991

SIGNIFICANCE: The Sacramento River pesticide spill resulted in the deaths of more than one million fish and thousands of birds and other animals that shared the river habitat. Although the chemical contamination was short-lived, the ecological consequences were long lasting.

On the night of July 14, 1991, a Southern Pacific train derailed at the Cantara Loop north of Dunsmuir, California, causing a chemical tank car to rupture and spill 19,000 gallons of the pesticide metam sodium into the upper Sacramento River. This event, known as the Sacramento River spill or the Cantara spill, was the largest inland ecological disaster that had taken place in California up to that date. As the metam sodium rapidly mixed with the river's water, it released highly toxic compounds, virtually sterilizing what had been one of the premier trout streams in California. The green contaminant plume eventually flowed 58 kilometers (36 miles) downstream into California's largest reservoir, Lake Shasta, where a string of air pipes placed at the bottom of the lake aerated the chemical. The aeration project accelerated the breakdown of the metam sodium, reducing toxic components to undetectable levels by July 29, 1991.

Studies were launched to identify and quantify the spill's damage to natural resources. Exposure to the pesticide had killed all the aquatic life in the Sacramento River between the Cantara Loop and Lake Shasta. More than one million fish died, including more than 300,000 trout. Millions of insects, snails, and clams perished, along with thousands of crayfish and salamanders. Hundreds of thousands of trees, particularly willows, alders, and cottonwoods, eventually died as a result of the pollution of the water, and many more were severely injured. The vegetative damage caused a corresponding dramatic loss of many wildlife species that depended on the river's vegetation for food and shelter. Birds, bats, otters, and mink either starved or were forced to relocate to other areas because their food sources were no longer available. In addition to the devastating effects on wildlife and plant life, the spill produced many reported negative effects on human health. It also halted recreational activities for miles along the river, resulting in substantial economic losses for the residents of the Dunsmuir area.

Although virtually no trace of the metam sodium remained in the river about one month after the spill, it became clear that full recovery remained years away. In 1992 the U.S. Department of Fish and Game planted more than three thousand trees to accelerate the recovery of severely injured vegetation along the river. Some plants, such as elephant ears and torrent sedge, recovered after two growing seasons. During 1994 the trout population reached about one-half of what it was prior to the spill, and trout angling was again allowed on the river. However, strict regulations were established that would allow protection for the recovering wild trout fishery.

By late 1995 ospreys, dippers, sandpipers, and mergansers were all making good progress toward recovery. By 1996 many aquatic and insect populations were nearing numbers that existed prior to the spill; some species, however, particularly clams, snails, crayfish, and salamanders, were struggling to make a comeback. To accelerate recovery, which continued into the twenty-first century, state and federal trustee agencies spent several million dollars to fund specific projects, including research, recovery monitoring, habitat acquisition and restoration, resource protection, and public education.

Alvin K. Benson

FURTHER READING

Elmer-Dewitt, Philip. "Death of a River: An Ecological Catastrophe in California Points to the Need for New Rules on the Transport of Toxic Compounds." *Time,* July 29, 1991, 24.

Friis, Robert H. "Pesticides and Other Organic Chemicals." In *Essentials of Environmental Health.* Sudbury, Mass.: Jones and Bartlett, 2007.

Hill, Marquita K. "Water Pollution." In *Understanding Environmental Pollution*. 3d ed. New York: Cambridge University Press, 2010.

SEE ALSO: Pesticides and herbicides; Restoration ecology; Water pollution.

Safe Drinking Water Act

CATEGORIES: Treaties, laws, and court cases; water and water pollution

THE LAW: U.S. federal law concerning standards for safe public drinking water

DATE: Enacted on December 16, 1974

SIGNIFICANCE: The Safe Drinking Water Act was the first law passed in the United States to set standards regarding acceptable levels of certain pollutants in the more than 170,000 public drinking-water supplies across the nation.

Under the Safe Drinking Water Act of 1974, the U.S. Environmental Protection Agency (EPA) is required to set standards regarding the maximum amounts of certain materials allowed in public drinking water; these include harmful inorganic and organic substances, radioactive substances, microorganisms, and suspended materials. The act requires the individual U.S. states to enforce the EPA's standards, and it requires each public drinking-water supplier to monitor the quality of its water sent to home users. Initially the EPA included twenty-five materials on its list of contaminants, but over time it has added many others.

DRINKING-WATER CONTAMINANTS

The EPA sets maximum levels of contaminants in public drinking water for microorganisms (such as viruses, colliform bacteria, *Giardia*, and *Cryptosporidium*), disinfection by-products (such as bromate and chlorite), disinfectants (such as chloramines and chlorine), inorganic chemicals (such as arsenic, asbestos, chromium, cyanide, fluoride, lead, mercury, and nitrate), organic chemicals (such as atrazine, benzene, dichlorobenzene, and dioxin), and radionuclides. In addition to setting the maximum allowable amount of each contaminant (the maximum contaminant level, or MCL) in drinking water (usually in milligrams of contaminant per liter of water), the EPA states the

ideal goal level of the contaminant (usually also in milligrams per liter). The EPA also provides information on the typical sources of the contaminants listed and their possible harmful health effects on humans.

Nitrate, for example, is a common pollutant in natural waters that may come from sewage, animal waste, fertilizer runoff, or the erosion of natural deposits. Nitrate levels thus can become high in groundwater near feedlots for cattle or in areas where large amounts of fertilizers are used, such as agricultural areas. Nitrogen compounds formed from nitrate may bond with hemoglobin in the blood of humans so that less oxygen can be transported through the body by hemoglobin. Humans, especially babies, can become seriously ill from drinking nitrate-polluted waters. The EPA sets the MCL for nitrate (measured as nitrogen) in public drinking water at 10 milligrams per liter.

Harmful microorganisms in water often come from human and animal fecal material. They can produce diarrhea, vomiting, and cramps. Harmful organic materials found in water supplies often come from chemical plants, herbicides, sewage, or insecticides. These can have a variety of negative health impacts, including increased risk of cancer, liver and kidney damage, anemia, reproductive problems, and nervous system problems. Harmful radioactive materials, such as uranium or radium, may enter water supplies from natural materials or as the result of improper disposal of radioactive wastes. Humans and animals exposed to such materials may have increased risk of cancer.

MUNICIPAL WATER SYSTEM PROBLEMS

Individual U.S. states have often failed in their enforcement of the EPA drinking-water standards. In 1985, for example, more than eighteen hundred cases were reported in which water in public supplies contained contaminants at levels higher than the maximum allowed. The most common problems involved levels of microorganisms, nitrate, and fluoride that exceeded EPA-required levels.

Because of such problems, amendments to the Safe Drinking Water Act were passed by Congress in 1986. This legislation gave the EPA deadlines by which it was to set and enforce reasonable standards for more than eighty potentially dangerous contaminants in public water systems. In setting the MCLs, the EPA was required to take into consideration not only the danger of the contaminants but also the costs of meeting these standards in public waters. Also included in this law was a ban on the use of lead solder and pipes in public water systems. In addition, the 1986 amendments required the EPA to monitor materials injected under the ground, such as oil field brines, to ensure that the injected materials do not contaminate groundwater supplies.

Robert L. Cullers

FURTHER READING

Cech, Thomas V. "Water Quality." In *Principles of Water Resources: History, Development, Management, and Policy.* 3d ed. New York: John Wiley & Sons, 2010.

Dorsheimer, Wesley T. "Removing Nitrate from Groundwater." *Water Engineering and Management* 144, no.12 (1997): 20-24.

Gray, N. F. *Drinking Water Quality: Problems and Solutions.* 2d ed. New York: Cambridge University Press, 2008.

Ketcham-Colwill, J. "Safe Drinking Water Law Toughened." *Environment* 28, no. 7 (1986): 42-43.

Royte, Elizabeth. *Bottlemania: Big Business, Local Springs, and the Battle over America's Drinking Water.* New York: Bloomsbury, 2009.

SEE ALSO: Acid mine drainage; Agricultural chemicals; Drinking water; Environmental health; Hazardous and toxic substance regulation; Heavy metals; Lead; Mercury; Pesticides and herbicides; Polychlorinated biphenyls; Water pollution; Water-pollution policy; Water quality.

Sagebrush Rebellion

CATEGORY: Land and land use

DEFINITION: A U.S. movement that called for state, rather than federal, management of public lands

SIGNIFICANCE: The Sagebrush Rebellion evidenced the growing conflict between environmentalism and the economic exploitation of public lands. Although the movement did not survive past the early 1980's, it succeeded in changing the policy debate regarding public lands.

The Sagebrush Rebellion, which originated in the western United States during the late 1970's and early 1980's, was part of a larger historical trend that began during the nineteenth century as settlers in the West struggled to gain control over public lands.

Movements opposing federal management of these lands had arisen periodically, as they would again during the 1990's in the form of the wise-use movement. The Sagebrush Rebellion received national attention during the 1980's in part because of President Ronald Reagan's support for the movement and because it was part of a larger movement that demanded the reduction of the federal government's involvement in American life. The Sagebrush Rebellion represented a response to the environmental movement that gained ground during the 1970's.

The federal government owns nearly 50 percent of the land in the western United States. In response to growing environmental activism, government regulations concerning the use of these public lands increased significantly during the 1970's. Several of these regulations threatened lumbering, cattle grazing, and other commercial activities that were permitted on public lands. Many westerners regarded the regulations as unfair, claiming that they stifled development and slowed economic growth in the region. Some maintained that the federal government was limiting opportunities in the West in order to protect established industries in the eastern United States.

In response, these westerners advocated the transfer of land management from the federal government to state governments, which were more amenable to development. They formed organizations such as the League for the Advancement of States' Equal Rights to advance their cause. During the late 1970's two members of Congress from western states offered legislation at the federal level to transfer land management to the states; however, these initiatives gained little support.

The movement garnered national attention in 1979 when the Nevada legislature passed Assembly Bill 413, commonly known as the Sagebrush Rebellion Act. Nevada's legislators maintained that the huge expanse of federal lands within the state's borders (83 percent of the land in Nevada belonged to the federal government) put the state at a disadvantage in comparison with other states. They claimed the right to manage the 19.4 million hectares (48 million acres) of land then under the authority of the U.S. Bureau of Land Management (BLM), which constituted almost all the federal lands in the state. After passage of the act, Nevadans made no effort to seize control of the public lands. Instead, the state appropriated funds to initiate a lawsuit against the federal government.

e so-called Sagebrush Rebels soon counted several victories throughout the region. Four western states enacted similar measures, with the Arizona legislature overriding Governor Bruce Babbitt's veto. In 1982 Alaskans showed their support for the so-called Tundra Rebellion when more than 70 percent of voters supported efforts to take control of that state's BLM lands. Success was not guaranteed—the Sagebrush Rebels saw legislative or executive measures supporting their cause fail in seven western states—but the notion of state control of public lands was clearly popular with western voters.

The movement gained a powerful boost in 1980 when president-elect Reagan declared himself to be a Sagebrush Rebel. As president he followed through on his rhetoric, nominating another "rebel," James Watt, for the cabinet position of secretary of the interior. Emboldened by the changed political climate in Washington, D.C., in May, 1981, Senator Orrin Hatch of Utah again offered legislation requiring state management of BLM lands and lands managed by the U.S. Forest Service. As in the past, the proposal did not gain sufficient support in Congress. The failure of Hatch's legislation indicated that the Sagebrush Rebellion lacked support from key politicians. Although some government officials agreed with the westerners' complaints, others merely made campaign promises that they had no intention of fulfilling.

WEAKNESSES OF THE MOVEMENT

The Sagebrush Rebellion suffered from several weaknesses. First, the state initiatives offered a potent rallying point for the movement, but few politicians or other observers took them seriously. The courts clearly would not approve measures that intruded on the powers of the federal government. Land transfers had to take place at the federal level, and despite Reagan's stated enthusiasm for the rebellion, key members of his administration opposed the notion of state or federal government ownership of lands. They favored the sale of public lands to private concerns. In the early 1980's many Sagebrush Rebels allied themselves with environmentalists to thwart the privatization movement.

Observers also noted that the rebellion faced opposition from the very groups that it was intended to assist. In many western states ranchers complained about federal regulation but benefited from the low cost of grazing on public lands. Mining companies also feared that the transfer of public land to state control might raise their costs and threaten their mineral rights. Some western state governments estimated that the costs of managing the lands would exceed the income the lands would generate while under state control. Thus powerful western interests were arrayed against the Sagebrush Rebellion.

The movement was short-lived, lasting only from the late 1970's to 1983, when the debate over privatization effectively killed it. If it is assessed by its goals, the Sagebrush Rebellion failed because federal lands remained in the public domain. However, the movement did change the policy debate regarding public lands. It advanced the ethic of multiple-use management, which remained an important issue in public land policy from the 1980's onward. It also provided public support for the Reagan administration's efforts to scale back environmental regulations.

Thomas Clarkin

FURTHER READING

Cawley, R. McGreggor. *Federal Land, Western Anger: The Sagebrush Rebellion and Environmental Politics.* Lawrence: University Press of Kansas, 1993.

Davis, Charles, ed. *Western Public Lands and Environmental Politics.* 2d ed. Boulder, Colo.: Westview Press, 2001.

Echeverria, John D., and Raymond Booth Eby, eds. *Let the People Judge: Wise Use and the Private Property Rights Movement.* Washington, D.C.: Island Press, 1995.

Graf, William L. *Wilderness Preservation and the Sagebrush Rebellions.* Savage, Md.: Rowman & Littlefield, 1990.

Nelson, Robert H. *Public Lands and Private Rights: The Failure of Scientific Management.* Lanham, Md.: Rowman & Littlefield, 1995.

Platt, Rutherford H. *Land Use and Society: Geography, Law, and Public Policy.* Rev. ed. Washington, D.C.: Island Press, 2004.

Robbins, William G., and James C. Foster, eds. *Land in the American West: Private Claims and the Common Good.* Seattle: University of Washington Press, 2000.

SEE ALSO: Antienvironmentalism; Land-use policy; Privatization movements; Wise-use movement.

St. Lawrence Seaway

CATEGORIES: Places; ecology and ecosystems

IDENTIFICATION: International waterway built to connect Lake Superior to the Atlantic Ocean

DATE: Opened on April 25, 1959

SIGNIFICANCE: The St. Lawrence Seaway has been the site of various events of environmental significance, including oil spills and the introduction of exotic, or invasive, species of marine life, in large part because the seaway allows for the mixing of the Great Lakes ecosystem with oceanic and other ecosystems. Existence of the seaway has also encouraged increased development in the surrounding region, which has led to pollution and loss of wetlands.

The St. Lawrence Seaway, a joint venture between the United States and Canada, provides deep-draft ocean ships with access to the Great Lakes. It is the world's longest waterway, extending 3,769 kilometers (2,342 miles) from Lake Superior to the Atlantic Ocean. The seaway, which follows an ancient trade route known to Native Americans as "the river without end," uses a system of locks to raise or lower ships to the appropriate water levels along the route. Each lock is at least 24 meters (80 feet) wide and 8 meters (26 feet) deep. The route from Montreal, Quebec, to Lake Ontario contains seven locks. Another eight locks make up the Welland Ship Canal route from Lake Ontario to Lake Erie. The Soo Canals, which connect Lake Huron and Lake Superior, have four locks. The total lift from the mouth of the St. Lawrence River to Lake Superior is approximately 177 meters (581 feet).

The seaway was built before there was much interest in collecting baseline environmental data, making it difficult to assess the effects of its operation since it opened on April 25, 1959. The massive project involved the excavation of more than 360 million tons of rock, construction of three dams, and completion of more than 105 kilometers (65 miles) of canals. The

The St. Lawrence Seaway, which opened on April 25, 1959, has allowed freshwater and oceanic ecosystems to intermix, resulting in drastic changes to the ecology of the Great Lakes. (National Archives)

seaway can accommodate ships up to 35,000 tons. The typical operating season is 270 days between April and December, with an average of 2,800 vessel transits. Ice-breaking equipment is used to enhance winter navigation. All tankers using the system must be double-hulled.

During the 1970's two sizable oil spills provided the impetus for development of an oil-spill contingency plan and led to closer monitoring of environmental impacts stemming from operation of the seaway. In 1989 the oil-spill plan enabled a successful response after a tanker spilled 14,000 gallons of xylene, a volatile and hazardous solvent, in the seaway.

The St. Lawrence Seaway allows the mixing of the Great Lakes ecosystems with oceanic and other ecosystems. Sea life lurking in cargo holds can enter the river and lake systems. Perhaps the most significant example of an intruding species is the zebra mussel, which arrived in the ballast water of an Eastern European ship during the 1980's and quickly spread throughout the Great Lakes region. Other invaders simply swim unaided into the Great Lakes from the Gulf of St. Lawrence, as in the case of sea lampreys. The seaway has also affected the riverine ecosystem through changes in river flow rates and replacement of natural channel features with canals.

With more than fifty associated ports, the seaway has spurred regional development, which has resulted in increased navigation on the river, increased recreational activity, shoreline development, tourism, point and nonpoint water pollution discharge, and loss of wetlands. Changes in operation of the seaway could have additional environmental consequences. Increased winter navigation would affect icing patterns, which in turn affect riparian habitat, archaeological resources, other natural resources, and the generation of power from dams. Pressure waves from ships disturb the bottom of the river and cause ice damage along the shoreline, including the shearing of plants caught in the ice. Other issues associated with increased operation range from potential effects on wildlife migration patterns to the consequences of shipping accidents.

Robert M. Sanford

FURTHER READING

Beck, Gregor Gilpin, and Bruce Litteljohn, eds. *Voices for the Watershed: Environmental Issues in the Great Lakes-St. Lawrence Drainage Basin.* Montreal: McGill-Queen's University Press, 2000.

Parham, Claire Puccia. *The St. Lawrence Seaway and Power Project: An Oral History of the Greatest Construction Show on Earth.* Syracuse, N.Y.: Syracuse University Press, 2009.

SEE ALSO: Introduced and exotic species; Lake Erie; Oil spills; Water pollution; Zebra mussels.

Sale, Kirkpatrick

CATEGORY: Activism and advocacy
IDENTIFICATION: American journalist, historian, and environmental writer
BORN: June 27, 1937; Ithaca, New York
SIGNIFICANCE: In a long career as journalist and activist, Sale has helped shape the modern environmental movement through his writings emphasizing human scale, bioregionalism, decentralization, and a thoroughgoing critique of technology and the idea of progress.

Kirkpatrick Sale was educated at Quaker-associated Swarthmore College and at Cornell University, where he majored in history and journalism. This educational background underlies Sale's long-standing efforts to forge links between environmental and social justice concerns—efforts reflected in the many books he has produced in his career as journalist and historian, from *The Land and People of Ghana* (1963) and *SDS* (1973) to *Rebels Against the Future: The Luddites and Their War on the Industrial Revolution—Lessons for the Computer Age* (1995) and *After Eden: The Evolution of Human Domination* (2006). Although sometimes portrayed as a simple Luddite, anarchist, and secessionist, he has served as editor with *The New York Times Magazine* and *The Nation* and is a member of both the E. F. Schumacher Society and the American Association for the Advancement of Science.

The trajectory of Sale's career as environmental writer and activist can perhaps be best understood in terms of the importance of a Schumacher-influenced "appropriateness" in Sale's thought and writings—most important, appropriateness in tools, scales, and relations among humans and of humans with other species and the nonhuman environment generally. His works trend toward a critique of what might be called the technoglobal dominance of the human species, the dangerous failings of which are, at root, all ultimately failures of appropriateness.

Although for Sale technology is never neutral, tools in themselves are less an issue than the prior history built into them by the culture that made them. Cultures that increasingly elevate the technological and material—to the exclusion of environmental, social, civic, and other "irrelevant" values—produce increasingly inappropriate and unsustainable technologies. As a bioregionalist who believes that human beings live best when they are aware of and live within the constraints of the regions in which they find themselves, Sale views the increasingly global scale of human technological culture also as increasingly inappropriate to long-term sustainability. The increasingly dominant (and increasingly inappropriate) power position of a relatively few human beings within the human species and the increasingly dominant position of *Homo sapiens* as the "crown of creation" in the natural world are at odds with horizontal and decentralized power relations among human beings and between the human species and the rest of nature—these are key elements of Sale's thought concerning what is appropriate and sustainable.

The broad "revolutions" of the past seventy thousand years—the introduction of big-game hunting among hunter-gatherers, the agricultural revolution, the Industrial Revolution, the digital revolution—are, for Sale, all intensifications of the same pattern of increasing inappropriateness in human tools, scales, and relations, a vast "boom" leading inevitably to a catastrophic "bust" on the same scale—unless action is taken in time.

Whether he is discussing the industrial capitalism of the past few centuries or the growing technoglobal dominance of *Homo sapiens* over the past seventy thousand years, Sale presents a message that mingles apocalypse and optimism. The forces he presents as arrayed against the change to a more appropriate way of living always seem to be unstoppably and invincibly leading humankind to global catastrophe—but this is precisely where the relevance of history to the future comes into play. Although humanity may be suffering a second, digital revolution in the present, we still have before us the model of the nineteenth century Luddite revolt against the first industrial revolution. Although we live in a technoglobally dominant *Homo sapiens* world, we still have the nearly two-million-year run of the more appropriate world of *Homo erectus* to inspire us. Sale, as historian, reminds us that the past

was not the same as the present, and this makes legitimate the hope for appropriate change, since the future need not be merely a continuation of the present.

Howard V. Hendrix

FURTHER READING

Katz, Eric, Andrew Light, and David Rothenberg, eds. *Beneath the Surface: Critical Essays in the Philosophy of Deep Ecology.* Cambridge, Mass.: MIT Press, 2000.

Sale, Kirkpatrick. *Dwellers in the Land: The Bioregional Vision.* Athens: University of Georgia Press, 2000.

Schumacher, E. F. *Small Is Beautiful: Economics As If People Mattered.* 1973. Reprint. Point Roberts, Wash.: Hartley & Marks, 1999.

Thayer, Robert L., Jr. *LifePlace: Bioregional Thought and Practice.* Berkeley: University of California Press, 2003.

SEE ALSO: Agricultural revolution; Antinuclear movement; Appropriate technology; Balance of nature; Biophilia; Bioregionalism; Carrying capacity; Deep ecology; Environmental ethics; Indigenous peoples and nature preservation; Industrial Revolution; Overconsumption; Schumacher, E. F.; Snyder, Gary; Subsistence use; Sustainable agriculture; Sustainable development.

SANE

CATEGORIES: Organizations and agencies; activism and advocacy; nuclear power and radiation

IDENTIFICATION: Organization of environmental activists, scientists, and pacifists formed with the aim of protesting nuclear weapons testing

DATE: Founded in 1957

SIGNIFICANCE: SANE experienced some success in increasing public awareness regarding the effects of nuclear testing on the environment.

In April, 1957, prominent antinuclear activists and pacifists, led by *Saturday Review* editor Norman Cousins, met in Philadelphia; the ad hoc organization formed at this meeting evolved into SANE. In September, the group's officials named their organization the National Committee for a Sane Nuclear Policy. The formation of what soon came to be known simply as SANE was announced in *The New York Times* on November 15, 1957, in an advertisement with the headline "We Are Facing a Danger Unlike Any Danger

That Has Ever Existed." The advertisement, which was sponsored by such notables as theologian Paul Tillich, social critic Lewis Mumford, novelist James Jones, and humanitarian Eleanor Roosevelt, brought the group many new members and donations.

SANE was at first an informal venture meant to make the public aware of the dangers of nuclear weapons testing, but it proved so popular that it became a permanent organization. The group's initial membership consisted largely of scientists, writers, and other professionals, but as the antinuclear movement grew, thousands of other people joined. By mid-1958, SANE had approximately twenty-five thousand members and more than one hundred chapters.

Despite Cold War opposition, SANE and other groups pressed their campaign against nuclear testing. In August, 1958, the United States followed the lead of the Soviet Union and voluntarily suspended nuclear tests. The two countries further agreed to meet in Geneva, Switzerland, to begin negotiations for a test ban treaty.

In the early 1960's, SANE came under attack by members of the U.S. Congress for the organization's alleged harboring of communists. Investigations by SANE leaders established that some members did in fact have communist affiliations, and though congressional investigations eventually exonerated SANE's leadership of wrongdoing, the effects of the investigations on SANE were disastrous. Membership declined drastically, and several prominent members and sponsors resigned.

In 1961, when the Soviet Union resumed the testing of nuclear weapons, SANE condemned the action and called for international protests. The organization also urged the United States to refrain from following the Soviet example, but to no avail.

Such concerns became secondary during the Cuban Missile Crisis of October, 1962, when the world stood on the brink of nuclear war. In the aftermath, support for a treaty between the United States and the Soviet Union gained new life, and the Limited Test Ban Treaty was signed on August 5, 1963. With the signing of the treaty, SANE's initial goals were partially achieved.

SANE leaders subsequently chose to direct the group's energies to opposing the Vietnam War, but this approach proved divisive; in 1967, many key SANE officials, including Cousins and the executive director Donald Keys, resigned from the organization. SANE's membership and influence subsequently declined precipitately, and in 1969 the group changed its focus to campaign against the construction of antiballistic missile systems. The organization did not experience a resurgence until the early 1980's, when the nuclear freeze movement gathered momentum. In the late 1980's, SANE merged with another organization, the Nuclear Weapons Freeze Campaign, to become SANE/Freeze, the largest peace-promoting organization in American history. SANE/Freeze was renamed Peace Action in 1993.

Alexander Scott

FURTHER READING

Giugni, Marco. *Social Protest and Policy Change: Ecology, Antinuclear, and Peace Movements in Comparative Perspective.* Lanham, Md.: Rowman & Littlefield, 2004.
Wittner, Lawrence S. *Confronting the Bomb: A Short History of the World Nuclear Disarmament Movement.* Stanford, Calif.: Stanford University Press, 2009.

SEE ALSO: Antinuclear movement; Comprehensive Nuclear-Test-Ban Treaty; Greenpeace; Limited Test Ban Treaty; Nuclear and radioactive waste; Nuclear testing; Nuclear weapons production.

Santa Barbara oil spill

CATEGORIES: Disasters; water and water pollution

THE EVENT: Blowout in a drilling well in the Santa Barbara Channel off the California coast that resulted in a massive oil spill

DATE: January, 1969

SIGNIFICANCE: The oil that spilled off the coast of the resort city of Santa Barbara threatened marine life and coated beaches, endangering the area's tourism industry. As a result of the spill, the public demanded more stringent regulations on oil companies drilling in offshore areas.

Santa Barbara is an old Spanish mission town situated between the Pacific Ocean and the Santa Ynez Mountains in Southern California. Residents of the city have long been familiar with the effects of minor oil pollution at places along the coast; oil has been escaping from natural fractures in the ocean floor for thousands of years. Hundreds of years ago, Native Americans reportedly caulked their canoes with oil-based substances found near the Santa Barbara shoreline.

In January, 1969, Santa Barbara and other nearby communities along the Pacific coast were confronted by a devastating oil slick. The source of the oil was a blowout at Union Oil Company's drilling platform A, which was less than 13 kilometers (8 miles) offshore in the Santa Barbara Channel. Gas-charged oil escaped from below the metal well casing (a pipe set in the well bore) before the blowout preventers could be closed. According to observers, ocean water east of the platform "boiled" violently for several hours after the preventers were closed.

Some reports suggested that the oil leaked to the surface along an unmapped fault. In a period of ten days, more than 11,000 tons of free oil reached the surface of the water and rapidly spread over an area of 200 square kilometers (80 square miles). An estimated 10,000 barrels of crude oil eventually reached the shoreline, where it coated gravel-sized beach rocks and rapidly infiltrated the sand. The light-colored Goleta Cliffs a short distance north of Santa Barbara were marked by a black band of gooey oil.

The spreading oil slick threatened marine life in the area, including porpoises, seals, whales, birds, and fish. Numerous oil-soaked birds perished, but many were saved by volunteers who removed the oil with a solvent. U.S. Navy personnel expressed concern that the oil would harm porpoises near the naval base at Point Mugu, California.

Remediation commenced immediately; methods included skimming the oil off the water's surface, burning the oil, applying chemical dispersants, and steam cleaning and vacuuming of beach areas. The most effective technique was the distribution of straw and other plant material along the beaches to absorb the oil. Many residents of Santa Barbara, including student volunteers from the city's campus of the University of California, worked many hours spreading and collecting the oil-soaked plant material, which was then taken away to be burned or buried.

As a result of the spill, Union Oil Company and three partners (Gulf, Mobil, and Texaco)—as well as the U.S. Department of the Interior—were sued for damages by the state of California and several coastal communities. The amounts of the claims ranged from $500 million to approximately $1.3 billion. The accident and the ensuing pollution also had major political repercussions. Public pressure resulted in a tem-

Work crews rake and shovel oil-soaked straw on the beach in Santa Barbara as part of cleanup efforts following the January, 1969, offshore oil spill. (AP/Wide World Photos)

porary halt on drilling in the channel, and more stringent regulations were imposed on oil companies drilling in offshore areas.

Donald F. Reaser

FURTHER READING

Fingas, Merv. *The Basics of Oil Spill Cleanup*. 2d ed. Boca Raton, Fla.: CRC Press, 2001.

Fitzgerald, Edward A. *The Seaweed Rebellion: Federal-State Conflicts over Offshore Energy Development*. Lanham, Md.: Lexington Books, 2001.

SEE ALSO: *Amoco Cadiz* oil spill; *Argo Merchant* oil spill; Bioremediation; BP *Deepwater Horizon* oil spill; *Braer* oil spill; *Exxon Valdez* oil spill; Ocean pollution; PEMEX oil well leak; *Sea Empress* oil spill; Tobago oil spill; *Torrey Canyon* oil spill.

Savannas

CATEGORY: Ecology and ecosystems

DEFINITION: Tropical or subtropical grassland environments with wet and dry seasons

SIGNIFICANCE: Savanna environments host large populations of insects, birds, herbivorous grazing animals, and predatory animals. Many nations have established nature preserves to protect valuable savanna resources, as hunting and poaching have rapidly reduced the populations of some once-numerous savanna inhabitants and agricultural encroachment has shrunk the land area of savannas worldwide.

Savannas form in tropical or subtropical wet-dry climates, with moisture from the intertropical convergence zone dominant in wet months. Savannas have an average temperature of at least 16 degrees Celsius (64.4 degrees Fahrenheit) during the coolest months and receive between 50 and 255 centimeters (20 and 100 inches) of rain yearly. Savannas have poor drainage; the vegetation in savanna regions is dominated by drought-resistant grasses and a few trees. Fires, which occur during the dry season and prevent forests from developing, are essential in maintaining these grasslands.

Savannas form at the borders of tropical rain forests. In Africa, the savanna of the Sahel lies to the north of the rain forest, and northern and southern savannas meet in the East African savanna. The Colombian and Venezuelan llanos of the Orinoquía region are north of the Amazon rain forest, and the Argentine pampas and Brazilian *cerrado* are south of the rain forest. Almost the entire country of Uruguay is savanna. Savannas occupy roughly 25 percent of the continent of Australia, running across the northern tropics. South of the Himalaya Mountains, the Terai-Duar savanna in northern India, Nepal, and Bhutan is maintained by the yearly monsoon.

Gazelles graze on an African savanna. (©Birute Vijeikiene/Dreamstime.com)

Savannas support large numbers of herbivorous and predator animals, birds, reptiles, and insects. Termites play an important role in savanna ecosystems; their mounds dot savannas worldwide. Termites consume cellulose and lignin from living grasses and trees, as well as vegetative litter. Primates (including human beings) exploit savanna termites as a high-protein source of food.

During the wet season (which may arise from monsoonal rains), large animal herds graze over the grasslands; during the dry season, herds move toward lakes and rivers. Predators follow the herbivores. Migratory birds from Asia, Europe, and North America winter in Southern Hemisphere savannas, joining large numbers of ground-living sedentary birds. Historically, sleeping sickness in Africa, spread by the tsetse fly, and malaria, spread by mosquitoes, kept human populations from exploiting many savannas, but modern insecticides and modern medicines have mitigated these disease factors somewhat, and human occupation of savannas has climbed steadily since the mid-twentieth century.

Human activities are the major threat to savannas worldwide. Illegal trade in endangered species of savanna birds, such as South American parrots, and African and Asian animals, including tigers, lions, and primates, is a multibillion-dollar industry worldwide. Some endangered birds and animals are prized as pets; others are killed for their pelts or for use of their body parts in foods or medicines. Efforts to stop the trade in endangered species and their products have had limited results. For example, hunting of the endangered African bush elephant (also known as the Savanna elephant) was prohibited in 1989, but poachers continue to kill the elephants for their ivory tusks. The population of these elephants stood at only an estimated ten thousand in 2010.

It has been estimated that if conversion of the Brazilian *cerrado* savanna to croplands continues at the rates seen in the late twentieth century, the savanna will disappear by 2030. Oil production in the Venezuelan llanos and Brazilian *cerrado* may also alter the savanna biome. In Australia, pressure from the cattle industry to alter the natural savanna to provide for optimal beef production has changed the savanna ecosystem. Problems have arisen there from the introduction of nonnative plant species, land clearance, and pastoral grazing, which has changed the natural pattern of fires in the savanna.

Anita Baker-Blocker

FURTHER READING

Bassett, Thomas J., and Donald Crummey, eds. *African Savannas: Global Narratives and Local Knowledge of Environmental Change*. Portsmouth, N.H.: Heinemann, 2003.

Mistry, Jayalaxshmi, and Andrea Berardi, eds. *Savannas and Dry Forests: Linking People with Nature*. Burlington, Vt.: Ashgate, 2006.

Pennington, R. Toby, and James A. Ratter. *Neotropical Savannas and Seasonally Dry Forests: Plant Diversity, Biogeography, and Conservation*. Boca Raton, Fla.: CRC Press, 2006.

Roach, Mary. "Almost Human." *National Geographic*, April, 2008, 126-145.

SEE ALSO: Africa; Asia; Australia and New Zealand; Elephants, African; Grazing and grasslands; Habitat destruction; Land clearance; Oil drilling; Poaching; Serengeti National Park; South America; Wildlife refuges.

Save the Whales Campaign

CATEGORIES: Activism and advocacy; animals and endangered species

IDENTIFICATION: A sustained advocacy effort on behalf of the protection of whales

DATE: Initiated in 1971

SIGNIFICANCE: The efforts of the variety of individuals and organizations involved in the Save the Whales Campaign have been influential in bringing about a ban on commercial whaling and in arousing international condemnation of the activities of pirate whalers and whale meat smugglers.

The unchecked commercial hunting of whales over several centuries in both open oceans and coastal waters caused the depletion of whale populations worldwide. Once harvested for their oil, whales later became prized for their meat, considered to be a delicacy in Japan and a few other countries. Many whale species approached extinction because of rampant overhunting.

The International Whaling Commission (IWC) was founded in 1946 to prevent further exploitation of whale populations, but instead it presided over some of the worst excesses in whaling's history. In 1971 the Animal Welfare Institute (AWI) joined with several other organizations to launch the Save the Whales Campaign in an effort to end the harvesting of whales. Pursuing the theme of this campaign, the United Nations Conference on the Human Environment in 1972 called for a ten-year moratorium on commercial whaling. In 1974 the IWC decided to regulate whaling according to the principle of maximum sustainable yield. Whenever a species of whale dropped below the optimal population for such a yield, the IWC instituted a ban on hunting that species so that the population could recover. In 1974 the blue whale, bowhead whale, and right whale populations reached low levels and were immediately protected. Because of the difficulty of obtaining reliable data and enforcing catch limits, the Save the Whales Campaign encouraged the IWC in 1982 to adopt an indefinite moratorium on commercial whaling; such a moratorium took effect in 1986 when the IWC instituted its international whaling ban.

During the 1990's the philosophy of the Save the Whales Campaign expanded to include not only the conservation of the whale population but also the issues of animal rights and the aesthetic value of observing whales. Many people worldwide simply believe that it is wrong to kill and eat such large, unique animals, and the World Wide Fund for Nature and others have pointed out that whales must be preserved for observation because of their intrinsic value as mammals of great intelligence.

In opposition to the Save the Whales Campaign and the IWC, Japan, Norway, Russia, and Iceland resumed whaling of the minke species in the early 1990's, citing their rights to refuse specific IWC rulings. These countries based their decisions on information submitted by the Scientific Committee of the IWC, which stated that the minke population was large enough to absorb sustainable exploitation. However, a deoxyribonucleic acid (DNA) test of whale meat imposed by the IWC in 1995 revealed that Japan was also harvesting some fin and humpback whales. At its May, 1995, meeting, the IWC strongly censured the continued whaling activities of these countries and took serious steps toward stopping pirate whalers and the illegal trade of whale meat.

Alvin K. Benson

FURTHER READING

Heazle, Michael. *Scientific Uncertainty and the Politics of Whaling*. Seattle: University of Washington Press, 2006.

Kalland, Arne. *Unveiling the Whale: Discourses on Whales and Whaling*. New York: Berghahn Books, 2009.

Vaughn, Jacqueline. *Environmental Politics: Domestic and Global Dimensions*. 5th ed. Belmont, Calif.: Thomson/Wadsworth, 2007.

SEE ALSO: Animal rights; Greenpeace; International whaling ban; International Whaling Commission; Marine Mammal Protection Act; Whaling.

Scenic Hudson Preservation Conference v. Federal Power Commission

CATEGORIES: Treaties, laws, and court cases; preservation and wilderness issues

THE CASE: U.S. Court of Appeals decision regarding basic principles of standing in U.S. environmental law

DATE: Decided on December 29, 1965

SIGNIFICANCE: The *Scenic Hudson* decision was a declaration of judicial authority that granted environmental groups with noneconomic interests the right to intervene in federal regulatory agency licensing decisions and held a federal regulatory agency responsible for engaging in its own evidentiary inquiry, considering alternatives to granting a license, and weighing the environmental impacts of its decisions.

In the early 1960's the Scenic Hudson Preservation Conference, a coalition of conservation and environmental groups, and the towns near Storm King Mountain in New York challenged the license that Consolidated Edison Company (Con Ed) had obtained to construct a storage reservoir, a powerhouse, and transmission lines on the Hudson River in the region. However, the Federal Power Commission (FPC) denied their motions to intervene and reopen the hearings to consider evidence on alternatives to the Storm King facility, the cost and practicality of underground transmission lines, and the feasibility of fish protection devices.

In the case *Scenic Hudson Preservation Conference v. Federal Power Commission*, the U.S. Court of Appeals dismissed the FPC's argument that Scenic Hudson lacked standing to seek judicial review of Con Ed's license because the organization had made no claim of personal economic injury. Scenic Hudson clearly had economic interests to protect. The Storm King reservoir would inundate 27 kilometers (17 miles) of hiking trails, and the planned aboveground transmission lines would decrease the value of public land and the tax revenues of private land and would interfere with community planning. Scenic Hudson also had noneconomic interests that would be affected by the Storm King project that were not barred from judicial review. The Court of Appeals held that the U.S. Constitution does not require an aggrieved party to have a personal economic interest, nor does the Federal Power Act (the FPC governing legislation), which permits "any aggrieved party" to obtain judicial review of an FPC order.

In terms of the licensing decision, the FPC had acknowledged that the "principal issue . . . is whether the project's effects on the scenic, historical and recreational values are such that we should deny the application." Since the FPC's standing argument was at odds with its recognition of the importance of noneconomic issues in its licensing decision, the court broadly read the statute to grant standing to organizations, such as Scenic Hudson, that had special interests in "aesthetic, conservational, and recreational" issues to act as private attorneys general to ensure that the FPC would protect the public interest. Since Scenic Hudson was an aggrieved party, the court read the statute's language, which gave the FPC the discretion to admit as a party "any person whose participation in the proceeding may be in the public interest," to create "an absolute right of intervention."

After the appellate court settled the standing issue, it found that the Federal Power Act gives the FPC a specific planning responsibility for the development of waterways for commerce, water power, and recreation. The statute's planning responsibility requires the agency to weigh these three factors in making a licensing decision. Then the court broadly read the FPC's responsibility to consider recreational objectives to include "the preservation of natural resources, the maintenance of natural beauty, and the preservation of historic sights." In fact, the FPC regulations require a recreation plan.

The FPC, in granting Con Ed a license for the Storm King project, did not fulfill its statutorily mandated planning duties because it failed to develop a complete record. First, the FPC did not explore alternatives to the Storm King project, including the use of gas turbines or the purchase of power from intercon-nections with other electric utilities, nor did the agency require Con Ed to supply this information. Therefore, the agency was unable to determine whether Con Ed needed Storm King to meet its future power needs. Second, the FPC did not seriously consider the cost of underground transmission routes; it simply accepted Con Ed's estimates of their cost even though the estimates had been questioned by the FPC's staff and hearing examiner. As a consequence, the agency was unable to weigh "the aesthetic advantages of underground transmission lines against the economic disadvantages." Third, there was sufficient evidence before the FPC on the danger of the Storm King project to fish that the FPC should have made further inquiries, but the agency failed to do so. It even dismissed as "untimely" the petitions of several groups to intervene and present evidence on Storm King's destructive impact on the spawning grounds for striped bass and shad and the unavailability of any powerhouse screening devices to protect fish eggs and larvae.

The Court of Appeals set aside the FPC's licensing order because the record had failed to support the agency's decision and then remanded the case with instructions to reexamine the alternatives to Storm King, the cost of underground transmission lines, and the fish spawning issue. In sum, the Scenic Hudson opinion was a bold declaration of judicial authority that granted environmental groups with noneconomic interests the right to intervene in federal regulatory agency licensing decisions and held a federal regulatory agency responsible for engaging in its own evidentiary inquiry, considering alternatives to granting a license, and weighing the environmental impacts of its decisions.

The FPC subsequently held hearings and issued Con Ed a license to build the Storm King project, which the federal Court of Appeals affirmed. However, the Scenic Hudson case was only the beginning of a two-decade controversy over the Storm King project. Further litigation on the project's threat to fish life expanded to include six other power stations along the Hudson River once the newly created Environmental Protection Agency conducted extensive hearings under the Federal Water Pollution Control Act amendments of 1972. The controversy was not resolved until 1981, when seventeen months of mediation led to Con Ed's agreement to forfeit its Storm King license and donate the site as a public park.

William Crawford Green

FURTHER READING

Dunwell, Frances F. "The 1960s: Scenic Hudson, Riverkeeper, Clearwater, and the Nature Conservancy Campaign to Save a Mountain and Revive a 'Dead' River." In *The Hudson: America's River.* New York: Columbia University Press, 2008.

Hoban, Thomas, and Richard Brooks. *Green Justice: The Environment and the Courts.* 2d ed. Boulder, Colo: Westview Press, 1996.

Houck, Oliver A. "Storm King." In *Taking Back Eden: Eight Environmental Cases That Changed the World.* Washington, D.C.: Island Press, 2010.

Lewis, Tom. *The Hudson: A History.* New Haven, Conn.: Yale University Press, 2005.

Talbot, Allan. *Power Along the Hudson: The Storm King Case and the Birth of Environmentalism.* New York: Dutton, 1972.

SEE ALSO: Dams and reservoirs; Environmental law, U.S.; Hydroelectricity; Power plants; *Sierra Club v. Morton*; *Tennessee Valley Authority v. Hill.*

Schumacher, E. F.

CATEGORY: Activism and advocacy

IDENTIFICATION: German British economist

BORN: August 16, 1911; Bonn, Germany

DIED: September 4, 1977; on a train en route to Zurich, Switzerland

SIGNIFICANCE: Schumacher's promotion of nonmaterialist values and his writings emphasizing the importance of protecting resources while attending to the needs of humans had a great influence on the environmental movement in the 1970's and 1980's.

Economist E. F. Schumacher became an important voice in environmental discourse when he published *Small Is Beautiful: Economics As If People Mattered* in 1973. Appearing first in Great Britain, the book found only a small audience in England and Europe, but its U.S. publication in 1974 led to Schumacher's becoming a guru to millions of environmentalists. The work's central theme, that people must replace materialistic values with nonmaterialistic ones, resonated with many Americans, especially college students. By the end of 1974 Schumacher had scheduled hundreds of lecture appearances, most of them in the United States.

Much of what is contained in *Small Is Beautiful* was foreshadowed by developments in Schumacher's life during the 1960's. At the time, Schumacher was the principal economic consultant to the National Coal Board in England. The German-born Schumacher had served as an adviser to British government officials during the 1940's. From 1946 to 1950 he had assisted with economic planning in British-occupied West Germany, and his work there gained the attention of famed economist John Maynard Keynes, who recommended him to the National Coal Board in 1950.

Schumacher remained with the Coal Board for more than twenty years and served as its statistical director from 1963 to 1970. In the midst of this fairly routine career as an expert economist, Schumacher experienced a spiritual crisis that changed the focus of his life. While visiting Burma as an economic consultant in the early 1960's, Schumacher became enchanted with Buddhism. He began to speak of "Buddhist economics," by which he meant the combination of spiritual harmony and material well-being. Although he eventually came to the opinion that Buddhism is not applicable to Western culture (he became a Catholic), the influence of Buddhist philosophy is evident in *Small Is Beautiful.*

The main theme of *Small Is Beautiful* is that modern industrial production, with its emphasis on the most advanced technology, destroys the creativity and dignity of the worker. Schumacher therefore urges the use of intermediate or "appropriate technology" as a way of giving workers a greater sense of satisfaction. He advocates small-scale, regional industry rather than huge national or international corporations.

Throughout *Small Is Beautiful* Schumacher concentrates on the needs of people as opposed to an exclusive concern about the environment. Therefore, he is usually categorized as a social ecologist rather than as a pure or "deep" environmentalist. *Small Is Beautiful* particularly encourages a regional approach to job creation for developing countries. Schumacher argues that such an approach can give a nation's people a sense of achievement and encourage them to protect their resources while gradually increasing the country's pool of skilled workers. In addition, it can slow the drift to urban areas, which leads to greater human misery and more pollution of the air and water. Schumacher's recommendations proved unrealistic for many of the world's less developed nations, but his ideas inspired greater regionalization of industry in the more environmentally conscious industrialized economies during the 1970's and 1980's.

Ronald K. Huch

FURTHER READING

De Steiguer, J. E. "E. F. Schumacher's *Small Is Beautiful*." In *The Origins of Modern Environmental Thought*. Tucson: University of Arizona Press, 2006.

Pearce, Joseph. *Small Is Still Beautiful: Economics As If Families Mattered*. Wilmington, Del.: ISI Books, 2006.

Schumacher, E. F. *Small Is Beautiful: Economics As If People Mattered*. 1973. Reprint. Point Roberts, Wash.: Hartley & Marks, 1999.

SEE ALSO: Deep ecology; Environmental economics; Renewable resources; Social ecology; Sustainable development.

Sea Empress oil spill

CATEGORIES: Disasters; water and water pollution

THE EVENT: Grounding of an oil tanker off the coast of Wales, resulting in the release of thousands of tons of crude oil into the sea

DATE: February 15, 1996

SIGNIFICANCE: The *Sea Empress* grounding resulted in the third-largest tanker spill in United Kingdom waters, causing adverse effects to a number of wildlife species and considerable difficulties for the fishing industry within the region.

On February 15, 1996, the *Sea Empress* oil tanker ran aground on a wave-exposed, current-scoured section of coastline near Milford Haven in southwest Wales. Over the next six days, the vessel released more than 75,000 tons of North Seas light crude oil and about 400 tons of heavy fuel oil. Still leaking oil, the *Sea Empress* was recovered and towed to Milford Haven on February 21.

Weather and wave conditions that prevailed at the time of the spill facilitated the spread of oil slicks well beyond the immediate area of the grounding. Heavy oil slicks drifted into Milford Haven and also flowed north and south along the open Pembrokeshore coast. During the first weeks after the incident, oil was observed across a wide area of the Bristol Channel. More distant shores that were affected included those around Lundy Island and the southeast coast of Ireland.

Initially, the three main concerns were to establish the size of the area affected by oil, plan cleanup measures, and determine how badly various populations of shellfish, finfish, and other wildlife had been contaminated. Shortly after the spill, a fishing exclusion order was applied to the affected region, banning the catching of any fish within a designated area. Some of the spilled oil was mechanically sucked up at sea. Between February 17 and 25, large amounts of chemical

Workers clean up oil spilled by the Sea Empress *from the beach at West Angle Bay in western Wales almost two weeks after the supertanker ran aground.* (AP/Wide World Photos)

dispersants were used to break up the oil into small droplets in order to reduce the risk to the coastline and to birds at sea. Mechanical methods were employed to clean beaches for the most part, but some dispersants were used to remove weathered oil from rocks next to selected beaches. The main recreational beaches were cleaned up by mid-April, allowing visitors to again enjoy them.

As local finfish were found to have little to no contamination, the ban on catching salmon and sea trout was lifted in May, 1996. The shellfish, however, were more heavily contaminated and recovered more slowly. Between 1996 and 1998, research was conducted to assess the impacts of the oil spill and the recovery of a range of key commercial fish species, particularly those that are important for food chains. By June, 1996, more than 6,900 oiled birds of at least twenty-eight species had been recovered dead or alive, and more than 3,000 birds had been cleaned and released.

Approximately one-third of the spilled oil evaporated from the sea surface, but because of a combination of natural and chemical dispersion, approximately 50 percent of the spill volume dispersed into the water column. The ultimate fate of this dispersed oil was unclear for some time, but water samples analyzed in 1997 showed low levels of total hydrocarbons. The impacts of the spill on the marine life in the area lasted for some years, but by 2001 the various populations appeared to have recovered.

Alvin K. Benson

FURTHER READING

Clark, R. B. *Marine Pollution*. 5th ed. New York: Oxford University Press, 2001.
Fingas, Merv. *The Basics of Oil Spill Cleanup*. 2d ed. Boca Raton, Fla.: CRC Press, 2001.
Speight, Martin, and Peter Henderson. "Threats to Marine Ecosystems: The Effects of Man." In *Marine Ecology: Concepts and Applications*. Hoboken, N.J.: John Wiley & Sons, 2010.

SEE ALSO: *Amoco Cadiz* oil spill; *Braer* oil spill; *Exxon Valdez* oil spill; Oil spills; Tobago oil spill; *Torrey Canyon* oil spill.

Sea-level changes

CATEGORIES: Water and water pollution; weather and climate

DEFINITION: Long-term rises and falls in mean sea level resulting from natural or human-caused alterations in climate or from deformations of the land

SIGNIFICANCE: Because a significant percentage of the human population lives near the oceans and most of them in large cities, even a modest rise in sea level could wreak great economic damage and displace hundreds of millions of people. It would also change sensitive coastal ecosystems, such as coral reefs and mangrove swamps.

The world's oceans and seas constantly rise and fall over short periods because of tides, currents, winds, air pressure, seasonal heating, and weather phenomena, such as El Niño and La Niña events. Moreover, because cold water is denser than warm water, the sea levels in polar oceans are generally lower than sea levels in tropical oceans. Nevertheless, scientists use data gathered by tidal gauges and satellites to determine an overall mean sea level (MSL), purely a statistical value. In some areas available tidal gauge data extend back more than one hundred years. These data and more recent data sets show that the MSL has been rising since the mid-nineteenth century after about six thousand years of relative stability. The twentieth century saw an average rise of 1-2 millimeters (0.04-0.08 inch) per year, with an increase to 3 millimeters (0.12 inch) per year at the end of the century. If that rate accelerates because of human-caused global warming, as many scientists have concluded will happen, sea levels could become high enough to imperil coastal cities, agriculture, ecosystems, and fisheries during the twenty-first century.

LONG-TERM CHANGE

Scientists distinguish two types of long-term change in sea level. The first, secular (or isostatic) change, comes from local movement of the land and seabed: land rebounding after the ice sheets of the last ice age melted (glacial isostatic adjustment), land subsiding as groundwater is removed, the deposit of sediments, and the effects of volcanoes, earthquakes, and tectonic plate movement. The second type, eustatic change, involves the volume of water in the oceans.

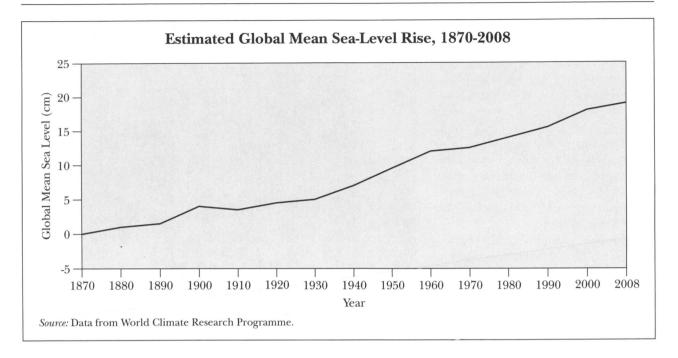

Estimated Global Mean Sea-Level Rise, 1870-2008

Source: Data from World Climate Research Programme.

The volume varies based on water temperature and on how much of the world's water supply is land-bound in lakes, aquifers, and ice (glaciers, ice caps, and ice sheets).

About half of the rise in sea level during the twentieth century came from warming water (thermal expansion). Water added to the oceans from melting glaciers and ice caps contributed most of the rest, a small amount coming from the ice sheets covering Greenland and Antarctica. These changes arise from temperature change caused by several factors influencing the amount of heat energy in the atmosphere, land surface, and water, principally orbital cycles varying the earth's distance from and attitude toward the sun and the concentration of greenhouse gases. For example, when global cooling locked greater volumes of water in land ice during the depths of the last ice age, twenty thousand years ago, the MSL was about 120 meters (394 feet) lower than that of the early twenty-first century. During the warm epoch of the Pliocene, about three million years ago, levels were 20 to 25 meters (66 to 82 feet) higher.

Future Change

If all the icebound water on the earth melted and ran into the oceans, the MSL would increase an estimated 64-70 meters (210-230 feet) above the contribution of thermal expansion. This is not likely to happen. Thermal expansion and melting will continue,

however, accelerated by an anticipated increase in global temperatures of 3 degrees Celsius (5.4 degrees Fahrenheit) because of greenhouse gas emissions.

Projections about sea-level changes in the near future are controversial, but a majority of climate scientists agree that the MSL will rise faster in the twenty-first century than it did during the twentieth century. In 2007, the Intergovernmental Panel on Climate Change estimated that the rise will be somewhere between 18 and 59 centimeters (7 and 23 inches) by 2100. However, that projection did not include possible contributions from the Antarctic and Greenland ice sheets, because scientists did not have sufficient understanding of the ice sheets' dynamics. Subsequent studies found increasing instability in parts of these ice sheets, and in 2009 the Scientific Committee on Antarctic Research forecast accelerated melting in Greenland and the Antarctic; the committee estimated that the MSL could rise as much as 1.4 meters (4.6 feet) by the end of the twenty-first century. Such a rise could bring serious disruptions to both agriculture and cities in coastal areas, principally because low-lying lan will be much more vulnerable to storm surges and to salinization of groundwater and wetlands. In 2000, some 600 million people were living along the coasts of the world, the majority in large cities.

Sea levels will continue to rise past 2100 because of further recovery from the last ice age, long-period climate variability, and accumulation of greenhouse gases. Scientists calculate that the MSL will come to equilibrium at about 50 meters (164 feet) higher than the MSL of the early twenty-first century in a few thousand years.

Roger Smith

FURTHER READING

Archer, David. *The Long Thaw: How Humans Are Changing the Next 100,000 Years of Earth's Climate.* Princeton, N.J.: Princeton University Press, 2009.

Langwith, Jacqueline, ed. *Water.* Farmington Hills, Mich.: Greenhaven Press, 2010.

McGranahan, Gordon, Deborah Balk, and Bridget Anderson. "The Rising Tide: Assessing the Risks of Climate Change and Human Settlements in Low Elevation Coastal Zones." *Environment and Urbanization* 19, no. 1 (2007): 17-37.

Pilkey, Orrin H., and Rob Young. *The Rising Sea.* Washington, D.C.: Island Press, 2009.

Pugh, David. *Changing Sea Levels: Effects of Tides, Weather, and Climate.* New York: Cambridge University Press, 2004.

SEE ALSO: Alliance of Small Island States; Beaches; Climate change and oceans; Coral reefs; Glacial melting; Groundwater pollution; Positive feedback and tipping points.

In 2008, the Steve Irwin, *the Sea Shepherd Conservation Society's flagship, prepares to embark from Brisbane, Australia, into Antarctic waters to interfere with Japanese whaling operations.* (AP/Wide World Photos)

Sea Shepherd Conservation Society

CATEGORIES: Organizations and agencies; activism and advocacy; animals and endangered species

IDENTIFICATION: International nonprofit organization devoted to the conservation of marine wildlife

DATE: Founded in 1977

SIGNIFICANCE: The Sea Shepherd Conservation Society's use of direct action and confrontation to disrupt fishing, seal hunting, and whaling operations have been the subject of controversy, but the group's activities have nonetheless focused public and policy attention on overexploitation of the seas.

In 1977 Paul Watson, a founding member of the environmental organization Greenpeace, left that organization after a disagreement with other members and founded the Earthforce Society, the group that later became the Sea Shepherd Conservation Society (SSCS). One of the first interventions carried out by Earthforce was the ramming of a vessel hunting whales in contravention of the worldwide whaling moratorium, an incident that set the tone for subsequent campaigns. Other actions the group has carried out since then have included marking baby seals with an organic dye so their pelts would be worthless to hunters, exposing illegal shark finners, capturing and destroying drift nets, and harassing the Japanese whaling fleet in the Southern Ocean Whale Sanctuary.

SSCS has been controversial since its inception. Detractors often label it a vigilante organization (a label SSCS proudly adopts) or an ecoterrorism group, whereas supporters note that SSCS enforces marine conservation laws and treaties that would otherwise go unenforced. On occasion SSCS has partnered with governments unable to police their own waters for illegal fishing, but it also operates on its own in national waters and on the high seas. SSCS cites Article 21 of the United Nations World Charter for Nature as the legal grounding for its activities, although legal scholars note that the organization's interpretation is probably outside the spirit of the text. Despite efforts by governments of targeted operations to disenfranchise SSCS by impounding its vessels, arresting members, and protesting to the nations in which SSCS branches are located, the organization remains largely unhindered. In fact, it has often used its encounters with law-enforcement agencies and upset citizenry to increase its profile in the news media.

Adam B. Smith

SEE ALSO: Animal rights movement; Earth First!; Ecotage; Ecoterrorism; Greenpeace; Monkeywrenching; Shark finning; Watson, Paul; Whaling.

Sea turtles

CATEGORY: Animals and endangered species
DEFINITION: Large marine reptiles found in all of the world's oceans except the Arctic
SIGNIFICANCE: All sea turtle species have long been listed as threatened or endangered, and they have been protected under international law. Nevertheless, their numbers have continued to decline, primarily because of human activities. Conservation efforts have helped stabilize the numbers of some species, but many populations remain at risk of extinction.

The seven species of sea turtles—leatherback, Kemp's ridley, olive ridley, green, hawksbill, loggerhead, and flatback—differ substantially in size, food preferences, and nesting and feeding locations. For example, the smallest, the Kemp's ridley, has a maximum weight of about 45 kilograms (100 pounds) and eats primarily crabs, whereas the leatherback can reach 907 kilograms (2,000 pounds) and feeds on jellyfish and various other marine life. All of these species (except the flatback, for which limited data are available) are listed as threatened or endangered under the U.S. Endangered Species Act of 1973 and exist at a small fraction of their historical population size.

Female sea turtles lay eggs on warm oceanic beaches, and it is there that the animals are most vulnerable. Egg clutches and hatchlings are eaten by natural and introduced predators, and eggs and adult turtles are poached by humans. Human-made alterations along nesting beachfronts, including buildings, seawalls, beach extensions, and other barriers, pose additional challenges to successful sea turtle reproduction. Further, hatchlings instinctively use reflected starlight or moonlight to find the ocean, and they can become confused by artificial light on land, migrate toward it, and perish.

New threats appear as the animals mature. Cold stunning, a change in temperature that disorients sea turtles, is a natural phenomenon that leaves their immune systems compromised and can lead to beach stranding. Human influences are the most perilous, however. Polluted waters, particularly in oceans with low water turnover, can cause the growth of potentially lethal fibropapilloma tumors on the eyes, skin, and internal organs of sea turtles. Turtles may mistake marine debris for prey, ingesting plastics and other materials that cause digestive failure; leatherback turtles, for example, sometimes ingest plastic bags that they mistake for their primary prey, jellyfish. Global warming also affects the turtles, as the temperature of the sand in which eggs are laid strongly influences the sex of the developing turtles.

Sea turtles must surface regularly to breathe, and they often drown when they are trapped in fishing nets or hooked as they take long-line fishing bait. Some efforts have been made to reduce sea turtle bycatch, including the use of turtle excluder devices

A green sea turtle. (©Pat Barry/Dreamstime.com)

in shrimp trawling nets, but such endeavors are not required by all governments. In fact, several nations still have active sea turtle fisheries in spite of the long-standing listing of these species as endangered by the Convention on International Trade in Endangered Species of Wild Fauna and Flora (CITES).

Sea turtles are also poached for their meat and even for their body parts. The shells of hawksbill turtles are used to make decorative items such as hair combs, jewelry, and handbags. The skin of some species is turned into shoes and purses. In many instances, unknowing tourists buy these illegal items only to have them confiscated when they return home.

All of these factors contributed to a drastic decline in sea turtle numbers, from historical millions to, in some instances, a few thousand (a greater than 99 percent reduction) by the early twenty-first century. Conservation efforts—including international programs aimed at reducing the accidental taking of turtles during fishing operations and at preserving, restoring, and monitoring nesting sites—have reversed this trend for some populations. For example, Kemp's ridley numbers, while still extremely low, have increased since the early 1990's, and local enforcement efforts in the Dominican Republic have almost eliminated the once-prevalent sale of turtle shell items there. Sea turtle products are in still in high demand in many Asian countries, however, most notably Indonesia and Vietnam, and millions of turtles continue to be caught and killed as fishing bycatch. Given the global distribution of sea turtles, a worldwide effort will be required to stem and reverse the decline of their population numbers.

Elizabeth F. Shattuck and David R. Foran

Further Reading

Safina, Carl. *Voyage of the Turtle: In Pursuit of Earth's Last Dinosaur.* New York: Owl Books, 2007.

Spotila, James R. *Sea Turtles: A Complete Guide to Their Biology, Behavior, and Conservation.* Baltimore: The Johns Hopkins University Press, 2004.

See also: Beaches; Conservation biology; Convention on International Trade in Endangered Species; Endangered Species Act; Endangered species and species protection policy; Global warming; Habitat destruction; Ocean pollution; Poaching; Turtle excluder devices.

Seabed disposal

CATEGORIES: Waste and waste management; water and water pollution

DEFINITION: Dumping of waste materials into the open sea or burial of such materials in the ocean floor

SIGNIFICANCE: The seabed disposal of debris, containers of toxic or radioactive waste, discarded chemicals, heavy metals, and other harmful wastes can jeopardize marine life by disrupting the balance of delicate ocean ecosystems.

Oceans cover more than 75 percent of the earth's surface. People have long believed that the vast ocean waters can dilute, redistribute, and render harmless any garbage, sewage, or other debris dumped into them. As landfill space becomes increasingly scarce, many have looked to the oceans and their tremendous beds as alternative sites for waste disposal. However, marine biologists and other scientists have learned that the oceans can be harmed by waste disposal. Like the land, which is a network of many different ecosystems, the oceans are made up of many ecosystems serving a variety of marine life. Many marine creatures and their habitats exist in a delicate ecological balance; any change in water quality, temperature, or food source is harmful.

In an effort to protect marine creatures and the balance of ocean ecosystems, many countries have worked together to stop ocean pollution and open-ocean dumping. The International Convention for the Prevention of Pollution from Ships (known as MARPOL, for "marine pollution") outlaws the dumping of plastic items anywhere in the oceans. It also regulates how far from shore ships must be in order to dump other kinds of debris legally. Much of this debris eventually settles on the seabed.

Also scattered across the seabed are containers of hazardous waste, chemicals, and radioactive waste. At one time scientists believed that the ocean floor would be a safe place for the disposal of sealed containers of these wastes; they argued that the materials would not come into contact with marine life at such depths and that the extremely cold temperatures of the deep ocean would keep the wastes safely contained. Dump sites for radioactive waste are located in the far northern parts of the Atlantic Ocean and in the Arctic Ocean. After World War II, several thousand tons of German chemical weapons, undetonated bombs, and other equipment were dumped in the waters off the coasts of Germany, Denmark, Norway, Sweden, and Poland. In 1972 the international Convention on the Prevention of Marine Pollution by Dumping of Wastes and Other Matter (known as the London Convention) banned ocean dumping of high-level radioactive wastes. In 1994 this ban was expanded to cover low-level radioactive wastes. Despite this agreement, Russia continued to dump large amounts of radioactive waste into the Barents and Kara seas.

Research has shown that the dumping of sealed containers of chemicals into the ocean, even in deep waters, poses threats to the environment. For example, several sites for ocean dumping of industrial wastes were established in the middle of fishing grounds off the coast of New England and used for twenty years. Although toxic and radioactive wastes are no longer dumped in these areas, years later they are still marked as hazardous because of the incredible staying power of the chemicals dumped. Unlike organic matter, chemicals do not break down quickly over time. Despite warnings, some people still harvest fish, shrimp, and shellfish from these areas. In a joint effort among several state and federal agencies, seafood from these areas and the surrounding waters was collected and tested for dangerous levels of hazardous or toxic chemicals. Tests performed by the U.S. Food and Drug Administration (FDA) showed only trace contaminants in the seafood—none of the samples tested had levels high enough to pose threats. When the anchor was pulled aboard a boat at one location, however, its tip had traces of waste that was high enough in radioactivity to set off the sensors worn by the study participants.

A video survey was taken of one site, known as the Foul Area, 29 kilometers (18 miles) off the coast of Boston, Massachusetts. The 3.2-kilometer (2-mile) expanse is officially named the Massachusetts Bay Industrial Waste Site. From 1953 to 1976 it served as a dumping ground for toxic and radioactive materials. The survey, taken in 1991, found almost one hundred objects scattered across the ocean floor at eighteen separate sites. Of these objects, sixty-four were identified as cement containers, and more than one-half of them had broken open over time.

Some scientists now advocate subseabed burial as an alternative to seabed dumping. Proposed mostly for high-level radioactive materials, subseabed disposal is argued to be safer than core-drilled burial on land. Proponents argue that the deep seabed is one of the most geologically stable places on the earth. They also argue that the sticky mud and clays found in the mid- and deep-ocean basins cause radioactive particles to cling to or bind with them, keeping them from migrating throughout the ocean waters. The results of tests conducted between 1974 and 1986 by an international team of scientists support these arguments. Opponents note that if any of the materials disposed of deep in the seabed should need to be retrieved, this process would be extremely difficult and costly. They also question the types of containers to be used and how the wastes can be safely transported to the seabed. While proponents were making progress in addressing these concerns, research funding was cut off in 1986 and the focus of decisions regarding disposal of radioactive wastes returned to land-based solutions. Although the London Convention has prohibited the dumping of radioactive waste at sea since 1994, subseabed disposal remains ambiguous, as do its effects on the ocean environment.

Lisa A. Wroble

FURTHER READING

Gorman, Martha. *Environmental Hazards: Marine Pollution.* Santa Barbara, Calif.: ABC-CLIO, 1993.

Hamblin, Jacob Darwin. *Poison in the Well: Radioactive Waste in the Oceans at the Dawn of the Nuclear Age.* New Brunswick, N.J.: Rutgers University Press, 2008.

Koslow, Tony. *The Silent Deep: The Discovery, Ecology, and Conservation of the Deep Sea.* Chicago: University of Chicago Press, 2007.

Laws, Edward A. *Aquatic Pollution: An Introductory Text.* 3d ed. New York: John Wiley & Sons, 2000.

Ringius, Lasse. *Radioactive Waste Disposal at Sea: Public Ideas, Transnational Policy Entrepreneurs, and Environmental Regimes.* Cambridge, Mass.: MIT Press, 2001.

Sindermann, Carl J. *Ocean Pollution: Effects on Living Resources and Humans.* Boca Raton, Fla.: CRC Press, 1996.

SEE ALSO: Hazardous waste; London Convention on the Prevention of Marine Pollution; Nuclear and radioactive waste; Ocean dumping; Ocean pollution; Waste management.

Seabed mining

CATEGORIES: Ecology and ecosystems; resources and resource management

DEFINITION: Extraction of minerals from the ocean floor and beneath it

SIGNIFICANCE: Seabed mining has the potential to yield an abundant supply of valuable minerals, but the processes involved in such mining also have the potential to destroy unique marine life-forms, change the composition of the ocean waters, and seriously disturb marine ecosystems.

Until the discovery of hydrothermal vents with rich deposits of minerals in the deep seabed in 1977, seabed mining operations were relatively simple, done close to seacoasts and with the permission of coastal states, which exercised their rights to explore and exploit the oceans within 200 nautical miles of their coasts—that is, within their exclusive economic zones. (A nation's exclusive economic zone, or EEZ, is the area of the sea and the seabed to which that nation claims exclusive rights.) Because of technological limitations of the processes used for seabed mining, such mining was possible only up to depths of 70 meters (230 feet). The first seabed mining operations involved dragging buckets along the seafloor to collect mineral-rich sand and then hauling the buckets up to boats, where iron ore was separated out through the use of magnets and other minerals were separated out through repeated sieving of the slurry. The waste remaining after the minerals were extracted was returned to the sea. Improved technology eventually replaced the buckets with suction pumps.

During the 1960's scientists found excessively warm deepwater temperatures in the Red Sea and a sediment sample that resembled hot tar. In 1966 the National Science Foundation funded a research voyage by the Woods Hole Oceanographic Institution to obtain sediment samples from the Red Sea, and the resulting samples were found to be filled with various metals, including iron, copper, and manganese. Given the fact that the Red Sea has a mid-ocean ridge where the seafloor is separating, the scientists concluded that the mineral-rich sediment and warm temperatures were most likely not a phenomenon unique to the Red Sea but might exist in other oceans where the seafloor was separating. As a result, research was carried out in the Atlantic, Pacific, and Indian oceans, as well as others, and researchers found sediment containing large amounts of minerals. Further research led to discovery of a link between volcanic action on the seafloor and the formation of mineral-rich sediment and the possible existence of hydrothermal vents.

In February, 1977, scientists on the Galápagos Hydrothermal Expedition discovered the first hydrothermal vent. They also made an unexpected discovery: An abundance of unusual marine life-forms were living in the water gushing from the vent. More exploration led to the discovery of more hydrothermal vents wherever the seafloor was separating. The discovery of these mineral-rich deposits of sediment in the deep sea revolutionized the concept of seabed mining and simultaneously raised myriad complex issues. Questions such as the following arose: Who would have the legal right to exploit these mineral deposits in the open ocean? What types of technologies would need to be developed? Could the deposits be mined commercially on a cost-effective basis? What impacts would mining have on the environment?

ENVIRONMENTAL CONCERNS

Conventional coastal seabed mining at depths of 25-70 meters (82-230 feet) has always created environmental concerns, especially in regard to waste disposal in the ocean. Deep-seabed mining raises much greater and more complex problems of environmental damage. The marine life discovered at the hydrothermal vents is composed of species and populations about which scientists have little knowledge; therefore they are unable to assess the effects of disturbances to the environment on these life-forms. Most scientists agree that if the environments of these newly discovered species are disturbed or destroyed before the species can be studied adequately, a vast amount of knowledge about how the planet formed will be lost. The species living in or near the hydrothermal vents are existing in darkness and in an atmosphere of extremely high temperature permeated with toxic fluids—an environment totally opposite to that which is necessary to the life-forms with which scientists are familiar. Not all of the hydrothermal vents are active, and some proponents of seabed mining have suggested that only inactive vents be mined; however, inactive vents are also surrounded by many unusual life-forms, so any mining activity will inevitably have impacts on the biodiversity of the seabed.

Another major concern is the disposal of the waste from the mining, or tailings. Minerals compose a small percentage of the deposits, and when the minerals are brought up to the surface, as much nonmineral waste as possible is separated from the minerals to reduce the bulk. One way in which this is done is by grinding the rock, or nodules, into slurry on the seabed floor. Depending on the method used, the tailings are discarded into the ocean at various depths, introducing foreign pollutant matter into the environments of the marine life at these various levels.

The potential for seabed mining to pollute the ocean waters, both at lower depths and at the surface, remains a serious concern. The minerals are brought up in a wet sulfide slurry, which is transported on barges. If oxidized, the slurry can quickly become sulfuric acid, which is harmful to the environment. In addition, the chemicals, fuels, and other materials used in the mining process pose the threat of accidental spilling, which could cause further harm to the environment.

Mining Technology

Although some companies are attempting to develop more environmentally friendly equipment and methods for mining the deep seabed, the technologies in use thus far pose significant environmental threats to the life-forms at both active and inactive hydrothermal vents, to other marine life, and to the oceans. The primary technique used in deep-seabed mining is one of scraping, either with a tow sled or a self-powered unit. This process raises clouds of sediment and buries large numbers of life-forms around the hydrothermal vents. Because new populations quickly establish themselves at the vent sites, some mining interests have argued that disturbances on the seabed are not significant. The same species do not reappear at the sites, however; studies of the new populations have revealed that the replacement populations may be of hardier yet less desirable types.

Both mining interests and scientific research groups have conducted experiments aimed at the development of mining technologies that could minimize environmental risks. However, tests and experiments to determine the environmental safety of various mining technologies and the types of damage associated with seabed mining cannot accurately depict what will occur in actual mining operations, as commercial mining activities are of much longer duration than any cost-effective experiment can be, and the physical area of seabed actually mined is much larger than can be devoted to an experiment. Nevertheless, scientists and seabed mining companies have gained some insights from these studies.

Legal Issues and Voluntary Codes

Although some deep-seabed mining can actually be done in the EEZs of island nations where hydrothermal vents occur in back-arc basins, most of the world's hydrothermal vents and seabed mineral deposits are located in open areas of the oceans that are not under the control of individual nations. Because of the possibility for conflict and environmental dam-

Life-Forms of the Hydrothermal Vents

Although animal life exists on the ocean floor, much of that zone is barren, with relatively few animals distributed at large intervals. One of the most exciting discoveries in marine biology during the last quarter of the twentieth century occurred in 1977 with the discovery of abundant animal life in the regions immediately surrounding four hot water geysers on the ocean floor. These hydrothermal vents, located in the eastern Pacific Ocean at a depth of approximately 2,700 meters, boasted a spectacular variety of species all located within a few meters of the vent. The geyser is caused when water is heated by Earth's mantle and rises rapidly through the cooler waters above. Water in the geyser reaches temperatures of 8 to 16 degrees Celsius, as opposed to the surrounding water, which is about 2 degrees Celsius. This warm water is also rich in dissolved hydrogen sulfide gas, which serves as a source of energy for bacteria at the vent. The bacteria, in turn, act as food for the various animal species in the area of the vent. Similar hydrothermal vents, each with a population of animals surrounding it, have since been discovered at other sites in the Atlantic and Pacific Oceans.

The largest animals located at these vents include large clams and mussels and several species of tube (vestimentiferan) worms. Other species include crabs, snails, segmented worms and spaghetti worms, so called because of their appearance. Many of these species had not been previously observed or described. All vent communities appear to have similar types of organisms in association, but not always the same species. Current estimates indicate that the life span of the vents is quite short (probably in the tens of years) and that when the vent dies, so do the animal communities surrounding it. How these vents are located and colonized by organisms living on the deep ocean floor is just one of many interesting questions concerning these deep-sea oases that remain to be answered.

age in the mining of the deep seabed, in 1994 the United Nations Convention on the Law of the Sea created the International Seabed Authority to set environmental standards and issue licenses and permits for mining in those areas of the world's oceans not under jurisdiction of EEZs.

Deep-seabed mining interests have also made efforts to minimize damage to the oceans and marine biodiversity by mining the seabed in ways that avoid severely affecting marine ecosystems. As part of this effort, in 1987 the industry established the International Marine Minerals Society, which worked to develop its Code for Environmental Management of Marine Mining to provide guidance to mining companies. The code was adopted in 2001 and continues to undergo revisions; compliance with the code's provisions is voluntary.

All of the parties involved in or concerned about deep-seabed mining—individual nations, international bodies, scientific research groups, and mining companies and engineers—are significantly aware of the potential for environmental damage to the oceans associated with the retrieval of seabed mineral deposits. For the most part, these parties have taken into account the marine environment and its biodiversity, as well as the benefits gained from exploitation of the seabeds' mineral resources, in their decision making and actions.

Shawncey Webb

FURTHER READING

Cronan, David Spencer, ed. *Handbook of Marine Mineral Deposits.* Boca Raton, Fla.: CRC Press, 1999.

Friedham, Robert R. *Negotiating the New Ocean Regime.* Columbia: University of South Carolina Press, 1992.

Nelson, Jason C. "The Contemporary Seabed Mining Regime: A Critical Analysis of the Mining Regulations Promulgated by the International Seabed Authority." *Colorado Journal of International Environmental Law and Policy* 16, no. 1 (2005): 27-76.

Sohn, Louis B., et al. *The Law of the Sea in a Nutshell.* 2d ed. St. Paul, Minn.: West, 2010.

Van Dover, Cindy Lee. *The Ecology of Deep-Sea Hydrothermal Vents.* Princeton, N.J.: Princeton University Press, 2000.

SEE ALSO: Best available technologies; Biodiversity; Ecosystems; Marine debris; Ocean currents; Ocean dumping; Ocean pollution; United Nations Convention on the Law of the Sea.

Seal hunting

CATEGORY: Animals and endangered species

DEFINITION: Personal and commercial hunting of seals for their fur, food, and body parts

SIGNIFICANCE: Commercial seal hunting involves the wholesale slaughter of the animals in ways many people consider cruel. Public awareness of the practices used in seal hunting and concerns about dwindling seal populations have led to greater regulation of seal hunting, but many environmental rights activists do not believe that such regulation has gone far enough.

The Inuit and other residents of the northern latitudes have hunted seals for millennia as sources of food and clothing. Commercial seal hunting, or sealing, began in 1689, and by the middle of the eighteenth century some 20,000 seals were being killed annually. The cooking oil refined from adult seal fat was valuable at the time, and sealing produced 500 tons of the fat per year. According to official Canadian records, more than 35 million seals were killed from 1805 to 1936. After World War II the demand for seal oil declined, but the demand for the animals' fur increased; this meant that hunters' emphasis moved from adults seals to newborns, the fur of which was more desirable. Seal hunting still takes place in Greenland, Norway, Russia, and Canada, where most of the world's harp seals live.

The female harp seals bear one pup annually beginning in late February. When born, the pup, called a whitecoat, has snow-white fur, the color of which begins to change when the pup reaches the age of two weeks. This white fur was fashionable for use in clothing during the 1960's, especially in Germany; eventually, however, the killing of harp seal pups less than two weeks old became illegal.

Harp seal populations in Canadian waters declined to approximately 1.75 million by 1961, when the regulation of seal hunting began. In 1964 a documentary film about seal hunting caused such an outcry among the public that eventually hunting quotas, seasons, and licensing procedures were established. These regulations had so many loopholes, however, that the harp seal population continued to drop, reaching approximately 1 million by 1975.

In 1972 the United States banned the importation of harp seal pelts, but demand had never been high there. The European Economic Community, where

A harp seal pup. (©Dreamstime.com)

demand had been greater, banned harp seal pelts from its markets in 1983, causing the annual harvest of seals to decline to 60,000. By 1998 the worldwide harp seal population was estimated to be somewhere between 2.6 and 3.8 million.

Despite these gains in protection for the seals, the annual kill actually increased in the late twentieth century as demand rose in the Far East for the body parts of adult male seals, which were believed to have aphrodisiac qualities and to provide a cure for impotence. The Canadian Department of Fisheries and Oceans (DFO) set the total allowable catch of harp seals at 270,000 per year in 2007. The 2006 catch was 354,000, and in 2007 it was 234,000. The official position of the DFO is that these levels do not harm the harp seal population and actually help preserve the cod population in the harp seals's habitat, as the seals feed on the cod. Animal rights activists assert that these figures grossly underestimate the total number of seals killed, because many seals' bodies fall into the water or under the ice and seals with pelt damage are discarded and therefore not counted. They also argue that the quotas do not take into account the number of illegal kills that are likely to take place as hunters fill their quotas. In addition, animal rights activists generally consider the process of clubbing or shooting seals itself to be a form of animal cruelty.

Thomas R. Feller

FURTHER READING

Mowat, Farley. *Sea of Slaughter.* 1984. Reprint. Mechanicsburg, Pa.: Stackpole Books, 2004.

Schultz, Stacey, and Julian E. Barnes. "Red Tide Rising." *U.S. News & World Report,* May 6, 2002, 56.

Watson, Paul. *Seal Wars: Twenty-five Years on the Front Lines with the Harp Seals.* Buffalo, N.Y.: Firefly Books, 2003.

SEE ALSO: Amory, Cleveland; Animal rights; Animal rights movement; Commercial fishing; Greenpeace; People for the Ethical Treatment of Animals; Sea Shepherd Conservation Society; Watson, Paul.

Secondhand smoke

CATEGORIES: Human health and the environment; atmosphere and air pollution

DEFINITION: Tobacco smoke emitted from burning cigarettes, cigars, and pipes, as well as that exhaled by smokers

SIGNIFICANCE: A strong respiratory irritant, secondhand smoke has been shown to contain thousands of chemicals, more than two hundred of which are harmful to human health and fifty of which are known cancer-causing agents. When concentrated indoors, secondhand smoke becomes a significant pollutant that greatly reduces air quality.

Since the 1950's research has shown that tobacco smoke poses serious health risks to smokers. The World Health Organization has identified smoking as the single largest preventable cause of disease and premature death. From 1972 onward U.S. surgeons general have cited exposure to secondhand smoke (also called passive or involuntary smoking) as a possible threat to nonsmokers.

In 1986 the U.S. surgeon general reviewed the work of more than sixty physicians and scientists from the United States and elsewhere relating to secondhand smoke exposure. The resulting report, published by the U.S. Department of Health and Human Services and titled *The Health Consequences of Involuntary Smoking: A Report of the Surgeon General* (1986), listed three conclusions: Involuntary smoking causes disease in nonsmokers; children exposed to secondhand smoke have an increased frequency of respiratory infections and respiratory symptoms, as well as a reduced rate of lung function capacity as the lung matures; and separating smokers from nonsmokers sharing the same airspace may reduce the exposure of nonsmokers to secondhand smoke but does not eliminate their exposure.

HEALTH EFFECTS

An examination of the contents of secondhand smoke reveals a complex mixture of more than 4,000 chemicals, at least 250 of which are known to be harmful to human health. These chemicals are present in two forms: gases and particles. The exact chemical makeup of secondhand smoke depends on the type of tobacco burned, the additives it contains, the way it is smoked, and (in the case of cigarettes) the paper in which it is wrapped.

The major gaseous toxins in secondhand smoke include carbon monoxide, carbonyl sulfide, benzene, formaldehyde, hydrogen cyanide, and nitrogen oxides. The presence of carbon monoxide, which is also a component of car exhaust, is a cause for particular concern. When inhaled, carbon monoxide interferes with the blood's ability to carry oxygen to the cells of the body. Red blood cells normally carry oxygen from the lungs to the cells. When carbon monoxide is present, it binds to red blood cells and makes them incapable of taking on the oxygen. The brain, heart, and other tissues do not get the oxygen they need.

The chemicals in secondhand smoke that are in particle form include tar, nicotine, and phenol. Tar is considered to be carcinogenic (cancer causing). Nicotine is toxic, and phenol promotes tumor growth. Many other particulate chemicals in secondhand smoke are carcinogenic, including catechol, benz(a)anthracene, quinoline, arsenic, beryllium, cadmium, chromium, nickel, and polonium 210. So many cancer-causing agents are found in secondhand smoke that the U.S. Environmental Protection Agency, the U.S. National Toxicology Program, and the International Agency for Research on Cancer all classify secondhand smoke as a known cancer-causing agent in humans.

Being in the presence of secondhand smoke produces immediate consequences. Common reactions include burning sensations in the eyes, nose, and throat; increased phlegm; rises in heart rate and blood pressure; and headaches and stomachaches. Exposure to secondhand smoke over time increases the risk not only of lung cancer but also of other lung conditions, heart attacks, and strokes. According to a 2005 report by the California Environmental Protection Agency, secondhand smoke causes approximately 3,400 deaths from lung cancer and 46,000 deaths from heart disease every year among adult nonsmokers. According to a 2006 surgeon general's report, the chances of developing lung cancer increase 20 to 30 percent for a nonsmoker living with a smoker.

Infants and young children are especially susceptible to the dangers of secondhand smoke. Childhood health problems related to smoke exposure range from ear infections to sudden infant death syndrome. Children whose parents smoke in the home experience a significantly higher risk of developing lower respiratory tract infections and asthma. Bronchitis, tracheitis, and laryngitis are three acute respiratory illnesses that children of smokers suffer more frequently than do the children of nonsmokers. Chil-

Health Effects of Secondhand Smoke

In remarks delivered at a press conference held on the occasion of the publication of the 2006 report Health Consequences of Involuntary Exposure to Tobacco Smoke: A Report of the Surgeon General, *U.S. surgeon general Vice Admiral Richard H. Carmona emphasized the dangers of secondhand smoke:*

Secondhand smoke is a health hazard for all people: It is harmful to both children and adults, and to both women and men. It is harmful to nonsmokers whether they are exposed in their homes, their vehicles, their workplaces, or in enclosed public places. We have found that certain populations are especially susceptible to the health effects of secondhand smoke, including infants and children, pregnant women, older persons, and persons with preexisting respiratory conditions and heart disease.

It is not surprising that secondhand smoke is so harmful. Nonsmokers who are exposed to secondhand smoke inhale the same toxins and cancer-causing substances as smokers. Secondhand smoke has been found to contain more than 50 carcinogens and at least 250 chemicals that are known to be toxic or carcinogenic. This helps explain why nonsmokers who are exposed to secondhand smoke develop some of the same diseases that smokers do....

We know that secondhand smoke harms people's health, but many people assume that exposure to secondhand smoke in small doses does not do any significant damage to one's health. However, science has proven that there is NO risk-free level of exposure to secondhand smoke. Let me say that again: There is no safe level of exposure to secondhand smoke.

dren whose parents smoke also have more difficulty with chronic (persistent and ongoing) coughs and increased phlegm than do the children of nonsmokers. Children who suffer from asthma and are exposed to secondhand smoke show an increase in the severity of their symptoms and the frequency of their attacks.

EXPOSURE REDUCTION

The social and economic consequences of secondhand smoke are far-reaching. Exposure to secondhand smoke leads to increases in human illness, suffering, and medical expenses. Lost wages caused by illnesses related to secondhand smoke exposure also contribute to the economic costs. The overwhelming body of evidence pointing to the dangers of secondhand smoke for nonsmokers has led to a flurry of public outcry and regulatory action throughout the developed world and, to a lesser extent, among developing nations.

In some countries, measures to reduce nonsmokers' exposure to secondhand smoke limit smoking to designated areas in enclosed public spaces such as airplanes, public transportation vehicles, airports, bars, and restaurants. The obvious problem with such measures—smoke does not confine itself to a designated smoking area—has given rise in other countries to outright bans on smoking in many enclosed public spaces. Concerns regarding the health of such workers as restaurant staff, bartenders, casino staff, and flight attendants, whose jobs require them to spend long hours in smoke-filled environments, has fueled arguments supporting bans.

Many countries have forbidden smoking in all workplaces, including restaurants and bars. In the United States, smoking is banned on interstate buses, most trains, airplanes flying domestic routes, and most airplanes flying between the United States and international destinations. Facilities in the United States that provide federally funded services to children are required to be smoke-free under the Pro-Children Act of 1994.

Many state and local governments in the United States have banned smoking in public facilities such as schools, hospitals, bus terminals, and airports. Some have also forbidden smoking in restaurants, bars, and other workplaces. Some communities within the United States and in other nations have even declared outdoor public spaces—such as university campuses, stadiums, parks, and beaches—to be official smoke-free zones. Multiple-family dwellings such as apartment buildings and condominiums have become additional targets for smoking bans, as shared ventilation systems make it impossible for residents to smoke without affecting others living in the same complex.

Secondhand smoke exposure can also be reduced through a decrease in the number of smokers. Common methods that governments and private organizations have used to encourage smokers to quit and discourage nonsmokers from starting include conducting antismoking education campaigns, limiting minors' access to tobacco products, increasing the prices of or taxes on those products, and providing health insurance coverage for treatments that help tobacco users quit.

Louise Magoon
Updated by Karen N. Kähler

FURTHER READING

Boyle, Peter, et al., eds. *Tobacco: Science, Policy, and Public Health.* 2d ed. New York: Oxford University Press, 2010.

California Environmental Protection Agency. *Proposed Identification of Environmental Tobacco Smoke as a Toxic Air Contaminant: Part B, Health Effects.* Sacramento: California EPA Office of Environmental Health Hazard Assessment, 2005.

Dean, Michael. *Empty Cribs: The Impact of Smoking on Child Health.* Frederick, Md.: Arts and Sciences, 2006.

Kluger, Richard. *Ashes to Ashes: America's Hundred-Year Cigarette War, the Public Health, and the Unabashed Triumph of Philip Morris.* New York: Vintage Books, 1997.

U.S. Department of Health and Human Services. *The Health Consequences of Involuntary Exposure to Tobacco Smoke: A Report of the Surgeon General.* Atlanta, Ga.: Author, 2006.

Watson, Ronald R., and Mark Witten, eds. *Environmental Tobacco Smoke.* Boca Raton, Fla.: CRC Press, 2001.

World Health Organization. *Protection from Exposure to Second-Hand Tobacco Smoke: Policy Recommendations.* Geneva: Author, 2007.

SEE ALSO: Carbon monoxide; Carcinogens; Environmental illnesses; Indoor air pollution.

Sedimentation

CATEGORY: Water and water pollution

DEFINITION: Deposition of particulate matter by wind, water, chemical precipitation, gravity, or ice

SIGNIFICANCE: Sedimentation, whether the result of natural forces or human influences on nature, can have both negative environmental effects, such as the degradation of water quality, and positive effects, such as beach rejuvenation and growth.

Particulate matter accumulates as sediment through transport and subsequent deposition of materials. Transport mechanisms include wind, running water, gravity, and glaciers. Materials that have been dissolved and transported in solution may be deposited as sediment through chemical precipitation. Sedimentation is an ongoing process, and, in a natural system, a delicate balance exists between positive and negative effects.

Sedimentation takes place in low-energy environments. Particle transport distance is a function of particle shape, size, and mass, and the available energy for transport. High-energy transport mechanisms will transport larger particles than low-energy mechanisms, and less energy is required to transport small particles the same distance as larger particles. Sedimentation can result in a degradation of the quality of land or water, or a loss of use of these resources. Degradation can be measured in terms of the identifiable adverse affects on aquatic organisms, wildlife, and humans.

Sedimentation can have negative environmental effects as a result of natural processes or as a result of processes induced by humans. Negative effects from natural processes can result from landslides, mudslides, the migration of dunes in developed areas, and the in-filling of lagoons or lakes that might otherwise provide recreational or commercial opportunities. Natural degradation of water quality results from sedimentation in areas such as north-central Oklahoma, where large accumulations of salt have precipitated from saline groundwater to form the Great Salt Plains along the Salt Fork of the Arkansas River, thereby affecting the chemistry of the Arkansas River.

The discharge of dredged or fill material, industrial wastes, or other materials into a natural system can affect the chemical, physical, and biological integrity of the system. Industrial wastewaters are a major source of chemicals that may include highly stable organic compounds that are capable of accumulating in high concentrations through sedimentary processes. Both the sedimentation rate and the individual compound's rate of solubility in water will affect the concentration of accumulated materials.

Adult freshwater organisms are generally tolerant of the normal extremes of suspended solids, but the introduction of excess materials and resulting sedimentation will kill eggs, larvae, and insect fauna while altering the characteristics of the aquatic bottom environment. Studies have also shown that high concentrations of suspended solids can interfere with the filter mechanisms of aquatic organisms and the ability of sight feeders to locate prey.

The positive effects of sedimentation can be seen in many natural processes. Beach rejuvenation and growth take place as the ocean's waves and currents deposit materials in quiet-water environments. Similar land growth takes place in the quiet-water environments of lakes and rivers. In many instances, this process has resulted in substantial land growth that has allowed development of commercial or recreational facilities. River floodplains are frequently rejuvenated with nutrients needed for agricultural land use through sedimentation. Salt harvested from brine ponds relies on the natural process of chemical precipitation and sedimentation. Sedimentary processes have also been responsible for creating accumulations of materials such as gold and platinum in river placer deposits.

Kyle L. Kayler

FURTHER READING

Chiras, Daniel D. "Water Pollution: Sustainably Managing a Renewable Resource." In *Environmental Science*. 8th ed. Sudbury, Mass.: Jones and Bartlett, 2010.

Taylor, Kevin G. "Sediments and Sedimentation." In *An Introduction to Physical Geography and the Environment*, edited by Joseph Holden. 2d ed. Harlow, England: Pearson Education, 2008.

SEE ALSO: Dams and reservoirs; Dredging; Erosion and erosion control; Soil conservation.

Seed banks

CATEGORY: Agriculture and food

DEFINITION: Repositories for the genetic material contained within seeds of agricultural and other plant species

SIGNIFICANCE: Because modern agricultural practices focus on only a limited number of crops, seed banks have been established to help promote biodiversity and protect the world's future food supply by preserving the seeds, or genetic material, of thousands of plant species that might otherwise become extinct. Seed banks also provide a backup of genetic plant material in the event of wars, accidents, or environmental disasters on a local or global scale.

Modern-day seed banks, also referred to as gene banks, are based on an agricultural tradition that has persisted for thousands of years: saving seeds from one season to ensure a supply that can be used to plant crops the next season. Seeds have also been passed

An armed guard walks through the Svalbard Global Seed Vault in Norway. This facility, which stores 4.5 million seed samples from around the world, is nicknamed the "Doomsday Vault" and is designed to withstand nuclear warfare and terrorist attacks as well as natural disasters. (AP/Wide World Photos)

down from one generation of farmers to the next, but as agricultural mass production has increased around the world, thousands of plant species have been irrevocably lost simply because no farmers or agricultural corporations chose to grow them over a period of time.

By the early twenty-first century, approximately 1,400 organizations identifying themselves as seed or gene banks were in operation worldwide. These banks can range from small, informal facilities that specialize in certain types of seeds to larger organizations such as the Seed Savers Exchange, a nonprofit organization established in 1975 to preserve not only the genetic but also the cultural and historical heritages of plants. Located on 360 hectares (890 acres) in Decorah, Iowa, the Seed Savers Exchange maintains more than 25,000 varieties of fruit, flower, vegetable, and herb seeds. This group goes beyond the activities traditionally conducted by seed banks in that it makes many uncommon seed varieties available for sale, and it plants its seed stock on a rotating basis in order to generate new seeds and ensure that the plant species remain part of the planet's active ecosystem.

One of the world's most impressive seed banks is the vast Svalbard Global Seed Vault, a facility opened by the Norwegian government in 2008 with assistance from the Global Crop Diversity Trust. Built inside a mountain on an island within the Arctic Circle, this high-tech seed bank has the capacity to store 4.5 million seed samples; it maintains a constant interior climate in spite of temperature variations in the surrounding permafrost. Nicknamed the "Doomsday Vault" by the media, it is designed to withstand nuclear warfare and terrorist attacks as well as natural disasters such as flooding, and is therefore considered to be a global "insurance policy" in the event of agricultural disaster. The Svalbard Global Seed Vault is actually a bank for banks; most of its "depositors" are smaller seed banks around the world that are located in areas that are geographically or politically less stable. It thus represents a vital step in efforts to protect the earth's plant biodiversity. The Global Crop Diversity Trust has the further goal of providing funding and other support for multiple crop repositories around the world.

Amy Sisson

FURTHER READING
Fenner, Michael, and Ken Thompson. "Soil Seed Banks." In *The Ecology of Seeds*. New York: Cambridge University Press, 2005.
Fowler, Cary. "The Svalbard Seed Vault and Crop Security." *BioScience* 58, no. 3 (2008): 190-191.
Rosner, Hillary. "The Gatherers." *Popular Science*, January, 2008, 60-64, 66, 91.
Shea, Neil. "Norway's Ark." *National Geographic*, June, 2007, 14-21.

SEE ALSO: Biodiversity; Extinctions and species loss; Intensive farming; Monoculture; Organic gardening and farming; Sustainable agriculture.

Selective breeding

CATEGORIES: Agriculture and food; animals and endangered species
DEFINITION: Deliberate introduction or propagation of human-desired characteristics into populations of wild or domesticated species of microorganisms, fungi, plants, or animals
SIGNIFICANCE: Artificial selection of desirable traits by human beings and their introduction and propagation within particular species is the basis of global agribusiness and the domesticated pet industry. It is also an important component of many other fields, including pharmaceuticals, biotechnology, and bioengineering.

Selective breeding is a form of artificial selection exercised by human breeders. In its classical form, it involves selecting from a population those individuals that have evident physical features that are specially desired by the breeder, such as a wider leaf, greater-than-average height, resistance to drought or diseases, greater yield of fruit or milk, or passivity. The breeder deliberately cross-breeds individuals with desired traits to create a new form of the species (or even a new species) that has pronounced, desirable, and predictable differences when compared to the original stock from which its ancestors came. Through selective breeding, over time the underlying average genotype (the genetic makeup) of a species changes as certain characteristics are directionally selected across generations.

Human beings in Paleolithic and Neolithic times regularly harvested wild grains. By as early as twelve thousand years ago, a series of plants were domesticated for human use. Both nonintentional and intentional selection were occurring as plants or animals were harvested or hunted. Often in hunting, slower animals were killed, meaning that over time, on average, wild animals became faster; this is an example of nonintentional selection. There is evidence that prehistoric peoples intentionally favored certain traits of plants—such as size, vigor, and yield—and replanted annually, as crops, the seeds preserved from desirable plants.

By the nineteenth century, plant and animal breeding was a reasonably advanced endeavor, even though it awaited work in genetics before the underlying mechanisms by which these traits were inherited were fully understood. Charles Darwin used artificial selection of plants and animals as analogues in conceiving and explaining his theory of evolution by means of natural selection. Both types of selection depend on variation among individuals of a population and the survival of certain traits among subsequent populations—the former undertaken by breeders consciously breeding for these traits, while in nature the struggle for existence "naturally" culls out individuals with traits not as well adapted to present or future environmental conditions.

In the twenty-first century selective breeding is a highly refined science involving insights from a host of disciplines, including genetics, genetic engineering, conservation biology, soil sciences, animal nutrition, biotechnology, agronomy, veterinary medicine, and economics. Perhaps the biggest impact on selective breeding has been the Green Revolution in agriculture, which has made food plentiful even if there are still enormous problems with global distribution systems. Humans have the capability to breed for particular desired traits across microorganisms, fungi, plants, and animals. Much of the work with microorganisms, for example, results in artificial production of pharmaceuticals and critical biological products for manufacturing on an unprecedented scale.

The development of selective breeding has introduced a variety of ethical issues, including issues related to the rise of monocultures of plants and animals that are vulnerable to single threats. Other issues raised by environmentalists and others involve the crossing of natural biological barriers between completely separate organisms, which can result in unpredictable results, and the accelerated extinguishment of natural biological diversity as selective breeding removes traits not currently prized but that may be invaluable in the future.

Dennis W. Cheek

FURTHER READING

Chiras, Daniel D. "Creating a Sustainable System of Agriculture to Feed the World's People." In *Environmental Science*. 8th ed. Sudbury, Mass.: Jones and Bartlett, 2010.

Kidd, J. S., and Renee A. Kidd. *Agricultural Versus Environmental Science: A Green Revolution*. New York: Chelsea House, 2006.

Varner, Gary E. *In Nature's Interests? Interests, Animal Rights, and Environmental Ethics*. New York: Oxford University Press, 1998.

SEE ALSO: Captive breeding; Dolly the sheep; Genetically modified foods; Genetically modified organisms; Green Revolution; High-yield wheat; Monoculture.

Septic systems

CATEGORY: Waste and waste management

DEFINITION: Systems for the on-site treatment of sewage and wastewater

SIGNIFICANCE: Although septic systems work well in isolation and in the appropriate soil environments, increased development in rural areas can cause environmental problems from improperly sited or inadequately separated septic systems.

Septic systems are the most common form of on-site wastewater treatment used in the United States. The conventional septic system is a model of simplicity and economy. It consists of a large chamber—the septic tank—into which domestic wastewater flows and where anaerobic digestion of the waste occurs. Nondegradable or slowly degradable compounds fall to the bottom of the septic tank, where they accumulate as sludge. The effluent then flows by gravity to a distribution box, where it is distributed by pipes through one or more gravel-filled ditches buried in the soil. These ditches constitute a lateral field where wastes slowly percolate through the gravel and underlying soil. During leaching, pathogenic organisms and residual organic matter are consumed. The percolating water eventually finds its way to underlying groundwater.

One of the assumptions underlying the use of septic systems is that there will be some pathogen reduction as the wastes are anaerobically digested and as

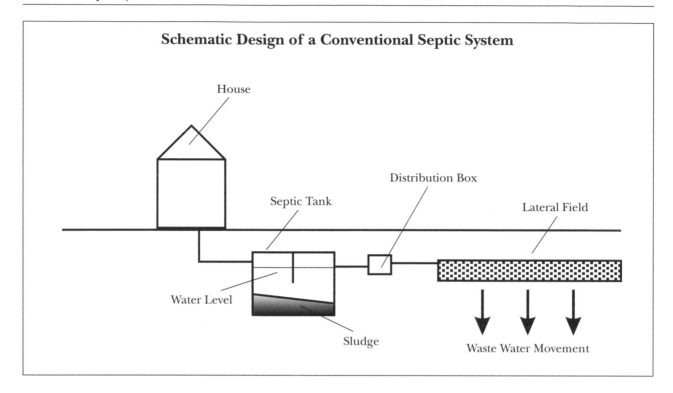

Schematic Design of a Conventional Septic System

they pass through the lateral field. This is normally the case. Anaerobic digestion reduces the pathogens in waste by about 10 percent, and up to a 99 percent reduction occurs as the pathogens pass through the biological mat that forms at the bottom of the lateral field. The biological mat is a rich collection of bacteria, fungi, and protozoa that consume harmful bacteria and viruses in domestic waste.

A second key assumption underlying the use of septic systems is that such systems will be properly sited and maintained. The soil is generally an effective biological filter, so most septic systems are designed to have 45 to 60 centimeters (18 to 24 inches) of permeable soil beneath the lateral field before wastewater encounters groundwater and is diluted. Likewise, the nondegradable materials in sludge must periodically be removed and either disposed of in a landfill or further processed in a municipal wastewater treatment system. If the sludge is not removed from the septic tank, it will eventually accumulate to such an extent that it will reduce the wastewater residence time in the tank. This causes inadequately treated waste and some solids to flow into the lateral field, which is subsequently unable to ensure proper wastewater treatment. The solids may also clog the leach field and cause wastewater to pond.

The dominant environmental problems associated with septic systems are the failure of some systems to reduce pathogens adequately and the systems' potential delivery of nutrients into groundwater. Domestic wastewater consists of gray water—water from showers, sinks, dishwashers, and clothes washers—and black water, which comes from toilets. Gray water is relatively free of nutrients and pathogens. Black water, in contrast, is the major source of environmental concerns regarding septic systems because it is both nutrient-rich and laden with pathogens that are potentially dangerous to humans. Because black water is commingled with gray water in most septic systems, all of the water leaving a house must be treated thoroughly.

In isolation, septic systems work well if properly sited and maintained because there is usually adequate distance between the domestic wastewater source and receiving waters. However, as development finds its way to rural environments that are not served by either municipal sewage treatment or municipal water facilities, the congregation of septic systems in limited areas can overwhelm the environment's capacity to filter and dilute wastewater effectively before it reaches drinking-water wells or recreational water supplies.

Properly locating septic systems is critical, and home sites may not always be located on soils that are suitable for conventional septic systems. The soils may be slowly permeable, the topography of the land may be too steep to permit an adequate lateral field, or there may be too little soil beneath the lateral field to allow proper filtration before impermeable layers and groundwater are reached. The consequence of these limitations to conventional septic systems is that soluble nutrients such as ammonium and phosphate, which come from the digestion of domestic wastes, may increasingly appear in underlying groundwater or shallow surface water and lead to eutrophication. Protozoan pathogens such as *Cryptosporidium* and *Giardia* may be inadequately treated or filtered. The potential for contamination by bacterial pathogens such as salmonella and viral pathogens such as hepatitis A, B, and C is a greater concern because their escape from septic systems into drinking-water or recreational water supplies represents a serious public health crisis.

As a result of these concerns regarding conventional septic systems, many U.S. states now require alternative forms of on-site wastewater treatment if either the soil or the density of habitation does not allow for the proper use of conventional septic systems. These alternative treatments include wetlands, in which domestic waste from a septic tank first passes through a bed of vegetation before entering a lateral field. Aerobic digestion occurs in the wetland, and soluble nutrients are taken up by the wetland plants. Mound systems are often used where shallow soils exist. In a mound system, a raised bed of soil is constructed to which domestic waste effluent is pumped and then allowed to percolate. A variety of other on-site systems use percolation through sand or peat filters to help process wastewater and reduce the impact of the nutrients and pathogens in domestic waste.

Mark Coyne

FURTHER READING

Hill, Marquita K. "Water Pollution." In *Understanding Environmental Pollution.* 3d ed. New York: Cambridge University Press, 2010.

Kahn, Lloyd. *The Septic System Owner's Manual.* Rev. ed. Bolinas, Calif.: Shelter, 2007.

Miller, G. Tyler, Jr., and Scott Spoolman. "Water Pollution." In *Living in the Environment: Principles, Connections, and Solutions.* 16th ed. Belmont, Calif.: Brooks/Cole, 2009.

Sullivan, Patrick J., Franklin J. Agardy, and James J. J. Clark. "Water Pollution." In *The Environmental Science of Drinking Water.* Burlington, Mass.: Elsevier Butterworth-Heinemann, 2005.

U.S. Environmental Protection Agency. *Water on Tap: What You Need to Know.* Washington, D.C.: Author, 2003.

SEE ALSO: Cultural eutrophication; Environmental health; Eutrophication; Groundwater pollution; Sewage treatment and disposal; Soil contamination; Waste treatment; Water pollution; Water quality.

Serengeti National Park

CATEGORIES: Places; preservation and wilderness issues

IDENTIFICATION: Large national safari park in northern Tanzania, East Africa

SIGNIFICANCE: The Serengeti National Park provides protection to vast numbers of animal, insect, and plant species. In addition to being a popular tourist destination, the park is active in conservation and research efforts.

The name Serengeti, meaning "endless plain," derives from the language of the Masai, a seminomadic people native to East Africa who have grazed their cattle on the Serengeti Plain for more than two thousand years. The first European to see the area was Oskar Baumann, an Austrian explorer and cartographer, in 1892. Germany colonized the Serengeti until the end of World War I, when Tanganyika (now Tanzania) became a British protectorate. In 1921 the British designated 324 hectares (800 acres) of the plain as a partial game reserve to protect the lions; this area became a full reserve in 1929.

The British established the whole of the Serengeti as a national park in 1951, before Tanganyika achieved independence. The Masai tribespeople on the plain were relocated to the southeastern portion of the park, in the Ngorongoro highlands. In the 1950's a book on the Serengeti by Bernhard Grzimek, president of the Frankfurt Zoological Society, and a documentary film based on the book that Grzimek wrote and directed brought worldwide attention to the Serengeti National Park. In the early twenty-first century, the Frankfurt Zoological Society works closely with Tanzania's national park service, provid-

Wildebeest cross the Mara River in Serengeti National Park during their annual migration. (©Jannie Nikola Laursen/iStockphoto.com)

ing funding for research, training, and conservation at the Serengeti National Park.

The park is roughly 14,760 square kilometers (5,700 square miles) in area. To the north it borders Kenya, to the west Lake Victoria, and to the east the Great Rift Valley. On the south it is surrounded by the buffer zones of the Ngorongoro Conservation Area, set up in 1959, and the Masai Mara National Reserve in Kenya. Within the park, tourists are accommodated in five lodges, seven tented camps, and a number of camp sites. Administration of the park is centered in Seronera, a small settlement within its boundaries.

The topography of the park is widely varied. The southern and central areas consist of savanna interspersed with rocky outbreaks known as kopjes or koppies. To the west is a corridor of more fertile ground watered by the Grumeti River. To the north are the permanent watercourses of the River Mara in the Lobo area. One of the great sights in the park is the annual migration of millions of wildebeest, joined by some 200,000 zebras and 300,000 Thomson's gazelles, from south to north, a distance of some 966 kilometers (600 miles), and then their return. Such migrations are attended by many predators: lions, cheetahs, jackals, leopards, hyenas, and African wild dogs.

In addition to these animals, the park is home to large herds of elephants, buffalo, impalas, giraffes, waterbucks, and other types of gazelles, as well as crocodiles that live in the rivers. The large population of black rhinoceros that was once present in the park has been decimated by poaching; the numbers of this spe-

cies had become dangerously low by the early years of the twenty-first century. Also native to the Serengeti are some five hundred bird species, including the secretary bird, ostrich, and black eagle. At the insect level species are abundant as well. One hundred varieties of dung beetle have been identified, for example. After the rains, the ground in many parts of the park is carpeted with a wide variety of wildflowers. Owing to the great biodiversity found there, the Serengeti National Park has been designated a World Heritage Site by the United Nations Educational, Scientific, and Cultural Organization (UNESCO).

Environmental challenges for the park include the comparative fragility of its ecosystems for such large numbers of animals; drought and erosion have posed particular problems. In addition, some conflicts have arisen on the borders of the park because animals from the park have killed the livestock of local herdsmen; also, diseases are sometimes spread between the wild animals of the park and domesticated animals. Efforts to reduce such tensions have included the establishment of four Wildlife Management Areas that encompass some twenty-three villages near the park's borders.

David Barratt

FURTHER READING

Holmern, Tomas, Julius Nyahongo, and Eivin Røskaft. "Livestock Loss Caused by Predators Outside the Serengeti National Park, Tanzania." *Biological Conservation* 135, no. 4 (April, 2007): 518-526.

Homewood, K. W., and W. A. Rodgers. *Maasailand Ecology: Pastoralist Development and Wildlife Conservation in Ngorongoro, Tanzania.* 1991. Reprint. New York: Cambridge University Press, 2004.

Roodt, Veronica. *The Tourist Travel and Field Guide of the Serengeti National Park.* 4th ed. Hartebeesport, South Africa: Papyrus, 2005.

Turner, Myles. *My Serengeti Years: The Memoirs of an African Game Warden.* New York: W. W. Norton, 1987.

SEE ALSO: Africa; Ecotourism; Grazing and grasslands; Poaching; Rhinoceroses; World Heritage Convention.

Sewage treatment and disposal

CATEGORY: Waste and waste management

DEFINITION: Processes applied to the domestic and industrial effluent collected by a sewer system

SIGNIFICANCE: The proper treatment and disposal of wastewater is a critical element in environmental planning because improper disposal or inadequate treatment can result in the contamination of groundwater and of drinking-water supplies.

The Minoan civilization on the island of Crete near Greece had one of the earliest known sewers in the world (c. 1600 B.C.E.). A large sewer known as the Cloaca Maxima was built during the sixth century B.C.E. in ancient Rome to drain the Forum. The Romans also reused public bathing water to flush public toilets. London, England, had a drainage system by the thirteenth century, but effluent could not be discharged into it until 1815. Sewers were constructed in Paris, France, before the sixteenth century, but less than 5 percent of the homes were connected by 1893. In general, the widespread introduction of sewers in densely populated areas did not occur until the mid-nineteenth century.

A typical wastewater disposal system consists of a network of pipes, a treatment plant, and an outfall to the ground or, more commonly, to a stream or the ocean. Older wastewater systems are generally combined; that is, domestic, industrial, and stormwater runoff are conveyed in the same pipes to a treatment plant. Although initially cheaper to build, combined systems are less desirable than separated systems because most of the effluent must bypass the treatment plant during storms, when street runoff rapidly increases. Newer wastewater systems are designed so that separate pipes handle wastewater and storm runoff.

About 60 to 75 percent of the water supplied to a community will wind up as effluent that must be treated and disposed. The remaining water is used in industrial processes, lawn sprinkling, and other types of consumptive use. Domestic sewage contains varying proportions of human excrement, paper, soap, dirt, food waste, and other substances. Much of the waste substance is organic and is decayed by bacteria. Accordingly, domestic sewage is biodegradable and capable of producing offensive odors. The composition of industrial waste varies from relatively clean rinse water to effluent that can contain corrosive, toxic, flammable, or even explosive materials. Therefore, communities usually require pretreatment of industrial effluent.

The organic material in sewage is decomposed by aerobic (oxygen-requiring) bacteria. However, the dissolved oxygen (DO) in water can be used up in the process of microbial decomposition. If too much organic waste enters the water body, the biochemical oxygen demand (BOD) can exhaust the DO in the water, thereby damaging the aquatic ecosystem. Indeed, most species of fish die in water in which the DO falls below 4 milligrams per liter for extended periods of time.

The function of a wastewater treatment plant is to produce a discharge that is free of odors, suspended solids, and objectionable bacteria. The processes of wastewater treatment are categorized as primary, secondary, or tertiary. Primary treatment is mostly mechanical, as it involves the removal of floating and suspended solids through screening and sedimentation in settling basins. This form of treatment can remove 40 to 90 percent of the suspended solids and 25 to 85 percent of the BOD.

Secondary treatment involves biological processing in addition to mechanical treatment. One form of biological processing is a trickling filter, where wastewater is sprayed over crushed stone and allowed to flow in thin films over biologic growths that cover the stone. The organisms in the biologic growths—which include bacteria, fungi, and protozoa—decompose the dissolved organic materials in the wastewater. These growths eventually slough off and are carried to settling tanks by the wastewater flow. Another type of secondary treatment is the activated sludge process. In this procedure, flocs of bacteria, fungi, and protozoa are stirred in the wastewater with results that are about the same as those achieved with trickling filters. Depending on the efficiency of the plant and the nature of the incoming wastewater, both types of biological processes can remove 50 to 95 percent of the suspended solids and BOD. The efficiency of secondary treatment can be seriously lowered if the design capacity of the plant is overloaded with excessive effluent coming from stormwater runoff in combined sewers. This is one important reason public health officials favor separate sewers, even though they are more expensive. The biologic processes can also be severely affected by toxic industrial waste, which can kill the "good" bacteria that are crucial in the treatment pro-

A sewage treatment plant in eastern China. (AP/Wide World Photos)

cess. Accordingly, many communities require pretreatment of industrial wastes.

Tertiary treatment is the most advanced method and consequently the most expensive. It includes several procedures such as the use of ozone, which is a strong oxidizing agent, to remove most of the remaining BOD, odor, and taste; adding alum to remove phosphate; and denitrification. The final effluent from any treatment level is usually chlorinated prior to release.

In areas where population densities are lower than about 1,000 people per square kilometer (2,600 per square mile), the costs of a sewer system and treatment plant are difficult to justify. Accordingly, septic systems are commonly used for wastewater disposal in low-density residential areas. In such a system, household effluent is piped to a buried septic tank, which acts as a small sedimentation basin and anaerobic (without oxygen) sludge-digestion facility. The effluent exits from this tank into a disposal field, where aerobic biologic breakdown of dissolved and solid organic compounds occurs. In order for a septic system to operate effectively, the soil must be of sufficient depth and permeability so that microbial decomposition can take place before the effluent reaches the water table. The Environmental Protection Agency estimates that 25 percent of the homes in the United States use septic systems.

Robert M. Hordon

FURTHER READING

Hill, Marquita K. "Water Pollution." In *Understanding Environmental Pollution.* 3d ed. New York: Cambridge University Press, 2010.

Lester, J., and D. Edge. "Sewage and Sewage Sludge Treatment." In *Pollution: Causes, Effects, and Control,* edited by Roy M. Harrison. 4th ed. Cambridge, England: Royal Society of Chemistry, 2001.

McGhee, Terrence. *Water Supply and Sewerage.* New York: McGraw-Hill, 1991.

Miller, G. Tyler, Jr., and Scott Spoolman. "Water Pollution." In *Living in the Environment: Principles, Connections, and Solutions.* 16th ed. Belmont, Calif.: Brooks/Cole, 2009.

Qasim, Syed A. *Wastewater Treatment Plants: Planning, Design, and Operation.* 2d ed. Lancaster, Pa.: Technomic, 1999.

Roseland, Mark. "Water and Sewage." In *Toward Sustainable Communities: Resources for Citizens and Their Governments.* Rev. ed. Gabriola Island, B.C.: New Society, 2005.

Salvato, Joseph A. *Environmental Engineering and Sanitation.* 4th ed. New York: John Wiley & Sons, 1992.

SEE ALSO: Environmental engineering; Septic systems; Sludge treatment and disposal; Stormwater management; Wastewater management; Water pollution; Water treatment.